Sustainability and Sustainable Development

An Introduction

Lisa Benton-Short

George Washington University

ROWMAN & LITTLEFIELD

Lanham • Boulder • New York • London

Executive Acquisitions Editor: Michael Kerns
Assistant Acquisitions Editor: Elizabeth Von Buhr
Sales and Marketing Inquiries: textbooks@rowman.com

Credits and acknowledgments for material borrowed from other sources, and reproduced with permission, appear in the Credits section.

Published by Rowman & Littlefield
An imprint of The Rowman & Littlefield Publishing Group, Inc.
4501 Forbes Boulevard, Suite 200, Lanham, Maryland 20706
www.rowman.com

86-90 Paul Street, London EC2A 4NE

British Library Cataloguing in Publication Information Available

Library of Congress Cataloging-in-Publication Data

Names: Benton-Short, Lisa, author.
Title: Sustainability and sustainable development : an introduction / Lisa Benton-Short.
Description: Lanham : Rowman & Littlefield, [2023] | Includes bibliographical references and index.
Identifiers: LCCN 2022043871 (print) | LCCN 2022043872 (ebook) | ISBN 9781538135358 (cloth) | ISBN 9781538135365 (paperback) | ISBN 9781538135372 (epub)
Subjects: LCSH: Sustainable development. | Environmental management—Social aspects.
Classification: LCC HC79.E5 B4527 2023 (print) | LCC HC79.E5 (ebook) | DDC 338.9—dc23/eng/20220909
LC record available at https://lccn.loc.gov/2022043871
LC ebook record available at https://lccn.loc.gov/2022043872

∞™ The paper used in this publication meets the minimum requirements of American National Standard for Information Sciences—Permanence of Paper for Printed Library Materials, ANSI/NISO Z39.48-1992.

This book is dedicated to my father, Robert Bruce Benton (1941–2021). He knew the earth was alive and precious: He discovered spirits living in wishing stones, sea glass, and driftwood, he meditated on philosophy, and found peace in the waters of Friday Harbor.

BRIEF CONTENTS

CONTENTS

BOXES

MAPS

FIGURES

TABLES

PHOTOS

This book's aim is to introduce students to sustainability—an ambitious task. The book gives readers a sense of the complexity and many dimensions of sustainability in the contemporary world. The book focuses on a global perspective while discussing examples from the United States, as a base of comparison. The writing style is accessible to the general reader and the scholarship is comprehensive, so that different interpretations are presented. Broad arguments are enlivened with detailed case studies in text boxes. Maps, graphs, photos, and infographics help reinforce important trends in advancing sustainability.

The challenge in teaching an Introduction to Sustainability course is there are many ways to teach such a course, and many issues to cover. There is no consensus on what topics to include nor a general framework from which to structure the course. There are many introductory-level books that explore sustainability from a primarily environmental perspective; there are those that focus more on human development issues such as health or education, but not many that cover the diverse and comprehensive range of issues that are fundamental to the concept of sustainability. This book attempts to provide an introduction to sustainability and sustainable development that covers the diverse range of environmental, economic, and human health and well-being challenges to a more sustainable future.

This book is structured around the seventeen UN Sustainable Development Goals (SDGs), adopted in 2015. The seventeen SDGs confront the causes of poverty and environmental degradation in an integrated fashion and tackle a broad set of issues including ending hunger, improving health, reducing inequalities, combating climate change, and protecting ocean and forest resources.

The SDGs are now *the* guiding framework for planning and implementing sustainability through 2030. They are the focus of international development efforts, and the "lingua franca" of sustainability. They are the international consensus on "what is sustainability?" As such the UN SDGs present an ideal framework for an introductory-level course textbook because they genuinely integrate the "Three Es"—environment, economic development, and equity—that are the core definition of sustainability. Beyond the intellectual merits of understanding the SDGs, benchmarking, tracking, and assessing the seventeen SDGs means jobs for our students, so introducing them to the UN SDGs gives them a competitive advantage in the job market.

This book synthesizes UN reports with current international scholarship to provide the context for understanding the SDGs. It provides a global perspective, with attention given equally to the highest-income countries of the Global North, and to the moderate- and low-income countries of the Global South. The book integrates basic environmental science, policy, and interdisciplinary perspectives while investigating key challenges to developing a more sustainable future through the SDG framework.

This textbook is structured around the seventeen SDGs with one chapter for each goal. The aim is to provide a general introduction to the goal by first introducing the issue by defining key concepts and describing trends and geographic patterns (e.g., where is poverty most prevalent in the world and why). Then each chapter explores strategies for achieving the goals, challenges to achieving these targets, and documents progress toward these goals. Each chapter ends with a brief summary of progress to date.

Each chapter is approximately 9,000 words: Long enough to provide an introduction to the topic, yet concise enough to allow instructors to supplement the text with scholarly articles that explore the topic in more depth should they choose to do so. Each chapter begins with learning objectives. Each chapter ends with a discussion on the interconnections among the SDGs to reinforce the fact that these goals are interdependent in many ways. Discussion questions are at the end of each chapter. Key terms are also located at the end of each chapter and are bolded in the text and defined in a glossary. Finally, each chapter has a list of further readings.

To unify the book, text boxes in each of the chapters share common themes that include:

- *Key Terms and Concepts*: define and explain one or two terms or concepts;
- *Solutions*: highlight approaches to specific SDG targets;
- *A Critical Perspective*: offers a critique on the SDG in particular, its targets, or the topic more broadly;
- *An Expert Voice*: features scholars, policy makers, practitioners, thought leaders, and so forth who discuss critical trends and successes; and
- *SDGs and the Law*: provides examples of important ways we govern for sustainability.

The introduction sets the context and provides a brief introduction to the concept of sustainability, and the launch of the UN SDGs. The introduction also briefly explores the theme of uneven development, an underlying driving force for the sustainability agenda. The remaining chapters explore the SDGs in numerical order (Chapter 1 is SDG 1 Poverty, Chapter 2 is SDG 2 Hunger and Food Insecurity, and so on). Tackling each chapter in order is certainly one way to proceed. Another way to consider moving through this textbook is by grouping chapters by the "Three Es": environment, economics, and equity. In this case, a sequence could look like:

- *Environment*: SDGs 6, 7, 13, 14, 15 (water, energy, climate, oceans, land ecosystems);
- *Economics*: SDGs 1, 8, 9, 12 (poverty, decent work, economic growth, industry, infrastructure, responsible consumption, and production); and
- *Equity*: SDGs 2, 3, 4, 5, 10, 11, 16, 17 (food insecurity, health, education, gender equality, reduced inequalities, peace, justice, and good governance).

There are likely other ways to organize and move through the book to best suit the structure of your course. The structure of the book, while providing a coherent whole, also allows sectional choices to meet different needs, time constraints, and interests of individual instructors. The entire book may be used, but some may want to focus on certain chapters and topic areas.

Because there are eighteen chapters in this text, and most courses on a semester schedule are only fifteen weeks, some instructors may elect to group together some chapters to cover in the same week. In my experience teaching the course in a fifteen-week term, we have grouped some SDGs together, for example:

- SDG 3 Good Health and Well-Being and SDG 4 Quality Education are complementary;
- SDG 5 Gender Equality and SDG 10 Reduced Inequalities both focus on inequalities (SDG 5 focuses on gender; SDG 10 focuses on other forms of inequalities such as discrimination based on race, age, ethnic identity, and disability);
- SDG 6 Clean Water and Sanitation and SDG 14 Life Below Water both focus on water (fresh water in SDG 6, oceans in SDG 14—and build on an understanding of water quality, common pollutants, and impacts of water contamination);
- SDG 7 Affordable and Clean Energy and SDG 13 Climate Action are complementary, as contemporary energy use of nonrenewable energy is a precursor to understanding the major forces driving climate change;
- SDGs 8 and 9 also pair well together with a focus on economic dimensions including the infrastructure needed for economic growth; and
- SDG 16 Peace, Justice and Strong Institutions and SDG 17 Partnerships and Governance are considered to be highly interconnected—many practitioners have referred to SDG 16 and SDG 17 as the "governance goals." Some have observed SDG 16 represents good governance and SDG 17 the means to implement good governance.

Textbooks are written at specific times in specific places by specific people. My own teaching of sustainability and my background as a geographer has shaped the material covered. This book has many maps, which is not surprising given that I am a geographer, but the maps also reinforce the global nature of the sustainability agenda, and highlight regions and countries where the issue is most pronounced, and where it is not. Graphs and charts also provide a general sense of trends over the last twenty years. Where possible, I have used the most recent data in maps, figures, and charts. However, a word of caution about data: Not all data is collected annually and not all countries collect data for all topics. As a result, some of the most recent data may be five or six years old by the time of publication. To counter the issue with missing, inaccurate, or out-of-date data, instructors should stress that the goal of the course, and of the book, is to provide a general understanding of trends over time, not a snapshot of the most recent year. This is because few of these important issues—poverty, hunger, water pollution, and carbon dioxide emissions—change dramatically from one year to the next. Rather, getting to sustainability will be about persistent and determined change over time. It will not happen overnight.

The book in your hands is the product of one author, but one who is part of a collaborative effort. I have had the privilege of teaching Introduction to Sustainability with five other instructors since 2012. The team of professors represents different disciplines (geography, biology, public health, law, creative writing, and landscape design), and I have learned much from them over the years. A very big thank you to Adele Ashkar, Peter LaPuma, Lee Paddock, Tara Scully, and Michael Svoboda. We have been collaborative in the truest sense—coteaching in the same lecture and sharing how we each approach solving the challenges and advancing sustainability. Much of my approach to this book integrates their knowledge, insight, and approaches to sustainability. I also want to thank Brooke Iacone, the talented cartographer that produced all of the maps, charts, graphs, and infographics. Her work was supported by the generosity of the Geography Department at George Washington University. This book is a synthesis of the expertise of many scholars and development practitioners, and I relied extensively on their research, ideas and critiques. I have made every attempt to acknowledge these sources in the section "Resources Used and Suggested Readings."

Lastly, I want to thank the many reviewers who looked at the proposal and draft chapters and offered both encouraging and constructive feedback. They include:

Karen Allen, Furman University
Hussein A. Amery, Colorado School of Mines
Katherine Amey, Kent State University
Eleanor Andrews, Cornell University
Pratyusha Basu, University of Texas, El Paso
Richard Biehl, University of Central Florida
Carolin Boules, University of Maryland
William Brown, State University of New York at Fredonia
Alec Foster, Illinois State University
Camille Gaskin-Reyes, Georgetown University
Bruce Gervais, California State University, Sacramento
David Himmelfarb, Eckerd College
Robert Kirkman, Georgia Institute of Technology
Jonathan S. Krones, Boston College
Smita Malpani, Washtenaw Community College (MI)
Basilio G. Monteiro, St. John's University
Meghan Mordy, Colorado State University
Michelle A. Myers, Pennsylvania Highlands Community College
Velvet Nelson, Sam Houston State University
K. J. Pataki-Schweizer, Portland Community College (OR)
David R. Perkins IV, Missouri State University
Debbie Roberts, Troy University
Jens Rudbeck, New York University
Phil Schoenberg, Western New Mexico University
Charles Scruggs, Genesee Community College (NY)
Steven E. Silvern, Salem State University
Scott Singeisen, Savannah College of Art and Design
C. N. Steacy, University of Georgia
Marylynn Steckley, Carleton University
Jennifer Titanski-Hooper, Francis Marion University
Michael Tlusty, University of Massachusetts, Boston
Jeremy VanAntwerp, Calvin University
Daniel Vivian, University of Kentucky
Edward Weisband, Virginia Tech

Sustainability and Sustainable Development

LEARNING OBJECTIVES

After reading this chapter, you should be able to:

- Define sustainability
- Describe the "three Es" of sustainability
- Explain the importance of the UN Sustainable Development Goals
- Discuss the context and impact of uneven development
- Define and discuss environmental justice
- Explain the importance and limitations of the Human Development Index

We will live in a world where nobody anywhere lives in extreme poverty
 Where no one goes hungry
 Where no one wakes in the morning asking if there will be food today
 We will live in a world where no child will die of diseases we know we can cure and where proper health care is a lifelong right for us all
 We will live in a world where everyone goes to school and education gives us the knowledge and skills for a fulfilling life
 We will live in a world where all girls and all women have equal opportunities to thrive and be powerful and safe
 We will live in a world where all people can get clean water and proper toilets at home, at school, and at work
 We will live in a world where there is sustainable energy for everyone, with heat, light, and power for the whole planet without destroying the planet
 We will live in a world where economies prosper and new wealth leads to decent jobs for everyone
 We will live in a world where our industry, our infrastructure, and our best innovations are not just used to make money but to make all of our lives better
 We will live in a world where our prejudices and extremes of in equality are defeated inside our countries and between different countries
 We will live in a world where people who live in cities and communities are safe, progressive, and support everyone who lives there
 Where we replace what we consume and we put back what we take out of the earth
 We will live in a world that is decisively rolling back from the threat of climate change
 Where we restore and protect the life in our oceans and seas
 And where we restore and protect life on land and forests, animals and the earth itself

With peace between and inside countries
Where governments are open and answer to us for what they do at home and abroad
And justice rules with everyone equal before the law
Where all countries and we—their people—work together in partnerships of all kinds
To make these goals a reality
For everyone
Everywhere*

DEFINING SUSTAINABILITY

Sustainability and sustainable development are often used interchangeably, but there are nuanced distinctions. Both concepts speak to the impacts and consequences of consuming or degrading resources faster than they can be replenished. The idea of **sustainability** emerged in the 1960s primarily out of concern about environmental degradation. It is a broad term that describes managing resources without depleting them for future generations.

Sustainable development, while closely related, is the process for improving long-term economic well-being and quality of life for people now and in the future. Sustainable development emerged in the 1980s out of discussions about the persistence of poverty, poor health, and resource scarcity. In 1987 the UN World Commission on Environment and Development issued a report called *Our Common Future*. The report was the product of a commission of foreign ministers, finance and planning officials, economists, and policy makers in agriculture, science, and technology. The report is often referred to as the **Brundtland Report**, after Gro Harlem Brundtland, Chair of the Commission. The Brundtland Report was an effort by the United Nations to take the lead in coordinating international efforts to implement sustainability. The report defined sustainable development as "development that meets the needs of the present without compromising the ability of future generations to meet their own needs." This definition is a fairly abstract concept, but its core is about finding a way to support human populations and their economies, without ultimately threatening the health of humans, animals, and plants. It also takes a long-range perspective noting that current decisions should not impair the prospects for maintaining or improving future living standards. The landmark Brundtland Report suggests that environmental protection and environmental institutions are an integral part of all development policies and practices.

Sustainability expert Kent Portney notes the most important distinction between earlier ideas of environmental protection and the more modern concept of sustainability is that the former tends to focus on environmental remediation and preservation/conservation, whereas sustainability has a more holistic and broader approach. Sustainability and sustainable development are integrative concepts. Hayden Washington called sustainability the destination, while sustainable development is the journey. Some see sustainability as the next major revolution in human history. Is it?

Conditions have worsened: Water scarcity and droughts threaten food supplies; climate change has brought firestorms that burn hundreds of thousands of acres and often destroy communities; and polluted water causes diseases that affect the health of millions of people. At least a million species on land

---

*"We the People for the Global Goals." United Nations, September 24, 2015. https://www.youtube.com/watch?v=RpqVmvMCmp0

and in the oceans are now threatened with extinction. Every indicator of global environmental degradation is on the rise. There are also crises beyond the physical environment: Women, minorities, and indigenous groups confront discrimination in nearly all countries, and war and conflict over the last decade have forced some sixty-eight million people to flee their countries as refugees. An estimated 767 million people remain in poverty. The current state of the world is *not* sustainable.

The concepts of sustainability and sustainable development have become global buzzwords. Political leaders speak of a green economy or community resilience; corporations pledge to become carbon neutral or to enact corporate social responsibility; academics research ecological footprints, carrying capacity, and systems thinking; and the international development community advances programs to reduce poverty and end hunger.

As concepts, sustainability and sustainable development consist of three pillars: the environment, the economy, and equity. These are often referred to as the **three Es** and can be visualized as three overlapping concentric circles (Figure I.1). The argument is that sustainability can only be achieved when all three pillars are balanced and rejects the notion that achievement in one pillar can be accomplished by sacrificing either of the other two. Genuine sustainability requires us to consider our actions in relation to others and to pursue a more democratic civic politics. It involves ensuring environmental and economic benefits are distributed equitably among all citizens. Achieving equity and environmental justice are one of the biggest challenges to a more sustainable future.

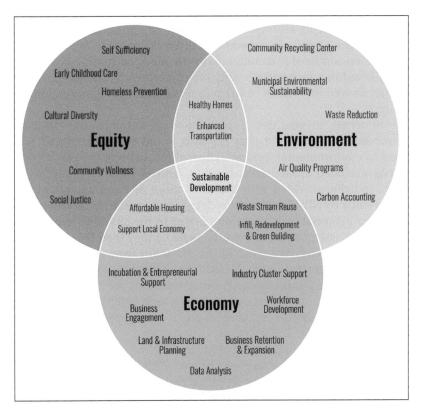

FIGURE I.1 The Three Es of Sustainability
They can be visualized as three overlapping concentric circles. Where the environment, the economy, and equity overlap is sustainability.

The Challenge of Equity

Most Americans know something about water pollution, air pollution, climate change, and the use of fossil fuels. However, there is more to sustainability and sustainable development than simply minimizing our environmental impact, developing policies that promote recycling, or passing legislation to prevent environmental degradation. It is important that we pay equal attention to the third "E" of sustainability—**equity**. The distribution of environmental "goods" such as parks and street trees, and environmental "bads" such as polluting industries and waste facilities, should be just. Environmental inequity disproportionately impacts minorities and the poor.

In many countries environmental "bads" such as toxic facilities are predominately concentrated in lower-income and minority-dominated communities. Marco Martuzzi and colleagues found that in Europe waste facilities were disproportionally located in lower-income areas and in places where ethnic minorities lived. In many countries, poorer communities have less pleasant environments and often bear the brunt of pollution or exposure to hazards. It is through their neighborhoods that highways are constructed; it is in their neighborhoods where heavy vehicular traffic can cause elevated lead levels in the local soil and water. There is a direct correlation between socioeconomic status and the quality of one's environment.

In the United States, study after study reveals a correlation between negative environmental impacts and the presence of racial or ethnic minorities. Race continues to be the most significant variable associated with the location of hazardous waste sites. The greatest number of commercial hazardous facilities are located in areas with the highest composition of racial and ethnic minorities. One study showed that three out of every five Black and Hispanic Americans lived in communities with one or more toxic waste sites. The study noted that race and ethnicity intertwined with access to power to produce an uneven experience of environmental quality at home and in the workplace.

Environmental inequalities occur not only through the presence of environmental "bads" but they are also found in absences and lack of access to environmental "goods." Throughout history, people of color have been excluded from parks, beaches, and swimming pools, while Whites have been provided with access to more of these amenities. For example, a study by Nik Heynen and colleagues mapped trees in Milwaukee and found an inequitable distribution of trees in the city. The more affluent White areas had more extensive urban tree canopy than the poorer and blacker areas of the city. As a result of the city's shift from public to private funding of the urban tree canopy, wealthier residents could afford to plant more trees than the poorer areas. Because trees can positively affect the quality of neighborhoods, through aesthetic and pollution diminishing properties, the distribution of uneven green space reproduced uneven social space.

It is important to distinguish between equality and equity. As Figure I.2 shows, equality is treating everyone the same. Equity, however, is about giving everyone what they need to be successful. An example is school funding. Advocating for equality would mean ensuring that all schools had the same amount of resources per pupil, which could be an improvement for some schools. However, advocating for equity would mean

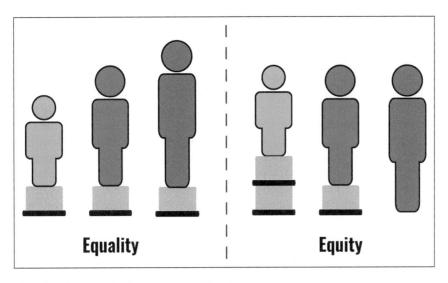

Equality   **Equity**

FIGURE I.2 Equity Compared to Equality
Equality is treating everyone the same. Equity, however, is about giving
everyone what they need to be successful. For example, imagine the image
is showing three people trying to look over a fence to see a baseball game.
Only the person in green can see over the fence. Equality would be providing a step of the same height to all three. Equity would be providing a step
for each that allows each individual to look over the fence.

recognizing that some schools—like those serving students in low-income
or minority communities—may need more resources (funding, experienced teachers, relevant curriculum, etc.) to reduce existing educational
disparities.

Despite equity's prominence in definitions of sustainability and sustainable
development, it is often the least defined and operationalized of the three Es
and frequently takes a back seat in sustainability efforts.

A society is stronger when more members have the education, health, and
other support they need to reach their full potential and contribute fully to
society. Such a community would be more stable, require fewer support goods
and services, and potentially generate more revenue to pay for sustainability
initiatives. Conversely, inequities are a barrier to sustainability.

Part of facing this reality and working to unravel this complex challenge
of discrimination, mistreatment, and disempowerment is to take action on
equity. It is crucial to elevate goals of equity alongside those of economics and
environmental protection as we work to advance sustainability.

WHY DO WE NEED SUSTAINABILITY AND SUSTAINABLE DEVELOPMENT?

Our world is marked by tremendous inequality across many dimensions,
including economic, political, social, and environmental inequalities. Despite
decades of attempts to reduce inequalities, they persist and are the reason
for the emergence of equity considerations in sustainability and sustainable
development.

Global Inequality

Global inequality is marked by disparities in living standards, resource access and use, and quality of life. There are places of wealth and places of poverty. There are places where people enjoy access to health care, jobs, and social services; there are places of deep inequality, discrimination, and poor health. According to estimates made by the UN Development Program (UNDP), the wealthiest 20 percent of the global population earns 82 percent of the total global income. The richest 1 percent has 50 percent of the world's wealth. Table I.1 highlights that uneven development varies over space. For example, there is a huge disparity in income between Germany and Malawi. In Malawi, many people do not live beyond their early sixties, while in Germany, many live beyond eighty years. Hunger and undernourishment are not prevalent in Germany, but they impact nearly half of Haiti's population. Women struggle to find employment in Haiti and Brazil, while those in Germany and the United States are more likely to be employed.

It is also important to consider spatial scale in inequality. Disparities are evident not only between countries but also *within* a country. For example, in the United Kingdom, wealth and economic growth has been concentrated in London, while areas such as Wales and Scotland have experienced less investment and economic growth. In China, rapidly industrializing urban areas along the east coast have created jobs and generated robust economic growth, yet the western regions remain poorer. In Malaysia, the richest 10 percent of the population claim 32 percent of national income.

Spatial scale is also evident in Moldova, where 69 percent of the urban population has access to safe drinking water, compared with only 23 percent of the rural population, signaling a gap between rural and urban development that exists within many countries.

Disparities also exist socially. Globally, women earn less than men and hold only 25 percent of administrative and managerial positions in the business world. Inequality and/or discrimination creates imbalances that leave women at a particular disadvantage in many parts of the world. Another area of disparity is found in the existence of environmental inequalities, which also disproportionately impact women, minorities, and the poor.

Environmental Inequalities and Environmental Justice

Environmental inequalities are pervasive. At the global scale, we can apply the concept to scenarios such as wealthy countries exporting their waste to lower-income countries, where many of those living near waste sites are exposed to hazards and low wages. In countries including Colombia, Indonesia, and

TABLE I.1 Global Differences in Development of Selected Countries					
	Brazil	**Germany**	**Vietnam**	**Malawi**	**USA**
GNI per capita (USD)	15,200	51,680	2,380	1,180	60,200
Life expectancy	75	80	75.5	62	78
Percentage population undernourished	2.5	2.5	6.7	26	2.5
Percentage female unemployment	13.8	3.3	1.9	7.07	3.99

Source: World Bank, *World Development Indicators* (2018).

Honduras, farmers have been forced to sell their land for the production of palm oil, the production of which causes deforestation and water pollution. Another example of environmental injustice occurs around mining; the extraction of iron in Guinea or gold in Burkina Faso generates highly toxic chemicals that affect land and water resources. It also poses harm to miners. A concerning trend is the increased number of cases of antimining activists that have died due to police or army repression. Around the world, the poor and the marginal more often than not live in the areas of the worst environmental quality.

The existence of environmental inequalities, whether in terms of siting of waste plants or the absence of green spaces, raises the issue of **environmental justice (EJ)**. The US Environmental Protection Agency (EPA) defined the term in 1994 as the "fair treatment and meaningful involvement of all people regardless of race, color, national origin, or income with respect to the development, implementation, and enforcement of environmental laws, regulations, and policies."

"Fair treatment" implies that no single group should bear a disproportionate share of negative environmental consequences or a disproportionate lack of positive environmental consequences. "Meaningful involvement" implies that people should have the opportunity to participate in environment decisions and effect regulatory bodies. The international community has largely adopted the US EPA's definition of environmental justice.

Environmental justice issues arise from the obvious fact that there is a correlation between the siting of hazardous facilities and low-income communities and/or minority communities. However, legal redress is difficult if procedures were correctly followed or permits were legally obtained. In many cases, environmental injustices are created and maintained through the everyday operation of the market.

Many countries face challenges associated with environmental injustice. An interactive database called the *Environmental Justice Atlas*, also known as *EJAtlas*, provides information on major global issues including nuclear; mineral ores and building extractives; waste management; biomass and land conflicts; fossil fuels and climate justice; water management; infrastructure and built environment; tourism recreation; biodiversity conservation conflicts; and industrial and utilities conflicts. *EJAtlas* offers an inventory of data as well as information about communities struggling for environmental justice around the world (https://ejatlas.org/).

If the aim of sustainability is to produce more equitable outcomes, we need more positive interventions. Environmental impact statements, for example, need a more explicit assessment of equity and justice issues. Low-income communities facing environmental challenges should also receive greater resources from government, in addition to equal treatment under the law. Unless more positive outcomes are engineered, the system will continue to produce inequitable results.

How can we explain these global inequalities? One way is to understand the theory and practice of development since 1950.

Development 1950–87

A long history of economic development has led to unsustainable practices, behaviors, and cultural norms that have caused environmental degradation and exploitation, and an uneven distribution of wealth and resources. It has

KEY TERMS AND CONCEPTS I.1

The Global North and the Global South

How we refer to the places in the world is enmeshed in wider value judgments. The geographical terms used to describe the world are constantly changing. In the 1990s, the term "developed" was used to describe the previous First World, and some of the Second World countries. The terms "developing" or "underdeveloped" replaced the term "Third World." Yet even these terms are problematic.

First, no one single definition is widely agreed upon. The United Nations does not have an official definition of a "developing country." It is highly problematic to see the world in such binary terms. For example, China, Peru, and Malawi, which have very different incomes, are all lumped under the term "developing."

Second, using the term "developing" or "underdeveloped" indicates some type of failure of progress, and further implies that many countries lack the ability to be prosperous, trapped in a permanent state of poverty and misery.

Third, it belies the tremendous progress that many countries have seen over the last twenty years.

Fourth, and most importantly, these categories reflect an amnesia about how that gap came into existence. It is not the case that "developing" countries have been slower in catching up to the rest of the developed countries, but that the wealth and prosperity of the developed countries was based on the underdevelopment or exploitation of the

so-called developing countries. Colonial and imperial histories are vitally important in setting the context for levels of development and the "gap" between the richest and the poorest. These legacies still impact economic, political, and social structures and institutions, even half a century after decolonialization. The term "developing country" is not due to lack of progress but an unfair distribution of wealth and the transfer of resources from colonies to imperial centers. It is vital to understand that levels of development and levels of income are the outcome of such processes and relationships, some of which have been operating for a century or longer, and not the "failings" of a people or its government.

In 2016, the World Bank eliminated the term "developing country" from its data indicators. The World Bank noted that this was important to do, not only because the terms were problematic but also because these terms shaped people's mental models of the world. This marks an evolution in thinking about the geographic distribution of poverty and prosperity. The World Bank has since adopted four levels of income classification: high income, upper-middle income, lower-middle income, and low income for its data collection, as shown in Table I.2. Research has shown that measures of GNI per capita (income) are a useful proxy for social indicators.

Today, the terms "developed" and "developing" have been replaced by the terms

Category	GNI per capita (Gross National Income)	Population	Examples
High income	$39,820	1.36 billion	Australia, USA, UK, Russia
Upper-middle	$7,598	2.4 billion	Argentina, China, Mexico
Lower-middle	$4,370	2.56 billion	Bolivia, Nigeria, Pakistan
Low income	$709	848 million	Afghanistan, Haiti, Rwanda, Malawi, North Korea

TABLE I.2   The World Bank's World Income Classification

Source: World Bank, "How Does the World Bank Classify Countries?" (May 23, 2021). https://datahelpdesk.worldbank.org/knowledgebase/articles/378834-how-does-the-world-bank-classify-countries%20May%2023

"Global North" and "Global South." Most of the world's richest countries are in the Global North. Many of the world's lowest-income countries are located in the Global South. But even this new division is flawed. There has been significant progress in the last several decades and the global economy has created more of a continuum of development. For example, Brazil, Indonesia, and Vietnam, countries in the Global South, are considered good investments because of a strong base of natural resources, demographic factors, and projections for future economic growth.

This textbook will use the terms **Global North** and **Global South**. It will also refer to "high-income countries" and "low- or lowest-income countries" as a way to underscore the biggest "gaps" among countries. However, even this categorization should be treated with caution: How we divide up the world and the terms we apply are not innocent of wider political, social, and economic implications.

also resulted in uneven economic development, where glaring economic inequalities exist among countries and even within countries. Uneven development is both a material condition and a theory of how this material condition came to be. To explain uneven development and its consequences in the contemporary world is a difficult and complex task, particularly because many of the causes have been unfolding over centuries. In this section, we look briefly at the way development was institutionalized starting in 1950.

In the modern era a predominate worldview was based on the idea that nature was something to subjugate, defeat, and use. The onset of the industrial revolution in the 1850s in Europe and the United States profoundly and irrevocably shifted human relationships with their physical environment. The economic foundations of the industrial age were the exploitation of the coal mine, the vastly increased production of iron, and the steam engine. This **anthropocentric** worldview saw the world as little more than a "resource" for human use. Factories and smokestacks were seen as indicators of progress, and economic growth was not only good, but it was also imperative. Countries in western Europe and the United States and Canada experienced economic growth and improved quality of life. Although this prosperity was based on accelerating resource use and resource waste, few people attached importance to preserving the quality of the environment.

After World War II, and in the context of decolonialization, the world was categorized and divided into the First World of capitalist economies with the United States in a leading role, a Second World of communist countries headed by the Soviet Union, and the Third World, which included a wide range of countries from Argentina to Cameroon and India. The categorization of Third World implied a description of poor countries. A few decades later, the term Third World would be substituted by "underdeveloped" or "developing" (*Key Terms and Concepts I.1* discusses the problematic nature of these terms and introduces the terms "Global North" and "Global South"). There was a general agreement among Western countries that developing countries needed assistance. The establishment of institutions such as the World Bank was designed to assist in "development." A major goal of **development** was the eradication of poverty so prevalent in the Global South. The World Bank funded projects such as roads, railways, dams, and port facilities to construct the infrastructure needed to industrialize and modernize economies in the Global South.

From the 1950s until the 1980s, neoclassical or conservative economics dominated international development theory and practices. A key tenet of neoclassical economics is a progrowth agenda and the belief that the free market will control and correct all that is needed for human benefit. Broadly, the theories behind neoclassical development consisted of the idea that developed countries were better, were harder working, and had stronger institutions and policies, while poor countries or those "less developed" just needed to catch up. The international community encouraged poorer countries to accept the advice of experts such as the World Bank and follow the West's path to modernization. It was assumed following this strategy would result in economic growth, which in turn would reduce a country's level of poverty and improve the health and living standards for its people. The effectiveness of development projects was measured by assessing growth in macroeconomic variables, such as GDP, exports, or Foreign Direct Investment (*Key Terms and Concepts I.2*). Measurements of development were limited to measurements of economic growth. Factors such as health, crime, discrimination, and environmental degradation were not factored into measures of success.

By the 1980s decades of development had not eradicated global poverty or resulted in the promise of economic growth and prosperity for many countries in the Global South. This failure was the impetus behind the Brundtland

 KEY TERMS AND CONCEPTS I.2

Economic Development and GDP, GNP, and GNI

There are many different ways to measure economic development. For much of the twentieth century, economists relied heavily on a single macro-level measurement: the gross domestic product (GDP) of a country. The GDP accounts for the total production occurring with in the geographic boundaries of a country, in a given year. The gross national income (GNI) also counts all goods and services produced. Gross national product (GNP) includes the earnings from all assets owned by residents, but it also includes earnings that do not flow back into the country. And it omits the earnings of all foreigners living in the country, even if they spend it within the country.

The more preferred measurement is now GNI. GNI measures all income of a country's residents and businesses, regardless of where it is produced. GNI is preferred because it captures more of the economic dynamics happening in developing economies. Consider the Philippines, where many of its citizens have migrated to other countries so they can earn a higher wage. Many Filipino immigrants send money, or remittances, back to their families in the Philippines. The income from these remittances can drive economic growth. This would be counted in GNI, but not in GDP. The World Bank now uses GNI rather than GNP or GDP (although it is still common to see the other terms in older publications and reports).

GDP, GNP, and GNI are often equated with a country's standard of living but each has been criticized because these measures:

- Fail to account for the distribution of wealth within a country;
- Do not show how that income is spent;
- Fail to measure human well-being or happiness of life satisfaction;
- Fail to account for negative externalities such as the generation of pollution or the use of funds to combat crime, which can result in an increase of GNI.

For these reasons, we should be highly cautious about using these measurements to assess sustainability.

Commission's charge to critically assess whether development, as practiced between 1950 and 1980, was effective. The Brundtland Report concluded that development as was then theorized and practiced was not delivering on its promise to reduce poverty, improve human health, and protect the environment. In some places, poverty had increased, human health had worsened, and the environment had been further exploited. There had been economic growth without "development." Traditional development, the report concluded, was not sustainable. What was needed was a shift to sustainable development.

Since the Brundtland Report, the concept of development and of sustainable development has been the subject of scholars for decades and it is difficult to do justice to the vast research on this topic here in the introduction. Although development research is multifaceted, complex, and continually advancing, we can place contemporary development scholarship into a continuum within three main theoretical schools. The first is conservative free-market economists, who adhere to neoclassical or conservative economics. The second are the liberal reformers, who believe economic growth is the way to sustainability. They argue that what is needed are reforms to the existing political and capitalist economic system. The third is the critical political economists who see reform as ineffectual and instead advocate for systemic transformation of both the economy and our value systems.

Conservative Free-Market Economists

Conservative free-market economists have their historical links in Adam Smith and John Maynard Keynes and modern concepts of free trade and the free market and development. They do not all adhere to an explicitly "endless growth" directive, but they do believe that the cure for poverty, unemployment, debt, crime, and other woes is to encourage economic growth through the private market. In other words, capitalist economic growth is still the main way to reduce global inequalities.

Starting in the 1990s, neoliberalism characterizes conservative free-market economics. Neoliberalism promotes policies that allow capitalism to generate increased development. Its trademark policies of privatization, deregulation, tax cuts, and free trade deals are strategies to facilitate economic growth. Neoliberals tend to see human–environmental relationships as opposites: Environmental protection comes at the expense of jobs and economic growth. Because of their embrace of the private market, neoliberals generally argue for less environmental regulations (not more), and in the United States neoliberalism has been behind recent efforts to scale back on air-quality standards, shrink public lands, and pressure for withdrawal from the Paris Climate Agreement, while maintaining fossil fuel subsidies and increasing tax cuts for corporations. In some instances, neoliberal policies have increased economic inequality, exacerbated poverty, and created the enduring patterns of uneven development that have been caused by the Global North.

Liberal Reformers

For many economists, neoliberalism does not sufficiently address global inequalities. Some economists are not fully aligned with a free-market ideology but are not so radical that they see capitalism as the problem. Liberal reformers seek to understand and shape macro- and microeconomic policies to lift poor countries out of poverty. Liberal reformers are not convinced (as neoliberals are) about the beneficial effects of free markets and see a need for public policies to produce more equitable outcomes. They are more critical in their viewpoints and inclined to support "interventionist" policies. Government taxes and expenditures that reduce income inequalities would be examples of such public policies.

Even within the liberal reformers framework, there are many offshoots. Economists may differ on the extent to which regulation should occur and how, and the extent to which institutions, such as the World Bank, need reform. There is also considerable disagreement about what types of institutional reforms should occur. Milton Friedman, for example, argued for free trade, smaller government, and a slow, steady increase of the money supply in a growing economy. His emphasis on monetary policy became known as monetarism.

Reformists have a long history of studying economic inequality and include economists such as J. K. Galbraith, Joan Robinson, and Gunnar Myrdal; they also include contemporary economists such Joseph Stiglitz, Jeffrey Sachs, Amaytra Sen, and William Easterly. Stiglitz has advocated for a limited but important role of government, noting that unrestrained or unregulated markets do not often work well. He has criticized the World Bank and IMF about its conventional approaches to assisting low-income countries. So too have William Easterly and Dambisa Moyo. William Easterly, in his book *The Tyranny of Experts*, states that the World Bank has failed to address the core of the problem of inequitable development. Dambisa Moyo has been critical of foreign aid in Africa, noting that development assistance has fostered dependency, encouraged corruption, and perpetuated the cycle of poverty. These economists have been highly critical of a neoliberal free market and of institutions such as the World Bank, but they believe that reform is possible through better interventions.

The economist Jeffrey Sachs, who was once considered a driving force for neoliberalism, has more recently reinvented himself as a kinder, liberal reformer. Sachs advocates for making governments more responsible global partners and has openly rejected neoliberalism. In his book *The Age of Sustainable Development*, Sachs describes the history of world economic development and the factors that contribute to some nations being more impoverished than others. He identifies numerous complex and interrelated factors that have resulted in inequitable development including poorly conceived economic policies in the short term and long term, a lack of industry and services, low educational levels, and cultural barriers that discriminate against women and minorities. Sachs also states that inequalities and a lack of sustainability are caused by problems with governance—corruption, inefficiency, and incompetence. The problems can be overcome through economic, political, and social reforms. Sachs suggests several strategies for sustainability. One is that richer countries increase their foreign aid and assistance. This would allow poorer countries to pay for essential needs and would reduce poverty. Another is to reduce political corruption that will lead to more foreign investment, which will in turn help grow the economy and reduce poverty. Sachs has also proposed cancelling the debt of many of the poorest countries in the Global South and advocates for the establishment of social funds to improve public health care and education systems. His type of liberal reform has dominated development economics—in both theory and practice—for the better part of the past two decades, as evidenced by the fact that he has served as one of the thought

KEY TERMS AND CONCEPTS I.3

The Human Development Index

In 1990, the United Nations created the **Human Development Index (HDI)** to broadly assess a country's level of development and to move beyond measuring solely economic factors. The HDI is a summary measure of the achievement in three key dimensions of human development:

1. A decent standard of living, measured by GNI per capita;
2. A long and healthy life, measured by life expectancy at birth; and
3. Being knowledgeable, measured by mean years of schooling.

The HDI is the geometric mean of normalized indices for each of the three dimensions. The index is on a scale of 0–1.0, with 1.0 being the highest level of development. Currently, no country has received a 1.0 on the HDI.

Map I.1 shows the most recent HDI rankings and suggests that although some countries are enjoying periods of exceptional wealth and success, the gap between them and the poorest countries is still considerable. It remains an inequitable world. Figure I.3 shows the improvements in HDI over time.

Table I.3 highlights the highest and lowest HDI countries. The highest-ranking countries are in Europe; the lowest-ranking countries are concentrated in sub-Saharan Africa. Interestingly, the United States is not ranked in the top 10 in the HDI.

What explains Norway's high HDI ranking? First, Norwegians have a relatively high life expectancy of almost eighty-two years, in part due to Norway's accessible and affordable public health care systems. Second, Norwegians spend an average of 17.7 years in school, a measurement that reflects on levels of knowledge as well as of choice, both of which indicate a high level of human development. Third, Norway has the seventh-highest GNI per capita in the world, but most experts agree that that alone is not sufficient to explain Norway's ranking. The country also

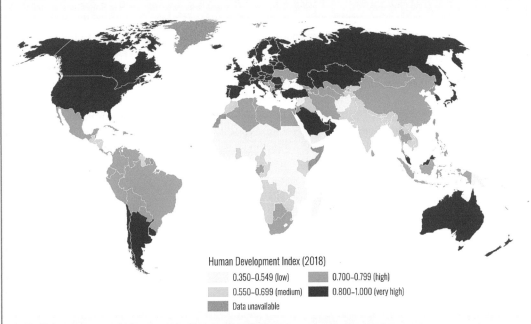

Human Development Index (2018)

- 0.350–0.549 (low)
- 0.550–0.699 (medium)
- Data unavailable
- 0.700–0.799 (high)
- 0.800–1.000 (very high)

MAP I.1 The Human Development Index (2018)
Geographically, countries in the Global North have higher levels of human development than most countries in the Global South. There are exceptions, such as Chile, Argentina, and Uruguay in South America, and the rich oil states of Saudi Arabia and the UAE, which have very high HDIs comparable to the Global North. Regionally, Africa has among the lowest levels of human development.

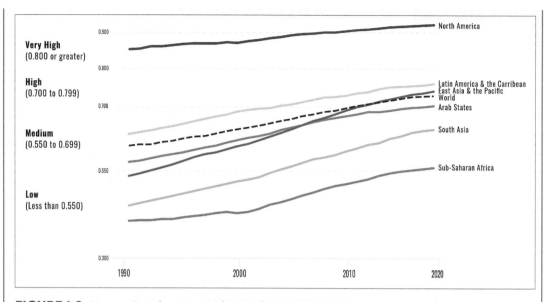

FIGURE I.3 Human Development Index Ranking, 1990–2020
Human development has been steadily improving for all countries, although there remain significant differences between regions, and between the Global North and Global South.

has strong institutions and a holistic approach to development. It has made strides in environmental sustainability and climate mitigation, and it also strives for gender parity in parliament.

It is noteworthy that HDI scores have improved for *all* countries over the last twenty years. This is true even for Niger, the lowest-ranked country in the HDI. Two factors appear to contribute the most to low rankings: gender inequality and armed conflicts. Other recurring factors include

extreme egalitarianism, high inequity and inequality of distribution of wealth, surplus labor and low productivity, and excessive government spending and corruption.

The HDI simplifies and captures only part of what human development entails. It does not reflect on inequalities, poverty, human security, or empowerment. For example, it does not include factors that can have a significant influence on quality of life, such as environmental

TABLE I.3 Human Development Index for Selected Countries

Highest-ranking HDI countries		Lowest-ranking HDI countries	
1. Norway	.957	180. Eritrea	.459
2. Ireland	.955	181. Mozambique	.456
3. Switzerland	.955	182. Burkina Faso	.452
4. Hong Kong, China (SAR)	.949	183. Sierra Leone	.452
5. Iceland	.949	184. Mali	.434
6. Germany	.947	185. Burundi	.433
7. Sweden	.945	186. South Sudan	.433
8. Australia	.944	187. Chad	.398
9. Netherlands	.944	188. Central Africa Republic	.397
10. Denmark	.940	189. Niger	.394

Source: United Nations, "Human Development Index" (2020). http://hdr.undp.org/en/content/latest-human-development-index-ranking

degradation, which in turn can impact human health and mortality rates.

Although the HDI may not fully capture all of the complexity of development, it is not realistic to expect any index that accommodates the diversity of all countries' development to do so. For now, the HDI will serve as a reference point for new indices as we move toward understanding human development in a more comprehensive and capabilities-focused way. It remains the global standard measure of sustainable development.

leaders behind the creation of the UN Millennium Development Goals and the Sustainable Development Goals.

One impact liberal reformers have had is in changing the way institutions such as the United Nations and World Bank measure development. No longer do these institutions solely rely on economic indicators; instead most have adopted the Human Development Index as their measure (*Key Terms and Concepts* I.3).

One glaring problem with either the conservative/neoliberalism or liberal reform approach is that the free market has proven itself unable so far to protect the environment, eradicate poverty, ensure political and social freedom, and deliver genuinely sustainable economic systems. For this reason, critical development scholars reject the free market as a solution for sustainable development.

Critical Political Economy

Critiques of development materialized in the 1980s, in part, led by scholars who were analyzing the consequences of development under free-market approaches. This view regards liberal redistributive reforms as "band-aids" and sees this as insufficient to resolve the system injustices that arise from extreme inequalities. "Critical political economy" is a broad term given to frameworks that have their intellectual roots in Marxian economics. In contrast to free markets, liberalization, and privatization, critical development places class, race, gender, and empire at the center of the discussion on development. It views capitalism as the problem rather than the solution. Theirs is a more radical framework.

Critical political economy scholarship examines the processes, policies, and practices that reproduce conditions of poverty, inequality, and oppression around the world. There is a focus on the poor and less powerful who are most adversely affected by exclusionary power relations and practices. At the heart of critical development studies is analyzing the systemic changes needed to transform the current world to one where economic and social justice and environmental integrity prevail.

Scholars such as Richard Peet, Elaine Hartwick, and David Harvey have helped advance a political economy perspective. Political economy attempts to explain the relationships between economic production and political processes. Political economists argue that global inequalities and uneven development were an *inevitable* consequence of the accumulation of capital in a capitalist system. Capitalism created **uneven development**. The term "political economy" has expanded in meaning beyond uneven economies, to include uneven resource distribution and use, uneven sociocultural indicators (gender, class, race, ethnic group, age), and more. The term also implies uneven geographical development that involves a number of different metrics (employment rates, income levels, rates of economic growth, and so on), and it has been described at all geographical scales—from intraurban disparities all the way through subnational regional differences to uneven international development.

A political economy framework provides an explanation for uneven development and global inequality. Jason Hickel, in his book *The Divide: A Brief Guide to Global Inequality and Its Solutions*, asserts that poor countries are poor because they have been integrated into the global economic system on unequal terms. Colonial and imperial histories are vitally important in understanding the context for levels of development and the "gap" between the richest and the poorest. These legacies still impact economic, political, and social structures and institutions, even half a century after decolonialization. Political economy scholars argue these structures have caused economic development that favored some countries more than others, and that some countries are rich not because they are further along some development trajectory, but because they had an initial advantage that was then reinforced by colonialism and economic imperialism.

Political Ecology

The phrase "political ecology" combines the concerns of ecology and a broadly defined political economy. It links critical social theory with the environment. Starting in the 1990s scholars such as Piers Blaikie, Murray Bookchin, Farhana Sultana, Michael Watts, and Erik Swynegedouw advanced **political ecology** as a framework for understanding economic inequalities and environmental degradation. Like political economy, political ecology has a focus on issues of power, but in relation to explaining environmental impacts. Political ecology studies the relationships between political, economic, and social factors with environmental issues and changes. Many environmental problems are less a problem of poor management, inappropriate technology, or overpopulation, and instead are social in origin. Michael Watts' research on the rapid deforestation in eastern Amazonia, argued that deforestation must be understood in terms of why those who were clearing tropical rainforests did so in pursuit of economically inefficient and environmentally destructive cattle ranching, and how the social forces of ranchers, peasants, workers, and transnational companies were shaped by larger political-economic forces, such as the role of Brazilian government subsides, class alliances, and corruption, as well as a global food system.

Other research has examined the causes of land degradation and erosion in the Global South. These studies challenged the conventional thinking that degradation was the result of the ignorance or indifference of peasant farmers, and instead argued it was the outcome of fundamental economic and political inequalities at the local, national, and even global levels, which concentrated landownership in the hands of a small group of elite.

There have emerged other fields of study as political ecology has deepened its concerns and areas of research. One example is Jennifer Wolch, who has studied how animal rights can be understood within a political ecology framework. Another example is the emergence of feminist political ecology. Scholars such as Yaffa Truelove, Vandana Shiva, and Ariel Salleh focus their attention on the intersection of capitalism, patriarchy, and the environment. Their work sees gender as a crucial variable in understanding both the environment and development. Vandana Shiva, for example, examined how rural Indian women experience and perceive ecological destruction and its causes, and how they have conceived and initiated processes to arrest the destruction of nature and begin its restoration. In 2021, the Feminist Economic Justice for People & Planet Action Nexus released a report about gender inequality. Their central message is that at the root of gender inequalities and many of the violations of women's human rights are economic policies that have failed most of the world's population and, most acutely, women and girls. It concludes that the entrenched patriarchy and the neoliberal economic model makes it difficult for anyone, especially women and other marginalized persons, to critique or challenge it.

The political ecology framework sees an inherent incompatibility between the current world economy and the ideal of sustainability. The current form of capitalism only exacerbates inequalities between genders, races, the Global North, and the Global South. Political ecologists criticize free-market reformers for their inattention to inequalities in wealth distribution and for remaining locked in a mantra of free-market progrowth. These experts argue that sustainability will not come from *more* growth. Only a fundamental transformation in economic and political systems will ensure real change.

A critical/political economy framework is useful for understanding the flaws in and limitations of the contemporary capitalist system: It is not yet as effective at providing an alternative way forward. How would a systemic transformation occur? What would a postcapitalist economy look like? Giorgos Kallis and colleagues have recently written about the case for "degrowth," but transforming economic systems so radically remains conceptually vague and hazy. The reality is that capitalism is not going away any time soon.

While there are competing theories of development and sustainable development, this textbook leans toward a critical/political ecology perspective, while also being mindful that sustainable development is currently operationalized and implemented in a world that remains oriented to capitalism. In the short term, sustainable development will be led by existing institutions and systems. It is important for anyone interested in sustainability to understand how this is implemented within the existing political, economic, and social systems, and to what degree progress is realized, all while retaining a critical perspective.

OPERATIONALIZING SUSTAINABILITY: THE MILLENNIUM DEVELOPMENT GOALS

A key challenge in sustainability is that it is so broad-ranging and ambitious that it is not easy to create a set of underlying principles that establishes immediate practical relevance and action. How is sustainable development to be achieved? How do we operationalize it?

Since the Brundtland Report, there have been numerous international conferences and summits attempting to do so. One of the most significant milestones was the global "Earth Summit" held in Rio de Janeiro in 1992. More than 172 governments were represented, and a total of 17,000 delegates participated in the conference sessions. Its major contribution was to reassert that concerns about the environment and development were equal. The success of the Rio Earth Summit helped to establish further global gatherings hosted by the United Nations, but by the late 1990s, there was a call to reinvigorate political will for action and implementation.

At the beginning of the new millennium, world leaders gathered at the United Nations to shape a broad vision to fight poverty in its many dimensions. That vision was translated into eight **Millennium Development Goals (MDGs)** and set a global framework between 2000–2015. This gathering also galvanized unprecedented efforts and set concrete goals and steps toward reaching the goals (Table I.4).

The MDGs helped to lift more than one billion people out of extreme poverty, make inroads against hunger, enable more girls to attend school than ever before, and address environmental issues. They generated new and innovative partnerships, motivated public opinion, and showed the value of setting ambitious goals.

Although the MDGs have been called a "success," they have also been widely criticized. Some critiqued the MDGs for being generated in a non-transparent way, driven by development ministers and heads of development agencies seeking a new rationale for aid. Economist Samir Amin criticized

TABLE I.4   The Millennium Development Goals Achievements

Goal	Progress made by 2015
Goal 1: Eradicate extreme poverty and hunger	• The number of people living in extreme poverty declined by more than half, falling from 1.9 billion in 1990 to 836 million in 2015.
Goal 2: Achieve universal primary education	• The primary school net enrollment rate in the developing regions reached 91 percent in 2015, up from 83 percent in 2000. • Sub-Saharan Africa had the best record of improvement in primary education of any region since the MDGs were established. The region achieved a 20-percentage-point increase in the net enrollment rate from 2000 to 2015.
Goal 3: Promote gender equality and empower women	• Many more girls are now in school compared to 2000. • The developing regions as a whole have achieved the target to eliminate gender disparity in primary, secondary, and tertiary education.
Goal 4: Reduce child mortality	• The global under-five mortality rate has declined by more than half, dropping from 90 to 43 deaths per 1,000 live births between 1990 and 2015. • Measles vaccination helped prevent nearly 15.6 million deaths between 2000 and 2015. The number of globally reported measles cases declined by 67 percent for the same period.
Goal 5: Improve maternal health	• Since 1990, the maternal mortality ratio has declined by 45 percent worldwide and most of the reduction has occurred since 2000.
Goal 6: Combat HIV/AIDS, malaria, and other diseases	• New HIV infections fell by approximately 40 percent between 2000 and 2013, from an estimated 3.5 million cases to 2.1 million. • By 2015, 13.6 million people living with HIV were receiving antiretroviral therapy (ART) globally, an immense increase from just 800,000 in 2003. ART averted 7.6 million deaths from AIDS between 1995 and 2015.
Goal 7: Ensure environmental sustainability	• In 2015, 91 percent of the global population was using an improved drinking water source, compared to 76 percent in 1990. • Of the 2.6 billion people who have gained access to improved drinking water since 1990, 1.9 billion gained access to piped drinking water on premises. • Globally, 147 countries met the drinking water target, 95 countries have met the sanitation target, and 77 countries have met both.
Goal 8: Develop a global partnership for development	• Official development assistance from developed countries increased by 66 percent in real terms between 2000 and 2015, reaching $135.2 billion.

Source: United Nations. *The Millennium Development Goals Report: 2015.* New York: United Nations (2015).

the MDGs for being an initiative pushed primarily by the Global North onto the Global South. Human rights scholars and practitioners criticized the MDGs for not adequately aligning with human rights standards and principles. Scholar Sakiko Fukuda-Parr, who has written widely on human development and human rights, noted that many civil society groups found the MDGs problematic because of the omission of inequality, weak goals on global "partnership" that lacked quantitative targets, and the lack of ambition in the targets. Other omissions from the MDGs include women's reproductive health issues, governance, conflicts, economic growth and employment, and many other important objectives. These critiques have led some to disparage the MDGs as the "Minimum Development Goals." Jason Hickel also takes issue with the true extent to which hunger and poverty were reduced with the MDGs, noting that what counts as the poverty line shifted from $1.08 per day in 2000 to $1.25 in 2008, thus it appeared that fewer people were poor, even though nothing had really changed.

Additionally, the MDGs were criticized for being too simple and contextualizing the concept of development as the elimination of poverty. By equating development with poverty reduction, the MDGs failed to focus on enlarging the productive capacity of economies to improve living standards, or to address much needed institutional reforms. Some development organizations and other stakeholders were frustrated that the targets were not in synch with their agendas and visions. Other critics said such a short list of goals could not sufficiently address the complexity of development challenges.

Some critics pointed out that the MDG goals and targets—such as universal primary education—were mostly relevant for low-income countries of the Global South and gave the impression that only these countries needed investment and strategies.

Despite the progress made, most of the MDGs were not fully achieved, and there was a consensus that more collective, long-term effort was needed. The MDGs needed a sequel that would address some of these criticisms.

THE SUSTAINABLE DEVELOPMENT GOALS

In 2013 the United Nations created a thirty-member Open Working Group of the General Assembly to prepare a proposal for a new global agenda. In addition to the thirty countries representing the Working Group, the United Nations also assembled experts and representatives from stakeholders that represented specific social groups with expertise on issues relating to women, human rights and indigenous peoples, farmers, trade unions, and children and youth. There were also twenty-five thematic working groups whose experts provided input on conflict, violence, and disaster; education; energy; environmental sustainability; food security and nutrition; governance; growth and employment; health; addressing inequalities; population dynamics; and water. In addition, the United Nations held national consultations with more than 100 countries that included governments and civil society groups. During this process, numerous governments, nongovernmental organizations' (NGOs') grassroots activists, scholars, and researchers, and other stakeholders were deeply engaged in developing, discussing, and deliberating on the goals, especially when compared to the process of creating the MDGs.

In 2015 after several years of negotiations and debate, 193 countries agreed to adopt the Sustainable Development Goals (SDGs), officially known as

Transforming Our World: The 2030 Agenda for Sustainable Development. Some have referred to the consensus achieved by UN member states in formulating and agreeing to the SDGs as nothing short of miraculous. The **Sustainable Development Goals** build on the success of the MDGs and aim to go further to end all forms of poverty. The SDG agenda also includes issues that were not in the MDGs such as climate change, sustainable consumption, innovation, and the importance of peace and justice for all (Figure I.4). Whereas the MDGs focused on Global South countries, the SDGs are viewed as more universal in that they call for action by all countries—high-income, middle-income and low-income—to promote prosperity while protecting the planet. Some experts argue the SDGs differ from the MDGs by incorporating a broader and more transformative agenda and taking into consideration the ongoing debates about development priorities. There is also more balance among the three Es of sustainability within the SDGs compared to the MDGs. The seventeen SDGs set out to tackle a diverse range of issues including ending hunger, improving health, combating climate change, and protecting oceans and forests. They recognize that ending poverty must go hand-in-hand with strategies that build economic growth and address a range of social needs including education, health, social protection, and job opportunities, while tackling climate change and environmental protection.

Within each SDG is a set of targets. Combined, there are 169 targets to achieve by 2030. These targets will direct efforts to implement the three Es of sustainability and are the guiding agenda for the protection of both the local and the global commons. The SDGs set an agenda for investment in advancing sustainability; as a result, it is estimated that there has been a $200 billion increase in development investment efforts since the SDGs were launched.

FIGURE I.4 The Seventeen Sustainable Development Goals

CRITIQUING THE SUSTAINABLE DEVELOPMENT GOALS

Critical political economy scholars and some practitioners have critiqued the SDG agenda, the goals, and the negotiation process. And rightfully so. Some have been disparaging of the SDG creation process, noting that the goals and targets were the result of compromise. For example, SDG 16 on governance and human rights was a contentious goal, and its inclusion was initially resisted by some over the inclusion of LGBTQ+ rights; ultimately it was included, but with weaker language. This has led some to say that using the concept of sustainability as a "catchall" solution for global problems may be fraught, particularly because any sustainability agenda can also be politicized.

Some critics believe the seventeen SDGs lack coherence by being too vague and too numerous. One issue is with the wording, with some claiming that some targets should have been constructed more clearly. Another criticism is that the SDG targets are a result of international negotiation, and as such may not reflect local or national sustainable development priorities and problems. While the SDGs and targets are considerably broad, there may still be gaps from local and national perspectives.

An additional concern is that countries could neglect some of the goals through loopholes of selectivity or oversimplification. While defenders say these goals better reflect the complexity of development, opponents say that too many targets make it difficult to prioritize. While some critics say there are too many goals, others counter that there are "missing" goals, including a specific goal around population growth. Still others criticize the goals for being focused on metrics, whereas genuine sustainability is a broader theory. They worry that those countries that meet their targets will consider themselves "sustainable."

There has also been concern raised that the targets included in the SDGs are not the right ones. For example, the Copenhagen Consensus Centre wrote a cost-benefit analysis report on the SDG targets and concluded that fulfilling some of the targets would provide less value for the money spent. For example, expanding immunization programs would yield better results than doubling the amount of HIV medications available for the sickest. Similarly, there would be greater value in preventing coral reef loss than in increasing the number of designated protected areas.

Political economy scholars have written that the SDGs do not explicitly challenge individuals or society to transform fundamental cultural values that influence how we perceive our relationship with the natural environment. Yet many political ecology scholars argue that genuine sustainability will require changing the anthropocentric mindset that has dominated the world for the past several hundred years. These critics label the SDGs and the global agenda as "greenwashing"—noting that most efforts to address sustainability will not be successful without a radical transformation of the market-oriented global capitalist system and the cultural value system that places economic growth ahead of other interests. They contend that the SDGs are a simplistic vision of meeting basic needs for all without recognizing the root causes of poverty embedded in power relations and exacerbated by current economic models that prioritize globalization and corporate profit over human rights.

Political economy and political ecology scholars also say that the United Nations and the SDGs take a flawed pro-economic-growth approach. For example, Hayden Washington and Paul Twomey contend that the SDGs need to be decoupled from economic-growth targets; otherwise they are only partial solutions. The SDGs appear to have contradictory targets related to economic growth. Jason Hickel sees the economic growth targets in Goals 8 and 9 to be in contradiction to the goals to protect the environment in Goals 6, 13, 14, and 15. For these scholars a progrowth agenda fails to create transformational change toward **ecocentrism**, where we acknowledge that ecological conditions set the foundation for human life.

It is too early to evaluate whether the SDGs will live up to their potential and promise. Indicators and monitoring processes are still being developed and national strategies are still being drafted. Of course, the real litmus test will be implementation.

Despite these legitimate criticisms, the SDGs represent a clear shift in development theory and practice from seeing poverty and development as separate from environmental concerns, to recognizing that the two are intimately bound together. The SDGs are considered the global consensus on defining and implementing sustainability and sustainable development through 2030. And despite their limitations, they are an important vantage point from which to explore the concepts of sustainability and sustainable development. This is because the diversity of topics covered in the SDGs, as well as the attention and financial commitments of governments, civic organizations, and the private sector, make understanding the SDGs a crucial part of understanding current efforts in sustainable development. It is the reason the SDGs form the framework for this book.

SUMMARY AND PROGRESS

The SDGs are a global agenda that attempts to coordinate efforts, strategies, and priorities. The SDGs are first and foremost a public investment program—in core infrastructure (roads, power, communications, water and sanitation), human capital (health, education), and the environment. They are also part of a system of international governance (see *SDGs and the Law* I.1).

The lead is the United Nations, an international organization consisting of numerous programs, funds, and initiatives, all of which constitute the "United Nations system." The United Nations is an international organization founded in 1945, currently made up of 193 member states. The United Nations consists of six principal organs (The General Assembly, Security Council, Economic and Social Council, International Course of Justice, Trusteeship Council, and the UN Secretariat), and numerous separately administered funds and programs. Throughout its history the UN General Assembly has established a number of programs and funds to address particular humanitarian and development concerns, referred to as the UN system.

Some of these offices, specialized agencies, funds, and programs of the UN system include UN Children's Fund (UNICEF), Food and Agriculture Organization of the United Nations (FAO), World Health Organization (WHO), and the UN Educational, Scientific and Cultural Organization (UNESCO). External partners include the World Bank, International Monetary Fund (IMF), Organization for Economic Cooperation and Development (OECD), and the World Trade Organization (WTO).

SDGs AND THE LAW I.1

There are many policy instruments that can be used to advance sustainability at many scales. Local, state/province, and national governments can use regulation, legislation, and incentives.

There are also international policy instruments. At the global scale, the United Nations has more than 560 multilateral treaties that cover a broad range of subject matters such as human rights, disarmament, and protection of the environment. A treaty may also be known as an international agreement, protocol, covenant, convention, pact, or exchange of letters, among other terms. Some examples of important UN treaties include:

- The Paris Climate Agreement
- Convention on the Rights of the Child
- The Convention on Biological Diversity
- Chemical Weapons Convention
- International Convention on the Elimination of All Forms of Racial Discrimination

A declaration is a document stating standards or principles but is not legally binding. In contrast, a treaty, convention, covenant, or charter is a formal, legally binding written agreement between actors in international law. It is usually entered into by sovereign states and international organizations, but can sometimes include individuals, or business entities. Generally, the United Nations writes a draft of the treaty. Then parties negotiate the terms of the treaty. It is then open to ratification by member states who can also make reservations (exceptions that will not apply to them). A convention becomes legally binding to a particular state when that state ratifies it. Signing does not make a convention binding, but it indicates support for the principles of the convention and the country's intention to ratify it.

In international law and international relations, a protocol is generally a treaty or international agreement that supplements a previous treaty or international agreement. A protocol can amend the previous treaty or add additional provisions. Parties to the earlier agreement are not required to adopt the protocol. Sometimes this is made clearer by calling it an "optional protocol," especially where many parties to the first agreement do not support the protocol. A good example is the UN Framework Convention on Climate Change (UNFCCC), which established a framework to develop greenhouse gas emissions, while the Kyoto Protocol contained the specific provisions and regulations that were later agreed upon.

The international system may have implications at the national level. For example, a treaty may affect a country's legal system being incorporated into national law, or through direct application through constitutional provisions to national law or due to legal reforms.

NGOs and national human rights institutions have also used treaty standards in relation to proposed government legislation and policies.

Finally, it is important to distinguish between hard law and soft law. Hard law is legally binding, legally enforceable, and includes most international treaties. However, soft law is not legally binding. Soft law includes principles, declarations, and resolutions. Confusing the issue is that sometimes hard laws are really soft laws. For example, the 1992 Rio Declaration is a soft law, but the 2011 UN Conventional on Biological Diversity and the 2016 Paris Climate Agreement are considered hard laws.

Like their predecessors, the UN SDGs are considered a soft law because the agenda is a voluntary agreement rather than a binding treaty. On the one hand, this may present a drawback as countries may be more tempted to avoid their commitments. On the other hand, a country may be willing to adopt a more ambitious agenda when there are no legally binding obligations. In short, the power of a soft law is to encourage leaders and decision makers to think more imaginatively about the future.

Many of the chapters that follow will highlight a notable or impactful law or treaty because these are important instruments of international governance around sustainability.

In addition to the UN system, numerous other international organizations, as well as national and local organizations, are engaged in implementing the targets around the SDGs. A 2021 Brookings Institution Report noted that while the UN system has the foremost official responsibility for shepherding SDG advances, it forms only one piece of a broader puzzle of action required for SDG success. Most of the practical actions for the SDGs will be developed and implemented far outside the walls of UN conference rooms.

In many countries, governments are developing a plan of action, establishing budgets, and working with individuals, institutions, and organizations of all kinds to adopt sustainable practices and achieve the SDGs. The private sector is also undergoing a sustainability transformation. Sustainability thinking has permeated financial decision making, institutional investors such as pension plans, key Wall Street players, and an emerging industry of analytic groups. Beyond government and business, there is growing interest in communities and grassroots support for many sustainability initiatives, which have launched many local, municipal, and regional initiatives. The most recent progress reports note there are more major challenges than achievements, and there is much work that remains to be done.

The good news is that there are a number of actors stepping in to assist sustainability efforts. Within the United Nations, numerous divisions are working toward implementation and to date there are more than 4,000 partners ranging from small nonprofits to large businesses.

There are many NGOs involved in data gathering and reporting, such as the World Health Organization, Save the Children, and Local Governments for Sustainability (ICLEI). ICLEI is a global network of more than 1,500 cities, towns, and regions that offers networking and technical assistance on SDG targets (and more). In addition, companies and organizations have created a wide variety of networks and benchmarking systems that collect data, measure and rank sustainability, or otherwise help countries develop and apply practical strategies, tools, and methodologies, and share best practices and lessons learned.

However, the most recent progress report notes that overall, the transformations required to meet the SDGs by 2030 are not yet advancing at the speed or scale required to meet the Goals. A 2020 UNESCO Report outlined that progress continued to be made in some areas: Global poverty continued to decline, albeit at a slower pace; maternal and child mortality rates were reduced; more people gained access to electricity; and countries were developing national policies to support sustainable development and signing international environmental protection agreements. In other areas, however, progress had either stalled or been reversed: The number of people suffering from hunger was on the rise, climate change was occurring much faster than anticipated, and inequality continued to increase within and among countries.

Some of the reasons for the slow progress include a deteriorated commitment to multilateral cooperation, economic losses from disasters, and of course, the impacts of the 2020 novel coronavirus, COVID-19. It is important to note that before the COVID-19 pandemic, progress was uneven. However, due to COVID-19, an unprecedented health, economic, and social crisis is making the achievement of the SDGs even more challenging. No country has been spared infections and deaths. Health systems in many countries have been undermined, and unemployment rates have affected the livelihood of hundreds of millions of people in the workforce. Tens of millions of people may be pushed back into extreme poverty and hunger, erasing the progress made in recent years.

⧉ INTERCONNECTIONS I.1

Interconnections among the SDGs

The seventeen SDGs depend on each other and are interconnected but documenting how action on one goal could positively or negatively affect another goal was initially less understood when the SDGs were first launch. More recently, there has been more research on the interlinkages across goals and targets in a more integrated and holistic way. There are now different approaches to understanding the interlinkages across the SDG goals and targets.

Understanding the interconnections among the SDGs is important because often policy makers and planners operate in silos. Different ministries handle energy, agriculture, and health. Policy makers also lack tools to identify which interactions are the most important to tackle, and evidence to show how particular interventions and policies help or hinder progress toward the goals. Many preconceptions that influence decisions are outdated or wrong, such as the belief that rising inequalities are necessary for economic growth, or that mitigating climate change is bad for economic growth in the long term. A better understanding of interconnections can provide insight into potential synergies or trade-offs: Models and scenarios that incorporate these can be useful in assessing alternative paths to the SDGs.

Researchers Mans Nilsson, Dave Griggs, and Martin Visbeck have assessed relationships between targets to highlight priorities for integrated policy and action. Nilsson and colleagues note that if countries ignore the interconnections and simply start trying to tick off targets one by one, they risk contradictory outcomes. For example, using coal to improve energy access (SDG 7 Affordable and Clean Energy) would accelerate climate change and acidify the oceans (undermining SDGs 13 Climate Action and 14 Life Below Water), as well as exacerbate other problems such as damage to health from air pollution (disrupting SDG 3 Good Health and Well-Being).

Nilsson and colleagues say there are four key questions decision makers should ask. First, is the interaction reversible or not? For example, failing on education (SDG 4 Quality Education) could irreversibly damage social inclusion (SDG 8 Decent Work and Economic Growth). Loss of species due to lack of action on climate change (SDG 13) is another irreversible interaction. Conversely, converting land use from agriculture to bioenergy production (SDG 7) might counteract food security (SDG 2 Zero Hunger) and poverty reduction (SDG 1 No Poverty) but could be reversed.

Second, does the interaction go in both directions? For instance, providing energy to people's homes benefits education, but improving education does not directly provide energy.

A third consideration is the strength of the interaction: Does an action on one goal have a large or small impact on another? Negative interactions can be tolerable if they are weak, such as the constraints that land resources might put on the development of transport infrastructure.

Fourth, how certain or uncertain is the interaction: Is there evidence that it will definitely happen or is it only possible?

Nilsson and colleagues recommend that for practical policy making, the process should start from a specific SDG and map out, score, and qualify interactions in relation to the other sixteen goals and their targets. It is not clear if many decision makers fully understand the interconnections among the seventeen SDGs.

Sources: Griggs, D. J, Nilsson, M., Stevance, A., and McCollum, D. (eds.). *International Council for Science (ICSUL): A Guide to SDG Interactions: From Science to Implementation*. Paris: International Council for Science (2017). https://doi.org/10.24948/2017.01 and https://council.science/publications/a-guide-to-sdg-interactions-from-science-to-implementation/

Nilsson, M., Griggs, D., and Visbeck, M. "Policy: Map the Interactions between Sustainable Development Goals." *Nature* 534: 320–22 (June 16, 2016). https://doi.org/10.1038/534320a

How This Book Is Organized

This book will introduce sustainability challenges and practices from a variety of angles and highlight examples of best practices. Each chapter will provide a general introduction to the goal by first outlining the issues, including key terms, concepts, and trends. Next, each chapter will explore strategies for achieving that goal and provide some examples of progress. Finally, each chapter ends with a brief summary of progress to date.

This book is organized with each chapter focused on a different SDG, but I recognize the irony of a book that stresses the importance of holistic thinking yet features topical chapters. It is important to realize that the SDGs are highly interdependent—objectives overlap; policies compete and complement; and countries have to set priorities, sequence their efforts, and manage trade-offs. The *Interconnections* I.1 box highlights the connectivity among SDGs. SDGs should not be tackled in isolation given the many interactions and interdependencies across SDGs. Actions in one SDG area have multiple ripple effects, synergistic as well as conflicting ones, across many other SDGs that need to be accounted for to ensure efficient and effective implementation. Working on targets that improve gender equality (SDG 4 Gender Equality) may improve women's health (SDG 3 Good Health and Well-Being) and can also improve job opportunities for women (SDG 8 Decent Work and Economic Growth). Similarly, action to increase the use of renewable energy resources (SDG 7 Affordable and Clean Energy) can reduce air pollution, which in turn improves health (SDG 3) and can reduce the impact of climate change (SDG 13 Climate Action). Yet, if actions to increase access to energy (SDG 7) rely on fossil fuels, this counteracts targets in climate change (SDG 13). Each chapter will also discuss some of the interconnections and synergies among the SDGs.

QUESTIONS FOR DISCUSSION AND ACTIVITIES

1. Is a definition of sustainability possible? Or is it too broad a concept to be useful?

2. Can the United States have economic success and still fail as a country?

3. Describe the three Es of sustainability. Why do you think the "equity" aspect of sustainability is overlooked?

4. Discuss an example of environmental injustice in the Global South.

5. Explain the importance of the UN SDGs.

6. Explain one or two limitations of using the UN SDGs as a way to understand sustainability.

7. Discuss whether the UN SDGs are sufficient to overcome inequalities and uneven development.

8. How would your world change if sustainability became the norm? How long would it take?

9. Which of the seventeen SDGs are the biggest challenge in your community and why?

10. What, in your opinion, will happen to the United States in the next twenty-five years if sustainability does not become the norm?

11. Why is uneven development a challenge to sustainability?

12. What theories of uneven development do you find most convincing and why? Discuss ways you could take action based on the development theory you find most convincing.

13. Why is it important to measure and track progress in sustainable development? What limitations are there in using the HDI to assess progress in sustainability?

TERMS

anthropocentrism
Brundtland Report
development
ecocentrism
environmental justice
equity
Global North
Global South
Human Development Index (HDI)

Millennium Development Goals (MDGs)
political ecology
sustainability
sustainable development
Sustainable Development Goals (SDGs)
three "Es"
uneven development

RESOURCES USED AND SUGGESTED READINGS

Adams, B. *Green Development: Environment and Sustainability in a Developing World.* 3rd ed. London: Routledge (2008).

Brinkman, R. *Introduction to Sustainability.* Hoboken, NJ: Wiley/Blackwell (2016).

Carson, R. *Silent Spring.* New York: Houghton Mifflin (1962).

Copenhagen Consensus Centre. *Weighing the World: Cost-Benefit Analyses of the Sustainable Development Goals* (2015). https://www.copenhagenconsensus.com/sites/default/files/weighing_the_world-_cost-benefit_analyses_of_the_sustainable_development_goals_-_markus_anderljung_1.pdf

Dodds, F., Donoghue, D., and Rosech, J. L. *Negotiating the Sustainability Development Goals.* New York: Routledge (2016).

Dresner, S. *The Principles of Sustainability.* 2nd ed. New York: Routledge (2008).

Easterly, W. *The Tyranny of Experts: Economists, Dictators and the Forgotten Rights of the Poor.* New York: Basic Book (2014).

Edward, A., and Orr, D. *The Sustainability Revolution: Portrait of a Paradigm Shift.* Gabriola, British Columbia, Canada: New Society Publishers (2005).

Feminist Economic Justice for People and Planet Action Nexus, The. *A Feminist Agenda for People and Planet: Principles and Recommendations for a Global Feminist Economic Justice Agenda* (2021): 3–4. https://wedo.org/wp-content/uploads/2021/06/Blueprint_A-Feminist-Agenda-for-People-and-Planet.pdf?blm_aid=59404

Fukuda-Parr, S. "From the Millennium Development Goals to the Sustainable Development Goals: Shifts in Purpose, Concept, and Politics of Global Goal Setting for Development." *Gender and Development* (18 February 2016): 4. https://doi.org/10.1080/13552074.2016.1145895

Harvey, D. *A Brief History of Neoliberalism.* Oxford: Oxford University Press (2007).

Heynen, N., Perkins, H., and Roy, P. "The Political Ecology of Uneven Urban Green Space." *Urban Affairs Review* 42 (1): 3–25 (2006).

Hickel, J. *The Divide: A Brief Guide to Global Inequality and Its Solutions.* London: William Heinemann (2017).

Hickel, J. "The Contradiction of the Sustainable Development Goals: Growth Versus Ecology on a Finite Planet," *Sustainable Development* 27 (5): 873–84 (2019).

Kallis, G., Paulson, S., D'Alisa, G., and Demaria, F. *The Case for Degrowth.* Cambridge: Polity Press (2020).

Macekura, S. J. *The Mismeasure of Progress: Economic Growth and Its Critics.* Chicago: University of Chicago (2020).

Martuzzi, M., Mitis, F., and Forastiere, F. "Inequalities, Inequities, Environmental Justice in Waste Management and Health." *European Journal of Public Health* 20 (1): 21–26 (2010).

Moore, S. *Environment and Society: A Critical Introduction.* 2nd ed. Oxford: Blackwell (2014).

Moyo, D. *Dead Aid: Why Aid Is Not Working and How There Is a Better Way for Africa.* New York: Farrar, Straus and Giroux (2009).

Mulligan, M. *Introduction to Sustainability.* New York: Earthscan/Routledge (2018).

Munier, N. *Introduction to Sustainability: Road to a Better Future.* New York: Springer (2005).

Nicolai, S., Hoy, C., Berliner, T., and Aedy, T. "Projecting Progress: Reaching the SDGs by 2030." *A Development Progress Flagship Report (pp. 1–48).* London: Overseas Development Institute (September 2015).

Oreskes, N., and Conway, E. *Merchants of Doubt: How a Handful of Scientists Obscured the Truth on Issues from Tobacco Smoke to Global Warming.* New York: Bloomsbury Press (2011).

Peet, R., and Hartwick, E. *Theories of Development: Contentions, Arguments, Alternatives.* New York: Guilford (2009).

Portney, K. *Sustainability.* Cambridge, MA: The MIT Press Essential Knowledge Series (2015).

Remington-Doucette, S. *Sustainable World: Approaches to Analyzing and Resolving Wicked Problems.* Dubuque, IA: Kendall Hunt (2013).

Risse, M. *On Global Justice.* Princeton, NJ: Princeton University Press (2012).

Robbins, P., Hinz, J., and Gaia, V. *Adventures in the Anthropocene: A Journey to the Heart of the Planet We Made.* Minneapolis: Milkweed Editions (2014).

Rogers, P., Jalal, K., and Boyd, J. *An Introduction to Sustainable Development.* New York: Routledge (2007).

Roosa, S. *Sustainable Development Handbook.* Boca Raton, FL: CRC Press, Taylor & Francis (2008).

Sachs, J. *The End of Poverty: Economic Possibilities for Our Time.* New York: Penguin (2005).

Sachs, J. *The Age of Sustainable Development.* New York: Columbia University Press (2015).

Sen, A. *Development as Freedom.* New York: First Anchor Books (2000).

Sheppard, E., Porter, P., Faust, D. R., and Nagar, R. (eds.). *A World of Difference, Encountering and Contesting Development.* 2nd ed. New York: Guilford (2009).

Shiva, V. *Staying Alive: Women, Ecology and Development.* Atlantic Highlands, NJ: Zed Books (1989)

Stiglitz, J. *People, Power and Profits: Progressive Capitalism for an Age of Discontent.* New York: W. W. Norton & Company (2020).

Stillwell, F. *The Political Economy of Inequality.* Medford, MA: Polity Press (2019).

United Church of Christ Commission for Racial Justice. *Toxic Wastes and Race in the United States: A National Report on The Racial and Socio-economic Characteristics of Communities With Hazardous Waste Sites.* New York: Public Data Access (1987).

United Nations. "About the Sustainable Development Goals." United Nations Sustainable Development Goals (2020). http://www.un.org/sustainabledevelopment/sustainable-development-goals/

United Nations. "Sustainable Development Goals Partnership Platform." United Nations Partnerships Platform for the Sustainable Development Goals (2022). https://sdgs.un.org/partnerships

United Nations. *The Sustainable Development Goals Report: 2020.* New York: The United Nations (2020). https://unstats.un.org/sdgs/report/2020/The-Sustainable-Development-Goals-Report-2020.pdf

United Nations General Assembly. *Transforming Our World: The 2030 Agenda for Sustainable Development* (September 2015). http://www.un.org/ga/search/view_doc.asp?symbol=A/RES/70/1&Lang=E

United Nations, World Commission on Environment and Development. *Report of the World Commission on Environment and Development: Our Common Future* (1987). https://sustainabledevelopment.un.org/content/documents/5987our-common-future.pdf

United States EPA (2017). https://www.epa.gov/environmentaljustice

United States EPA. "EJAtlas—Global Atlas of Environmental Justice." Environmental Justice Atlas (2019). https://ejatlas.org/

Wacquant, L. *Punishing the Poor: The Neoliberal Government of Social Insecurity*. Durham, NC: Duke University Press (2009).

Washington, H. "Is 'sustainability' the same as 'sustainable development'?" in H. Kopnina and E. Shoreman-Ouimet (eds). *Sustainability: Key Issues (pp. 356–76)*. New York: Routledge (2015).

Washington, H., and Twomey, P. *A Future beyond Growth: Towards a Steady State Economy*. London: Routledge (2016).

Watts, M. "Latin America: Implications of Land Use and Abuse," *Land Use Policy* 1(3): 243–51 (1984).

Poverty

LEARNING OBJECTIVES

After reading this chapter, you should be able to:

- Identify where in the world poverty is most pressing
- Identify how poverty is measured
- Explain the causes and consequences of poverty
- Identify and explain the various targets and goals associated with SDG 1
- Explain how the three Es are present in SDG 1
- Explain how poverty reduction strategies connect to other SDGs

Amudha is a 14-year-old student in 10th grade at a school in a small rural community near Madurai. She lives with her father, mother, sister, nephew, and niece. Her father's hand was broken while picking coconuts. This stopped him from working as a manual laborer. While he was recovering, he worked as a watchman and her mother became a construction worker. Both parents now work in construction. Her father earns Rs 400 per day, and her mother earns Rs 350 and has severe knee and back pain due to the strenuous work. Amudha's parents cook with wood, unable to afford a liquid petroleum gas cylinder. The family lives in a rented primitive shack next to a dried-up pond on wasteland owned by the local government. They have no drinking water or toilet facilities. They defecate in the open next to the pond and obtain electricity from a neighbor's supply. The meagre wages are not sufficient to maintain a family of six. Amudha's mother dreams of having a hut of her own before she dies.

Amudha's day starts at 6:00 AM. She helps her mother at home and then walks to school. Her government-provided bicycle is broken and there is no money to repair it. After school, she attends remedial classes until 9:00 PM. She then comes home for dinner. Later, she helps her mother wash dishes and goes to sleep by 10:00 PM. Amudha's ambition is to become a doctor. Her mother lost two babies, giving birth at home with no access to medical care.

This is what a day spent in poverty looks like for many of the world's estimated 770 million people living in poverty.[*]

---

[*]Oxford Poverty and Human Development Initiative. *Global Multidimensional Poverty Index 2018: The Most Detailed Picture to Date of the World's Poorest People.* Oxford: University Press (2018).

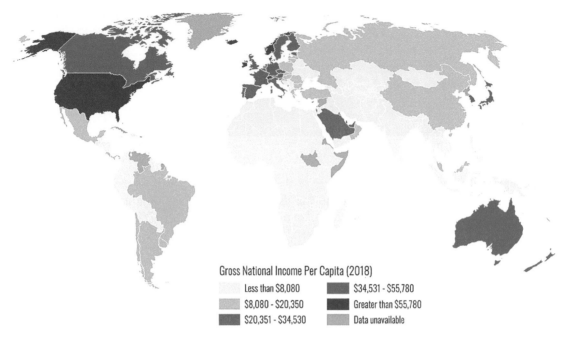

Gross National Income Per Capita (2018)

Less than $8,080	$34,531 - $55,780
$8,080 - $20,350	Greater than $55,780
$20,351 - $34,530	Data unavailable

MAP 1.1 Gross National Income Per Capita (2018)
There are patterns evident: Countries in the Global North have among the highest GNI per capita, while many countries in the Global South are lower. The lowest GNI per capita is found in Africa and South Asia.

POVERTY: UNDERSTANDING THE ISSUE

While there are many consequences of global inequality, perhaps the most striking is related to income and **poverty**. Map 1.1 highlights the GNI per capita for the most recent year. It shows relative wealth in the Global North (North America, Europe, Japan, Australia), middle incomes in South America and parts of East Asia, and low incomes in sub-Saharan Africa, North Africa, South Asia, and Central Asia. Given the wide disparity in income, and that 767 million live in poverty today, it is not surprising that the first SDG focuses on eradicating poverty. The causes of poverty are complex and multidimensional.

According to the World Bank, if an individual is living on $1.90 a day or less, they are living in **extreme poverty**. The 767 million people in that category have $1.90 a day or less in purchasing power to fulfill their daily needs—which include food and housing. Globally, extreme poverty has rapidly declined since 1900 (Figure 1.1). New poverty estimates by the World Bank suggest that the number of those in extreme poverty—has fallen from 1.9 billion in 1990 to about 700 million today. But there remain places where poverty is rising—notably sub-Saharan Africa, which comprises more than half the world's extreme poor.

Understanding the geography of poverty is crucial because two regions, sub-Saharan Africa and South Asia, account for 70 percent of the global total of extremely poor people. The World Bank forecasts predict that by 2030, nearly nine in ten extremely poor people will live in sub-Saharan Africa.

This is not to say poverty is absent in the highest-income countries. The United Nations estimates that there are 30 million children growing up poor in the world's richest countries.

Overall, the trend in global poverty has been decreasing, but not everywhere equally. There are regions and countries where poverty persists, but there are also regions that have seen tremendous progress. Sub-Saharan Africa

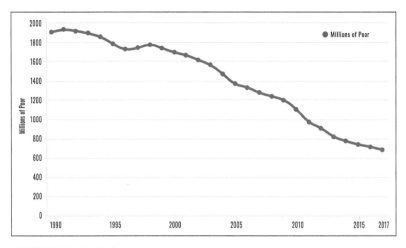

FIGURE 1.1 Global Poor Population at the US $1.90 Poverty Line (1990–2017)
Poverty has been declining since 1990, with declines accelerating considerably since 2000.

has made some advances in reducing poverty that have led to economic growth in countries such as Kenya, although there is still much to be done.

Measuring Poverty

The **World Bank**, whose core mission is to eradicate poverty, gathers data to provide statistics on poverty (See *Key Terms and Concepts* 1.1). They establish and maintain the global norms for measuring national wealth, but calculating poverty is not easy. The World Bank cannot simply add up the national poverty rates of each country because this would mean using a different yardstick to identify who is poor in each and every country. Instead, it created a **poverty line** that measures poverty in all countries by the same standard.

World Bank researchers examined national poverty lines from some of the poorest countries in the world and converted those lines to a common currency by using **purchasing power parity (PPP)** exchange rates. The PPP exchange rates calibrate the cost of the same quantity of goods and services to be priced equivalently across countries.

In 1990, the value of the national poverty line was about $1 per day per person, which formed the basis for the first dollar-a-day international poverty line. After a larger volume of internationally comparable prices were collected in 2005, the international poverty line was revised based on fifteen national poverty lines from some of the poorest countries in the world. The average of these fifteen poverty lines was $1.25 per person per day (again in PPP terms), and this became the revised international poverty line. In 2015, the World Bank used the poverty lines of those same fifteen poorest countries from 2005 (holding steady the yardstick against which they measure) to determine the new global poverty line of $1.90 per person per day in PPP terms. They also expanded their poverty line to include: $3.20 a day for lower-middle-income countries (like Egypt, India, and the Philippines) and $5.50 a day for upper-middle-income countries (like Brazil, Jamaica, and South Africa). The new numbers better account for differences in the cost of living across the world. For example, in the United States, a single individual making an income less than $12,490 per year is consider in poverty; for a family of four, poverty is a collective income of under $25,750 per year. Clearly these income levels would be considered "wealthy" in many poorer areas of the world;

🔑 KEY TERMS AND CONCEPTS 1.1

The World Bank Group

The World Bank Group was established in 1944 with the primary mission to end extreme poverty. Originally, its loans helped rebuild countries devastated by World War II. In time, the focus shifted from reconstruction to development, with a heavy emphasis on infrastructure such as dams, electrical grids, irrigation systems, and roads. Starting in 1960, there has been greater emphasis on the poorest countries, part of a steady shift toward the eradication of poverty becoming the World Bank Group's primary goal.

As the world's largest development institution, the World Bank has supported huge infrastructure projects across Asia, Africa, and South America. In 1947, the World Bank made four loans totaling $497 million; in 2015, it made 302 commitments totaling $60 billion.

The World Bank is not a bank in the traditional sense; it is a cooperative made up of 189 member countries to reduce poverty and support development. The bank provides low-interest loans, zero to low-interest credits, and grants to low-income countries. These loans and grants support a diversity of investments in such areas as education, health, public administration, infrastructure, financial and private-sector development, agriculture, and environmental and natural resource management. Some World Bank projects are cofinanced by governments, other multilateral institutions, commercial banks, and private-sector investors.

In the 1980s, critical development scholars criticized the World Bank for financing a series of environmentally destructive projects, including dams on the Narmada River in India and road building into the Brazilian Amazon, which resulted in deforestation and displacement of indigenous peoples. In the early 1990s, the World Bank adopted a set of environmental policy reforms and committed to sustainability and environmental protection in development.

More recently, critical development scholars have criticized the bank for following a progrowth business-friendly agenda that allows Western companies to make profits from lower-income countries without paying a fair share of tax in those countries. Others have pointed out the inherent imbalance of power of the bank being run by one of the most economically powerful countries (the United States), despite the main borrowers from the World Bank being lower-income countries.

The World Bank has worked closely with the United Nations since the adoption of the MDGs and the SDGs. The World Bank sees the SDGs as aligned with their twin goals of ending extreme poverty and boosting shared prosperity. The World Bank is working with client countries to deliver on the 2030 agenda in three critical areas: (1) finance, (2) data, and (3) implementation—supporting country-led and country-owned policies to attain the SDGs.

Source: The World Bank Group, "The World Bank Group—Home Page." http://www.worldbank.org

---

yet the cost of living in the United States is high and an income of under $13,000 is unlikely to provide adequate food, shelter, and other basic necessities (Photo 1.1).

For this reason, economists often distinguish between **relative poverty** and **absolute poverty**. Absolute poverty is the threshold defined by the World Bank and people in absolute poverty lack even the basic necessities. Relative poverty is a state of living in which people can afford necessities but are unable to meet their society's average standard of living, and they may not have any of the extras beyond just making ends meet. While perhaps not quite equated with a "keeping up with the Joneses" perspective, it is true that people may feel "poor" in the United States if they are living without a car to drive to and from work, or without any money in savings should a family member fall ill.

There has been some criticism of the World Bank's theoretical framework for poverty assessments. One criticism is that setting the poverty line

PHOTO 1.1 Poverty in the United States
Poverty is evident in even the highest income countries such as the United States. There are an estimated 10,000 individuals who are unhoused in the San Diego metro region, making the city home to the fourth-largest number of homeless in the United States. Some critics say that the city—and the country—has neglected and ignored its increasing homeless population.

is somewhat arbitrary. There are also problems with reliable and comparable data. Some economists have also noted that few countries collect statistics on the income poverty of women, who are thought to be disproportionately affected by poverty, and that any poverty assessment should be handled with caution.

Most experts call poverty a "multidimensional problem" that involves numerous factors. The World Bank's measure of poverty, while widely used, does not accurately capture poverty's many nonmonetary impacts on education, health, sanitation, water, electricity, and so forth. These indicators are extremely important for understanding the many dimensions of poverty that people experience. These indicators are a key complement to monetary measures of poverty and are crucial to effectively improving the lives of the poorest. One attempt to measure these nonmonetary indicators has been the Global Multidimensional Poverty Index (MPI).

The Global Multidimensional Poverty Index

The Global Multidimensional Poverty Index (MPI) was developed in 2010 by the Oxford Poverty & Human Development Initiative to identify the most vulnerable people—the poorest among those in poverty. Multidimensional poverty considers the many overlapping deprivations that people in poverty experience, and Table 1.1 outlines the indicators that comprise the MPI. The MPI assesses poverty at the individual level. If someone is deprived in a third or more of ten (weighted) indicators, the global index identifies them as "MPI poor," and the extent—or intensity—of their poverty is measured by the number of deprivations they are experiencing. The MPI can be used to create

TABLE 1.1 The Dimensions of Poverty in the Global Multidimensional Poverty Index

Dimensions of Poverty	Indicators	SDG Area	A person is deprived if
Health	Child mortality Nutrition	SDG 2 SDG 3	Any child has died in the family in past five years Any adult or child, for whom there is nutritional information, is stunted
Education	Years of schooling School attendance	SDG 4 SDG 4	No household member has completed six years of schooling
Living Standards	Cooking fuel Sanitation Drinking water Electricity Housing Assets	SDG 7 SDG 11 SDG 6 SDG 7 SDG 11 SDG 1	The household cooks with dung, wood, or charcoal The household's sanitation facility is not improved or it is improved but shared with other households The household does not have access to safe drinking water or safe drinking water is more than a thirty-minute walk from home roundtrip The household has no electricity The household has a dirt, sand, or dung floor The household does not own more than one of: radio, TV, telephone, bike, motorbike, or refrigerator and does not own a car or truck

Source: Oxford Poverty and Human Development Initiative. *Global Multidimensional Poverty Index 2018: The Most Detailed Picture to Date of the World's Poorest People.* Oxford: University Oxford Press (2018): 6.

a more comprehensive picture of people living in poverty and allows comparisons across countries and regions, and within countries by ethnic group, urban/rural location, and other community characteristics. Table 1.2 details the results from the most recent MPI calculations.

Although the UN Human Development Index (discussed in the "Introduction") remains more widely used to measure "development," its use of only three

TABLE 1.2 How Many People Are Multidimensional Poor and Deprived in...

How many people are MPI poor and deprived in	Million	Share of MPI poor
Nutrition	827	62%
Child mortality	173	13%
Years of schooling	671	50%
School attendance	493	37%
Cooking fuel	1,218	91%
Sanitation	1,058	79%
Water	602	45%
Electricity	740	56%
Housing	1,064	80%
Assets	585	44%

Source: Oxford Poverty and Human Development Initiative. *Global Multidimensional Poverty Index 2018: The Most Detailed Picture to Date of the World's Poorest People.* Oxford: University Oxford Press (2018): 19.

indicators makes it more susceptible to bias. Multiple existing measures of poverty make things more complicated because it is possible for an individual to be earning above $1.90 a day, but still be considered in poverty by the MPI because he or she is deprived in one-third or more of the indicators. Those in poverty generally lack not only income, but education, health, justice, credit, and other productive resources and opportunities. Many economists and development experts prefer to use the MPI because it better captures the true reality of poverty. The MPI accounts for severe deprivations in traditional areas like education, health, and living standards, but also includes other indicators (e.g., child mortality) that have a daily impact on a person's poverty status. The MPI is valuable for highlighting the different ways that people experience poverty and has become an important way to understand what poverty really looks like for many of the most vulnerable people.

The most recent MPI report measures acute poverty for 105 countries, covering 5.7 billion people, approximately 77 percent of the global population (Map 1.2). A key finding was that 1.3 billion people live in multidimensional poverty in 105 countries, nearly all located in the Global South (see Table 1.3). Half of all poor people are children under the age of eighteen years; more than 665 million children (one out of every three children) are spending their childhood in multidimensional poverty. Another finding was that multidimensional poverty is found everywhere in the Global South, but it is particularly acute in sub-Saharan Africa and South Asia where more than 1.1. billion live (Figure 1.2). In India, about 611 million people—46 percent of those who are multidimensionally poor—live in severe poverty; that is, they are deprived in at least half of the weighted indicators in health, education, and living standards. Sub-Saharan Africa, with 342 million people living in severe poverty, accounts for 56 percent of the world's severely poor.

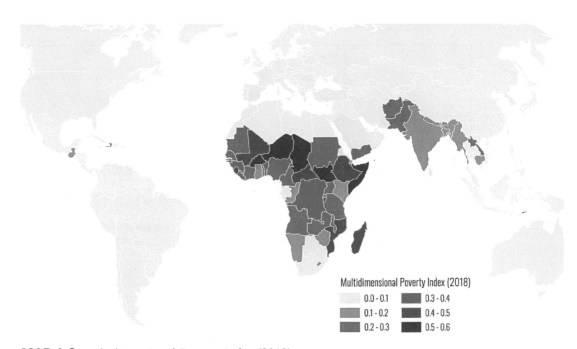

Multidimensional Poverty Index (2018)

0.0 - 0.1	0.3 - 0.4
0.1 - 0.2	0.4 - 0.5
0.2 - 0.3	0.5 - 0.6

MAP 1.2 Multidimensional Poverty Index (2018)
There are regional clusters of poverty: South Asia and sub-Saharan Africa have much higher rates of poverty than other regions of the world.

TABLE 1.3 Poverty Is Highest in...	
Largest number of people living in multidimensional poverty	Highest percentage of people living in multidimensional poverty
India—364 million	South Sudan—85%
Nigeria—97 million	Niger—81%
Ethiopia—86 million	Chad—80%
Pakistan—85 million	Burkina Faso—78%
Bangladesh—67 million	Ethiopia—78%

Source: Oxford Poverty and Human Development Initiative. *Global Multidimensional Poverty Index 2018: The Most Detailed Picture to Date of the World's Poorest People.* Oxford: University Oxford Press (2018).

When examining trends and patterns, it is essential to pay attention to scale as there can be significant variability within regions. Consider Figure 1.3 that shows the MPI for the region of South Asia. Overall, there is variability among countries such as India, Bangladesh, and Afghanistan. Although India has the highest number of poor, Afghanistan has the highest percentage of its population living in poverty. In addition, there is variability in which indicators of poverty are more pressing in which countries. Nepal, for example, is challenged by nutrition, while Afghanistan is more challenged by higher child mortality and education attainment. Understanding this level of detail in poverty is crucial when investing in poverty reduction strategies and reminds us that there is no "one-size-fits-all" model for ending poverty.

There is also variability within a country. There remains a significant rural-urban inequity in many Global South countries. Rural-urban differences are particularly pronounced in sub-Saharan Africa, South Asia, and East Asia and the Pacific. The Global MPI provides important data and understanding

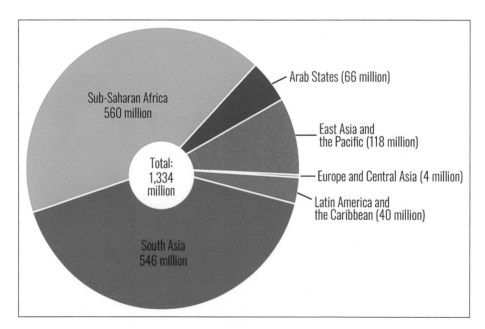

FIGURE 1.2 Multidimensional Poverty Index Estimates, by Region (2018)
More than 80 percent of those experiencing multidimensional poverty live in two regions: sub-Saharan Africa and South Asia.

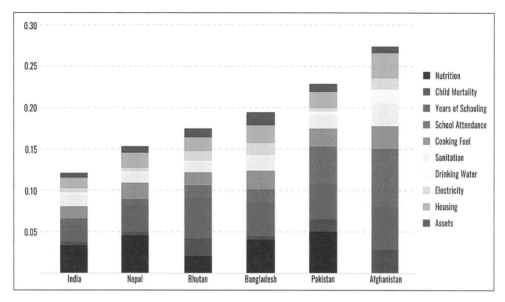

FIGURE 1.3 South Asia Value of the MPI and Its Composition (2018)
Although India has the largest number of poor according to the MPI, other countries such as Bangladesh, Pakistan, and Afghanistan have a larger percentage of their populations in poverty. Also note the variation in indicators. Some countries such as Bhutan are challenged more by educational issues, while Pakistan has high rates of malnutrition, and Afghanistan is dealing with high levels of child mortality.

for SDG 1. Rather than viewing challenges one by one, in silos, the MPI shows how deprivations are concretely interlinked in poor people's lives.

THE IMPACTS OF POVERTY

Poverty is associated with many adverse outcomes. Poverty is more than the lack of income and resources to ensure a sustainable livelihood. The most obvious aspect of poverty is physical: hunger, disease, substandard housing, and the lack of basic necessities. Its manifestations include malnutrition, limited access to education and other basic services, social discrimination and exclusion, as well as the lack of participation in decision making. There is also an emotional aspect of poverty: shame, the humiliation of dependency, and desperation. Poverty often creates moral issues, as those in poverty struggle to decide how to use the little money they have. Should limited funds be used to send children to school or to buy medicine for a sick child?

Impact of Poverty on Children

Poverty has a direct impact on children's development from infancy to early education. Poverty is a major factor of high infant mortality and child mortality rates. Low family income can impede children's cognitive development and their ability to learn. It can contribute to behavioral, social, and emotional problems, and it can cause and exacerbate poor health as well. For example, poverty leads to malnutrition, which can negatively affect a child's physical and mental development. Other manifestations of poverty such as distance, lack of transportation, and financial resources make it very difficult for poor children in low-income countries to get an education. Poor children that are able to obtain an education typically attend schools with inadequate facilities and receive the kind of education that barely provides them with the tools to further their studies or seek employment, thereby restricting them and their children to poverty and generating a vicious cycle of poverty across generations. Low-income parents also struggle to provide the developmentally enriching books, toys, activities, and early care and learning environments that are abundant in the lives of more affluent children.

Impact of Poverty on Health

Health is perhaps the one area where poor people suffer the most. A disproportionately large percentage of diseases in low-income countries are caused by the consequences of poverty such as poor nutrition, indoor air pollution, and lack of access to proper sanitation and health education. According to World Health Organization estimates, poverty-related diseases account for 45 percent of the disease burden in the lowest-income countries. For example, HIV/AIDS disproportionately impacts the poorer African countries such as Zambia and Zimbabwe where one in five adults live with HIV or AIDS. A recent study indicated that poverty is the root cause of tuberculosis and malaria. More than ten million people fall ill with tuberculosis each year and 1.6 million die, although there is treatment.

In many urban areas, poorer communities bear a disproportionate share of air pollution generated by traffic, industrial pollution, and waste disposal. It is not uncommon for cities such as Bangkok, Beijing, Kolkata, and Delhi to experience 30–100 days or more of poor air quality. Those who live in slums have increased exposure. The lack of infrastructure such as electricity and vented stoves means people turn to other forms of energy such as wood, biomass, and coal, all of which emit high levels of particulate matter and other pollutants.

Impact of Poverty on Other Social Indicators

Growing inequality is detrimental to economic growth and undermines social cohesion, increases political and social tensions, and, in some circumstances, drives instability and conflicts. Women may be forced to take any job, no matter how demeaning, dangerous, or illegal it may be. Often households will disintegrate and the social fabric in poor communities is frayed.

Inadequate family income and economic uncertainty can compromise parenting, which in turn can cause depression and anxiety and increase the risk of substance abuse and domestic violence within an impoverished family. Crime and violence against women are also exacerbated in communities living in poverty from unemployment and lack of income.

Poverty and the Environment

The Brundtland Report advanced the idea that there is a strong relationship between poverty and environmental degradation and contended that poverty often leads to unsustainable agriculture, food insecurity, and deforestation through inefficient practices. The perceived tendency of poor societies for environmental degradation is explained as the poor are "short-term maximizers" who are forced to degrade to survive. For those experiencing poverty, short-term survival may trump long-term conservation. To survive, the poor may overgraze the land, be forced to hunt endangered species for food, or slash and burn rainforests to grow crops for subsistence or export. Slash-and-burn agriculture may reduce long-term soil fertility, damages animal habitats, can lead to water pollution, and contributes to climate change. Lacking electricity or other forms of fuel, the impoverished may scavenge wood from already threatened forests. "To care about the environment," said anthropologist and conservationist Richard Leakey, "requires at least one square meal a day." Basic needs must be met to implement environmental protection and sustainability.

Critical development scholars such as Peter Rogers, Kazi Jalal, and John Boyd counter that it is a myth that most environmental degradation is caused by those in poverty. Rather, most serious forms of degradation are caused by the wealthier populations (who consume far more resources per capita). They also point out that poverty reduction does not lead to environmental degradation; rather it is often environmental degradation that transforms low-income people into poor people. Once poverty is reduced, the stress on the environment is also reduced. There are many indications that the poor are concerned about the environment and often possess indigenous knowledge that can be better suited to support their local environment. Scholars have documented many case studies of the poor protecting the environment. Vandana Shiva has noted that conventional literature that pits the poor against the environment is distorted; instead we should be asking why are the poor poor?

No doubt the connection between poverty, development, and the environment is a complex one. But one thing is clear: Poverty and sustainability are not compatible.

REASONS FOR POVERTY

Not surprisingly, because poverty has many dimensions it has many interconnected causes. It is impossible in this chapter to detail the many and often interrelated causes because the root of poverty is the legacy of inequitable development, as briefly discussed in the "Introduction." It is important to understand how poverty is related to the way in which modern economic growth diffused throughout the world over the last 250 years and created uneven development. In the process of economic transformation, many things can go wrong. Some experts point to government corruption, a lack of economic infrastructure, poor access to education, and poor access to health care. There is no single explanation of the persistence of extreme poverty.

For free-market reformers, the traditional argument has been that free markets open up trade opportunities and when a country becomes a global partner in trade, they are able to import and export their products to a larger market, thus increasing their economic wealth. Higher levels of economic wealth are associated with lower rates of poverty. Free-market reformers believe creating competitive markets increases the effort people put forward. Under a free-market system incentives are created to reward those who are productive, work a lot of hours, and have good skills and brilliant ideas. At the one end of this framework, neoliberals tend to criticize government regulation that impedes or stifles incentives for the free market. Neoliberals argue that wealth is the personal responsibility of the individual to work hard and see government intervention as action that increases the dependence of the poor on government aid and makes the situation worse. Therefore, efforts to alleviate poverty should come from a system that rewards economic productivity and industriousness. Free markets and free trade function best when government avoids any intervention that may distort the operation of that free market. Free-market reformers point out that, thirty years ago, 50 percent of people in poor nations were living in extreme poverty. Since the development of global markets, however, 21 percent of people in the lowest-income nations around the world are considered to be living in extreme poverty.

It is true that the historical reduction of extreme poverty around the world happened as markets liberalized and capitalism flourished. As Esteban Ortiz-Ospina has noted, however, it is also true that this reduction of poverty and improvement of living conditions happened at the time that governments increased spending on social protections such as unemployment, family benefits, public and subsidized housing, and social security benefits. Free markets do not necessarily deliver living wages, worker health and safety, or environmental protection, all of which can exacerbate conditions for those in poverty.

Many development economists now criticize the way that the international community—and specifically the World Bank and International Monetary Fund (IMF)—handled a series of economic crises in the Global South during the 1980s and 1990s. Structural adjustment programs consist of loans provided by the World Bank or the IMF to countries that experience an economic crisis. Their purpose is to adjust the country's economic structure, improve international competitiveness, and restore its balance of payments. Structural adjustment programs were introduced in the 1980s as a way to "restructure" the debt that many countries in Latin America were experiencing. During this debt crisis, many Latin American countries were unable to repay their loans, and the IMF and World Bank made access to new loans and debt restructuring conditional on a set of economic policies. In the 1990s, structural adjustment programs were created for other highly indebted poor countries in other regions, including sub-Saharan Africa.

Structural adjustment programs followed neoclassical or neoliberal principles that included privatizing state-owned enterprises such as water, electricity, or phone; easing regulations and lowering taxes to attract investment by foreign business; and reducing public-sector employment and other public spending to reduce deficits. In many low-income countries, the implementation of structural adjustments involved reducing government expenditures through cuts in social services—such as health care and education. These policies were aimed at opening up economies to international trade, which in turn, would create growth allowing debt to be repaid.

Structural adjustment programs have attracted sharp criticism, however, for imposing austerity policies on already-poor nations. Although some macro-level economic indicators such as inflation or GNI growth were stabilized or improved, many social and environmental indicators worsened. Bolivia, for example, saw child malnutrition increase, educational attainment decrease, incomes stagnate, and inequality increase. In Ghana poverty rates were higher after structural adjustment than before its implementation in 1983. Many refer to this as "development reversal." Critics argued that the burden of structural adjustments fell most heavily on those in poverty, women, children, and other vulnerable groups. Instead of reducing poverty, structural adjustment programs contributed to it by forcing layoffs in both the government and the private sector. The legacy of structural adjustment has continued to trap many countries in poverty.

The failure of structural adjustment has led many, including those who believe in the free market, to rethink causes of poverty and move away from a neoclassical or neoliberal approach to poverty reduction. Instead, many free-market reformers see a crucial role for government investment and government regulation.

The continued legacy of structural adjustment on poverty in the Global South has been just one reason that many criticize the effectiveness of the World Bank, and of its progrowth framework. Critical development scholars, for example, say structural adjustment is a neocolonial instrument that led to economic growth without "development." Additionally, critical development scholars argue that poverty is not about individual "failings": The poor should not be blamed for being poor. Nor is poverty necessarily a problem of an overregulated market. Instead, they look to the systemic factors—deeply rooted in the political economic and social structure of a country. This may sound abstract, but this approach highlights the means by which the poor and marginalized gain access to and control over resources.

Consider Haiti. Haiti is the poorest country in the Western Hemisphere and one of the poorest countries in the world. Its per capita income—$765— is far below the regional average of $9,270. About 25 percent of Haitians live in extreme poverty, nearly 60 percent live on less than $2 a day, and some 80 percent of the rural Haitian population live in poverty. The staggering level of poverty in Haiti is associated with a profile of social indicators that is also shocking. Life expectancy is only sixty-three years compared with the Latin American average of sixty-nine years. Less than half of the population is literate. Only about 50 percent of children of secondary-school age attend secondary school. Health conditions are similarly poor; vaccination coverage for children, for example, is only about 25 percent. Only about 50 percent of the population has access to safe water. In addition, Haiti is also highly vulnerable to natural disasters, mainly hurricanes, floods, and earthquakes, which disproportionately affect the poor.

What accounts for the dire extent of poverty in Haiti? Political economy scholars acknowledge that contemporary political instability, woefully poor governance, and corruption, and long-term underinvestment in basic services (such as health care and education) are underlying factors of poverty. However, they also point to the legacy of colonialism and slavery in Haiti. When Haiti declared independence from France in 1804, it became the first modern state founded by Blacks who won their freedom by force of arms. But independence brought a new set of problems, called the Independence Debt. In 1825 the French government required Haiti to pay the price of 150 million francs to compensate the French colonists for their lost revenues from slavery. To do so the Haitian government took out loans. From 1825 to 1944, Haiti paid the French government; this enormous burden drained the Haitian treasury for the next century and created a vicious spiral of poverty.

Many political economists say the poverty of modern Haiti is inextricably linked to the Independence Debt. French prosperity was built on Haitian indebtedness and poverty. The annual income of a French family is $31,112, whereas the average annual income for a Haitian family is $450. To date, France has not confronted the morally dubious repayment forced upon the Haitians, nor the consequences of the brutal slavery it imposed on Africans who labored on plantations there. It is estimated that the Independence Debt would be worth as much as $21 billion in today's dollar. While many former

Haitian presidents have demanded recompense from France, there has been little movement on this issue.

SOLVING POVERTY

There is no single solution to reducing poverty. If there were, we would have implemented it by now. Poverty is a complex problem with many causes, many consequences, and many variables across multiple places.

Many of those living below the poverty line are in fragile and war-torn countries. Paul Collier describes how poor countries are more prone to civil conflict. He notes that conflict lowers incomes, destroys physical capital (equipment and infrastructure), and diverts resources to military spending. Despite this linkage, few solutions proposed for poverty reduction adequately focus on conflict resolution and it is not listed as a specific target for SDG 1.

Table 1.4 displays the targets for achieving SDG 1; it also lists indicators that will mark progress toward these targets. How these targets are implemented will vary from one country to another, from one region to another, as will the levels of success. This is because poverty is directly dependent on context, underlying causes, and priorities.

The international community, which includes the World Bank and the UN system, tends to frame poverty reduction solutions based on economic and political reform measures. Currently there are three categories of people that the UN system prioritizes for poverty reduction programs:

1. Those that live in the lowest-income countries, typically in rural and agricultural communities, which have been less able to benefit from economic globalization;

2. Those who live in conflict-affected areas, where conflict has led to massive displacements that in turn strains economies and resources; and

3. Those who live in "pockets" of extreme poverty in middle-income countries, usually in remote communities that are often disconnected or who belong to marginalized groups that face social and economic discrimination.

Generally, many experts say that to effectively reduce poverty, a government must adopt public policy interventions that help to modify the social, cultural, and economic conditions that created poverty in the first place. A 2017 UNDP progress report on SDG 1 said that to eradicate poverty by 2030, given current rates of population growth, it will be necessary to reduce by about 110 million *every year* the number of people living on less than $1.90 a day. The report identified several "priority" actions that, taken in combination, may reduce poverty:

- Provide opportunity to generate inclusive, sustainable, and sustained growth through international trade and foreign direct investment that can contribute to poverty reduction by creating job opportunities, enhancing productivity, lowering domestic prices, and transferring technology;
- Empower the poor to ensure that all citizens have a voice;
- Invest in rural development;
- Prevent and reduce vulnerability to conflict, natural disasters, and economic downturns that send poor households into a downward spiral;

TABLE 1.4   SDG 1 Targets and Indicators to Achieve by 2030

Target	Indicator of progress
1.1 Eradicate extreme poverty for all people everywhere, currently measured as people living on less than $1.25 a day	**1.1.1** Proportion of population below the international poverty line, by sex, age, employment status, and geographical location (urban/rural)
1.2 Reduce at least by half the proportion of men, women, and children of all ages living in poverty in all its dimensions according to national definitions	**1.2.1** Proportion of population living below the national poverty line, by sex and age **1.2.2** Proportion of men, women, and children of all ages living in poverty in all its dimensions according to national definitions
1.3 Implement nationally appropriate social protection systems and measures for all, including floors, and by 2030 achieve substantial coverage of the poor and the vulnerable	**1.3.1** Proportion of population covered by social protection floors/systems, by sex, distinguishing children, unemployed persons, older persons, persons with disabilities, pregnant women, newborns, work-injury victims, and the poor and the vulnerable
1.4 Ensure that all men and women, in particular the poor and the vulnerable, have equal rights to economic resources, as well as access to basic services, ownership, and control over land and other forms of property, inheritance, natural resources, appropriate new technology, and financial services, including microfinance	**1.4.1** Proportion of population living in households with access to basic services **1.4.2** Proportion of total adult population with secure tenure rights to land, with legally recognized documentation and who perceive their rights to land as secure, by sex and by type of tenure
1.5 Build the resilience of the poor and those in vulnerable situations and reduce their exposure and vulnerability to climate-related extreme events and other economic, social, and environmental shocks and disasters	**1.5.1** Number of deaths, missing persons, and persons affected by disaster per 100,000 people **1.5.2** Direct disaster economic loss in relation to global gross domestic product (GDP) **1.5.3** Number of countries with national and local disaster risk reduction strategies
1.A Ensure significant mobilization of resources from a variety of sources, including through enhanced development cooperation, to provide adequate and predictable means for developing countries, in particular least developed countries, to implement programs and policies to end poverty in all its dimensions	**1.A.1** Proportion of resources allocated by the government directly to poverty reduction programs **1.A.2** Proportion of total government spending on essential services (education, health, and social protection)
1.B Create sound policy frameworks at the national, regional, and international levels, based on propoor and gender-sensitive development strategies, to support accelerated investment in poverty eradication actions	**1.B.1** Proportion of government recurrent and capital spending to sectors that disproportionately benefit women, the poor, and vulnerable groups

Source: United Nations, "Sustainable Development Goals Knowledge Platform, Targets and Indicators Goal 1." https://sustainabledevelopment.un.org/sdg1

- Curb inequalities and exclusion that leave segments of society stuck in poverty by addressing the gap between the rich and the poor, inequality of opportunities in access to basic services such as health and education, and gender inequality; and
- Halt environmental degradation that causes and aggravates deprivation and increases the risks of setbacks by placing environmental sustainability at the center of policy making.

Provide Economic Opportunities and Create Decent Jobs

For many years, the focus of the World Bank and other UN organizations was a focus on increasing economic opportunity through economic growth. For the World Bank, providing opportunity means encouraging economic growth that employs the labor force of the poor. The experiences of countries that have reduced poverty show that an important driver of poverty eradication is economic growth accompanied by increases in decent employment and in labor income, as well as reductions in informal employment. This is a balance of the free market with government involvement in programs and policies.

Accelerating economic growth through sustainable industrialization may significantly reduce poverty. A 2018 UN Report found that between 1990 and 2013, East Asian countries that focused on industrialization were able to reduce the number of people living in extreme poverty from about 1 billion to 71 million.

At the same time, a growth boom that generates large gains for the richest but keeps stable the income of low- and middle-income people, would not be poverty reduction nor an expansion of human development.

To break the cycle of poverty and underdevelopment, a country should identify those economic sectors with potential for both an immediate payoff in terms of job creation and the possibility of sustainable long-term growth. Productive employment and decent work remain the basic and sustainable route out of poverty for individuals, while the corollary is also true: Underemployment, unemployment, informal work, poor-quality employment, vulnerable employment, and working poverty hinder efforts to achieve poverty eradication. However, the linkages between trade and investment, on one hand, and growth and poverty reduction, on the other, are complex and country specific, which means approaches may differ. Countries can also take advantage of Information and Communication Technologies (ICT) to open up new markets through e-commerce and other platforms (discussed in more detail in Chapter 9: "SDG 9 Innovation").

Use the Multidimensional Poverty Index to Identify Vulnerable Households

A challenge for many countries is the lack of timely collection and analysis of poverty-relevant data that can guide propoor policies and projects. Efforts to strengthen national statistical systems is a priority in poverty eradication. One promising strategy is to use the Global Multidimensional Poverty Index to identify the most vulnerable households for intervention.

In addition to using the Global Multidimensional Poverty Index, some experts suggest using the Poverty and Social Impact Analysis (PSIA) as a tool to assess both the economic and social impact of reforms on different social and income groups.

Implement Social Protection Programs

While economic growth is an essential factor for success in the fight against poverty, it is often insufficient. There is the need for growth to be augmented by policy interventions such as social protection to mitigate vulnerability. A 2015 UN Food and Agriculture Organization (FAO) report said that globally social protection intervention has helped in lifting about 216 million people out of hunger and vulnerability between 1990 and 2015.

Social protection refers to policies or actions directed at reducing weakness and shocks through alleviation of poverty perils by promoting effective labor markets, minimizing individuals' vulnerability to risks and building their capabilities to coordinate economic and social disturbances including old age and ill health, disability, unemployment, and financial exclusion. Some of the more effective social protection measures are public or government cash transfers, which is a form of income redistribution from the rich to those in poverty to bridge the gap of inequality. For example, in agricultural communities, social protection could include subsidizing the price of fertilizers sold to farmers and leasing of land to farmers at lower rates amongst others. This can be a critical intervention because in many instances agriculture is associated with various shocks and risk such as floods, pests, disease, and so forth that negatively affect crops and therefore livelihoods. In the United States, social protection includes Social Security, unemployment benefits, public or subsidized housing, reduced cost or free lunch in schools, and the Supplemental Nutrition Assistance Program (SNAP), sometimes called "food stamps," which has reduced child hunger. A 2019 study by the National Academies found that social protection programs that alleviate poverty directly (such as income transfer) or indirectly (by providing food, housing, or medical care) can substantially improve child well-being.

Social protection is considered a key strategy for countries in Africa. This is because 80 percent of the population has no social protection. Recent research recommends that the governments of African countries should implement effective social protection programs and policies in the agricultural sector in the form of insurance, in-kind and cash support, among others to make farming attractive, thereby increasing employment and productivity.

Social protection systems are fundamental not only to lifting people out of poverty but also in preventing them from falling back into poverty. A study by Oluwatoyin Matthew and her colleagues found that social protection programs, when properly earmarked and structured, can bridge the income shortcomings of families in poverty and protect health. Similarly Stephen Devereaux found that social protection incorporates investing for subsequent growth, as it helps households to pull out of poverty cycles through health and educational investment for their children.

According to the UNDP, effective poverty reduction also requires a focus on both the dynamics of "exiting" poverty and the dynamics of not "falling back." For those who have been able to climb out of poverty, progress is often temporary: persistent economic slowdown, food insecurity, and climate change threaten to rob them of their hard-won gains and force them back into the cycle of poverty. In some development circles, risk prevention or risk mitigation is crucial to alleviating possible economic "shocks" such as volatility of the financial markets, unemployment, or illness that can undermine poverty reduction. Social protection measures may also help keep people from falling back into poverty. In the vast majority of countries, the risk of losing social and economic gains is as important as tackling deprivations.

Invest in Rural and Agricultural Development

Two-thirds of the world's extreme poor live in rural areas and depend on agriculture, fishing, or forest resources for their livelihoods. To eradicate poverty, low-income countries must transform rural areas. This is because most of the people living in rural poverty have less access to health, education, and social protection, and basic infrastructure such as roads, water, and electricity.

One key strategy is to support projects that enhance agricultural productivity and encourage young people into agricultural jobs. Development experts suggest that agriculture is an important contributor to the alleviation of poverty because it is the biggest labor supplier, particularly in low-income countries. In addition, increasing agricultural productivity has multiple benefits: More food is sent to market and as a result food security may be improved for individuals and the wider community. The *Solutions* 1.1 box details a project in Tajikistan that has seen small-scale progress.

Investment in rural development can involve improving access to credit and markets, facilitating farm mechanization, revitalizing agricultural extension systems, strengthening land tenure rights, and formulating gender-sensitive agricultural and rural development policies. According to the FAO, investments in rural and agricultural areas can (1) increase small-scale farmers' productivity and income; (2) diversify farmers' income through the development of value chains; and (3) create more and better jobs for the rural poor. Some of these will require public sector investment, such as agricultural research and extension, infrastructure, and education. Other actions can involve the private sector, including creating markets and value-added chains, as well as bringing technologies to rural areas.

 ## SOLUTIONS 1.1

Rural Development in Tajikistan

Target 1.4: Ensure that all men and women, in particular the poor and the vulnerable, have equal rights to economic resources, as well as access to basic services, ownership, and control over land and other forms of property, inheritance, natural resources, appropriate new technology, and financial services, including microfinance.

"In the Khatlon region of Tajikistan, poor smallholders and family farmers dominate the livestock sector, but most struggle with the limited availability and high costs of feed, degraded pasture lands, acquisition of good quality fodder seed or animal health care, and access to financial services, markets, or technologies. In addition to these challenges, poor farmers generally do not practice proper animal husbandry. Collectively, this leads to low productivity of livestock.

FAO, together with IFAD, is supporting around 25,000 rural households in Tajikistan to increase the productivity of pastures and enhance women farmers' capacity to process and market livestock products.

By strengthening institutions and community organizations, including Pasture Users Unions, promoting private-sector services and improving pasture management through veterinary trainings and study tours, the Livestock and Pasture Development Project aims to reduce poverty and enhance the nutrition of rural households. This $15.8 million initiative also aims to increase income-generating opportunities of rural women, including youth, by providing them with job skills, productive inputs, and organizational capacities.

After only two years of implementation, about 92,200 men and 88,600 women have already benefitted from better management of natural resources and improved animal health. The project has also contributed to improving pasture lands through the distribution of mineral fertilizers, fodder crop seeds, and innovative technologies. Combined with enhanced infrastructure, pasture rotation techniques, and a supply of agricultural machinery and equipment, pasture lands have greatly improved.

So far, livestock productivity has increased by 15–20 percent, creating more than 300 jobs in the region. The initiative has also contributed to establishing around 200 Pasture Users Unions across the Khatlon region, and developing 200 Community Livestock and Pasture Management Plans, while fostering increased investments in infrastructure (with around twenty-four new veterinary clinics built), as well as agricultural technologies and machinery.

In the Muminobod district, thanks to the provision of agricultural machinery and organizational strengthening, rural women received training on milk processing, poultry keeping, small ruminants breeding, and bee-keeping activities. They are now able to access better income-generating opportunities and make a living from their labor."

Source: Food and Agricultural Organization of the United Nations. *Ending Poverty and Hunger by Investing in Agriculture and Rural Areas* (2017): 17. http://www.fao.org/3/i7556e/i7556e.pdf

Diversify Poverty Reduction Strategies

Given the complex nature of poverty, and its interconnectedness with development and inequality, implementing a range of policies may allow human development to advance in the long term. This includes building stronger institutions, supporting a strong and vocal civil society, and stronger national accountability through global governance. Undoubtedly, increased international assistance will be needed in some contexts. In addition, for a country to really be successful in reducing poverty, there are a few preconditions that need to be met. It is crucial for countries to find pathways out of conflict, work to mitigate climatic risks, eliminate discriminatory laws and policies, and empower women and girls, which will make efforts to reduce poverty far more effective.

There are many actions that may reduce extreme poverty. *Solutions* 1.2 profiles a set of strategies for achieving poverty reduction in Haiti. What is interesting is that many of them do not involve directly intervening in the economy.

Eradicating poverty may be the most difficult and challenging of all the SDGs because there are so many dimensions and no "one-size-fits-all" solution. A 2017 UN report acknowledged that poverty eradication can only be achieved when interconnected factors are also achieved, including inclusive growth, livelihoods and decent work, social protection, access to basic infrastructure and services, food security, nutrition, health, education, empowerment of women and girls, environmental sustainability, governance, and more equitable access to opportunities and distribution of income and wealth—targets found within many of the other SDGs. No doubt, poverty eradication will demand coordinated, cross-cutting initiatives.

SOLUTIONS 1.2

World Bank Solutions for Poverty in Haiti

Target 1.1: Eradicate extreme poverty for all people everywhere, currently measured as people living on less than $1.25 a day. Target 1.2: Reduce at least by half the proportion of men, women, and children of all ages living in poverty in all its dimensions according to national definitions.

What can be done to reduce poverty in Haiti? The World Bank has outlined the following strategies as priorities for Haiti:

- **Promote inclusive growth by creating greater economic opportunities,** particularly outside of the capital city of Port-au-Prince, by strengthening access to energy, developing renewable energy, facilitating access to financing, and promoting the competitiveness and productivity of the private sector through the development of public and private energy infrastructure.
- **Strengthen human capital and access to services,** by improving primary education and maternal and child health care, while extending access to water and sanitation in the communities most affected by cholera and implementing preventative health care and treatment measures.
- **Improve capacity to adapt to climate shocks,** by strengthening capacity to respond to disasters and protecting a greater number of Haitians through investments in mechanisms to combat flooding as well as in other climate-resilient infrastructure projects, including drainage systems, reinforced bridges, and all-weather roads.
- **Strengthen governance to improve state effectiveness,** by investing in mechanisms to promote transparency and accountability, including accountability within the framework of public financial management; strengthen institutions and government capacity to generate key data and adopt policies based on reliable data; and, finally, enhance government capacity to finance the provision of basic services.

Critical development scholars, however, say that more needs to be done to address the historical injustices and say that Haiti should be at the center of a global movement for recompensation from France.

Ending poverty in Haiti will not happen quickly, or easily, particularly because the country's political systems and economic institutions must undergo significant reform. The good news is that Haiti has made significant progress in several areas, including:

Education
- Grants for the enrollment of 9,500 students in forty-three nonpublic schools in rural areas;
- More than 23,000 students in public basic schools receive hot meals and snacks served in schools in rural areas;
- Support for sixty-one public basic schools and sixty-one community-run public schools in rural areas; and
- Construction and rehabilitation of ninety-one schools or semipermanent structures after Hurricane Matthew in October 2016.

Health
- Expansion of vaccination coverage for 50 percent of children under the age of five;
- Financing of all routine vaccines at the national level for the years 2016 and 2017 and the special diphtheria vaccination campaign for the years 2017 and 2018; and
- Training in cholera prevention for almost four million people and 6,000 health and hygiene officers.

Water and Sanitation
- Increased access to drinking water for more than 314,000 people through the construction, rehabilitation, and extension of drinking water supply systems; and
- Rehabilitation of sanitation and supply facilities in more than sixty schools and rehabilitation of around thirty markets.

Source: The World Bank, "The World Bank in Haiti" (April 2019). https://www.worldbank.org/en/country/haiti/overview#3

Is Development Aid a Solution?

One solution that has been debated is the role of foreign aid in poverty reduction. Free-market reformer Jeffrey Sachs in his book *The End of Poverty: Economic Possibilities for Our Times* draws on his experience dealing with poverty reduction efforts. He says the key to ending poverty is for wealthy countries to increase development aid to poorer countries. He prioritizes the following areas for development aid: agriculture (increasing food production); expanding health coverage; increasing educational attainment; reducing gender inequality; and investing in infrastructure (roads, bridges, electricity, and information technology access). These types of investments could focus on programs such as vocational training, antiretroviral therapy, and off-grid generators. Such target areas of investment are less about "growing an economy" directly and more about providing the conditions in people's lives that allow them to contribute to economic development in the future. However, Sachs' recommendations have been criticized for being "top-down" (poor countries are reliant on the rich countries for assistance) and "status quo" (the economy only needs to be reformed, not transformed).

Others have become more skeptical of foreign aid, noting that this approach has been a mainstay of development since the 1960s and yet has failed to eliminate poverty. Foreign aid has been shown to be ineffective in Somalia, Afghanistan, and Sudan, for example, raising the question of whether foreign aid is a part of the solution to end poverty. Economist Amartya Sen's ideas were instrumental in creating the Multidimensional Poverty Index. He disagrees with Sachs' ideas to increase development aid and instead calls for a "capability approach." This approach focuses more on the expansion of people's choices and their participation in the policies and programs designed to reduce poverty. Sen's approach focuses on changing the underlying economic, social, and political configuration of society, and is more multidimensional (see *Expert Voice* 1.1).

Economist and former World Bank consultant Dambisa Moyo has also written about why foreign aid has not worked in Africa. She notes that in the past fifty years, more than $1 trillion in development-related aid has been transferred to Africa, and only a few are better off while many are worse off. She argues overreliance on aid has trapped low-income nations in a vicious circle of aid dependency, corruption, market distortion, and further poverty, leaving them with nothing but the "need" for more aid. Aid, she says, does not stimulate economic growth but rather economic stagnation. Her solution is to cut aid to Africa. Doing so will force African leaders to make the necessary changes to their economies to lead them out of poverty.

A Critical Perspective on Poverty

Political economists and political ecologists, not surprisingly, are skeptical about the way the United Nations and World Bank have implemented poverty reduction programs. This is because they believe the underlying root of poverty is the distributional inequities that exist in almost all societies. More than 90 percent of those living on less than $1.25 a day are in fragile or vulnerable states. Clearly, there is more analysis needed on the role of conflict and political instability in causing or reinforcing poverty. For example, the largest recipients of global humanitarian assistance are countries in protracted or recurrent crisis, yet this assistance has not resulted in a meaningful reduction of poverty.

EXPERT VOICE 1.1

Amartya Kumar Sen on Development as Freedom

According to Indian economist and philosopher Amartya Kumar Sen, development—and poverty eradication—is the process of expanding human freedom and not only increasing economic growth and GDP. Sen argues development requires the removal of major sources of unfreedom: poverty as well as tyranny, poor economic opportunities as well as systemic social deprivation, and neglect of public facilities as well as intolerance or overactivity of repressive states. The following is an excerpt from Sen's 1999 book *Development as Freedom* and is based on his work that influenced the creation of the Human Development Index:

> We live in a world of unprecedented opulence of a kind that would have been hard even to imagine a century or two ago. There have also been remarkable changes beyond the economic sphere. The twentieth century has established democratic and participatory governance as the preeminent model of political organization. Concepts of human rights and political liberty are now very much a part of the prevailing rhetoric. People live much longer, on an average, than ever before. Also, the different regions of the globe are now more closely linked than they have ever been. This is so not only in the fields of trade, commerce and communication, but also in terms of interactive ideas and ideals.
>
> And yet we also live in a world with remarkable deprivation, destitution and oppression. There are many new problems as well as old ones, including persistence of poverty and unfulfilled elementary needs, occurrence of famines and widespread hunger, violation of elementary political freedoms as well as of basic liberties, extensive neglect of the interests and agency of women and worsening threats to our environment and to the sustainability of our economic and social lives. Many of these deprivations can be observed, in one form or another, in rich countries as well as poor ones.
>
> Overcoming these problems is a central part of the exercise of development. We have to recognize, it is argued here, the role of freedoms of different kinds in countering these afflictions. Indeed, individual agency is, ultimately, central to addressing these deprivations. On the other hand, the freedom of agency that we have individually is inescapably qualified and constrained by the social, political and economic opportunities that are available to us. There is a deep complementarity between individual agency and social arrangements. It is important to give simultaneous recognition to the centrality of individual freedom and to the force of social influences on the extent and reach of individual freedom. To counter the problems that we face, we have to see individual freedom as a social commitment.
>
> This work outlines the need for an integrated analysis of economic, social and political activities, involving a variety of institutions and many interactive agencies. It concentrates particularly on the roles and interconnections between certain crucial instrumental freedoms, including economic opportunities, political freedoms, social facilities, transparency guarantees, and protective security. Societal arrangements, involving many institutions (the state, the market, the legal system, political parties, the media, public interest groups and public discussion forums, among others), are investigated in terms of their contribution to enhancing and guaranteeing the substantive freedoms of individuals, seen as active agents of change, rather than as passive recipients of dispensed benefits.

Source: Sen, A. K. *Development as Freedom*. Oxford: Oxford University Press (1999): xi–xii.

Political economists also point to the need to understand the political economic scale when analyzing poverty reduction solutions. Global trends in poverty reduction have masked big national differences. The MDGs are touted for halving poverty (although some have noted that the definition of poverty changed midway through, so it may be misleading to say that the MDGs halved poverty). Despite making progress on poverty, in thirty countries extreme poverty went up during the time of the MDGs. The most rapid increase in extreme poverty can be seen in eighteen sub-Saharan African countries. To end poverty by 2030, sub-Saharan Africa would need to reduce poverty at a pace faster than South Asia has achieved over the last fifteen years, an unlikely scenario.

Further, national trends have masked subnational differences. Between 2005 and 2012 India saw poverty reduce by 10 percent a year in some states, but in eight smaller states available data shows that poverty increased. We must see a shift beyond national averages to subnational approaches to poverty reduction. This requires a change in how decisions are made about targeting resources and much more detailed and transparent data on people in poverty and how they are progressing. Scale also comes into play with regard to urban versus rural poverty. In urban areas poverty may be attributed to lack of employment opportunities and inequality in income across households. In rural areas, poverty may be a result of lack of access to factors that would increase agricultural production such as access to credit or technology. This research underscores the need to have a clearer understanding of what mechanisms influence the situation and dynamics of poverty in a country, and at various scales.

Political ecology scholars have criticized poverty alleviation programs that avoid or do not address certain historical and political realities surrounding resource ownership. For example, Ikubolajeh Logan and Bill Mosely examined Zimbabwe and found that many racially-based colonial resource ownership structures remained largely in place. As a result, land tenure is insecure, and this places serious limitations on poverty reduction that focuses on investments in agriculture. They make the case that rural poverty cannot be resolved in Zimbabwe without adequate attention to land reform.

Similarly Susan Stonich examined poverty in Honduras. She noted that poverty and food insecurity cannot be understood apart from the dominant development strategy designed to increase agricultural exports (such as bananas). These short-term policies often came at the expense of longer-term development needs. She argues poverty in Honduras was caused by extreme socioeconomic and political inequality, war and the large-scale displacement of people in the region, a debt crisis, the overemphasis on export agriculture, and poorly developed mechanisms for distributing food to those most in need. She suggests that poverty reduction must center around the needs of small rural farmers, including land reform, improved access to credit, greater government accountability, and community participation.

Political economists and political ecologists say that effective poverty reduction policies will require credible political and institutional transformations that would engage all economic agents or institutions in the country to improve the allocation and efficiency of economic resources. Consider Ethiopia, an economy still based primarily on agricultural production. Agriculture will remain a dominant force in the economy for the foreseeable future, so the most pressing challenge is to reduce rural poverty by increasing agricultural productivity. The way for farmers to do this is to build terraces, plow their lands well, attend to weeding, and improve irrigation. However, one of the major obstacles to poverty reduction is that smallholder farmers have no security of ownership of the land. Farmers without land tenure are reluctant to invest into their land if they think it may be taken from them. The solution is to ensure security of land by reforming the land tenure system. This, however, will require a major shift in Ethiopia's political economy structures, including renegotiating land distribution and ownership, which will require the political will for such changes.

Adopt Propoor Policies That Empower and Protect the Poor

Political economists advocate for more propoor policies. Economic growth will not produce jobs and cut poverty unless the needs of the poor and marginalized are at the center of development priorities. Countries need to plan, budget, and implement propoor policies. Political economists and political ecologists say a key strategy for reducing poverty is to recognize the poor and other excluded groups should be agents of their own development. This means they must have meaningful involvement and participation in all aspects of political, economic, and social life, especially in the design and implementation of policies that affect the poorest and most vulnerable groups of society. At the local scale, it is important to privilege local knowledge and practices, ensure the participation of the most poor and marginalized (bottom-up planning), and ensure social and environmental equity and justice. At the national scale, governments must align poverty reductions programs with local needs and aspirations and to empower representative local decision-making authority.

Empowering the poor also consists of providing basic social services such as education, health care, and family planning. To do so requires good governance and investment in social services. Examples of propoor investment include projects that promote financial literacy and management skills, or research that involves or is led by farmers. Because every country has a primary responsibility for poverty reduction, the role of national policies is crucial, as is public investment.

SUMMARY AND PROGRESS

No doubt, eradicating poverty is a difficult challenge. The conceptualization, experience, and determinants of poverty are multidimensional, deep, and resistant to change. The good news is that many countries have acted and many are making progress. Over the last thirty years, China lifted more than 800 million of its citizens out of poverty; India lifted 133 million out of poverty between 1994 and 2012; and Vietnam's poverty rate fell from nearly 60 percent to 20 percent in the past two decades (*Making Progress* 1.1). Reducing poverty is achievable.

Even before COVID-19, the world was off track to end poverty by 2030. Progress had slowed and the pace of global poverty reduction had been decelerating. The COVID-19 pandemic has deeply impeded progress. It is anticipated that tens of millions of people may be pushed back into poverty, undoing years of steady improvement. One estimate is that an additional seventy-one million people may be living in extreme poverty as a result of COVID-19, the first rise in global poverty since 1998. Southern Asia and sub-Saharan Africa

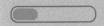 **MAKING PROGRESS 1.1**

Poverty Reduction

Over the last thirty years China lifted more than 800 million of its citizens out of poverty and is often cited as a success story. But other countries are making impressive progress on poverty too.

Uganda has recorded one of the fastest rates of extreme poverty reduction in sub-Saharan Africa. The percentage of the population living on less than $1.90 a day declined from 53 percent in 2006 to 34 percent in 2013 and currently stands at 22 percent. Experts attribute government investment in some fundamentals such as agriculture, infrastructure, and better market information for farmers and traders as key to poverty reduction in Uganda.

Another success story is Vietnam. Vietnam has integrated into global markets since it adopted an "open door" economic policy in 1986. Vietnam's success in promoting economic growth was widely attributed to the comprehensive reform that transformed the old central-planning economy into a new, dynamic market economy.

The government has pushed a series of economic reforms in conjunction with poverty-reduction policies. This has led to steady GNI growth of some 7–8 percent a year, while poverty rates have more than halved. Economic sectors such as textiles and apparel, agriculture, and seafood processing have provided jobs to millions of workers and farmers. As a result, Vietnam has been reclassified from one of the poorest countries to a lower-middle income country.

Government-supported universal education programs helped develop an educated labor force, while commitment to gender equity has enabled women to benefit from economic development. Other social indicators such as school enrollment rate, health-care insurance coverage, and access to amenities (electricity, clean water, and sanitary facilities) have also improved. Food security has provided a stable social and political foundation for Vietnam to reach the next rung of the development ladder: manufacturing. Recently, it has begun attracting the electronics and high-tech sectors as it seeks to diversify its industrial economy.

Every year in July, the Golden Cup is awarded in recognition of the contribution of individuals and organizations to humanitarian acts in poverty reduction and community development. This event is attended by government officials and broadcast live on national television. These movements and events serve to remind Vietnamese citizens and their government about the importance of poverty reduction and community development and encourage political leaders to pursue a more propoor agenda.

Many experts and international donors believe Vietnam has gradually improved the quality of governance and civil participation. Nonetheless, the government is still nondemocratic under the communist party leadership, and civil and political rights remain weak. Despite such success, there are still challenges: Pressure for political reform will likely continue.

Source: Binh, L. Q. "What Has Made Viet Nam a Poverty Reduction Success Story?" Working Paper for Oxfam (2008): 1–29. https://oxfamilibrary.openrepository.com/bitstream/handle/10546/112471/fp2p-cs-what-has-made-vietnam-poverty-reduction-success-story-140608-en.pdf?sequence=1&isAllowed=y

are expected to see the largest increases in extreme poverty. Women and young workers are being disproportionally affected by the pandemic and are expected to be the most vulnerable to unemployment. In many low-income countries, there are no unemployment benefits or other social protections to provide assistance. No doubt, the COVID-19 pandemic will have both immediate and long-term economic consequences.

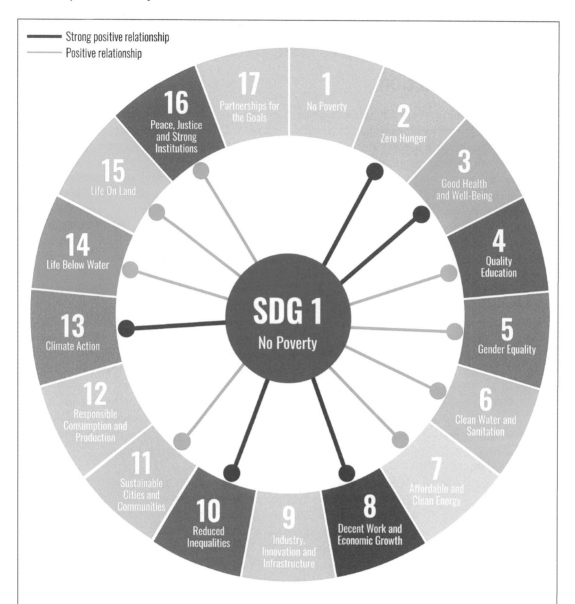

SDG 2: Access to food for the poor, ending malnutrition and supporting small scale farming are important for poverty alleviation.

SDG 3: Eradicating income poverty can improve health as poor people face substantial financial barriers to gaining access to the health services they need. In turn, good health plays a critical role in reducing poverty as it increases an individual's labor productivity and income generating potential.

SDG 6: Having access to clean and safe water and sanitation is important for breaking poverty cycles.

SDG 7: Access to modern and sustainable energy and increase energy efficiency is fundamental for eliminating poverty.

SDG 10: Without a significant reduction in inequality, especially in countries with high rates of poverty and inequality, the world will not meet its goal of ending extreme poverty.

FIGURE 1.4 Interconnections in SDG 1 and the Other Goals
Poverty connects to all the SDGs in some way; this figure highlights several of the stronger connections.

⟳ INTERCONNECTIONS 1.1

Disaster Risk Reduction

Target 1.5: Build the resilience of the poor and those in vulnerable situations and reduce their exposure and vulnerability to climate-related extreme events and other economic, social, and environmental shocks and disasters.

At first glance, disasters may not seem directly related to poverty. However, there are significant economic and social impacts associated with disasters. The impact of economic losses from disasters such as floods or fires varies greatly at the country level. Higher-income countries like the United States, Germany, and Japan, or larger countries such as China and India, experience the bulk of the absolute losses (such as buildings damaged), but countries on the lower end of the HDI like Mongolia, Haiti, Yemen, and Honduras experience harsh losses in percentages of GNI and lives lost. Small island countries are especially vulnerable: In 2004, Hurricane Ivan cost Grenada an estimated 200 percent of GNI, and in 2015, tropical storms may have cost the Commonwealth of Dominica an estimated equivalent of 50 percent of its GNI. Countries in sub-Saharan Africa face both environmental (drought, floods, cyclones, landslides, and wildfire) and social hazards (epidemics such as HIV/AIDS, malaria, and tuberculosis). Both types of hazards impact poor households and communities severely and are an impediment to sustainable development.

Disaster risk-reduction policies and institutional mechanisms do exist at various degrees of completeness in countries around the world. However, their effectiveness is limited. Disaster management structures often focus on only one or two key natural disasters and do not account for social disasters. Many plans do not focus on strengthening traditional copying strategies or cover small, localized disasters. In addition, many disaster management institutions suffer from inadequate financial support.

Target 1.5 synergizes with the recently adopted 2015 UN Sendai Framework for Disaster Risk Reduction. The Sendai Framework represents a shift from the traditional emphasis on disaster response to disaster reduction by prevention. Disaster risk reduction is the concept and practice of reducing disaster risks through systematic efforts to analyze and reduce the causal factors of disasters. Examples of disaster risk reduction include reducing exposure to hazards, lessening vulnerability of people and property, wise management of land and the environment, and improving preparedness and early warning for adverse events.

Questions for Discussion:

1. What other SDGs might benefit from attention to reducing risks from hazards and disasters?
2. In what ways could investing in disaster reduction have a negative impact on SDG 1?

Sources: United Nations Office for Disaster Risk Reduction. *2018 Annual Report.* Geneva: UNISDR (2019). https://www.unisdr.org/files/64454_unisdrannualreport2018eversionlight.pdf
United Nations Office for Disaster Risk Reduction. *Disaster Risk Profile: Namibia.* Geneva: UNISDR (2018). https://www.unisdr.org/we/inform/publications/63278
"Who We Are." United Nations Office for Disaster Risk Reduction. https://www.unisdr.org/who-we-are/international-strategy-for-disaster-reduction

As this chapter has discussed, ending poverty is not just about increasing people's income but also their access to basic needs, such as quality education and health care, clean water and sanitation, decent housing, and security. Figure 1.4 highlights those goals that have the strongest reinforcing synergies with SDG 1. The *Interconnections* 1.1 box challenges you to make connections between the reduction of risks due to disasters and hazards and poverty. These inherent interconnections remind us that poverty eradication needs to be understood as aiming for the well-being, welfare, and freedom of each individual, and this will involve a range of SDGs.

QUESTIONS FOR DISCUSSION AND ACTIVITIES

1. Describe how the three Es are present in SDG 1.

2. In which regions and countries in the world is poverty most pressing?

3. How is poverty measured?

4. Explain why the Global MPI provides a more comprehensive view of poverty than other ways it is measured.

5. Explain the causes and consequences of poverty.

6. Why is poverty so persistent?

7. Explain how poverty is connected to development that has resulted in inequalities.

8. What is the role of the World Bank in eliminating poverty? What are the limitations of the World Bank in eliminating poverty?

9. Explain how investing in agriculture, education, or health could result in reducing poverty.

10. Which solutions discussed do you believe hold good potential to reduce poverty and why? What are other solutions you have learned about?

11. Do high-income countries have an obligation to address poverty? If so, what would that involve?

12. Investigate what poverty "looks like" by selecting a range of countries by income at this interactive website: https://www.gapminder.org/dollar-street

13. Based on *Interconnections* 1.1, find an example of how poverty has been exacerbated in a country by hazards or disasters.

TERMS

absolute poverty
extreme poverty
Global Multidimensional Poverty
   Index (MPI)
poverty

poverty line
purchasing power parity (PPP)
relative poverty
social protection
World Bank

RESOURCES USED AND SUGGESTED READINGS

African Poverty Clock. (2021). https://www.africanpoverty.io/index.html

Alkire, S., and Jahan, S. "The Global MPI 2018: Aligning with the Sustainable Development Goals." *UNDP Human Development Report Office Occasional Paper* (2018).

Banerjee, A., and Duflo, E. *Poor Economics: A Radical Rethinking of the Way to Fight Global Poverty.* New York: Public Affairs (2012).

Collier, P. *The Bottom Billion: Why the Poorest Countries Are Failing and What Can Be Done About It.* New York: Oxford University Press (2007).

Collins, D. *Portfolios of the Poor: How the World's Poor Live on $2 a Day.* Princeton: Princeton University Press (2009).

Development Initiatives. "Five Reasons Why Ending Poverty Will Be Tougher Than Halving It Has Been." (October 5 2015). http://devinit.org/post/five-reasons-why-ending-poverty-will-be-tougher-than-halving-it-has-been/

Devereux, S. "Social Protection for Rural Poverty Reduction and Rural Transformations." *Technical Papers Series #1* (2016).

"Fragile States Index." (2021). https://fragilestatesindex.org//

Godinot, X., Cady, A. Hayes, A., and Stone, J. *Eradicating Extreme Poverty Democracy, Globalisation and Human Rights.* London: Pluto Press (2008).

International Monetary Fund. World Economic Outlook Database (April 2021). https://www.imf.org/en/Data

International Monetary Fund. Washington, DC (2021). www.imf.org/en/Publications/WEO/weo-database/2021/April

Leakey, R. as quoted in Schmidheiny, S. *Changing Course: A Global Business Perspective on Development and the Environment (p. 135).* Cambridge, MA: MIT Press (1992).

Logan B. I., and Moseley, W. "The Political Ecology of Poverty Alleviation in Zimbabwe's Communal Areas Management Programme for Indigenous Resources (CAMPFIRE)." *Geoforum* 33: 1–14 (2002).

Matthew, O., Osabohien, R., Fagbeminiyi, F., and Fasina, A. "Greenhouse Gas Emissions and Health Outcomes in Nigeria: Empirical Insight from ARDL Technique." *International Journal of Energy Economics and Policy* 8 (3): 43–50 (2018).

Moyo, D. *Dead Aid: Why Aid Is Not Working and How There Is a Better Way for Africa.* New York: Farrar, Straus and Giroux (2009).

National Academies of Sciences, Engineering and Medicine. *A Roadmap to Reducing Child Poverty.* Washington, DC: The National Academies Press (2019).

Ortiz-Ospina, E. "Historical Poverty Reductions: More Than a Story About 'Free Market Capitalism' in Our World In Data" (2017). https://ourworldindata.org/historical-poverty-reductions-more-than-a-story-about-free-market-capitalism

Oxford Poverty and Human Development Initiative. *Global Multidimensional Poverty Index 2018: The Most Detailed Picture to Date of the World's Poorest People.* Oxford: University Oxford Press (2018).

Rogers, P., Jalal, K., and Boyd, J. *An Introduction to Sustainable Development.* New York: Routledge (2007).

Sachs, J. *The End of Poverty: Economic Possibilities for Our Times.* New York: Penguin (2006).

Sen, A. *Development as Freedom*. New York: First Anchor Books (2000).

Sen, A. K. *Collective Choice and Social Welfare*. London: Penguin (2016).

Shipler, D. K. *The Working Poor: Invisible in America*. New York: Knopf (2006).

Shiva, V. "Recovering the Real Meaning of Sustainability." *Environment in Question* (pp. 195–201). New York: Routledge (2005).

Stonich, S. *I Am Destroying the Land: The Political Ecology of Poverty and Environmental Destruction in Honduras*. New York: Routledge (2020).

Thiele, L. P. *Sustainability*. Malden, MA: Polity Press (2013).

Thurow, R., and Kilman, S. *Enough: Why the World's Poorest Starve in an Age of Plenty*. New York: Public Affairs (2009).

United Nations. "Ending Poverty and Hunger" (2015). https://www.un.org/sustainabledevelopment/blog/2015/09/ending-poverty-and-hunger/

United Nations. *High-Level Panel Political Forum on Sustainable Development: HLPF Thematic Review of SDG 1: End Poverty in All Its Forms Everywhere*. Geneva, Switzerland: United Nations (2017). https://sustainabledevelopment.un.org/content/documents/14379SDG1format-final_OD.pdf

United Nations Food and Agricultural Organization. *The State of Food and Agriculture 2015 in Brief: Social Protection and Agriculture: Breaking the Cycle of Rural Poverty*. Geneva, Switzerland: United Nations (2015): 1–13.

United Nations General Assembly. *Implementation of the Third United Nations Decade for the Eradication of Poverty (2018–2027)* (2018). https://documents-dds-ny.un.org/doc/UNDOC/GEN/N18/250/39/PDF/N1825039.pdf?OpenElement

World Bank. *Monitoring Global Poverty: Report of the Commission on Global Poverty*. Washington, DC: World Bank (2017).

"World Bank World Development Indicators." The World Bank. http://datatopics.worldbank.org/world-development-indicators/

Zhan, P., Yangyang, S., and Li, S. "China's Poverty Alleviation Policies and Multidimensional Poverty: 1995–2013." *China Economist* 14 (2): 25–38 (2019). 0.19602/j .chinaeconomist.2019.3.03.

Hunger and Food Insecurity

LEARNING OBJECTIVES

After reading this chapter, you should be able to:

- Describe food insecurity trends
- Describe the geography of food insecurity and identify the most vulnerable populations
- Identify and understand the impacts of food insecurity and hunger
- Explain the causes of food insecurity
- Identify and explain the various targets and goals associated with SDG 2
- Explain how the three Es are present in SDG 2
- Explain how SDG 2 strategies connect to other SDGs

Andres Burgos wakes up around 3 AM every day to prepare arepas, the Venezuela staple of cornbread. After filling his backpack, he rides his bicycle through the streets of Caracas, Venezuela. He looks for people prying into trash bags for food and offers them this bread stuffed with ham, cheese, or vegetables.

Due to the economic situation in the country, the pattern of consumption has forced the fragile population to change diet habits. Individuals are forced to consume more carbohydrates such as rice, pasta, and beans. Items including meat, fish, eggs, cheese, and vegetables are often too expensive for many. However, a diet of mostly carbohydrates can lead to chronic malnutrition. Given the economic crisis, which has exacerbated poverty, it is possible that 74 percent of Venezuelan households face extreme poverty and food insecurity.* They represent a small part of the 821 million hungry today.

HUNGER AND FOOD INSECURITY: UNDERSTANDING THE ISSUE

Ending hunger and achieving food security are important goals around sustainable development. On the one hand, between 1990 and 2018, undernourishment declined from 19 percent to 11 percent and child stunting fell from 40 percent to 23 percent. On the other hand, the World Food Programme reports there has been a rise in world hunger for the last several years in a row. The most recent statistics count 821 million chronically undernourished people in the world.

This upward trend of world hunger follows decades of steady decline (Figure 2.1). It also underscores the challenge of achieving the Zero Hunger target by 2030. The World Food Programme points to several reasons behind the recent rise in

---

*Velarde Vásquez, C. "Food Insecurity in Venezuela." The Borgen Project (2021, May 19). https://borgenproject.org/food-insecurity-in-venezuela/

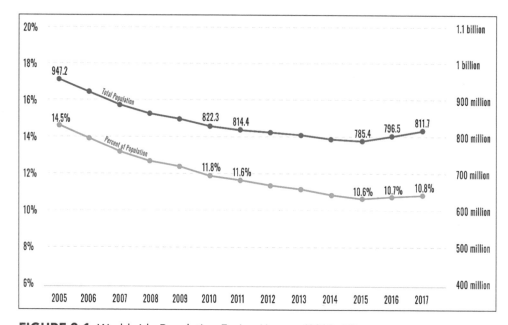

FIGURE 2.1 Worldwide Population Facing Hunger (2005–17)
This graph shows hunger—in both total numbers as well as percentages—declining steadily from 2005 through 2015. Since 2016, hunger has been increasing again, although it is more pronounced in regions such as sub-Saharan Africa.

hunger, including conflict, climate variability, and economic slowdowns. Food and hunger are intricately linked to environmental and economic change. Ending hunger and all forms of malnutrition by 2030 will require faster downward trends.

Food experts are also concerned about the unfolding obesity epidemic, which continues to increase in all regions. An estimated 40 million children under five are overweight, more than 300 million children between the ages of 5–18 are overweight, and about 2 billion adults are overweight.

Without good nutrition, human beings cannot achieve their full potential. Beyond the immense human costs of malnutrition, the economic costs of hunger are projected to be significant. Obesity costs the world $2 trillion annually and it is possible that food insecurity may reduce GNI by 11 percent in Africa and Asia. Food security is one of the most complicated unsolved problems of sustainable development, and there are numerous challenges to address to achieve SDG 2. Some challenges include a rising food demand coupled with continued population growth, and the projected negative impacts of climate change on agriculture in the most vulnerable countries. Additionally, by 2050, the planet may be home to 10 billion people. If we cannot feed everyone today, how will we be able to feed another 2.5 billion by 2050?

Defining Food Security

Food security is a concept that has evolved over time to reflect changes in thinking. The term originated in the mid-1970s when the World Food

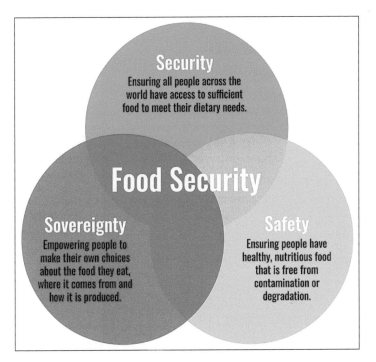

FIGURE 2.2 The Three Ss of Food Security
These include food sovereignty, food safety, and food security.

Conference defined food security in terms of food supply (assuring the availability and price stability of basic foodstuffs). However, the technical successes of the **Green Revolution** did not automatically lead to dramatic reductions in poverty and levels of malnutrition everywhere, prompting food experts to recognize food security was also a result of the lack of effective demand (*Critical Perspectives* 2.1). Further, during the 1970s and early 1980s, episodes of famine and other food crises prompted food experts to recognize that the behavior of potentially vulnerable and affected people was also a part of the bigger picture of food security. In 1996, the World Food Summit adopted a more comprehensive definition of food security: "food security exists when all people, at all times, have physical and economic access to sufficient safe and nutritious food that meets their dietary needs and food preferences for an active and healthy life." Conversely, food insecurity exists when people do not have adequate physical, social, or economic access to food. Food security can be conceptualized as an umbrella term for three interconnected concepts, often called the "Three Ss" (Figure 2.2).

1. *Security*: Food must be available and accessible at all times, supplied through domestic production or imports (including food aid).

2. *Sovereignty*: Food is always accessible and people are empowered to make their own choices about the food they eat, by purchasing food, growing food for their own consumption, or bartering for food.

3. *Safety*: Food must be safe and have a positive nutritional impact on the body.

CRITICAL PERSPECTIVES 2.1

The Green Revolution

In 1943, the Rockefeller Foundation sent agricultural expert Norman Borlaug to Mexico, a country experiencing food shortages, with the objective of exporting US agricultural technology as a solution. The aim was to improve traditional crops grown in Mexico, particularly maize and wheat. Borlaug and his colleagues used wheat from other areas of the world to breed a dwarf hybrid. By the 1960s, Mexico had tripled wheat production and had closed the gap between food production and food needs. The initiative was expanded to other parts of the world experiencing famines and food insecurity, launching a new era of agriculture called the Green Revolution.

The Green Revolution successfully increased agricultural production throughout the world from the 1950s to the 1990s. During this time, agricultural output equaled or stayed slightly ahead of population demands. Grain yields increased by more than 2 percent annually, resulting in nearly a threefold increase in world grain production during that forty-year period. It is estimated that the Green Revolution may have prevented one billion people from starvation.

The Green Revolution included three integrated strategies:

1. Developing hybrid varieties of maize, rice, and wheat (genetically engineered crops) that produced substantially higher yields;
2. Increasing use of synthetic pesticides and fertilizers; and
3. Increasing use of mechanization of equipment (tractors, harvesters, etc.) that required higher inputs of capital (money).

During the 1970s and 1980s, varieties of rice were widely planted in the paddies of Southeast Asia and China, and new varieties of wheat in the drier regions of Asia and Latin America.

In Latin America, more than 80 percent of the wheat grown are Green Revolution varieties.

However, political ecologists among others have documented numerous shortcomings of the Green Revolution that include:

- **Environmental impacts:** There is public concern about the long-term environmental implications of the use of synthetic pesticides and fertilizers. Studies have shown a reduction in soil fertility and increased erosion due to the application of pesticides and fertilizers. Second, expanding irrigation to meet increased water needs of new hybrid varieties has led to groundwater depletion and water shortages in some areas. Third, traditional farming methods such as crop rotation or intercropping were replaced by grain monocultures, decreasing agricultural diversity.
- **Social/Economic Impacts:** Increased mechanization of farming had unintended consequences. One was the consolidation of small farms and fields that resulted in displacing many tenant farmers and intensified rural poverty. This especially impacted small farmers and women. In turn, the resulting displacement caused massive rural to urban migration, thus creating problems in cities. Additionally, the inputs of mechanized equipment are costly, which tends to benefit only the larger landholders.
- **Left Out:** Not all countries adopted Green Revolution methods. Most of sub-Saharan Africa was initially bypassed by the Green Revolution. Many areas within the region did not traditionally grow either wheat or rice, lacked irrigation, and were dominated by small, subsistence farming, all of which did not suit Green Revolution methods.

Of particular note is food sovereignty, which began as a call for empowering farmers to make decisions about what to grow and when. In many countries in the Global South, the food sovereignty movement has become a bottom-up movement to protect the rights of local farmers and small-scale farmers, many of whom are indigenous. Food sovereignty is part of the food justice movement and has the potential to substantially transform political economic relationships around food, land, and markets.

Food security is a multidimensional phenomenon. The challenge to achieving food security is not just producing enough nutritious food but ensuring it reaches those who need it. To add complexity to the bigger picture, in many places people rely heavily on a handful of crops—all of which are susceptible to disease, climate change, and require ever-evolving technology for crop maintenance. In addition, in many countries where the population is growing, food insecurity is already a problem. Furthermore, the food system is a major contributor to climate change and other environmental changes.

Hunger Versus Food Insecurity

While hunger and food security are related, they are not the same. **Hunger** is physiological and reflects the physical pain and discomfort an individual may experience. Hunger is a potential consequence of food insecurity that results in discomfort, illness, weakness, or pain that goes beyond the usual uneasy sensation.

Food insecurity occurs in a broader socioeconomic context. Food insecurity is defined as "limited or uncertain availability of nutritionally adequate and safe foods or limited or uncertain ability to acquire acceptable foods in socially acceptable ways." Food insecurity is the result of food intake that is continuously insufficient to meet **minimum dietary energy requirements**. This is often measured at the household level and reflects economic or social conditions of limited or uncertain access to adequate food.

In the United States, for example, food security is measured by the US Department of Agriculture (USDA), which conducts surveys each year. The surveys ask about various indicators of food insecurity, from the least severe ("We worried whether our food would run out before we got money to buy more") to the most severe ("In the past 12 months did you or other adults in your household ever not eat for a whole day because there wasn't enough money for food?"). The USDA groups households into three classifications of food security: severe, moderate, and low (see Figure 2.3). For example, **moderate food security** is defined as uncertain access to food of sufficient quality and/or quantity, but not so extreme that it causes undernourishment. While severe food insecurity is associated with the concept of hunger, people experiencing moderate food insecurity face uncertainties about their ability to obtain food and have been forced to compromise on the quality and/or quantity of the food they consume. For this population, even if they are not necessarily suffering from hunger, they are at greater risk of various forms of malnutrition and poor health.

Food insecurity has replaced hunger as the primary focus of organization, action, and policy surrounding food access.

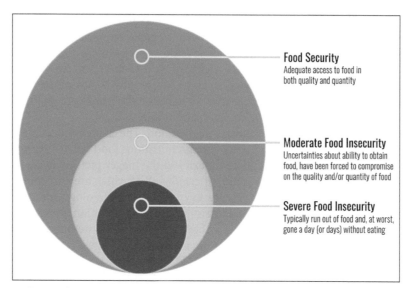

Food Security
Adequate access to food in both quality and quantity

Moderate Food Insecurity
Uncertainties about ability to obtain food, have been forced to compromise on the quality and/or quantity of food

Severe Food Insecurity
Typically run out of food and, at worst, gone a day (or days) without eating

FIGURE 2.3 The Levels of Food Security
The USDA groups households into three classifications of food security: high, moderate, and severe. Moderate food security is defined as uncertain access to food of sufficient quality and/or quantity, but not so extreme that it causes undernourishment. Severe food insecurity is associated with the concept of ongoing hunger.

The Geography of Food Insecurity

Map 2.1 shows the distribution of hunger and makes apparent that food insecurity is found in all world regions. Even in high-income countries, sizeable portions of the population face food insecurity and moderate to even severe food insecurity affects approximately 8 percent of the population in North America and Europe. For example, 2019 data shows that thirty-seven million people were food insecure in the United States, of which eleven million were children. Food insecurity affects all states in the United States, but food insecurity is highest in Alabama, Arkansas, Kansas, West Virginia, and Mississippi. Globally, food insecurity is primarily concentrated in low- and middle-income countries. More than two-thirds of all undernourished people live in South Asia and sub-Saharan Africa (Figure 2.4). In every continent, the prevalence rate is higher among women than men. Hunger is rising in almost all subregions of Africa and, to a lesser extent, in Latin America and the Middle East. There has been good progress in North Africa and the Middle East, but this has stagnated in recent years. South Asia has also seen improvement, but there remains a high level of child undernourishment in this region. Sub-Saharan Africa has the highest prevalence of child undernourishment at almost 20 percent. For example, while Malawi has made progress in recent years, more than 50 percent of the population lives in poverty and more than 47 percent of children under five are stunted. Ironically, agriculture is the main source of income for the majority of the population.

Importantly, the populations of both sub-Saharan Africa and South Asia are increasing faster than in other regions. Projected population growth in these regions over the next fifteen years will challenge food security, particularly as demand for food also increases along with population growth.

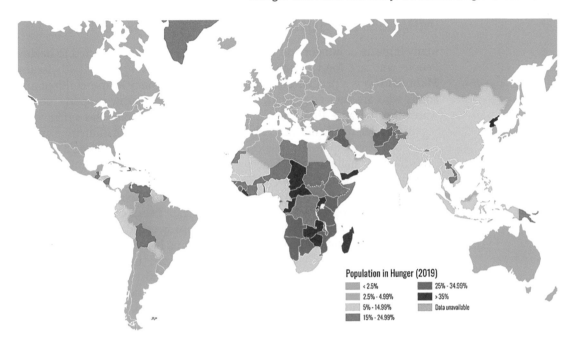

MAP 2.1 Population in Hunger (2019)
Most of the countries in the Global South are confronted by issues of food insecurity and hunger.
There are regional clusters where hunger is more pronounced: Many countries in Africa and South Asia
have significant percentages of the population hungry.

There have also been success stories. Bangladesh, Brazil, Colombia, Peru,
and Vietnam have made tremendous gains in addressing food insecurity.
Egypt, Ethiopia, Kenya, Nepal, Rwanda, Tanzania, and India have similarly
made good progress.

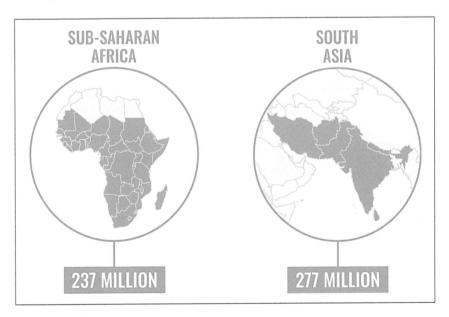

FIGURE 2.4 Regions with the Most Undernourished People
More than two-thirds of all undernourished people live in South Asia and
sub-Saharan Africa. Within these regions, the number of women and children
make up more than 80 percent of those who are undernourished.

IMPACTS OF HUNGER AND FOOD INSECURITY

Many impacts of hunger and food insecurity are deeply connected to health. People who are food insecure are impacted by a range of health issues. **Malnutrition** is a pervasive problem and can take many forms: children and adults who are skin and bones, children who do not grow properly, people who suffer because their diets are imbalanced, or people who suffer from nutrition-related noncommunicable diseases (Figure 2.5). Malnutrition affects all countries and approximately one in three people on the planet. There are three types of malnutrition. One type of malnutrition is chronic hunger, or **undernourishment**. The United Nations Food and Agriculture Organization (FAO) defines chronic hunger as the insufficient intake of energy (calories) and proteins. Hundreds of millions of people experience chronic hunger.

Another type of less visible malnourishment is called **hidden hunger**, or **micronutrient deficiency**. A person experiencing hidden hunger may be getting sufficient calories or proteins, but micronutrients such as vitamins or fatty acids are not adequately present in the diet. Such micronutrient deficiencies result in vulnerability to various kinds of ill health, infections, and diseases. Populations in many low-income countries confront micronutrient deficiencies such as vitamin A, vitamin B12, zinc, iron, folate, omega-3 fatty acids, and iodine. A deficiency in vitamin A can cause eye problems and stunted growth in children; a deficiency in vitamin B12 or iron may cause anemia. Insufficient zinc can lead to loss of appetite, hair loss, and impaired immune system, while insufficient omega-3 fatty acids can cause joint pain and skin lesions.

The third type of malnutrition is the excessive intake of calories leading to **obesity**. Obesity occurs in a person whose weight is far too high for his or her height. Obesity is also defined as a body mass index (BMI) greater than 309, where BMI equals the weight in pounds divided by the height measured in feet. The term "overweight" is defined as a BMI greater than 25. Obesity is on the rise in many countries, especially among school-age children and adults, and contributes to four million deaths globally (Figure 2.6).

Obesity is a challenge because the causes are not fully understood. Part of the cause of obesity is total caloric intake; another contributor may be physical inactivity. In many countries, high caloric intake can result from increased intake of the "wrong types of calories" such as highly processed foods (e.g., soft drinks, potatoes, rice, and baked goods).

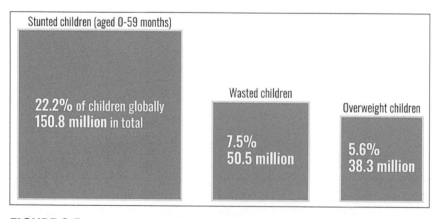

FIGURE 2.5 Malnutrition in Children

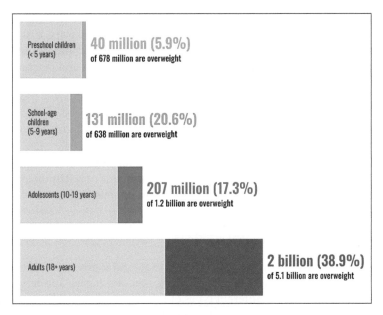

FIGURE 2.6 Obesity by Age Cohorts
The largest number of people experiencing obesity are adults
over the age of eighteen. However, obesity is on the rise
among school-age children. Obesity contributes to four
million deaths globally.

Undernourishment

When children are undernourished, their physical and mental development
may be damaged. For example, **undernourishment** can impair brain development. It can increase vulnerability to both communicable and noncommunicable diseases such as metabolic disorders. Chronic undernourishment of
young children is measured with several indicators. One is **stunting**. A child
who is stunted has very low height for his or her age. Stunting can be measured by assessing a child's height relative to the standard population distribution of height for age. Stunting is often the result of many factors: chronic
undernourishment, hidden hunger, infections (such as from worms), and a
lack of access to safe water and sanitation that can spread other pathogenic
diseases. Children who are more than two standard deviations below the
norm are considered stunted. A 2021 UNICEF study reported 149 million
children suffer from stunting. The highest stunting rates are found in sub-
Saharan Africa and South Asia; currently sub-Saharan Africa and South Asia
account for more than nine of out ten stunted children. On an encouraging
note, the number of stunted children has declined over the last several years,
although the extent of progress varies considerably.

The FAO considers **wasting**, another indicator of undernourishment,
even more urgent. Wasting is defined as a low weight for height and is
often a sign of life-threatening undernutrition. A child who is wasted
may require therapeutic foods, which are high-intensity nutritional foods
designed specifically to combat acute undernutrition. The 2021 UNICEF
report found that forty-five million children suffered from wasting. Combined, some 200 million children under the age of five suffer from stunting
or wasting.

There is also a distinction between **chronic undernutrition**—which is long term—and **acute undernutrition**, which may occur because of wars, disasters, droughts, and displacement of people. In situations of acute undernutrition, massive suffering and loss of life may arise from starvation and disease.

In other situations, people who are food insecure can be affected by diet-sensitive chronic diseases such as diabetes and high blood pressure. Children are at higher risk for asthma and anemia, and behavioral problems such as hyperactivity, anxiety, and aggression.

Tackling malnutrition in all its forms requires a country-specific plan of action and leveraging more general best practices from the international community. Countries must address the underlying determinants of nutrition and implement targeted nutrition interventions to prevent or treat malnutrition.

TRENDS IN FOOD PRODUCTION

A major challenge for addressing food security is that food is grown in different ways around the world. There are small farms, medium farms, and large, extensive farms. Farmers grow different types of food depending on the area's physical geography, as well as social, economic, and political influences. The following are the factors that affect food production:

Climate: Climate determines the suitability of a crop to a particular region and its yield potential. More than 50 percent of crop variation is determined by climate. The most important climatic factors that influence growth, development, and yield of crops are solar radiation, temperature, and rainfall.

Temperature: All plants have maximum, optimum, and minimum temperature limits. Temperature effects on plant growth and development are dependent on the type of plant species. The biological and chemical processes that take place in the soil are connected with air temperature. Temperatures that are too high or too low can negatively impact plant growth and cause the plants to die. For example, wheat grows most optimally at 25 degrees Celsius (77 degrees Fahrenheit), while rice grows optimally at around 30 degrees Celsius (86 degrees Fahrenheit). These crops will die if temperatures during the growing season exceed 50 degrees Celsius (120 degrees Fahrenheit). Even short periods of very hot weather can impede root growth or the growth of shoots.

Soil: Soil stores mineral nutrients and water used by plants, as well as housing their roots. Soil can make the difference between harvesting an abundant yield or a poor yield. However, the geography of soil classification varies greatly around the world, which in turn impacts agricultural productivity. The world's best agricultural soils are mollisols. Mollisols are fertile, dark soils found in the temperate grassland biomes of the midwestern United States, across temperate Ukraine, Russia, and Mongolia, and the pampas of Argentina. But other soil types such as oxisols—soils found in the tropical and subtropical rainforests—have limited fertility for agriculture. The types of soils and the nutrients in the soil are affected by both elevation and location.

Water/Precipitation: Water is critical to growing crops and livestock. When water is available and used effectively and safely, production and crop yield are positively affected. A decrease in applied water can cause production and yield to decrease. Water for agricultural purposes can come from fresh water sources (such as lakes, rivers, and streams), groundwater (such as aquifers), or collected directly from rainwater. The amount of fresh water available differs around the world, based on climate and other factors. Few areas of agricultural production have sufficient water supplies. In most agricultural areas, water is supplied to croplands by artificial means such as irrigation. For example, directing water from rivers through canals that are then used to provide water in fields.

Topography: Topography is the position or arrangement of both natural and artificial/human-made features and structures in an area. Modern farming practices rely on relatively flat or level fields for efficiency. Fields with a moderate slope are often contoured, a process that may involve added expense for the building of terraces and diversion ditches. Steeper slopes are also prone to erosion and rainwater runoff.

Pests: Pest species are cause for major concern due to the potential loss of or damage to crops. Pests can include rodents, birds, and insects. Insects are a significant threat. They may cause direct injury to plants by eating leaves and burrowing holes in stems, fruits, and/or roots. Insects can also cause indirect damage, where the insects do little or no harm, but transmit bacterial, viral, or fungal infection to a crop. Aphids are one of the main culprits in this regard, carrying diseases from plant to plant. Pests can also impact crops postharvest. Rodents such as rats and mice and insects such as moths and beetles can infest grain and commodities, causing damage to raw food materials and contaminating finished products and making them unfit for human consumption.

Elevation/Altitude: Elevation plays a large role in the health and growth of plants. Elevation may affect the type and amount of sunlight that plants receive, the amount of water that plants can absorb, and the nutrients that are available in the soil. As a result, certain plants grow very well in high elevations, whereas others can only grow in middle or lower elevations. Generally, high altitude locations are not ideal for agricultural production.

Labor: Agricultural labor maintains crops and tend to livestock. Every area under cultivation needs labor. Some crops must be harvested by hand (such as strawberries) while other crops require someone to operate equipment. In addition to harvesting, labor is involved in the production, transportation, and sale and distribution of food. In general, since 1800, the number of people employed in agriculture has declined everywhere around the world, although this decline has been highly pronounced in the highest-income countries. For example, in 1950, there were 6.6 million people employed in agriculture in the United States. Today that number is less than two million. In other regions, such as sub-Saharan Africa, more than 40 percent of the population is employed in agriculture, although this too varies from country to country.

Where Is Agricultural Production Optimal?

Most places in the world do not have the conditions for optimal food production, as shown in Map 2.2. The temperate midlatitude zones are among the most optimal, particularly for crops: They have good precipitation, good soils, moderate temperatures, all of which make these areas ideal for agriculture. These areas include the US Midwest, parts of western, central, and eastern Europe (such as Ukraine), Russia, China, and India. But most of the world is not a midlatitude temperate region. Desert areas, forests,

Optimal Areas for Growing Food

Highly fertile, deep and neutral pH soils

MAP 2.2 Optimal Areas for Growing Food
This map highlights the areas of the world that have optimal conditions for growing food. It is noteworthy that there are only a few countries that are optimal for growing food. Most of these areas are already under cultivation.

areas closer to the poles in both hemispheres, and mountainous areas lack many of the conditions for growing substantial amounts of food. Some of the semiarid regions are used for pasturelands. In fact, of the 130 million square kilometers of land on Earth, agriculture constitutes around fifty million square kilometers, or about 40 percent of the world's total land area. Of that fifty million square kilometers, about fourteen million is land that can be used for agricultural crops. This geography points to several important realities. First, not everywhere can grow food optimally, or even at sufficient levels. And second, many of the best areas for growing food are already under cultivation.

These geographic realities, combined with social, political, and economic factors provide some context for understanding why today forty-one countries need external assistance for food. Of these forty-one countries, thirty-one are in Africa.

Even within countries, there can be variations in agricultural land uses. For example, a country such as Ethiopia has deserts, and lowland and highland crops, while Mali has irrigated rice in the south and pastoralism in the north. In large parts of Niger and Chad, conditions for crops are poor and dependent on rainfall.

Given predictions for continued population growth, improving agricultural performance will be central to addressing both poverty and food insecurity. At the country scale, it is important to recognize that food self-sufficiency is not an essential prediction for food security in the future. However, given the geographically large regions where food insecurity is found (sub-Saharan Africa and South Asia), countries that currently face food insecurity will benefit tremendously from increased productivity. There are several strategies to improve productivity. One is to place more land under cultivation (called **land extensification**), but any new lands will likely be suboptimal for productivity. A second way is to increase the amount of yield from already existing land (called **land intensification**). This was the strategy of the Green Revolution of the 1960s. Finally, a third way is to reduce food waste.

Food Waste

Food waste is a more substantial problem than many recognize. Some estimates are that nearly one-third of food in the world is discarded or wasted for various reasons. Food loss refers to the unused product from the agricultural sector, such as unharvested crops. Food waste is defined as the loss of the edible amount of food available for human consumption (not, e.g., banana peels, bones, or eggshells).

In the United States, food waste is estimated at between 30 to 40 percent of the food supply. Food loss occurs for many reasons and at every stage of the production and supply chain. Some of the food is lost postharvest due to lack of safe or available storage for crops. Between the farm and the retail stage, food loss can be caused by problems during drying, milling, transporting, or processing that expose food to damage by insects, rodents, birds, molds, and bacteria. At the retail level, equipment malfunction, overordering, or discarding of blemished produce results in food loss. Once in the hands of consumers, there is still potential for food waste and spoilage. In the United States, 43 percent of wasted food comes from households. Many people tend to buy more food than they need, which can lead to more food waste. Consumer demand for flawless fruits and vegetables has led major grocery chains to buy only perfect food (in response, there has been an emerging movement to promote "ugly" produce that has physical imperfections but is absolutely safe and nutritious). Improperly storing food can lead to food waste. Many people are unsure of how to store fruits and vegetables, which can lead to premature ripening and eventually, rotten produce. In 2015, the US Department of Agriculture and the US Environmental Protection Agency set a goal to cut food waste by 50 percent by 2030. A 2021 report noted that progress was mixed: More food was being donated or composted, but more food was also making its way to landfills.

REASONS FOR HUNGER AND FOOD INSECURITY

There are many reasons for food insecurity, and these have been the subject of research and studies for many decades. At the most basic level, poverty and inequality is the most significant underlying cause of food insecurity. Other often-cited reasons include climate change, conflict and war, dietary preferences that impact the global food system (such as increased demand for meat), environmental disasters (droughts, hurricanes, floods, earthquakes), environmental problems (deforestation, the depletion of freshwater that threatens the irrigation of crops), food consumption patterns, and invasive species.

More recently, food insecurity has been increasing in many countries where economic growth has lagged. Some of these are middle-income countries. Economic shocks can prolong and worsen food insecurity. According to the World Food Programme, economic slowdowns and downturns disproportionately challenge food security and nutrition. As an example, the lack of economic growth or economic crises can adversely affect government investment in rural development, which in turn impacts agricultural credit services, and ultimately production. A country's economic resilience also has a direct effect on its nutritional resilience. Liberia's economy has struggled since the Ebola outbreak in 2014 and an estimated 50 percent of the population lives below the poverty line.

Conflict and Food Insecurity

Of the forty-one countries that need external food assistance, many are countries that are home to civil unrest or full-fledged conflict, while others face severe resource strains due to large influxes of refugees from those neighboring countries in conflict. In Afghanistan, 3.6 million people are at "emergency" levels of food insecurity, with another ten million at "crisis." In South Sudan, about 6.35 million people or 54 percent of the total population are estimated to be severely food insecure. And in war-torn Syria, some 6.5 million people are food insecure. Conflict and civil war are also key reasons for food insecurity in Burundi, Cameroon, Democratic Republic of Congo, Yemen, and parts of Myanmar and Nigeria. Still other countries are experiencing a political and economic crisis, such as that in Venezuela where hyperinflation has undermined people's ability to afford food. This underscores the need to understand the context-specific causes of food insecurity and malnutrition.

Environmental Disasters, Climate Change, and Food Insecurity

Environmental impacts of drought or flood have exacerbated food insecurity. Drought is ever present in many parts of the Horn of Africa and can lead to widespread periodic famine in the region. Floods also affect local areas. In Kenya and Somalia, recent droughts decreased cereal output and the price of maize and sorghum rose sharply. Adverse weather has also impacted food production in Zimbabwe, where the number of food insecure people increased from 30 percent in 2019 to 42 percent currently.

Climate change may compromise food production in countries and regions that are already highly food insecure. Some climate experts suggest that wheat, rice, and corn yields could decrease by 10 percent for every 1 degree (Celsius) rise in temperature. For example, North Africa and the Middle East are regions where climate evidence suggests there will be significant drying in the future, and therefore crop production will be stressed even more than it is today. In East Asia and Southeast Asia, higher temperatures may stress agricultural production. Beyond regions, climate change is likely to also affect food availability in the wider global food system.

Political Economy of Inequality and Food Insecurity

Critical development scholars such as political ecologists have documented how wider political economic systems can lead to hunger and food insecurity. For example, the loss of pastureland or cropland, or land-use policies such as government subsidies that divert corn production into ethanol for automobiles, have increased food insecurity in some countries. Consider the consequences of government policies. Development economists have pointed out that despite espousing the neoliberal principles of a free market, Europe and the United States maintain highly protectionist agriculture policies, enforced by measures such as farm subsidies and tariff and nontariff trade barriers. This has undermined the ability of the Global South to benefit as much as it could from agricultural exports. Additionally, food prices on a global market impacts food security in the Global South far more than in the Global North. The European Union currently spends $74 billion on farm subsidies as a result of its Common Agricultural Policy. In 2019, the United States increased farm subsidies to $28 billion, their highest level in fourteen years. These subsidies enable producers to keep prices artificially low, making it difficult for small farmers in the Global South to compete, even within their own markets. This may be one reason why some countries in Africa have become a net importer of food and agricultural products, despite the tremendous agricultural potential.

Food insecurity is also connected to social, political, and economic factors. Scholars have studied the impact of export agriculture on local food production for some time. Researchers recognize that in certain contexts, if export agriculture expands at the same time local food production declines, export production may undermine food crop cultivation. The allocation of land and labor also contributes to this problem. For many reasons, there are increased pressures for export-led production. Land-use policies at the national level may encourage producing export crops such as cotton in an attempt to increase foreign exchange earnings, while domestic crop production declines. The global food system means that many farmers in the developing world are increasingly enmeshed within commercial networks where production for export comes at the expense of domestic food security. For example, the increase in health awareness has increased demand for bananas in Europe and North America, with countries such as Ecuador, Costa Rica, and Guatemala scaling up banana exports. However, the production, purchase, transport, and marketing of bananas are under the control of just five big multinational trading companies such as Chiquita, Del Monte, and Dole.

Feminist political ecologists have documented that gender inequality exacerbates hunger and food insecurity. Research on the impacts of the Green Revolution showed that a substantial number of women farmers were driven into poverty as a result of the land-use changes that occurred, which in turn had wider implications for household nutritional levels because it is women who are primarily responsible for cooking and feeding the family. Female farmers are responsible for growing, harvesting, preparing, and selling the majority of food in poor countries, but often lack access to finances, credit, extension, and training programs, and they are frequently underrepresented at the forums where important decisions on policy and resources are made.

Critical development scholars have also documented ways in which food distribution and access are contextualized within a wider political economy system.

Food Distribution and Access

A food shortage may happen when not enough food is produced, such as when crops fail due to drought, pests, or too much moisture. But a major reason for food insecurity is the inequitable distribution of food at various scales: within countries, within states, and even within neighborhoods. If total calories from all the food produced were divided among all the people on earth, there would be 2,750 calories per person per day. Because the recommended daily minimum per person is 2,100 calories a day, there are currently enough

UNDERSTANDING THE ISSUE 2.1

Food Versus Feed

Some argue that on paper we have more than enough food to feed the current global population. But not if more than a third of edible crops are fed to livestock. Meat and dairy are highly resource intensive, requiring food, land, water, and energy, which amounts to far more crops than it would take to feed humans directly. Animals are extremely inefficient converters of food. Feed conversion ratios calculate how much more food each animal consumes than they produce as food for human consumption. Chickens consume two to five times their feed compared to their edible weight, while cows consume between six to twenty-five times their weight in feed. It takes twenty-five pounds of grain to yield just one pound of beef—while crops such as soy and lentils produce, pound for pound, as much protein as beef. In 2019, it was estimated that we fed and slaughtered some seventy-five billion land animals.

A large number of people eat a "Western diet" that is heavy in meat, dairy, and eggs. But few people calculate the resources required to produce their food.

- The United States uses fifty-six million acres of land for animal agriculture while dedicating only four million acres of land to growing produce.
- Seventy percent of grain in the United States is fed to farm animals rather than to people.
- It takes 4,200 gallons of water per day to produce a meat-eater's diet, while a plant-based diet uses only 300 gallons of water per day.

There is a tremendous amount of water in a typical hamburger; reducing meat in the diet will help to conserve water resources (Figure 2.7).

A majority of the crops fed to livestock are "calories" that are redistributed away from those who need it most to produce meat for those who can afford it most. The argument is that animal-based foods are a form of overconsumption and redistribution that reduces the amount of available food and increases the price of basic food staples, which is an issue of equity.

The United Nations predicts that global meat consumption will double over the next thirty years, particularly in lower- and middle-income

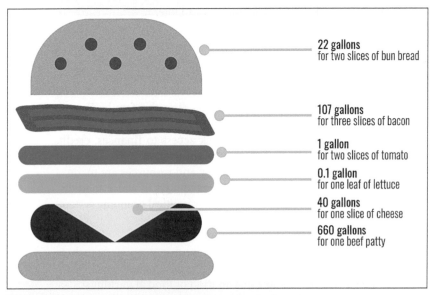

FIGURE 2.7 How Much Water Is in Your Burger?
As this figure illustrates there is a tremendous amount of water resources "hidden" in a food product such as a hamburger. A diet heavy in meats has both health and environmental consequences.

countries. Some contend that we could meet the increasing demand for meat and dairy through biotechnology. Others argue that we should be promoting dietary changes in the most affluent countries. A single individual may not feel he or she has the power to change government policies and business practices quickly to ensure food security, but we each have the power to control our lifestyle choices.

The debate about the environmental cost of livestock is an important one, yet we also need to be aware of crucial differences and inequities between consumers in the Global North versus the Global South. An estimated one billion smallholder farmers in the Global South depend on livestock for food and income; increasing their productivity may be important solutions to both poverty and food insecurity. Livestock provide food for families, generate an additional income, and act as a buffer where a crisis arises. Livestock also supply manure for crops, draft power for tilling, and transport for products and families. In addition, there are high numbers of malnourished women and children in the Global South who could benefit from more protein in their diet; increasing access to meat and dairy products is an important solution to reducing hunger.

There are two big questions. First, is reducing consumption of animal-based foods a necessary component of a sustainable food system? And if so, who should reduce their consumption?

Sources: Bittman, M. Food Matters: A Guide to Conscious Eating. New York: Simon and Schuster (2009). Cassidy, E., et al. "Redefining Agricultural Yields: From Tonnes to People Nourished Per Hectare." Environmental Research Letters 8 (3): 2–3 (September 2013). http://iopscience.iop.org/1748-9326/8/3/034015 Smil, V. Feeding the World: A Challenge for the Twenty-First Century. Cambridge, MA: MIT Press (2000). Yacoubou, J. "Factors Involved in Calculating Grain: Meat Conversion Ratios." Vegetarian Resource Group (2019). www.vrg.org/environment/grain_meat_conversion_ratios.php

calories to feed everyone in the world. But not everyone is getting the needed calories because food is not evenly distributed across the world.

As much as the production of food differs from place to place, so too do the distribution, access, and consumption of food. It is important not to oversimplify food insecurity as being just a problem of scarcity (not enough food) or just a problem of distribution (not enough access to food). More accurately, hunger and food insecurity result from a web of immensely complex and interconnecting factors, including both food supply and distribution issues.

Problems associated with food distribution can include food spoilage resulting from poor storage or pests; transportation obstacles such as poor/missing infrastructure; and supply/demand issues.

If we accept that the world is currently producing enough calories, protein, and micronutrients to sufficiently feed the world's population, problems in the supply chain, like food waste, distribution, and nutrient losses, result in distribution and access problems. For example, we lose most micronutrients, like Vitamins A and C, in postharvest waste of fruit and vegetables, while energy and protein are lost when crops end up being used as animal feed and biofuel (*Understanding the Issue* 2.1).

The issue of distribution and access is also a paradox—some people are consuming far too many calories (and are dealing with obesity), while others are dealing with chronic undernutrition.

Food insecurity is deeply related to income inequality, which impacts the consumption of resources, particularly energy and food. For example, the emergence of food deserts in wealthier countries such as the United States are connected to race and income (see *Key Terms and Concepts* 2.1). The global food system is complex and multifaceted and often disproportional.

🔑 KEY TERMS AND CONCEPTS 2.1

Food Deserts and Food Swamps

Issues of food equity and access affect every country. Even in the highest-income countries, there are people without access to food. **Food deserts**, a term coined by geographer Neil Wrigley, refers to an area with limited access to nutritious and affordable food. A food desert is an area lacking full-service grocery stores. Residents must either travel a long distance to get food or pay higher prices for food purchased at convenience stores with limited choice or lack of nutritional food. Food deserts can be found in rural areas, where the population density is too low or too poor to attract or support grocery stores within a close range. People living in rural food deserts live more than ten miles from a grocery store. Many food deserts are also located in cities and are found in poor or minority neighborhoods. Typically, in urban areas, food deserts are defined by populations that are a mile or more away from a grocery store.

In the United States, the US Department of Agriculture (USDA) estimates that about 2.3 million people live in a rural food desert and about 21.2 million people live in an urban food desert. The USDA has developed an interactive food desert locator that allows anyone to identify food deserts in their community. Research shows that food deserts reflect social exclusions and contribute to health inequalities. Areas with food deserts tend to have higher rates of obesity, asthma, and heart disease. In response to growing awareness of food deserts, the Obama administration launched the Healthy Food Financing Initiative (HFFI) in 2015 to eliminate food deserts. So far, HFFI has invested $500 million to increase fresh food access in food deserts. But eliminating food deserts is still a long way off.

Urban food deserts raise concerns about social justice and food insecurity. In Washington, DC, food deserts are concentrated in Wards 7 and 8, which not coincidentally have the lowest average incomes and are 94 percent Black. Of the city's forty-three full-service grocery stores, only three are located in Ward 8. By contrast, Ward 3, the highest-income ward, has eleven full-service stores. And despite a boom in farmers' markets, only three of the city's forty farmers' markets are

MAP 2.3 Washington, DC, Food Deserts (2015)
It is not a coincidence that the concentration of food deserts are greatest in Wards 7 and 8, as these are predominately low income and Black.

(map legend) Washington, D.C. Food Deserts (2015)

located in Ward 8. Map 2.3 shows the prevalence of food deserts in Wards 7 and 8. The opposite of a food desert is a **food swamp**, which is defined as an area with an abundance of fast food, junk food outlets, convenience stores, and liquor stores that outnumber affordable, nutritious food. Food swamps contribute to obesity. Wards 7 and 8 are simultaneously both food deserts and food swamps.

At the local level, there are many nonprofits working to solve food deserts in cities across the United States. DC Central Kitchen, for example, partners with local growers to provide fresh produce to corner stores in Ward 8. Their program is called "Healthy Corners." Recently the DC Department of Health teamed up with DC Farmers Market Collaborative to launch the "The Produce Plus Program." Through this program, any DC resident receiving food assistance gets a $10 voucher per family each week to spend on fresh fruit and vegetables at DC farmers' markets. Some farmers' markets double or even triple the voucher.

Despite these new initiatives, improving a community's health may prove more complicated than just bringing in grocery stores. Many working poor may lack access to a fully fitted kitchen, making "ready-made meals" the only option. And income isn't the only barrier to eating better: Sometimes people choose to eat unhealthy food even when nutritious food is available and affordable.

It is clear that cities will confront issues of food production and consumption as they plan for sustainability. Eliminating food deserts is just one example of how we are challenged to imagine a more sustainable and equitable future.

Sources: "Food Access Research Atlas." US Department of Agriculture. https://www.ers.usda.gov/data-products/food-access-research-atlas/go-to-the-atlas/
Wrigley, N. "Food Deserts in British Cities: Policy Context and Research." *Urban Studies* 39 (11): 2029–40 (2002).
For more information on DC Central Kitchen, see https://dccentralkitchen.org/healthy-corners/

Higher-income people disproportionally consume most of the world's meat and milk, for example. Solving food insecurity will require us to address poverty and wealth inequalities—among and within countries.

SOLVING HUNGER AND FOOD INSECURITY

Ending hunger will involve coordinating various scales of strategies: including local plans, national plans, and the complex global food system. Central to any genuine solution is to redress inequalities, among countries and within countries. Table 2.1 summarizes the targets for SDG 2.

The FAO of the United Nations has proposed Twenty Actions or strategies that integrate the many dimensions of agriculture and rural development and lay the foundations for resilient and sustainable agriculture as shown in Table 2.2. These twenty actions can be integrated into two larger strategies of investing in rural development and improving agricultural productivity.

Invest in Rural Development

The FAO suggests that one of the most important strategies involves reinvesting in rural areas. This is because in many countries agriculture remains the largest employer and main economic sector: both a major problem and major opportunity. More than two billion people are smallholder and family farmers (which includes family farmers, fishers, foresters, herders, and rural workers). In fact, 72 percent of farms on the planet are small farms, many of which have limited access to advanced production technologies, markets, credit, infrastructure, and basic services. The fact is there is little information on who smallholder farmers are, what they earn, and how much they produce.

In addition, the FAO suggests targeting investments for rural women who make up almost half the agricultural labor force in developing countries. In many areas, women own a much smaller share of land than men do and often have more limited rights of access and control over land. People without secure land rights are frequently excluded from access to key rural services, including loans. Secure land tenure is associated with higher levels of investment and productivity in agriculture, higher revenues, and improvements in both food and income security. Evidence shows that when women are given equal access

TABLE 2.1   SDG 2 Targets and Indicators to Achieve by 2030

Target	Indicators
2.1 End hunger and ensure access by all people, in particular the poor and people in vulnerable situations, including infants, to safe, nutritious, and sufficient food all year round	**2.1.1** Prevalence of undernourishment **2.1.2** Prevalence of moderate or severe food insecurity in the population, based on the Food Insecurity Experience Scale (FIES)
2.2 End all forms of malnutrition, including achieving, by 2025, the internationally agreed targets on stunting and wasting in children under five years of age, and address the nutritional needs of adolescent girls, pregnant and lactating women, and older persons	**2.2.1** Prevalence of stunting (height for age <–2 standard deviation from the median of the World Health Organization (WHO) Child Growth Standards) among children under five years of age **2.2.2** Prevalence of malnutrition (weight for height >+2 or <–2 standard deviation from the median of the WHO Child Growth Standards) among children under five years of age, by type (wasting and overweight)
2.3 Double the agricultural productivity and incomes of small-scale food producers, in particular women, indigenous peoples, family farmers, pastoralists, and fishers, including through secure and equal access to land, other productive resources and inputs, knowledge, financial services, markets, and opportunities for value addition and nonfarm employment	**2.3.1** Volume of production per labor unit by classes of farming/pastoral/forestry enterprise size **2.3.2** Average income of small-scale food producers, by sex and indigenous status
2.4 Ensure sustainable food production systems and implement resilient agricultural practices that increase productivity and production, that help maintain ecosystems, that strengthen capacity for adaptation to climate change, extreme weather, drought, flooding, and other disasters, and that progressively improve land and soil quality	**2.4.1** Proportion of agricultural area under productive and sustainable agriculture
2.5 Maintain the genetic diversity of seeds, cultivated plants, and farmed and domesticated animals and their related wild species, including through soundly managed and diversified seed and plant banks at the national, regional, and international levels, and promote access to and fair and equitable sharing of benefits arising from the utilization of genetic resources and associated traditional knowledge, as internationally agreed	**2.5.1** Number of plant and animal genetic resources for food and agriculture secured in either medium- or long-term conservation facilities **2.5.2** Proportion of local breeds classified as being at risk, not at-risk, or at unknown level of risk of extinction
2.A Increase investment, including through enhanced international cooperation, in rural infrastructure, agricultural research and extension services, technology development, and plant and livestock gene banks to enhance agricultural productive capacity in developing countries, in particular least developed countries	**2.A.1** The agriculture orientation index for government expenditures **2.A.2** Total official flows (official development assistance plus other official flows) to the agriculture sector

2.B Correct and prevent trade restrictions and distortions in world agricultural markets, including through the parallel elimination of all forms of agricultural export subsidies and all export measures with equivalent effect, in accordance with the mandate of the Doha Development Round	**2.B.1** Producer support estimate **2.B.2** Agricultural export subsidies
2.C Adopt measures to ensure the proper functioning of food commodity markets and their derivatives and facilitate timely access to market information, including on food reserves, to help limit extreme food price volatility	**2.C.1** Indicator of food price anomalies

Source: United Nations Sustainable Development Goals. https://sustainabledevelopment.un.org/sdg2

to resources, income opportunities, education, and social protection, agricultural output and food availability increases and the number of poor and hungry declines (Photo 2.1). Investment in rural youth is also critical, as the numbers of people aged fifteen to twenty-four is projected to increase sharply in rural sub-Saharan Africa and South Asia, areas that are already dealing with food insecurity. Improvements in nutrition can have multiple benefits that include reducing the risk of maternal mortality and low-birth-weight babies, reducing childhood stunting, and reducing the prevalence of infectious

PHOTO 2.1 USAID Supporting Small Farmers
USAID supports small farmers, like Matta in western Ghana, build better chicken coops, which increase the chickens' productivity and the families' food security. Often raising a few chickens can make a substantial difference in two ways: by increasing nutritional levels within a family and by increasing household income.

TABLE 2.2 The Food and Agriculture Organization of the United Nations "Twenty Actions"

1	Facilitate access to productive resources, finance, and services
2	Connect smallholders to markets
3	Encourage diversification of production and income
4	Build producers' knowledge and develop their capacities
5	Enhance soil health and restore land
6	Protect water and manage scarcity
7	Mainstream biodiversity conservation and protect ecosystem functions
8	Reduce losses and promote sustainable consumption
9	Empower people and fight inequalities
10	Promote secure tenure rights
11	Use social protection tools to enhance productivity and income
12	Improve nutrition and promote balanced diets
13	Prevent and protect against shocks: enhance resilience
14	Prepare for and respond to shocks
15	Address and adapt to climate change
16	Strengthen ecosystem resilience
17	Enhance policy dialogue and coordination
18	Strengthen innovation systems
19	Adopt and improve investment and finance
20	Strengthen the enabling environment and reform the institutional framework

Source: United Nations Food and Agriculture Organization. *Transforming Food and Agriculture to Achieve the SDGs* (2018). http://www.fao.org/3/I9900EN/i9900en.pdf

diseases. Improved nutrition can also empower women and girls, increasing their educational outcomes. Table 2.3 summarizes solutions that may improve nutrition.

Social protection systems open the door to a thriving economy and fairer society. They support small-scale producers, particularly rural women, by providing income generation opportunities. At the same time, they promote sustainable food systems and improve natural resource management. José Graziano da Silva, FAO Director-General concluded that "ultimately, political will is fundamental to achieving zero hunger, or indeed any of the SDGs."

Improve Agricultural Productivity

Critical to achieving SDG 2 is improving agricultural productivity. Although cereal yields have nearly doubled in sub-Saharan Africa since the 1990s, they are not rising fast enough to meet growing food demand. Experts say that cereal yields will need to increase by 3 percent per year each year until 2030, but currently they are only increasing by around 2.2 percent. Some growth in food demand could be met by expanding production to areas currently not under cultivation, but as the discussion on food production highlights, these areas will likely be suboptimal for productivity. Instead, there have been more

TABLE 2.3 Actions to Improve Nutrition

Evidence shows that multiple actions are needed to create sustainable improvements in diet. These may include:

Design programs to target nutritionally vulnerable groups and periods (e.g., children's first 1,000 days);

Ensure that monitoring and evaluation systems measure impact of programs and projects on nutrition;

Use cash transfers to enable households to purchase a nutritious diet;

Make cash transfers large enough and adjust them to reflect regional and urban and rural price differences;

Involve men alongside women in programs designed to raise awareness of nutrition;

Promote access to education;

Increase and promote access to health services;

Increase investment in health care system;

Provide energy-protein supplementation;

Implement therapeutic feeding for severely wasted children;

Implement biofortification of various foods for vitamin A, zinc, and other micronutrients;

Improve access to clean water;

Increased government spending on nutrition and ensure that nutrition interventions are included in local councils' annual budgets;

Recruit and empower local/district nutrition officers;

Counsel women during pregnancy about breastfeeding and healthy eating;

Targeted subsidies for nutritious foods;

Disseminate food-based dietary guidelines for infants, children, and adolescents;

Screen children for overweight as well as underweight yearly;

Pivot toward nutrition-conscious, sustainable agricultural practices;

Design and enact propoor growth policies;

Support the livelihoods of small-scale and family farmers;

Invest in rural women farmers by improving women's rights to land and tenure, participation in rural labor markets;

Establish legal frameworks to ensure smallholder access to productive resources;

Invest in rural life: programs in health, social protection, access to finances;

Encourage organic farming, agroecology, agroforestry, crop-aquaculture;

End conflicts;

Mitigate climate variability;

Reduce inequality;

Establish social protection programs to manage economic and social risks such as unemployment, sickness, disability, etc. (unemployment insurance, health insurance, cash or in-kind transfers);

Target social protection programs to include adolescence, especially for girls; and

Rethink personal diets.

Source: By author compiled from several sources.

efforts at intensification, which aims to increase yields per hectare. For example, Ethiopia has experienced unprecedented growth in the last fifteen years. In 1990, the average maize yield was 1.6 tons per harvested hectare; today, it has increased to around three tons per harvested hectare. The reasons for the increased yields include increased availability and use of modern inputs such as modern varieties and fertilizers, better educational extension services, and increasing demand. However, yields are still lower than in many other countries due to the limited water availability. One solution proposed has been A New Green Revolution for Africa.

A New Green Revolution for Africa

Africa has long been a continent of small farmers, half of them women. They often farm with little fertilizer, pesticide, or irrigation, on a tiny plot with a hoe. With pressure from population growth, poverty, and high rates of child malnutrition, Africa must increase food production, and do so in a more sustainable way. These solutions involve investments in rural infrastructure, improving technical training of farmers, leveraging new technologies, upgrading food processing, focusing on smallholder farmers, especially women, and expanding local market access.

In 2006, the Bill and Melinda Gates Foundation and the Rockefeller Foundation joined forces to launch the Alliance for a Green Revolution in Africa (AGRA). This effort promises to create a new Green Revolution uniquely for Africa, while avoiding some of the problems of the earlier initiatives. AGRA currently funds hundreds of projects that include research to develop and deliver better seeds, increase farm yields, improve soil fertility, upgrade storage facilities, improve market information systems, strengthen farmers' associations, expand access to credit for farmers and small suppliers, and advocate for national policies that benefit smallholder farmers. Credit programs are especially critical for poor farmers who lack collateral or creditworthiness to access traditional loans. AGRA carries out its activities in sixteen countries, with a special emphasis on Ghana, Mali, Mozambique, and Tanzania. The aim is to improve production of staple crops in "breadbasket" areas that have relatively good soil, adequate rainfall, and basic infrastructure, and then replicate successful approaches in other areas and other countries with similar conditions.

Projects underway that focus on Africa-specific solutions include improving yields of traditional crops such as cassava, sorghum, millet, and beans that are being bred for disease resistance and drought resistance. Cassava is especially important because it grows in poor soils and across the continent. Another crop plant, cowpea, is an indigenous African legume that grows well in semiarid regions. Cowpea is grown and marketed mostly by women and is considered an "insurance crop." Since the launch of AGRA, there has been progress; however, increasing productivity remains a challenge with African farms producing at less than a sixth of the yield of farms in the United States.

In many African countries, milk is a critical part of the strategy to eradicate malnutrition and to provide nutritious food to a growing population, and some of the AGRA initiatives have focused on this. For example, Rwanda, a country still challenged by food insecurity and high child malnutrition has been improving the dairy industry. Smallholder farmers often have one or two cows, as part of the country's "One Cow per Poor Family" initiative. A single cow can provide much-needed income, and also boost a family's nutrition. However, milk faces numerous obstacles on its journey from cow to consumer. Given that Rwanda's nickname is "Land of a Thousand Hills," getting milk to the community collection center is difficult. One challenge has been how to collect milk from so many individual farmers; most of the milk is collected in stainless steel containers, then bicycled to collection centers like that of Blessed Dairies shown in Photo 2.2. Blessed Dairies collects milk from more than 8,000 rural farmers in a system of collection centers.

PHOTO 2.2 Blessed Dairies Milk Cooperative
Blessed Dairies is a collection and distribution center in Rwanda that is an
example of a farmer cooperative. This collection center is one of many that
collects milk from more than 8,000 rural farmers. It then uses the milk to create
yogurt, cheese, and butter for domestic consumption.

This new Green Revolution for Africa is not without its critics who point to
a range of issues, including the potential social and environmental pitfalls of
mono-crop agriculture, the dangers of encouraging farmers to use genetically
modified seeds, and the likelihood that high-cost inputs will lead to growing inequalities within African farming communities. Political economists and
political ecologists are also concerned that African farmers may become more
dependent on foreign imports of seed and fertilizer. In addition, as was true
in the previous iterations of the Green Revolution, mechanization of farming
requires land consolidation, which could result in a rural exodus. Finally, some
critics wonder if the priorities of this new African Green Revolution align with
the aspirations of Africa's small farmers. For these reasons, some of the same
negative consequences of the 1960s Green Revolution could occur in Africa.

Consider the mixed results in Rwanda, where more than 85 percent of the
population farms on land that averages .76 hectares (the smallest farms in the
region). On the one hand, general aggregate data suggests that over the last
ten years agricultural productivity increased and conventionally measured
poverty rates have fallen. On the other hand, studies have shown that in many
local areas household subsistence practices were disrupted, poverty exacerbated, and land tenure security and autonomy were curtailed. All of which
underscores the complexities of increasing food security.

Practice Sustainable Agriculture

There has been considerable discussion on how to define "sustainable agriculture." In the past, sustainable agriculture focused primarily on environmental criteria. For example, if the soil was bad, or if water was used

SOLUTIONS 2.1

Biofortification

Target 2.2: End all forms of malnutrition, including achieving, by 2025, the internationally agreed targets on stunting and wasting in children under five years of age, and address the nutritional needs of adolescent girls, pregnant and lactating women, and older persons.

Fortification is the practice of deliberately increasing the content of a micronutrient such as vitamins or minerals in a food, so as to improve the nutritional quality of that food. **Biofortification** is the process of improving the nutritional quality of food during plant growth (rather than having nutrients added to the foods during processing of the crops—for example, adding vitamins to your breakfast cereal during production). Biofortification typically involves breeding crops to increase their nutritional value. This can be done through conventional selective breeding or through genetic engineering.

Biofortification may have the potential to improve nutritional outcomes in populations where supplementation and conventional fortification may be limited or unaffordable. This is one reason why biofortification is seen as an important strategy for dealing with malnutrition in low- and middle-income countries. Examples of biofortification projects include:

- Iron-biofortification of rice, beans, sweet potato, cassava, and legumes;
- Zinc-biofortification of wheat, rice, beans, sweet potato, and maize;
- Provitamin A biofortification of sweet potato, maize, and cassava; and
- Amino acids and protein-biofortification of sorghum and cassava.

The World Health Organization estimates that iron biofortification of crops could help cure the two billion people who suffer from iron deficiency anemia. Another example is Golden Rice, which is biofortified with beta-carotene that the body converts into Vitamin A. Other examples include iron-fortified biscuits, micronutrient fortification of flour, drinks that contain essential vitamins and minerals, fortification of edible food oil, and chickpea products.

In many parts of Africa, where rice is less commonly grown or consumed, biofortification of sweet potatoes has been effective. Mozambique, Uganda, and Rwanda introduced orange-fleshed sweet potatoes as a way to reduce Vitamin A inadequacies.

Biofortified sweet potatoes are part of a coordinated program consisting of (1) disseminating specific seeds and information about growing the sweet potato, (2) creating demand for consuming sweet potatoes, and (3) helping to create a market component. It has been a challenge to convince farmers and consumers that sweet potatoes are worth growing and consuming.

Rwanda has worked with USAID to promote the sweet potato as a "superfood." In 2015, Rwanda launched an initiative called SUSTAIN—Scale Up Sweetpotato through Agriculture and Nutrition. The program aims to increase production and consumption of sweet potatoes and provide more jobs for women in the value chain. The sweet potato grows well in Rwanda's poor soils and can produce many tubers in a relatively small area of land. The result is high yields in small spaces. Furthermore, sweet potatoes are high in calories and in Vitamin A. Vitamin A is essential to eye health in young children and a lack of Vitamin A can lead to blindness. Today, sweet potatoes are Rwanda's number one crop, accounting for 13 percent of all crop production while taking up only 5.2 percent of arable land according to UKAID.

The sweet potato is also used as a flour in biscuits and breads. The SUSTAIN program has supported numerous roadside stands in rural northern Rwanda. The stand promotes consumption of sweet potatoes and provides nutritional counseling cards, growing instructions, and cooking instructions.

Some critics worry that biofortification as a strategy could further simplify diets already overly dependent on a few carbohydrate staples, and that lack of access to a diverse and balanced diet is critical. Advocates of biofortification acknowledge this as a long-term concern but believe that in the short-term biofortification is an effective strategy to help reduce malnutrition.

For Rwanda, a country committed to ending malnutrition by 2030, the sweet potato is an important part of their plan to reduce child stunting from 44 percent to 18 percent by 2025. It is also a critical part of increasing the population's access to nutritious foods. How sweet is that?

Sources: "Achieving Agricultural Transformation: Rwanda Leads the Way." The World Bank Group (2015). http://www.worldbank.org/en/news/feature/2015/06/09/achieving-agricultural-transformation-rwanda-leads-the-way
"Agriculture and Food Security" (n.d.). USAID. https://www.usaid.gov/rwanda/agriculture-and-food-security
Feed the Future (2011). https://docs.igihe.com/IMG/pdf/rwandafeedthefuturemultiyearstrategy.pdf
"Scaling Up Sweetpotato through Agriculture and Nutrition (SUSTAIN) in Rwanda." UKAID (2015). http://www.sweetpotatoknowledge.org/wp-content/uploads/2015/12/Brief16_SPHI_SUSTAIN-Rwanda.pdf

inefficiently, then a farm might have been considered unsustainable. In recent years, there has been a realization that sustainable agriculture must also include the economic and social dimensions. If a farm cannot withstand a disaster, or if the well-being of farm workers is disregarded, then a farm cannot be sustainable. The FAO characterizes sustainable agriculture as an integrated system: mixed cropping, conservation agriculture, and agroforestry systems that aim at producing more food and feed from the same area of land with fewer inputs. One thing that many agree on is that we must improve the ability to grow food—we need to produce higher yields per unit of land. This may involve a range of solutions, including genetically engineered food. Second, communities must become more resilient to withstand the environmental changes that will come. Strategies here include introducing improved drought-resistance crops or making crop varieties more nutritious or biofortified (see *Solutions* 2.1). Another opportunity is to take advantage of "precision farming" that uses geospatial data and other types of information to economize on or reduce the use of water, nitrogen, and pesticides so that food can be produced with less environmental impact. There is a need for new advances in better harvesting, storage, and transport of crops, all of which reduce food loss and waste that occur from farm to plate. Finally, many agricultural communities may benefit from the formation of farming cooperatives, explored in the *Solutions* 2.2 box.

Agroecology

Another possible solution is agroecology. **Agroecology** is an alternative approach to agriculture that draws on a rich history of traditional farming systems that suggests that local farmers have developed long-standing livelihood strategies and adaptive responses to environmental change. Agroecology entails the use of ecological principles such as crop rotation, mixed cropping, livestock integration, agroforestry, and composting to design resilient, sustainable, low-input farming systems, drawing on local farmers' agricultural knowledge, and complex understanding of environmental conditions. The FAO reports there is evidence that restoring and scaling traditional agroecological practices through farmer-led education could help improve food security and build social resilience among the poor.

SOLUTIONS 2.2

Food Cooperatives

Target 2.3 Double the agricultural productivity and incomes of small-scale food producers, in particular women, indigenous peoples, family farmers, pastoralists, and fishers, including through secure and equal access to land, other productive resources and inputs, knowledge, financial services, markets and opportunities for value addition, and nonfarm employment.

Target 2.4 Ensure sustainable food production systems and implement resilient agricultural practices that increase productivity and production; that help maintain ecosystems; that strengthen capacity for adaptation to climate change, extreme weather, drought, flooding, and other disasters; and that progressively improve land and soil quality.

As communities and organizations work to achieve the targets of SDG 2, one established model that is attracting renewed attention is cooperatively owned businesses (or co-ops). Cooperatives exist in a diverse set of sectors, offering alternatives to traditional shareholder- or proprietor-owned business structures.

Cooperatives are member-owned, democratically controlled business enterprises. One of the main goals of cooperatives is economic inclusion. They are formed to help small players gain parity with large investor-owned competitors to address market failures where neither the private sector nor the government provides a needed service or to give consumers a deliberate choice of enterprise to better meet their common needs and aspirations. The structure of a cooperative requires that it be responsive to its member-owners and, in turn, to the local community. The nature of cooperatives is inherently both locally based and participatory, embodying a direct connection between member needs and the services provided. Because of this, cooperatives are well positioned to contribute directly to community vitality and stability, modeling equitable and inclusive economic practices. This essential aspect of cooperatives can affect community health through the support of stable, community-based institutions and the nurturing and developing of authentic local leaders and informed and empowered members.

Cooperatives exist across numerous types of businesses. Ace Hardware, REI, and Navy Federal Credit Union are examples. Farmer-owned cooperatives include Sunkist, Organic Valley, Cabot, Ocean Spray, Blue Diamond, Florida's Natural, and Land O'Lakes. They are not limited to the United States. AMUL is an Indian dairy cooperative jointly owned by 3.6 million milk producers. In Bolivia, El Ceibo is a cocoa cooperative that uses agroecology practices. There are hundreds of examples around the world; farmer cooperatives are expanding.

Farmer cooperatives market milk and dairy products, grains, livestock and poultry, fiber, nuts, produce, and other products on behalf of their member-owner producers, often adding value to raw products for greater economic benefit. Farmer cooperatives also sell farm supplies and provide services to their member producers at competitive prices.

Agricultural processing cooperatives add value to commodities through the collective ownership of processing equipment. Marketing cooperatives provide members access to regional, national, and international markets for their goods that would not be available to independent producers. Additionally, these cooperatives may give member-owners the benefit of national brand names that add value to products marketed through the cooperative.

The biggest name in organic dairy in the United States, Organic Valley, started out in 1988 with a small group of organic vegetables farmers in Western Wisconsin who wanted to build a brand and market for organic produce. Their effort soon expanded to other products; today, dairy products make up 88 percent of revenue. Before Organic Valley started, the only place an organic farmer could sell dairy products at the premium price it cost to produce them was at the farm stand or very small bottling operation.

To ensure that organic farming is both economically and environmentally sustainable, the co-op prices milk from the farmer up. This means they begin with the actual cost of production, and make sure that the price being charged will yield a fair and living wage for family farmers. Just as important is the concept of price stability. In the conventional dairy world, the price of milk fluctuates month to month, at times dramatically so, pushing constant risk and insecurity onto the individual producer. In contrast, Organic Valley opted to take on and share that market risk at the cooperative level, paying a steady and predictable price to farmer-members. With Organic Valley's price stability policy in place, a farmer knows what to expect, making it easier to make sound decisions about the farm operation, leading to greater business stability for co-op members. The typical Organic Valley dairy farmer has been farming for more than twenty years and has been certified organic for almost a decade. They know that milk prices eventually go down just as surely as they go up, and that Organic Valley's stable price philosophy benefits its members tremendously over the long term. By transparently setting prices based upon actual costs borne by certified organic farmers, Organic Valley made clear that undercutting that price would undercut the farmer. By taking firm leadership, setting the bar high, and clearly explaining why it needs to be there, the co-op has established an entire new industry that operates—much more than most—to the widespread advantage of consumers, small producers, and the environment.

Sources: The National Cooperative Business Association. "The ABCs of Cooperative Impacts." https://ncbaclusa.coop/resources/abcs-of-cooperative-impact/ The National Cooperative Business Association. "The ABCs of Farmer and Independent Small Business Co-Op Impact, Case Study: With Its Focus on Stability and Sustainability, Organic Valley Benefits Consumers, Small Producers and the Environment." https://ncbaclusa.coop/content/uploads/2020/09/Farmer-bifold-final-2.pdf

However, feminist political ecologists are cautious about adopting agroecology without the transformation of social and political roles and systems. Ruth Nyambura, a Kenyan ecofeminist and researcher, writes:

Food sovereignty is a deeply political project that those of us in the food justice movement take seriously. We are not just interested in a technologically different way of farming or providing food and jobs for people using alternatives like agroecology, but we realize that a technologically different way of farming from industrial agriculture will be useless if questions of power, by the class-gendered-racial, rural-urban, young-old are not constructively addressed.

She concludes that agroecology and family farming are limited in their revolutionary potential if women continue to face the violence of patriarchy in their immediate surroundings and especially with relation to access and control over the ecological resources and the exploitation of their reproductive and productive labor.

SUMMARY AND PROGRESS

SDG 2 is one of the few SDGs where progress has been reversed and not just because of COVID-19. Recall that food insecurity has been on the rise over the last four years notably in sub-Saharan Africa and Latin America. In Eastern Africa countries such as Kenya, an upsurge in the desert locust wreaked havoc on the food supply. In addition, conflict in Yemen has exacerbated food insecurity for millions. *Critical Perspectives* 2.2 offers some concerns on achieving the targets of SDG 2.

CRITICAL PERSPECTIVES 2.2

Critiquing SDG 2 Indicators

Some scholars have critiqued the UN set of indicators related to SDG 2 as not universally applicable and say their operationalization still requires fine-tuning. A recent article by Juliana Gil and colleagues highlighted several problems with some of the targets and indicators for SDG 2.

The researchers note that indicator 2.4.1, for example, refers to the percentage of agricultural area under sustainable practices, and while percentage is a quantifiable metric, agreeing on what sustainability is, when it is achieved, and what it translates into at different scales can be difficult. Targets 2.3 (agricultural productivity), 2.4 (sustainability of food production systems), and 2.5 (genetic diversity) are less clearly defined and not always universally relevant. How each country defines these terms could lead to a variety of interpretations due to the vagueness and lack of specificity of terms such as "sustainable" or "fair" regarding the scale of enforcement and monitoring or the boundaries of "food systems." They also note that goals to double agricultural productivity are not universally applicable. In some countries, the pursuit of agricultural intensification may collide with the pursuit of agricultural sustainability.

There are other criticisms. No set of indicators can fully capture the complex link between agricultural interventions, dietary change, and nutrition, which involves several complex factors within and beyond SDG 2 (e.g., food production, diet diversification, biofortification, food safety, gender empowerment, value chains, policy support).

Juliana Gil and colleagues caution that when implementing the SDGs, it is important to check the reliability of existing global figures and complement them with country-level data whenever possible. A great diversity may exist at the subnational level, making it necessary to use more detailed local information—in other words, be sensitive to geographic scale.

They also note that not all targets have the same degree of priority in different countries. Although SDG targets are the same for all countries, the pathways they will follow will differ for each country. Historical trends may offer hints on a country's future, but there is no guarantee that past trends will persist. Finally, synergies and trade-offs across SDGs and analytical scales must be examined to inform coherent policy design under different scenarios.

_____

Source: Gil, J. D. B., Reidsma, P., Giller, K., et al. "Sustainable Development Goal 2: Improved Targets and Indicators for Agriculture and Food Security." *Ambio* 48: 685–98 (2019). https://doi.org/10.1007/s13280-018-1101-4

Along with conflict, climate shocks, and the locust crisis, COVID-19 is likely to increase the numbers of people food insecure because of economic slowdowns and disruptions to food systems. The lockdown measures in place to prevent the spread of COVID-19 caused businesses and local markets to close and prevented many small-scale food producers from getting their products to consumers. It is estimated that an additional 132 million people may suffer undernourishment because of COVID-19 due to a loss of household income, a lack of available and affordable nutritious food, reduced physical activity, and disruptions in essential nutrition services. A 2021 UNICEF report on the state of food security estimates that by 2030, as many as 660 million people may still face hunger; this is thirty million more people than previously predicted because of the lasting effects of the pandemic. The report also presents new projections of potential additional cases of child stunting and wasting due to COVID-19. Based on a conservative scenario, it is projected that an additional twenty-two million children in low- and middle-income countries will be stunted, an additional forty million will be wasted between 2020 and 2030 due to the pandemic.

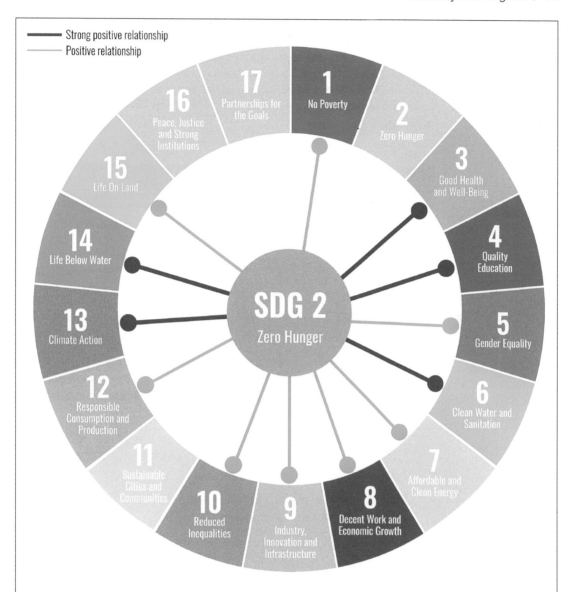

- Strong positive relationship
- Positive relationship

SDG 1: Achieving targets in SDG 2 can enable and reinforce through enhanced food and nutrition security – which are essential to reduce poverty and eradicate extreme poverty.

SDG 3: When people's nutrition status improves, it helps break the intergenerational cycle of poverty, generates broad-based economic growth and provides a foundation for improved health and well-being.

SDG 10: Supporting small-scale food producers can lead to substantial poverty reduction as rural people constitute the largest segment of the world's poorest, positively impacting both SDG 1 (No Poverty) and SDG 10 (Reduced Inequalities).

SDG 15: Agriculture impacts on the wellbeing of terrestrial ecosystems (sustainable food production system and agriculture practices), and if done sustainability, should reinforce the maintenance of terrestrial ecosystems and the prevention of land as well as biodiversity erosion. However, it is also possible that if increasing agricultural productivity relies on practices and technologies that contribute to land and soil degradation and high Greenhouse Gas emissions, targets focused on the conservation, restoration and sustainable use of terrestrial ecosystems, forests, soils and biodiversity might not be achieved.

FIGURE 2.8 Interconnections in SDG 2 and the Other Goals
Hunger and food insecurity connect to all the SDGs in some way; this figure highlights several of the stronger connections.

Solving hunger and creating sustainable food systems will require public awareness, political and individual changes, and the mobilization of new technologies. Some experts argue that a sustainable agricultural system must be tailored to local contexts. There will be no "one solution" fits all and advancing targets in other SDGs will also help meet targets of SDG 2. Figure 2.8 highlights the strongest synergies and interconnections between SDG 2 and the other goals.

QUESTIONS FOR DISCUSSION AND ACTIVITIES

1. Describe the geography of food insecurity and identify the most vulnerable populations.

2. Identify and explain the impacts of food insecurity and hunger.

3. Explain how the three Es are present in SDG 2.

4. Explain the causes of food insecurity; explain why some experts argue that the root of hunger is poverty.

5. Investigate whether there are food deserts in your community (to find out, go to the USDA food desert locator map at https://www.ers.usda.gov/data-products/food-access-research-atlas/). Discuss what strategies could be used to solve the problem of food deserts.

6. How could individual consumers influence the global food system through their food choices? How would you change your diet or approach to food in the future?

7. Of the solutions discussed for addressing food insecurity, which would you prioritize for investment and why?

8. Explain how SDG 2 strategies connect strongly to SDG 1, SDG 3, SDG 10, and SDG 15. Investigate real-world examples of these interconnections.

TERMS

acute undernutrition
agroecology
biofortification
chronic undernutrition
food deserts
food insecurity
food swamps
Green Revolution
hidden hunger
hunger
land extensification

land intensification
malnutrition
micronutrient deficiency
minimum dietary energy
   requirement
moderate food security
obesity
undernourishment
stunting
wasting

RESOURCES USED AND SUGGESTED READINGS

Altieri, M. A., Nicholls, C. I., Henao, A., & Lana, M. A. "Agroecology and the Design of Climate Change-Resilient Farming Systems." *Agronomy for Sustainable Development* 35 (3): 869–90 (2015).

Boucher, D. H. *The Paradox of Plenty: Hunger in a Bountiful World.* Oakland, CA: Food First (1999).

Chase, L. *Food, Farms, and Community: Exploring Food Systems.* Durham, NH: University of New Hampshire (2014).

Devereux, S., Vaitla, B. and Swan, Robert, S. H. Seasons of Hunger: Fighting Cycles of Quiet Starvation among the World's Rural Poor. London: Pluto in Association with Action Against Hunger, ACF International Network (2008).

Graziano da Silva, J. "Achieving SDG2: End Hunger, Achieve Food Security and Improved Nutrition and Promote Sustainable Agriculture." Food and Agricultural Organization of the United Nations (2016). https://www.fao.org/sustainable-development-goals/news/detail-news/en/c/424259/

Itieri, M. A., and Nicholls, C. I. "Agroecology and the Reconstruction of a Post-COVID-19 Agriculture." *The Journal of Peasant Studies* 47 (5): 881–98 (2020).

Kimura, Aya Hirata. *Hidden Hunger: Gender and the Politics of Smarter Foods.* Ithaca, NY: Cornell University Press (2013).

Merino, J. "Women Speak: Ruth Nyambura Insists on a Feminist Political Ecology," *Ms Magazine,* November 2017. https://msmagazine.com/2017/11/15/women-speak-ruth-nyambura-feminist-political-ecology/

Misra, M. "Moving Away from Technocratic Framing: Agroecology and Food Sovereignty as Possible Alternatives to Alleviate Rural Malnutrition in Bangladesh." *Agriculture and Human Values* 35: 473–87 (2018). https://doi.org/10.1007/s10460-017-9843-3

Palmer, L. *Hot, Hungry Planet: The Fight to Stop a Global Food Crisis in the Face of Climate Change.* New York: St. Martin's Press (2017).

Rieff, D. *The Reproach of Hunger: Food, Justice, and Money in the Twenty-First Century.* London: Verso (2016).

United Nations Children's Emergency Fund. *The State of Food Security and Nutrition in the World.* Geneva: United Nations (2021). https://data.unicef.org/resources/sofi2021/?utm_source=newsletter&utm_medium=email&utm_campaign=SOFI%202021

United Nations Children's Emergency Fund, World Health Organization, International Bank for Reconstruction and Development/The World Bank. *Levels and Trends in Child Malnutrition: Key Findings of the 2021 Edition of the Joint Child Malnutrition Estimates.* Geneva: World Health Organization (2021). https://data.unicef.org/resources/jme-report–2021/

United Nations Food and Agriculture Organization. *Transforming Food and Agriculture to Achieve the SDGs* (2018). http://www.fao.org/3/I9900EN/i9900en.pdf

United Nations Food and Agriculture Organization. *The Water-Energy-Nexus: A New Approach in Support of Food Security and Sustainable Development*. Rome: United Nations (2014). http://www.fao.org/3/bl496e/bl496e.pdf

World Food Programme. *The State of Food Security and Nutrition in the World* (2019). Wfp.org/publications/2019-state-food-security-and-nutrtion-world-sofi-safeguarding-against-economic

Health

LEARNING OBJECTIVES

After reading this chapter, you should be able to:

- Identify key threats to health and well-being
- Describe trends around health in various world regions
- Explain key health measures such as mortality rates
- Explain major obstacles to achieving good health
- Identify and explain the various targets and goals associated with SDG 9
- Explain how the three Es are present in SDG 9
- Explain how SDG 9 strategies connect to other SDGs

The first to die was her little boy, just five years old when he got sick. The following year, Jane Wamalwa lost a second child, this time her eighteen-month-old son. When her baby daughter fell ill two years later, Jane knew what was going to happen next and felt powerless to stop it. She watched in despair as her third child succumbed to an illness that is both deadly and easily preventable: diarrheal disease.* In western Kenya, where Jane lives, diarrhea is one of the top causes of death for children under five years old.

Tragically, diarrheal disease is both preventable and treatable.

HEALTH: UNDERSTANDING THE ISSUE

The COVID-19 **pandemic** that began in 2020 has commanded a tremendous amount of attention and deservedly so. It has caused loss of life and human suffering—5.5 million dead and 326 million infections worldwide at the start of 2022. The pandemic has destabilized the global economy and increased global unemployment dramatically. It has exposed and deepened existing economic, social, and institutional inequalities between regions and countries, and within countries, hitting the poorest and most vulnerable communities the hardest (*Understanding the Issue* 3.1). Refugees and migrants, as well as Indigenous peoples, older persons, and people with disabilities are particularly at risk. And hate speech targeting vulnerable groups is on the rise. The pandemic has also disrupted progress on all the other SDGs.

Although emerging and immediate infectious diseases—such as COVID-19—often dominate media attention and captivate much of the dialogue around global health threats, there remain persistent global health priorities the world has been addressing for several decades.

---

*"The Heartbreaking Truth about Diarrhea." PATH, March 22, 2013. https://www.path.org/articles/the-truth-about-diarrhea/

UNDERSTANDING THE ISSUE 3.1

The Diffusion and Disruption of COVID-19

In December 2019, a novel respiratory disease was identified in the city of Wuhan China. It was given the name SARS-CoV-2, later shortened to COVID-19. It spread quickly. By September 2020, it had infected thirty-two million people and caused more than one million deaths. By July 2021, it had infected 170 million and caused more than 3.5 million deaths. In the United States, more than thirty-three million people were infected and almost 600,000 died. Map 3.1 shows the cumulative cases of COVID-19 as of the summer of 2021.

More than 60 percent of new infectious diseases come from animals. When we transform forests or wild areas, we increase the possibility of more new viruses, for which we have no immunity because humans have had little or no previous exposure. The COVID-19 pandemic reveals the **zoonotic** threat caused by land conversion of wildlands or forests into farmland, pasture, and urban areas.

Another critical factor in the diffusion of the disease is globalization. Diseases spread more quickly in a shrinking world, turning local outbreaks into **epidemic diseases**. COVID-19 spread quickly around the world through air passengers. For example, in the first two weeks of January 2020, there were 1,300 flights from China to the United States, with 380,000 people arriving during that time frame. The virus spread through every country. However, its impact varied enormously over space and time. Figure 3.1 highlights these differences. Initially, in many US and European cities, it was the more affluent global travelers, the rich cosmopolitans, who brought the disease, first impacting affluent neighborhoods before moving into the lower-income minority communities. In the United States, early in the pandemic, states such as Washington, California, and New York were hit hardest; within several months, however, the virus had diffused into states in the South and Midwest. In the United States, Black, Native American, and Hispanics were most vulnerable as they were likely to be "essential workers" who live and work in environments where the virus was more likely to be present and more easily spread, such as in health care, food services, and public transportation.

More recently, middle- and lower-income countries have experienced significant rates of infections. Brazil, for example, was hard hit, with sixteen million infections and 460,000 deaths. Vaccination rates were far below where they need to be to create **herd immunity**. In early summer 2021, India experienced a second surge

COVID-19 Cases (As of 6/9/21)*
*Locations and case numbers (circle sizes) are approximate

MAP 3.1 Cumulative COVID-19 Cases (June 2021)
This map highlights the uneven global diffusion of COVID-19.

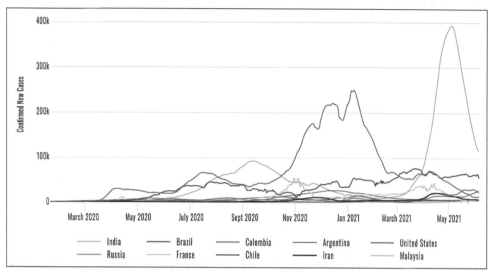

FIGURE 3.1 Daily Confirmed Cases of COVID-19 for Selected Countries
Every country has experienced cases of COVID-19, but the intensity and impact varies across time and space as social, economic, and political responses to the disease differ from one country to another.

in infections. More than twenty-seven million had been infected, with 325,000 deaths, which many say is severely understated. The deaths are second only to the United States at that time. The wave overwhelmed health-care systems, with people struggling to get hospital beds, oxygen, and medicines. Crematoriums ran out of space. Despite the fact that India is the world's largest vaccine manufacturer, the Indian government failed to purchase enough for their own population and only fourteen million doses had been administered to a country of 1.3 billion.

Global supply chains were also disrupted by the pandemic. In the first months of the pandemic in the United States many states were unable to access supplies and had to compete with each other on the open market to purchase vital lifesaving equipment. The lack of personal protection equipment (PPE) for frontline responders led the governor of Maryland to call in favors from South Korea to have PPE shipped by air from Seoul to Baltimore.

Worldwide, the pandemic created economic chaos. Lockdowns and restrictions closed down substantial parts of the global economy. In the lowest-income countries, with large informal economies, more people were simply left adrift. Higher-income countries were better placed to help their citizens. Across Europe, enhanced support allowed people to remain in employment and health-care provision was maintained. In the United States, in contrast, the lack of a safety net was exposed. In early spring 2020 more than thirty million people were collecting unemployment and many were applying for aid.

The pandemic was a giant "stress test" that revealed the cracks and strains of all countries. In the United States, the pandemic highlighted long-lasting issues including a health system that was geared to making money rather than providing health care for all, a decline of public health, hyperpartisanship, and attacks on science. It also revealed the enduring injustices of class and race. These effects will be long-lasting.

Source: Short, J. R. *Stress Testing the USA: Public Policy and Reaction to Disaster Events.* 2nd ed. Cham, Switzerland: Palgrave McMillan (2021): 143–52.

Health is a crucial social and economic asset, and a cornerstone of human development. SDG 3 aims to ensure healthy lives and promote well-being for all at all ages. However, there are significant challenges to achieving the goal and certainly the current pandemic reminds us that many people lack **access to health services** and that many health-care systems did not cope well with crisis.

TABLE 3.1 Examples of Infectious and Degenerative Diseases

Infectious Diseases	• Cholera • Diarrhea • Ebola • Influenza (pandemic, seasonal) • Meningitis • Novel coronavirus (COVID-19)	• Plague • Respiratory infections • Smallpox • Yellow fever • Zika virus
Degenerative Diseases	• Alzheimer's disease • Parkinson's disease • Cancers	• Coronary artery disease • Macular degeneration • Osteoarthritis

Source: By author and compiled from several sources.

There are three broad categories of threats to human life and well-being: biological, socioeconomic, and environmental. Biologically, two categories of diseases are among the leading causes of death. **Infectious diseases** or **communicable diseases** are disorders caused by **pathogenic** organisms—such as bacteria, viruses, fungi, or parasites. Many organisms live in and on our bodies. Typically, they are harmless or even helpful. But under certain conditions, some organisms may cause disease. Some infectious diseases can be passed from person to person.

Infectious diseases can be grouped in three categories: diseases that cause high levels of mortality; diseases that can cause short or long-term disability; and diseases that, owing to the rapid and unexpected nature of their spread, can have serious global repercussions.

Noncommunicable (chronic or **degenerative**) **disease** is the result of a continuous process based on degenerative cell changes, affecting tissues or organs, which will increasingly deteriorate over time. As a body ages, changes over time accumulate and there can be certain instances when any degenerative changes lead to symptoms and disease. Table 3.1 outlines examples of infectious and degenerative diseases.

There are also threats to human health and well-being that are socioeconomic in origin. These can be society-wide or individual-level factors. Examples include:

- Education level
- Poverty
- Health insurance status
- Employment status
- Racism
- Housing conditions
- Early childhood stressors (such as malnutrition)
- Drowning
- Income inequality
- Deteriorating built environment
- Racial segregation
- Crime and violence
- Social capital
- Availability of open or green spaces
- Violence or homicide

- Suicide or depression
- Unhealthy behaviors (tobacco use, alcohol or drug abuse)

Finally, there are also environmental factors that can cause ill health or death. These include exposure to pollution and natural disasters. Contact with or ingestion of polluted water, for example, can cause intestinal infections such as cholera or diarrheal disease, and air pollution is associated with increased **morbidity** from respiratory diseases. Today, environmental factors directly and indirectly contribute to 23 percent of all deaths globally.

Table 3.2 lists the leading causes of death currently. In 2019, the top ten causes of death accounted for 55 percent of the 55.4 million deaths worldwide that year. However, the causes of death vary greatly between the Global North and the Global South.

In the Global North, more deaths occur due to noncommunicable or degenerative diseases such as heart disease, stroke, and chronic obstructive

TABLE 3.2   Leading Cause of Death for All Ages 2019

Leading Cause of Death for All Ages for All Countries

Rank	Cause	% of total deaths
1	Ischemic heart disease	16.6
2	Stroke	10.2
3	Chronic obstructive pulmonary disease	5.4
4	Lower respiratory infections	5.2
5	Alzheimer's disease and other dementias	3.5
6	Trachea, bronchus, lung cancers	3.0
7	Diabetes mellitus	2.8
8	Road traffic injuries	2.5
9	Diarrheal diseases	2.4
10	Tuberculosis	2.3

Leading Cause of Death for All Ages for Low-Income Countries

Rank	Cause
1	Neonatal conditions
2	Lower respiratory infections
3	Ischemic heart disease
4	Stroke
5	Diarrheal diseases
6	Malaria
7	Road traffic injuries
8	Tuberculosis
9	HIV/AIDS
10	Cirrhosis of the liver

Source: World Health Organization. "The Top 10 Causes of Death" (2020). https://www.who.int/news-room/fact-sheets/detail/the-top-10-causes-of-death

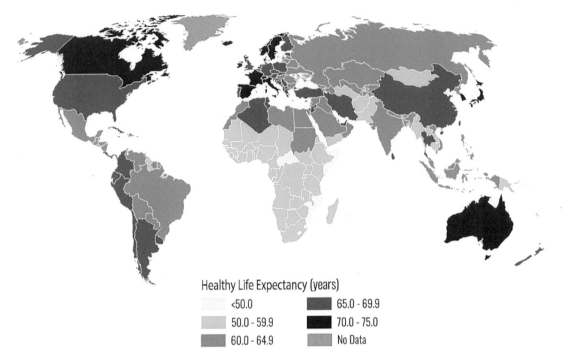

Healthy Life Expectancy (years)

<50.0	65.0 - 69.9
50.0 - 59.9	70.0 - 75.0
60.0 - 64.9	No Data

MAP 3.2 Healthy Life Expectancy at Birth, Both Sexes (2019)
As with many demographic measurements, countries in the Global North have higher life expectancy than those in the Global South.

pulmonary disease. In the highest-income countries, degenerative diseases such as lung cancer, diabetes, colon cancer, and kidney diseases are increasing, while heart disease and stroke have decreased in the last twenty years. In 2019, Alzheimer's disease and other forms of dementia ranked as the second leading cause of death in high-income countries. Women are disproportionately affected. Globally, 65 percent of deaths from Alzheimer's and other forms of dementia are women.

In contrast, in the Global South, people living in a low-income country are far more likely to die of a communicable disease than a noncommunicable disease. Table 3.1 highlights that in a low-income country, six of the top ten causes of death are infectious diseases such as malaria, tuberculosis, and HIV/AIDS. The contrast between high-income countries and low-income countries reminds us that where you live can greatly influence the types of risks to your health.

The good news is that over the last two decades, there has been progress in increasing life expectancy and reducing some of the common killers associated with child mortality, **infant mortality**, and maternal mortality (Map 3.2 shows life expectancy as of 2019). A focus on providing more efficient funding of health systems, improved sanitation and hygiene, and increased access to physicians has saved the lives of millions.

CHALLENGES TO GOOD HEALTH

There are many continuing challenges to good health. There is a need to fully eradicate a wide range of infectious diseases such as HIV/AIDS, malaria, and tuberculosis. More attention needs to be paid to neglected tropical diseases. More investments are needed to address depression or domestic violence.

Climate change is likely to increase vulnerability to many health complications, including heat stroke, malnutrition, and the spread of certain diseases. In addition, nearly half the world's people cannot afford or access quality health care and live in unhealthy environments. **Health systems** and **health services** in nearly all regions need to be improved. The following sections outline some of the more significant threats to human health and well-being.

Maternal Health

Complications during pregnancy and childbirth remain a cause of death in many of the lowest income countries. Today, nearly 300,000 women die each year from complications relating to pregnancy and childbirth, which is referred to as **maternal mortality**. More than 90 percent of them live in low- and middle-income countries in the Global South. The *Key Terms and Concepts* 3.1 box discusses maternal mortality trends. Many of those women at high risk for maternal mortality live in remote areas, are in poverty, and are the least likely to receive adequate health care. The majority of maternal deaths are preventable through appropriate management and care, including antenatal care by trained health providers, assistance during delivery by skilled health personnel, and care and support in the weeks after childbirth.

 ## KEY TERMS AND CONCEPTS 3.1

Maternal Mortality

Maternal death or **maternal mortality** is defined by the World Health Organization as "the death of a woman while pregnant or within 42 days of termination of pregnancy, irrespective of the duration and site of the pregnancy, from any cause related to or aggravated by the pregnancy or its management." Since 1990, the world maternal mortality rate has declined 44 precent, but every day 800 women still die from pregnancy or childbirth-related causes, from largely preventable causes before, during, and after the time of giving birth. That equates to about 300,000 women who die each year.

In most high-income countries, maternal mortality is now very low. The average rate in the European Union is six maternal deaths per 100,000 live births. The picture is very different in low-income countries. In sub-Saharan Africa, maternal mortality is 545 per 100,000 live births. More than 85 percent of maternal deaths are from impoverished communities in Africa and Asia (Map 3.3). In Sierra Leone a woman is 300 to 400 times more likely to die with each pregnancy. At an estimated rate of 1,360 deaths per 100,000 live births, around one in seventy-five pregnan-

cies end in the death of the mother. The five countries where a woman is most likely to die in a given pregnancy are Sierra Leone, Central African Republic, Chad, Nigeria, and South Sudan.

The effect of a mother's death results in vulnerable families. Their infants, if they survive childbirth, are more likely to die before reaching their second birthday.

There are many causes both direct and indirect. Direct causes can include infections, embolisms, or hemorrhage during birth, or complications from abortions. Other causes are indirect and can include poverty, access to health care, the overall health of the mother before and during pregnancy, and unmet need for family planning. One study found that maternal mortality rates would be almost two times higher without contraceptive use at current levels. Yet in some countries, women still cannot access family-planning methods. Other indirect factors include structural and social barriers, including the "3 delays":

1. Delays in seeking care are due to the decisions made by the women who are pregnant and/or other decision-making individuals. Decision-making

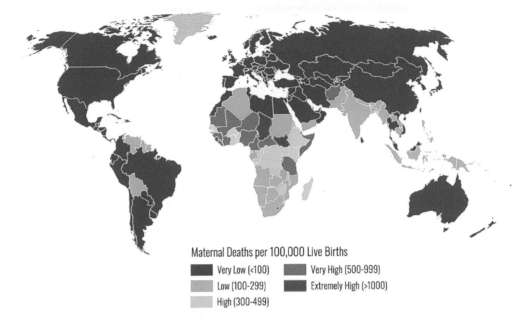

Maternal Deaths per 100,000 Live Births

- Very Low (<100)
- Low (100-299)
- High (300-499)
- Very High (500-999)
- Extremely High (>1000)

MAP 3.3 Maternal Mortality Ratio (2017)
Maternal mortality remains an issue in many countries in Africa where access to prenatal care and health services in general may be more limited.

individuals can include the woman's spouse and family members. Examples of reasons for delays in seeking care include lack of knowledge about when to seek care, inability to afford health care, and women needing permission from family members.
2. Delays in reaching care include factors such as limitations in transportation to a medical facility, lack of adequate medical facilities in the area, and lack of confidence in medicine.
3. Delays in receiving adequate and appropriate care may result from an inadequate number of trained providers, lack of appropriate supplies, and the lack of urgency or understanding of an emergency.

The positive news is that there are existing strategies for improving maternal mortality:

- Ensure that all women receive basic preventive and primary reproductive health care services, comprehensive sexuality education, family planning and contraception, as well as adequate skilled care during pregnancy, childbirth, and the postpartum period;
- Empower women, girls, and communities;
- Strengthen health systems to respond to the needs and priorities of women and girls;
- Improve water, sanitation, and hygiene (WASH) systems;

- Improve transportation and communication infrastructure;
- Support education at all levels for girls; and
- Focus on eliminating gender-based violence and discrimination.

For example, evidence shows that when girls exercise their rights to delay marriage and childbearing and choose to advance in school, maternal mortality goes down for each additional year of study they complete. Gender-based violence affects the reproductive health of women and girls throughout their lives. Its adverse consequences include unwanted pregnancies, pregnancy complications including low birth weight and miscarriage, injury and maternal death, and sexually transmitted infections such as HIV/AIDS.

If the world achieved European maternal mortality rates, only 11,000 women would die each year, instead of 300,000. If we can make maternal deaths as rare as they are in the healthiest countries in the world we can save almost 300,000 mothers each year.

Sources: Roser, M., and Ritchie, H. "Maternal Mortality." OurWorldInData.org (2013). https://ourworldindata.org/maternal-mortality
World Bank Development Indicators for Maternal Mortality. (2000 and 2017). https://databank.worldbank.org/source/world-development-indicators

Based on the latest data from 2013 to 2018, 81 percent of all births globally took place in the presence of skilled health personnel, a significant increase of 69 percent. In sub-Saharan Africa, where two-thirds of the world's maternal deaths occur, only 60 percent of births were assisted by skilled attendants. Projections show that substantially more resources will be required just to maintain current coverage rates in Africa, due to population growth.

Child Health

Children continue to die in many regions around the world. In 2019, approximately six million children and adolescents under the age of fifteen died, mostly from preventable causes. Leading causes of death in children under five years are preterm birth complications, birth asphyxia/trauma, pneumonia, congenital anomalies, diarrhea, and malaria, all of which can be prevented or treated. Access to affordable interventions including immunization, adequate nutrition, safe water and food, and health care by a trained health provider can prevent or treat these causes of death.

Since 2000 there has been progress made in child survival worldwide, and millions of children under five years of age are more likely to survive today than in 2000. The total number of under-five deaths dropped from 9.8 million in 2000 to 5.4 million in 2017. The under-five mortality rate has fallen by 49 percent—from 77 deaths per 1,000 live births in 2000 to 39 deaths currently. Yet the impacts of **child mortality** continue to be prevelant in certain regions and countries. Half of child deaths occur in sub-Saharan Africa, and another 30 percent in Southern Asia. Almost half (2.5 million) of the total number of under-five deaths took place in the first month of life—the most crucial period for child survival. In 2019 half of all under-five deaths occurred in just five countries: Nigeria, India, Pakistan, the Democratic Republic of the Congo, and Ethiopia. Nigeria and India alone account for almost a third of all child deaths.

The global **neonatal mortality** rate fell from 31 deaths per 1,000 live births in 2000 to eighteen deaths in 2017—a 41 percent reduction. However, progress will need to accelerate in about fifty countries, mostly in sub-Saharan Africa, to meet the target by 2030. According to UNICEF, a child born in sub-Saharan Africa was nine times more likely to die in the first month of life than a child from a high-income country. If neonatal mortality is reduced further, the lives of an additional ten million children under five years of age will be saved.

Infectious Diseases

Infectious diseases continue to significantly impact countries in the Global South, particularly the lowest-income countries. Three diseases in particular are noteworthy: **tuberculosis, HIV/AIDS,** and **malaria**. Combined, these three diseases kill more than 2.5 million people each year.

Tuberculosis

Tuberculosis (TB) is the world's leading cause of death from a single infectious agent and, according to World Health Organization (WHO), it is a worldwide **pandemic**. The symptoms of active TB disease include cough, weakness, weight loss, fever, loss of appetite, and night sweats. If left untreated, the mortality rate with this disease is more than 50 percent. In 2018, around ten million people became ill with TB and an estimated 1.5 million died. Another three million people with TB were not diagnosed and, as a result, were left behind without effective treatment and care.

Significant progress has been made over the last two decades. The incidence of tuberculosis declined by 21 percent since 2000—from 170 new and relapsed cases per 100,000 people in 2000 to 140 in 2015 and 134 in 2017. Still, there are regional differences. North Africa and sub-Saharan Africa are hit hardest.

In sub-Saharan Africa, Southern and Central African countries are among the most vulnerable. The TB epidemic in these countries is fueled by the HIV epidemic, with 50–80 percent of people with TB also living with HIV. The mining industry in this region plays a role; mining-related silicosis is a risk factor for TB and labor migration across international borders complicates the provision of treatment and care.

India is also impacted by TB and is the highest TB burden country. India is home to one in four people living with TB; approximately 220,000 people die of TB in India each year. Of the 2.2 million infected with TB in India, only half are able to afford the treatment drugs. Major challenges to controlling and treating TB in India include poor primary health-care infrastructure in rural areas of many states, high population density in urban slums, and the continued spread of HIV/AIDS infection. These factors often cause immunosuppression and may lead to a higher susceptibility to TB. In addition, the lack of political will and high corruption have hindered treatment.

Gender disparities in TB present huge challenges in providing access to services. Worldwide, men are much likelier than women to contract and die from TB, although women are more likely to provide care for people who are sick with TB, placing them at risk of exposure in caregiving situations.

Today large gaps in detection and treatment persist, and the current pace of progress is not fast enough to meet the target of ending the epidemic by 2030. Moreover, drug-resistant tuberculosis is a continuing threat. A much stronger focus on prevention is also essential to end TB. It had been widely acknowledged that the biggest gap in the fight against TB is political will.

HIV/AIDS

Despite significant progress, as of 2020, more than 37.9 million people globally were living with HIV (Map 3.4). Since the start of the epidemic in the early 1980s, almost seventy-six million people have become infected and thirty-three million have died from AIDS-related illnesses. Yet as Map 3.4 shows, the disease burdens some regions and countries far more than others. Sub-Saharan Africa and North Africa are two regions most severely affected, with nearly one in every twenty-five adults living with HIV and accounting for more than two-thirds of the people living with HIV worldwide.

No single prevention method or approach can stop the HIV epidemic on its own. Several methods and interventions have proved highly effective in reducing the risk of, and protecting against, HIV infection, including male and female condoms, the use of antiretroviral medicines such as pre-exposure prophylaxis (PrEP), behavior change interventions to reduce the number of sexual partners, the use of clean needles and syringes, and the treatment of people living with HIV to reduce viral load and prevent onward transmission.

According to UNAIDS, strong domestic and international commitment and funding for the AIDS response has accelerated programs for prevention,

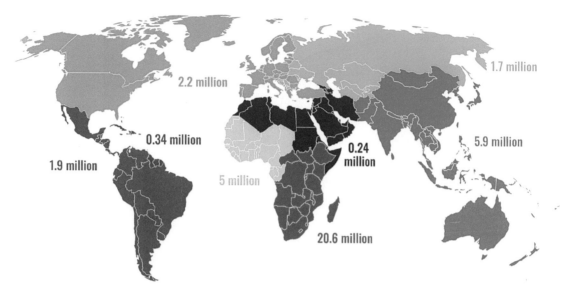

MAP 3.4 Estimated Number of People (All Ages) Living with HIV (2020)
Although HIV is found in all countries, those in sub-Saharan Africa have among the highest burden
of disease. Of the estimated 37.9 million living with HIV, almost 26 million live in sub-Saharan Africa.

testing, and treatment. Of the thirty-eight million people worldwide who
currently live with HIV, a record 23.3 million of them have access to antiretroviral therapy, which can control the infection. AIDS-related deaths have been
reduced by 60 percent since the peak in 2004.

Global HIV incidence among adults declined by 22 percent between 2010
and 2017, well short of the progress required to meet the 2030 targets. The
largest decline in incidence by age group was among children zero to fourteen
years old (37 percent from 2010 to 2017), reflecting increased provision of
antiretroviral medications to prevent mother-to-child HIV transmission.

The incidence of HIV among adults (fifteen to forty-nine years of age)
in sub-Saharan Africa declined by around 38 percent from 2010 to 2017.
Progress has been slower elsewhere, and some subregions have even seen an
increase in HIV incidence, including Western Asia (53 percent), Central Asia
(51 percent), and Europe (22 percent).

Despite reductions in HIV/AIDS infections there are still highly vulnerable
populations. Women and girls are particularly at risk in sub-Saharan Africa
and account for almost 60 percent of all new HIV infections. According to
UNAIDS, despite the fact that prevention exists, there are three reasons new
infections continue: first, a lack of political commitment, and as a result, inadequate investments; second, a reluctance to address sensitive issues related to
people's sexual and reproductive rights; and third, a lack of systematic prevention implementation, even when policy exists.

Malaria

Malaria is a disease that is transmitted from person to person through a
vector, in this case infected mosquitoes. This is known as a **vector-borne
disease**. The bite of an infected Anopheles mosquito transmits a parasite
that enters the victim's blood system and travels into the person's liver where
the parasite reproduces. There the parasite causes a high fever that involves
shaking chills and pain. In the worst cases, malaria leads to coma and death.

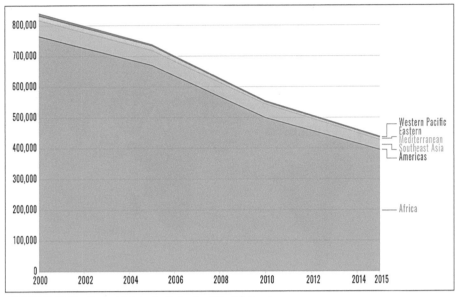

FIGURE 3.2 Global Malaria Rate by World Region
Malaria has steadily declined in all world regions between 2000 and 2015. However, Africa remains disproportionately impacted by the disease. In addition, progress has stalled since 2015.

At its peak in 2000 almost one million people died of malaria. Since then, progress on malaria was steady until 2015, but has subsequently stalled (see Figure 3.2).

Between 2000 and 2017, new malaria cases fell by 37 percent globally and by 42 percent in Africa. During this same period, malaria mortality rates fell by 60 percent globally and by 66 percent in the African regions. Other regions have achieved impressive reductions in their malaria burden. Since 2000, the malaria **mortality rate** declined by 72 percent in the region of the Americas, by 65 percent in the Western Pacific region, by 64 percent in the Eastern Mediterranean region, and by 49 percent in the Southeast Asia region. In 2017, an estimated 219 million cases of malaria and 435,000 deaths from the disease were reported (see Map 3.5). North Africa and sub-Saharan Africa account for nine out of ten malaria victims.

Much of the progress between 2000 and 2015 was the result of programs aimed at **vector control** as the main way to prevent and reduce malaria transmission. Two forms of vector control are effective in a wide range of circumstances: insecticide-treated mosquito nets and indoor residual spraying (Photo 3.1). Additionally, rapid diagnostic testing introduced widely has made it easier to quickly distinguish between malarial and nonmalarial fevers, enabling timely and appropriate treatment. An estimated 663 million cases of malaria have been averted in sub-Saharan Africa since 2001 as a direct result of the scale-up of these key interventions.

Despite the progress, sub-Saharan Africa continues to carry the heaviest **burden of disease**, accounting for more than 90 percent of global malaria cases, and the toll is rising. Children under five years of age are particularly susceptible to malaria illness, infection, and death, and account for 61 percent (266,000) of malaria deaths worldwide. As a result, malaria has had a significant economic burden on health systems. Malaria, like TB and HIV/AIDS, is concentrated in countries with low incomes, which struggle to bear the cost of treatment.

In 2021, WHO announced the E-2025 initiative, which focuses on twenty-five countries within reach of achieving zero malaria cases by 2025, including Botswana, Cabo Verde, the Dominican Republic, South Africa,

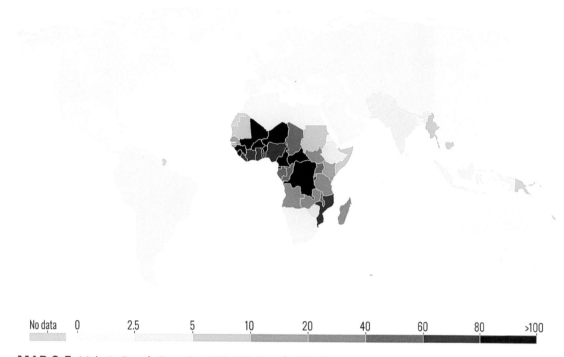

No data 0 2.5 5 10 20 40 60 80 >100

MAP 3.5 Malaria Death Rate Per 100,000 People (2017)
Malaria deaths are highest in Africa, although it should also be noted that malaria is concentrated in tropical climate countries.

and Thailand. According to WHO, eradicating malaria worldwide could save millions more lives and unlock trillions of dollars in economic potential, while strengthening countries' health systems and capacity to respond to both existing and emerging diseases. There is also a promising vaccine that was launched in a pilot program in Ghana, Kenya, and Malawi, where the malaria vaccine was administered to 650,000 children.

Neglected Tropical Diseases

Neglected tropical diseases also affect human health amongst the poorest, although they tend to be overshadowed by other public health issues such as the big three infectious diseases (TB, HIV/AIDS, and malaria). Neglected tropical diseases are a diverse group of about twenty communicable diseases found in 149 tropical and subtropical countries. They affect billions of people—particularly those who live in poverty, lack adequate sanitation, and are in close contact with infectious vectors and domestic animals. In addition, many of these neglected diseases are located in geographically isolated areas, making treatment and prevention much more difficult. Another reason for the "neglect" of these diseases is that they are not seen as commercial or large scale (e.g., compared with malaria) and the lack of profit may inhibit the pharmaceutical industry R&D investment. As a result, treatment or prevention for many of these diseases must come from either governments or philanthropy.

In 2017, 1.58 billion people required mass or individual treatment and care for neglected tropical diseases, down from 2.03 billion in 2010. Geographically, neglected tropical diseases are found in several countries in Africa, Asia, and Latin America (see Map 3.6). Neglected tropical diseases are especially common in tropical areas where people do not have access to clean water or safe ways to dispose of human waste. Neglected tropical diseases cost

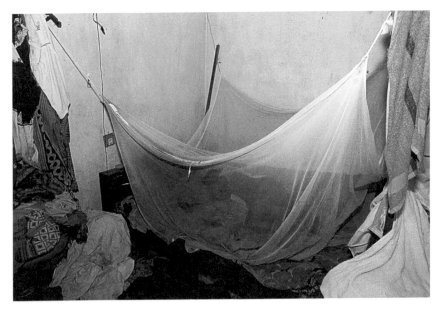

PHOTO 3.1 Mosquito Nets
Nets such as the one shown are effective at containing malaria, which is endemic among many populations in sub-Saharan Africa. Insecticide-treated mosquito nets are inexpensive (each net costs between $2–$5) and will last about a year. Unfortunately, the cost is often too high for the poorest families.

these countries the equivalent of billions of US dollars each year in direct health costs, loss of productivity, and reduced socioeconomic and educational attainment.

Some of the more common neglected tropical diseases are transmitted by mosquitoes or black flies and include Chagas disease, dengue fever, schistosomiasis, snakebite envenomation, and yaws.

More than 200,000 people die each year from snakebite venom, rabies, and dengue fever, but many more thousands are disabled, disfigured, or debilitated. Yaws, for example, is a bacterial infection that causes skin lesions and affects mostly children. It has a very low mortality rate but can cause disfigurement and disability if untreated.

Controlling the vectors that transmit these diseases and improving basic water, sanitation, and hygiene are highly effective strategies against these diseases. The good news is that there has been progress. According to WHO, today 600 million fewer people require interventions against neglected tropical diseases than in 2010, as a result of expanded preventive coverage and treatments. For example, pharmaceutical companies have donated nearly three billion tablets of safe, quality-assured medicines annually to support the control and elimination of neglected tropical diseases in countries where they are considered **endemic diseases**. Neglected tropical diseases are formally recognized as targets for global action in SDG target 3.3, which calls to "end the epidemics of . . . neglected tropical diseases" by 2030.

Other Major Causes of Death and Illness

Diseases are not the only major cause of poor health; socioeconomic and environmental factors also play a major role. For example, road accidents,

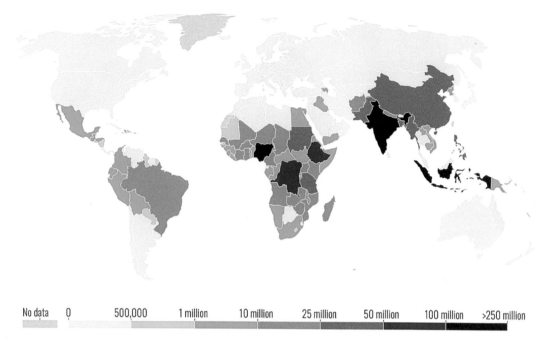

| No data | 0 | 500,000 | 1 million | 10 million | 25 million | 50 million | 100 million | >250 million |

MAP 3.6 Number of People Requiring Interventions against Neglected Tropical Diseases (2015) People in the Global South are disproportionately exposed to numerous tropical diseases due to a range of environmental/climate and social/economic/political factors.

contaminated water, and air pollution account for a significant number of deaths and illness each year.

Road Traffic Deaths

Road safety is not the first issue that comes to mind when thinking about good health. And yet it does not receive anywhere near the attention it deserves. A 2018 WHO report found that 1.35 million people were killed each year in traffic accidents, and the numbers could triple to 3.6 million per year by 2030. Tens of millions more are injured or disabled every year or may suffer life-altering injuries with long-term effects. These losses take a huge toll on families and communities. The cost of emergency response, health care, and human grief is immense. Road traffic injuries cause considerable economic losses to individuals, their families, and nations as a whole. This is one reason why reducing road traffic deaths and injuries is one of the SDG 3 goals (Goal 3.6). There are also targets to reduce road traffic deaths in SDG 11 because road congestion and traffic tend to be more pronounced in urban areas.

Low- and medium-income countries account for 93 percent of road traffic deaths, despite having only 60 percent of the world's vehicles. In the Global South, traffic accidents have become the fifth leading cause of death, outnumbering those who die of HIV/AIDS, malaria, and tuberculosis. Road traffic injury is the leading cause of death for people aged between five and twenty-nine years. The burden is disproportionately borne by pedestrians, cyclists, and motorcyclists, who represent more than half of all global road deaths. Pedestrians and cyclists represent 26 percent of all deaths, while those using motorized two- and three-wheelers comprise another 28 percent. Map 3.7 shows the distribution of deaths by road user type. The causes of road accidents and deaths are numerous and

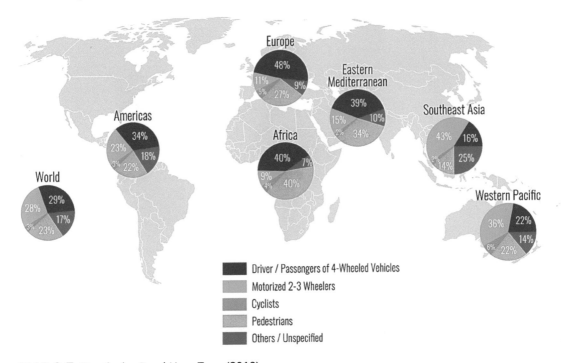

MAP 3.7 Deaths by Road User Type (2018)
In the Global North, more people die from accidents involving automobiles; in the Global South, many more are injured or killed by motorcycles. Road conditions can have an impact on road safety.

include speeding; driving under the influence; the lack of helmets, seat belts, and child restraints; unsafe road infrastructure; and distracted driving. Many experts question the term "road accidents," noting that many road deaths and injuries are preventable.

Nigeria has among the highest road accidents, with nearly thirty-four deaths for every 100,000 residents (Photo 3.2). In Indonesia, approximately ninety people die in accidents each day. There are more than sixty million motorcycles on Indonesia roads. Like many low-income or moderate-income countries, motorcycles costs around $1,000, far less than a car, which is one reason for the prevalence of two- and three-wheeled vehicles in countries such as Vietnam. Although wearing a helmet has been shown to prevent road traffic deaths, only 52 percent of passengers are reported to wear helmets. This is another reason why motorcycle riders account for more than 35 percent of the annual fatalities.

Contaminated Water
UNWater notes that 2.2 billion people do not have direct access to a clean source of drinking water, and instead are exposed to water contaminated with feces, putting them at risk of contracting other water-borne diseases such as cholera, dysentery, typhoid, and polio. The greatest increases in exposure to these occur in low- and lower-middle-income countries, primarily because of higher population and economic growth in these countries, especially those in Africa, and the lack of wastewater management systems. Contaminated water is particularly deadly in children: About 297,000 children under five die annually—more than 800 every day—from **diarrheal diseases** due to poor sanitation, poor hygiene, or unsafe drinking water. The connection between health and water is a clear and compelling one.

PHOTO 3.2 Bad Road Conditions in Nigeria
Nigeria has a substantial number of road deaths. A major reason is speeding,
but the country also has poorly maintained roads that are riddled with potholes,
making driving any vehicle dangerous.

Inadequate water, sanitation, and hygiene causes diseases such as diarrhea
and infection through soil-transmitted helminths (parasitic worms). These
conditions led to a total of 870,000 deaths in 2016. This large disease burden
could be significantly reduced if safely managed drinking water and sanitation
services were universally available, and good hygiene practices were followed,
a topic discussed in more detail in Chapter 6.

Air Pollution

Both indoor (household) and outdoor air pollution increase the risk of cardiovascular and respiratory diseases and are major risk factors for **noncommunicable diseases**. Air pollution kills an estimated seven million people
worldwide every year, largely as a result of increased mortality from stroke,
heart disease, chronic obstructive pulmonary disease, lung cancer, and acute
respiratory infections.

WHO data shows that nine out of ten people breathe air that exceeds WHO
guideline limits containing high levels of pollutants, with low- and middle-
income countries suffering from the highest exposures.

Children are particularly vulnerable to both indoor and outdoor air pollution because their lungs, brains, and immune systems are still developing
and their respiratory tracts are more permeable. Young children also breathe
more often than adults and take in more air relative to their body weight.
Air pollution harms children's developing lungs, but it can also cross the
blood-brain barrier and permanently damage brain development. UNICEF
estimates more than 300 million children are breathing toxic air and this is a
major contributing factor in the deaths of around 600,000 children under five
every year.

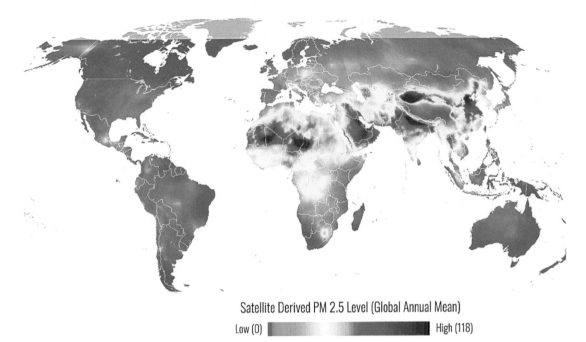

Satellite Derived PM 2.5 Level (Global Annual Mean)

Low (0) ▮▮▮▮▮▮▮▮▮▮▮▮▮▮▮▮▮▮ High (118)

MAP 3.8 Outdoor Air Pollution, Global Annual Average (2012–14)
This map shows areas in North Africa, the Middle East, and into South Asia are more exposed to high levels of particulate matter. Particulate matter can settle into lungs, scarring lung tissue and causing long-term respiratory issues. Not all particulate matter is manmade; some is from windborne soils and sands from deserts and agricultural areas.

Exposure to household air pollution causes around four million deaths each year. Indoor air pollution comes from polluting fuels and technologies for cooking, including wood, animal dung, and kerosene. Often low-income families do not have good ventilation indoors, magnifying exposure to airborne pollutants. Health risks from this type of pollution are particularly high among women and children, who typically spend the most time around the stove or open fire. Chapter 7 discusses strategies to introduce clean-cooking stoves in Africa and South Asia, which may significantly reduce exposure to indoor air pollutants.

Ambient air pollution from traffic, industry, power generation, waste burning, and residential fuel combustion resulted in around 4.2 million deaths in 2012–14 (Map 3.8). According to UNICEF, two billion children live in areas where outdoor air pollution, caused by factors such as vehicle emissions, heavy use of fossil fuels, dust, and burning of waste, exceeds minimum air quality guidelines set by WHO. South Asia has the largest number of children living in these areas, at 620 million, with Africa following at 520 million children. The region of East Asia has 450 million children living in areas that exceed guideline limits.

REASONS FOR POOR HEALTH: A CRITICAL PERSPECTIVE

Many of the most pressing health issues discussed previously must be analyzed within a broader political economy framework. For example, malaria, HIV/AIDS, TB, neglected tropical diseases, and even COVID-19 pandemic are not socially neutral diseases. Economic inequality between and within nations significantly contributes to the chances of contracting and dying from any of these.

These inequalities have been made stark with the COVID-19 pandemic. Low-income countries with weak health-care systems, workers whose jobs cannot be performed remotely, and the differences between those with and without access to soap and water to wash their hands also account for the unequal impact of the coronavirus. Racial and ethnic minorities experience higher death rates from COVID-19. Inequality is also embedded in international responses to COVID-19, as giving and receiving aid is often impacted by inequalities of national power and influence, resulting in global competition rather than the collaboration needed to end the pandemic.

Researchers Clare Bambra and colleagues suggest "the COVID-19 pandemic is occurring against a backdrop of social and economic inequalities in existing noncommunicable diseases as well as inequalities in the social determinants of health" (p. 965). They call this outbreak a **syndemic** when risk factors or comorbidities are intertwined, interactive, and cumulative—and exacerbate the disease burden and increase its negative effects.

Bambra and colleagues also point out that "minority ethnic groups, people living in areas of higher socioeconomic deprivation, those in poverty and other marginalized groups (such as homeless people, prisoners and street-based sex workers) generally have a greater number of coexisting noncommunicable diseases, which are more severe and experienced at a younger age" (ibid.). For example, people living in more socioeconomically disadvantaged neighborhoods and minority ethnic groups have higher rates of almost all the known underlying clinical risk factors that increase the severity and mortality of COVID-19, including hypertension, diabetes, asthma, obesity, and smoking.

Inequalities in chronic conditions arise as a result of inequalities in exposure to the social determinants of health: the conditions in which people "live, work, grow and age" including working conditions, unemployment, access to essential goods and services (e.g., water, sanitation, and food), housing, and access to health care. Bambra is not alone in her assessment of a syndemic. UN Secretary-General António Guterres admitted "COVID-19 cannot be seen in isolation from the fundamental problems with our health systems: inequality, underfunding; complacency, neglect."

The impact of COVID-19 on health inequalities will not just be in terms of virus-related infection and mortality; understanding the broader political economic context is crucial if we are to prevent the COVID-19 pandemic from increasing health inequalities for future generations.

SOLUTIONS FOR IMPROVING HEALTH AND WELL-BEING

There are many health issues to address, and many possible solutions involved. Table 3.3 lists the targets and indicators for SDG 3. Experts say there are four critical means to achieving the targets for SDG 3. These are (1) increasing health access and health systems by adopting universal health coverage; (2) reducing tobacco use; (3) supporting research, development, and universal access to vaccines and medicines; and (4) improving early warning systems for global health risks.

TABLE 3.3 SDG 3 Targets and Indicators to Achieve by 2030	
Target	**Indicators**
3.1 Reduce the global maternal mortality ratio to less than 70 per 100,000 live births	**3.1.1** Maternal mortality ratio **3.1.2** Proportion of births attended by skilled health personnel
3.2 End preventable deaths of newborns and children under five years of age, with all countries aiming to reduce neonatal mortality to at least as low as 12 per 1,000 live births and under-five mortality to at least as low as 25 per 1,000 live births	**3.2.1** Under-five mortality rate **3.2.2** Neonatal mortality rate
3.3 End the epidemics of AIDS, tuberculosis, malaria, and neglected tropical diseases and combat hepatitis, water-borne diseases, and other communicable diseases	**3.3.1** Number of new HIV infections per 1,000 uninfected population, by sex, age, and key populations **3.3.2** Tuberculosis incidence per 1,000 population **3.3.3** Malaria incidence per 1,000 population **3.3.4** Hepatitis B incidence per 100,000 population **3.3.5** Number of people requiring interventions against neglected tropical diseases
3.4 Reduce by one-third premature mortality from noncommunicable diseases through prevention and treatment and promote mental health and well-being	**3.4.1** Mortality rate attributed to cardiovascular disease, cancer, diabetes, or chronic respiratory disease **3.4.2** Suicide mortality rate
3.5 Strengthen the prevention and treatment of substance abuse, including narcotic drug abuse and harmful use of alcohol	**3.5.1** Coverage of treatment interventions (pharmacological, psychosocial and rehabilitation, and aftercare services) for substance use disorders **3.5.2** Harmful use of alcohol, defined according to the national context as alcohol per capita consumption (aged fifteen years and older) within a calendar year in liters of pure alcohol
3.6 Halve the number of global deaths and injuries from road traffic accidents	**3.6.1** Death rate due to road traffic injuries
3.7 Ensure universal access to sexual and reproductive health-care services, including for family planning, information and education, and the integration of reproductive health into national strategies and programs	**3.7.1** Proportion of women of reproductive age (aged 15–49 years) who have their need for family planning satisfied with modern methods **3.7.2** Adolescent birth rate (aged 10–14 years; aged 15–19 years) per 1,000 women in that age group

3.8 Achieve universal health coverage, including financial risk protection, access to quality essential health-care services, and access to safe, effective, quality and affordable essential medicines and vaccines for all	3.8.1 Coverage of essential health services (defined as the average coverage of essential services based on tracer interventions that include reproductive, maternal, newborn and child health, infectious diseases, noncommunicable diseases and service capacity and access, among the general and the most disadvantaged population) 3.8.2 Proportion of population with large household expenditures on health as a share of total household expenditure or income
3.9 Substantially reduce the number of deaths and illnesses from hazardous chemicals and air, water, and soil pollution and contamination	3.9.1 Mortality rate attributed to household and ambient air pollution 3.9.2 Mortality rate attributed to unsafe water, unsafe sanitation, and lack of hygiene (exposure to unsafe Water, Sanitation and Hygiene for All [WASH] services) 3.9.3 Mortality rate attributed to unintentional poisoning
3.A Strengthen the implementation of the World Health Organization Framework Convention on Tobacco Control in all countries, as appropriate	3.A.1 Age-standardized prevalence of current tobacco use among persons aged fifteen years and older
3.B Support the research and development of vaccines and medicines for the communicable and noncommunicable diseases that primarily affect developing countries, provide access to affordable essential medicines and vaccines, in accordance with the Doha Declaration on the TRIPS Agreement and Public Health, which affirms the right of developing countries to use to the full the provisions in the Agreement on Trade-Related Aspects of Intellectual Property Rights regarding flexibilities to protect public health, and, in particular, provide access to medicines for all	3.B.1 Proportion of the population with access to affordable medicines and vaccines on a sustainable basis 3.B.2 Total net official development assistance to medical research and basic health sectors
3.C Substantially increase health financing and the recruitment, development, training, and retention of the health workforce in developing countries, especially in least developed countries and small island developing states	3.C.1 Health worker density and distribution
3.D Strengthen the capacity of all countries, in particular developing countries, for early warning, risk reduction, and management of national and global health risks	3.D.1 International Health Regulations (IHR) capacity and health emergency preparedness

Source: United Nations Sustainable Development Goals. https://sustainabledevelopment.un.org/sdg3

Improve Health Access and Health Systems

The **accessibility of health services** is crucial for maintaining and improving health. Currently, at least half the people in the world do not receive the health services they need. About 100 million people are pushed into extreme poverty each year because of out-of-pocket spending on health.

WHO defines **universal health coverage (UHC)** as ensuring that all people have access to needed health services without suffering financial hardship. This includes prevention, promotion, treatment, rehabilitation, and palliative care. UHC requires adequate and competent health and care workers with an optimal skills mix at the facility, outreach, and community level, and who are equitably distributed, adequately paid, and supported. Moving toward UHC will also advance **equity in health**.

Achieving UHC is one of the targets the nations of the world set when adopting the SDGs in 2015. UHC has therefore become a major goal for health reform in many countries and is integrated within several of the targets of SDG 3 (such as 3.7 and 3.8). Achieving UHC requires multiple approaches. One is to increase health financing and investment in health systems. WHO says eighteen million additional health workers are needed by 2030 to meet the goals. Gaps in the supply of and demand for health workers are concentrated in low- and lower-middle-income countries. There is a shortfall of seven million skilled health workers in rural areas, exacerbating rural–urban inequities. Investments are needed from both public and private sectors in health worker education, as well as in the creation and filling of funded positions in the health sector and the health economy. Efforts are needed to improve working conditions for many health and care workers, including labor rights and adequate compensation. The COVID-19 pandemic also raised the importance of UHC, given the inequalities in access to health care for those infected, and even the inequalities in access to vaccines.

One solution has been to adopt **primary health care (PHC) or people-centered care.** A PHC approach focuses on organizing and strengthening health systems so that people can access services for their health and well-being based on their needs and preferences, at the earliest, and in their everyday environments. According to WHO, PHC entails three interrelated and synergistic components, including comprehensive integrated health services that embrace primary care as well as public health goods and functions as central pieces; multisectoral policies and actions to address the wider determinants of health; and engaging and empowering individuals, families, and communities for increased social participation and enhanced self-care and self-reliance in health. Another emerging strategy is health access is called **Health in All Policies**, profiled in *Solutions* 3.1.

Reduce Tobacco Use

Although tobacco use is not officially listed as a leading cause of death, there are a myriad of health issues that result from tobacco consumption, and it is indirectly responsible for lung cancers and respiratory diseases such as emphysema, asthma, and pneumonia. Globally, tobacco is estimated to eight million deaths a year. It kills around half of all smokers. Tobacco use is considered the leading cause of preventable death globally. WHO estimates there are 1.3 billion smokers in the world; 80 percent of them live in low- and middle-income countries in the Global South.

☰ SOLUTIONS 3.1

Health in All Policies

Target 3.7: Ensure universal access to sexual and reproductive health-care services, including for family planning, information and education, and the integration of reproductive health into national strategies and programs. Target 3.8: Achieve universal health coverage, including financial risk protection, access to quality essential health-care services, and access to safe, effective, quality, and affordable essential medicines and vaccines for all.

Health in All Policies, or "healthy public policy," is based on the idea that health starts where people live, work, learn, and play, and that community health is influenced by more than individual choices. One's physical and social environments, along with local government decisions and actions that shape these environments, have an impact on health outcomes.

Health in All Policies are a response to a variety of complex and often inextricably linked problems such as the chronic illness epidemic, growing inequality and health inequities, rising health-care costs, an aging population, climate change causing more frequent disasters, and the lack of efficient strategies for achieving governmental goals with shrinking resources. Such "wicked" problems are extremely challenging. Addressing them requires innovative solutions, a new policy paradigm, and structures that break down the silos of government to advance multidisciplinary and intersectoral thinking.

Health in All Policies are a collaborative approach to improving the health of all people by incorporating health considerations into decision making across sectors and policy areas.

Health is influenced by the combination of social, physical, and economic environments, collectively referred to as the "social determinants of health." Health in All Policies support improved health outcomes and health equity through collaboration between public health practitioners and those nontraditional partners who have influence over the social determinants of health. Health in All Policies approaches include five key elements:

1. Promoting health and equity,
2. Supporting intersectoral collaboration,
3. Creating multiple benefits to attract many partners,
4. Engaging stakeholders, and
5. Creating structural or process change.

Health in All Policies encompass a wide spectrum of activities and can be implemented in many different ways. It is increasingly emerging as a best practice in health that promises to have multiple benefits to a community.

Sources: American Public Health Association. "Health in All Policies" (2017). https://www.apha.org/topics-and-issues/health-in-all-policies Rudolph, L., Caplan, J., Ben-Moshe, K., and Dillon, L. "Health in All Policies: A Guide for State and Local Governments." Washington, DC, and Oakland, CA: American Public Health Association and Public Health Institute (2013). https://www.apha.org/~/media/files/pdf/factsheets/health_inall_policies_guide_169pages.ashx

Cigarette smoking is the most common form of tobacco use worldwide. Each year, more than 6.5 trillion cigarettes are sold around the world. Other tobacco products include waterpipe tobacco, various smokeless tobacco products, cigars, cigarillos, and pipe tobacco.

While the United States has significantly decreased its share of tobacco farming from more than 180,000 farms in the 1980s to just more than 10,000 today, it is still the fourth-largest producer in the world. This is despite the fact that smoking-related diseases cost the United States more than $300 billion per year. China, India, and Brazil are today the three largest tobacco-producing countries. Not surprisingly, it is within many of these nations that smoking awareness is at its lowest. For example, a survey in China showed that only 38 percent of smokers knew that smoking could cause heart disease, while only 27 percent knew that it could lead to a stroke.

The good news is that the number of young people taking up smoking has decreased in the Global North; the bad news is that many have taken up electronic nicotine delivery systems (ENDS), commonly referred to as e-cigarettes. The long-term effects of the use of these electronic versions are not well-understood.

One response has been the WHO Framework on Tobacco Control, an international treaty aimed at reducing tobacco use. This treaty is considered an important milestone in the promotion of public health as profiled in *SDGs and the Law* 3.1. Strengthening this framework is also a focus of SDG 3 Target 3.A.

Increase Research, Development, and Access to Vaccines and Medicines

Vaccines and immunization programs represent some of the most impactful public health advances seen to date, playing a critical role in reducing the spread of and, in some cases, eliminating the threat of the world's many devastating infectious diseases. Immunization is widely recognized as one of the world's most successful and cost-effective health interventions, saving millions of lives. Smallpox, once one of the deadliest diseases, has been eradicated around the world. Measles vaccinations have helped eliminate 99 percent of cases in the United States and WHO estimates that seventeen million lives have been saved globally since 2000 due to the measles vaccine. In 2017, 116.2 million children were immunized, the highest number ever reported. Over the last twenty years, infectious disease vaccine development and new medical breakthroughs in treatment have been dynamic and promising. Currently, there are nearly 260 vaccines in development by America's biopharmaceutical companies to both prevent and treat diseases including cancer, Alzheimer's disease, allergies, and autoimmune disorders. The biggest story of 2021, however, is the development of multiple and effective COVID-19 vaccines (see *Making Progress* 3.1).

However, outbreaks of measles and diphtheria, resulting in many deaths, have occurred in areas with lower immunization coverage. Additionally, many infectious diseases remain with a high unmet medical need and a growing list of vaccine preventable diseases.

In the United States, for example, the federal government supports vaccine research through several agencies including the National Institutes of Health (NIH). NIH-funded vaccine research totaled an estimated $2.4 billion in 2019, the year prior to the COVID-19 pandemic. NIH has funded research for vaccines against specific pandemic threats such as coronavirus, Zika, Ebola, and dengue fever. Experts note the rapid development of vaccines for COVID-19 relied on the application of existing vaccine technologies previously funded, confirming the importance of long-term and sustained federal funding.

Developing new vaccines and medicines for communicable and noncommunicable diseases will involve partnerships with the private sector pharmaceutical industry. Meeting the targets of SDG will not be possible without a significant and concerted effort from the pharmaceutical industry. There is a critical need for vaccines and medicines in low-income countries, but incentives for research and development investment can be at odds with business

SDGs AND THE LAW 3.1

The World Health Organization Framework Convention on Tobacco Control

The World Health Organization Framework Convention on Tobacco Control (WHO FCTC) is the first WHO treaty adopted under Article 19 of the WHO Constitution. The treaty came into force on February 27, 2005. The FCTC is one of the most quickly ratified treaties in UN history.

The WHO FCTC was developed in response to the globalization of the tobacco epidemic. The preamble to the convention reveals how countries viewed the need to develop such an international legal instrument. It cites their determination "to give priority to their right to protect public health" and the "concern of the international community about the devastating worldwide health, social, economic and environmental consequences of tobacco consumption and exposure to tobacco smoke." It then notes the scientific evidence for the harm caused by tobacco, the threat posed by advertising, promotion, and illicit trade, and the need for cooperative action to tackle these problems. Other paragraphs of the preamble note the role of civil society and the human rights that the convention aims to support.

The treaty's provision includes rules that govern the production, sale, distribution, advertisement, and taxation of tobacco. The treaty consists of the so-called six **MPOWER** measures or articles:

- **M**onitor tobacco use and prevention policies,
- **P**rotect people from tobacco use,
- **O**ffer help to quit tobacco use,
- **W**arn about the dangers of tobacco,
- **E**nforce bans on tobacco advertising, promotion, and sponsorship, and
- **R**aise taxes on tobacco.

As a result of commitments and attention to the issue, global tobacco production has been falling, especially over the last five years.

A 2018 progress report noted that good progress has been made in Article 8 (Protection from exposure to tobacco smoke), Article 11 (Packaging and labelling of tobacco products), Article 12 (Education, communication, training and public awareness), and Article 16 (Sales to and by minors) have been implemented successfully. More than 90 percent of the countries that signed the treaty indicated having implemented tax and/or price policies, and the same percentage declared having banned smoking in all public places. A considerable number of countries also shared their experience in extending or planning to extend smoking bans to outdoor environments, as well as on the inclusion of novel products in their existing smoke-free legislation. Education, communication, training, and public awareness campaigns have been carried out widely at the national and regional levels, often in conjunction with World No Tobacco Day. More than two-thirds of the countries reported enacting or strengthening legislation aimed at tackling illicit trade on the national level.

However, progress was less satisfactory on Article 18 (Protection of the environment and the health of persons), Article 19 (Liability), and Article 17 (Provision of support of economically viable alternative activities). The report cited a lack of human and financial resources. Additionally, technical assistance is still very much needed in the fields of taxation, policy development, research, and national cessation programs. Finally, tobacco industry interference, combined with the emergence of new and novel tobacco products, continues to be considered the most serious barrier to progress. One response was that WHO launched a Knowledge Hub to track and gather data on—and inform the public about—tobacco industry interference in public policy making.

Sources: World Health Organization. "2018 Global Progress Report on Implementation of the WHO Framework Convention on Tobacco Control" (2021). https://fctc.who.int/publications/m/item/2018-global-progress-report
World Health Organization. "The WHO Framework Convention on Tobacco Control: An Overview" (2021). https://www.who.int/fctc/WHO_FCTC_summary.pdf?ua=1 and https://fctc.who.int/who-fctc/overview
World Health Organization. "Tobacco: Key Facts" (2021). https://www.who.int/news-room/fact-sheets/detail/tobacco

MAKING PROGRESS 3.1

The COVID-19 Vaccines

Target 3B: Support the research and development of vaccines and medicines for the communicable and noncommunicable diseases that primarily affect developing countries, provide access to affordable essential medicines and vaccines, in accordance with the Doha Declaration on the TRIPS Agreement and Public Health, which affirms the right of developing countries to use to the full the provisions in the Agreement on Trade-Related Aspects of Intellectual Property Rights regarding flexibilities to protect public health, and, in particular, provide access to medicines for all.

Building on deep scientific knowledge gained from decades of experience with viruses such as MERS, SARS, influenza, HIV, and Hepatitis C, biopharmaceutical companies have made unprecedented progress in advancing treatments and vaccines to help fight COVID-19.

By spring of 2021 there were sixteen approved COVID-19 vaccinations that had been developed. In the United States, Pfizer, Moderna, and Johnson & Johnson developed approved vaccines. Others include the EU company AstraZeneca, Russia's Sputnik, India's Covaxin, and China's Sinopharm and CoronaVac. The rapid progress was remarkable.

A recent article in the *Wall Street Journal* examines how research conducted to develop the COVID-19 vaccines could lead to future breakthroughs for other infectious diseases and conditions. It noted:

- "The pandemic has opened a new era for vaccines developed with gene-based technologies, techniques that have long stumped scientists and pharmaceutical companies, suggesting the possibility of future protection against a range of infectious disease."

- "New vaccine technologies spurred by the pandemic are leading efforts to combat COVID-19 and herald a new arsenal of weapons for fighting lethal viruses in the future, infectious-disease researchers said, another example of how the fight against [the novel coronavirus] has supercharged technological development."

- "For years, vaccines for such infectious diseases as measles and polio were made from the viruses they targeted, in versions scientists rendered harmless. The shots rally the immune system by exposing people to the targeted virus. Yet such vaccines could take a decade or longer to develop, and manufacturing them took months…. The [COVID-19] vaccines offer several advantages over shots using older technology. They seem to activate not just the antibodies that neutralize a virus but also the memory and T-cells that keep the immune defense alert for the long-term."

The application of new technologies and approaches to the development of the COVID-19 vaccines may prove to be instrumental, as we are likely to see new infectious diseases emerge in the future.

Source: Loftus, P. "Covid-19 Vaccines Yield Breakthroughs in Long-Term Fight against Infectious Disease," *Wall Street Journal*, February 28, 2021. https://www.wsj.com/articles/covid-19-vaccines-yield-breakthroughs-in-long-term-fight-against-infections-disease-11614537238

models that strive for maximum profits. To advance this part of the SDG 3 agenda, pharmaceutical companies are challenged to rethink whether they also serve a social purpose in addition to their business purpose.

Access to Medicine Foundation, a Netherlands-based nonprofit, has developed an Access to Medicine Index. The index quantifies global pharmaceutical companies' contribution toward making medicines more accessible to the global population. Their 2021 index noted GlaxoSmithKline, Novartis, Pfizer, Johnson & Johnson, and Sanofi are among the top five for advancing affordable and accessible medicine in low- and moderate-income countries.

However, their 2021 report also noted that many of the world's lowest-income countries still do not benefit significantly from access strategies being implemented by the world's largest pharmaceutical companies. Less than half of key products controlled by twenty large companies are being offered through access strategies in countries classified by the World Bank as low- or moderate-income countries. The shortfall is particularly acute in the lowest-income countries, which are most consistently overlooked by companies despite being home to almost 700 million people.

Political economist Anil Hira has argued that the investment in pharmaceutical R&D has tended to focus on health issues of the aging middle- and upper-class populations of the Global North, with less emphasis on those that affect the Global South. A majority of pharmaceutical R&D is focused on cancers and a handful of diseases, while there has been less investment in R&D targeting maternal and neonatal health conditions such as neonatal sepsis and maternal hemorrhage. The capital-intensive nature of R&D makes it more likely that companies will focus on Global North, where there is an attractive enough market for profit. Hira says this is a market failure.

As a counter to the Global North pharmaceutical alliance, some countries have invested in biotechnology focused on the Global South. India, Cuba, China, and Brazil have growing pharmaceutical industries. Brazil, China, and India can manufacture most of their own vaccines and have been exporting these to other countries in the Global South. This is seen as a positive development for access and affordability, as well as R&D.

The Access to Medicine Foundation report concludes that although there has been an increase in access and affordability strategies by pharma companies, it is clear that progress is still only gradual and not quick enough. They note that solving the access to medicine problem is fundamentally a question of taking action at scale: Industry needs to reach more people with more products across a wider range of the world's poorest countries.

Establish Early Warning Systems for Global Health Crises

An early warning system is an instrument for communicating information about impending risks to vulnerable people before a hazardous event occurs, thereby enabling actions to be taken to reduce or avoid potential harm. Early warning systems are routinely used for environmental events such as hurricanes, tsunamis, and volcano eruptions. In contrast, there has been less attention on the development of such systems for infectious disease epidemics. The goal of a disease early warning system would be to provide public health officials and the general public with as much advance notice as possible about the likelihood of a disease outbreak in a particular location, thus widening the range of feasible response options. Ideally, an early warning system would shift the infectious disease paradigm from reactive—where first responders scramble to contain active threats—to preemptive management of risk.

Early warning systems are surveillance systems that collect information on epidemic-prone diseases. Most involve the coordination and analysis of numerous measures and indicators such as climate factors, demographics of an area (age, race, etc.), health systems in an area, and reports or data on infections or outbreaks. The more advanced early warning systems use new technologies such as big data, satellite imagery, and GIS/mapping, combined with qualitative reports and surveys to strengthen the existing surveillance systems. Such a system would use computer models to integrate environmental, epidemiological, and molecular data to forecast where disease risk is high and what actions could prevent outbreaks or contain an epidemic.

Some health early warning systems already exist. In the United States, the US EPA issues the Air Quality Index to predict ozone levels and other pollutants in most US cities; these are issued daily and provide information to those with respiratory conditions to take action to reduce health effects (such as avoiding outdoor exercise when ozone levels are high). Many countries in the European Union have early warning systems for heat waves. Early warning systems for heat waves involve forecasting the heatwave event, predicting possible health outcomes, triggering effective and timely response plans targeting vulnerable populations, notification of heatwave events, and communication of prevention responses.

In the Global South, some countries have health early warning systems for malaria. Health expert David Rogers notes that the mosquitos that spread malaria depend on suitable habitats to breed, which in turn depends heavily on climatic conditions. As a result, malaria transmission is highly seasonal, so knowing when the climatic conditions are optimal for transmission is critical. WHO established guidelines to implement long-range forecasting and early detection to monitor malaria situations that may escalate toward possible epidemics. Monitoring of climatic indicators, population vulnerability factors, and operational and environmental factors helps detect when conditions suitable for an epidemic have already appeared at a given time and place. Such early warning systems have the potential to help the health community foresee potential epidemics weeks to months in advance. The aim is to identify the beginning of an epidemic by measuring changes in the incidence of malaria cases. For example, Nigeria launched a satellite to give early warning signals of environmental disaster and to establish the relationship between malaria vectors and the environment that breeds malaria using remote sensing technology. At least eight African countries have malaria early warning systems, but many more countries could benefit from developing these early warning systems.

A recently developed mobile toolkit is WHO's Early Warning, Alert and Response System (EWARS). An "EWARS in a box" contains all the equipment needed to establish surveillance and response activities, particularly in difficult and remote field settings without reliable internet or electricity. The box contains sixty mobile phones, laptops, and a local server to collect, report, and manage disease data. A solar generator and solar chargers allow the phones and laptops to function without twenty-four-hour electricity. A single kit costs approximately $15,000 and can support surveillance for fifty fixed or mobile clinics serving roughly 500,000 people.

Most of these early warning systems are present in only a handful of countries and many are still in development. There is great potential to develop more health early warning systems and to expand the number of countries that adopt such systems.

Adopt Vision Zero Strategies

WHO has suggested several strategies to improve road safety and reduce road accidents. One is "Vision Zero," a strategy to eliminate all traffic fatalities and severe injuries, while increasing safe, healthy, equitable mobility for all. First implemented in Sweden in the 1990s, Vision Zero has proved successful across Europe—and now it is being implemented in other cities. The principles behind Vision Zero include the idea that traffic deaths are preventable, saving lives is not expensive, and that all of us—whether driving, walking, bicycling, using a wheelchair, or riding transit—have a right to safe mobility.

In Colombia, Bogotá recently adopted Vision Zero. The city reduced the number of traffic deaths by 50 percent between 1996 and 2006 by implementing an integrated approach to road safety and urban mobility. Under a strong mayoral leadership, the city integrated institutional, financial, and regulatory reforms, combined with messages that resonated with citizens. Bogotá invested about $2 billion in expanding Bus Rapid Transport (BRT) and building 300 kilometers of bikeways and improved pedestrian infrastructure. The investment was enabled by institutional and tax reform and a sustainable mobility-oriented response to public demand for improved commuting options. The city also invested in road safety improvements such as improved sidewalks and cycle lanes.

SUMMARY AND PROGRESS

Before the COVID-19 pandemic, there was good progress being made on many areas that impact health and well-being. Life expectancy was on the rise, maternal and child mortality was declining, and strong gains were made against leading communicable diseases such as tuberculosis, malaria, and HIV/AIDS. Progress was also being made on road traffic deaths, as some countries adopted Vision Zero, a strategy to eliminate all traffic fatalities and severe injuries, while increasing safe, healthy, equitable mobility for all.

However, progress in other areas had stalled and some of the targets were not likely to be met by 2030 without a recommitment to the targets and an infusion of investment. For example, the number of people covered by essential health services in 2017 was estimated to be between 2.5 billion and 3.7 billion—about one-third to one-half of the global population. Only 12 to 27 percent of the population in low-income countries were fully covered that year. If current trends continue, only 39 to 63 percent of the global population will be covered by such services by 2030.

Unfortunately, the COVID-19 crisis continues to cause health-care disruptions that may reverse decades of improvements. Illness and deaths will spike, and the pandemic interrupted childhood immunization programs in seventy countries. Together, these may lead to hundreds of thousands of additional under-five deaths than would have been expected in 2020 and 2021. The pandemic also highlighted the existing shortage of health professionals in many countries, particularly in regions with the highest burden of disease.

The COVID-19 crisis has crippled economies, reducing money available for other investments in health care and health systems. UNICEF reports that families are facing heightened stress under lockdown with many experiencing financial insecurity. And children are missing out on life-saving vaccines and much-needed free meals (such as school lunches) because of the suspension of services. Many children, especially the most vulnerable, even risk losing their lives to preventable diseases because access to health care is disrupted.

The road ahead is a challenging one: The world must recover from the COVID-19 crisis, and yet also continue to improve health-care systems and access to affordable vaccines and medicines. The current experience with the COVID-19 crisis has further highlighted the need for strengthening emergency preparedness as well as for rapidly scaling up response capacities and increasing international collaboration.

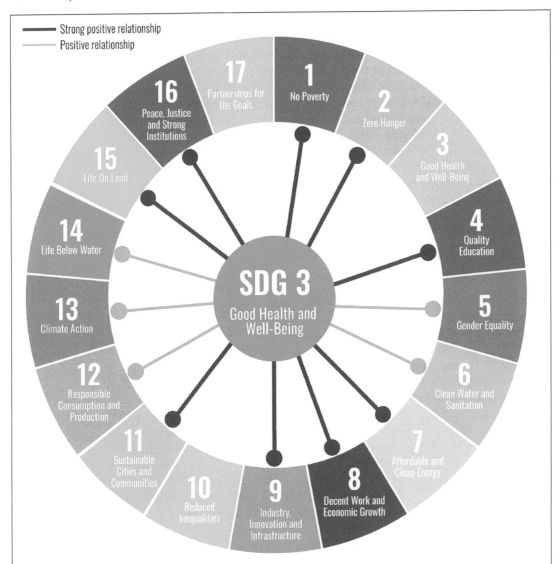

FIGURE 3.3 Interconnections in SDG 3 and the Other Goals

Health connects to all of the SDGs in some way; this figure highlights several of the stronger connections.

SDG 1: Poverty reduction leads to improved health and well-being, while good health is a strong enabling factor for effective poverty reduction.

SDG 4: Access to high-quality education is associated with better health, at both individual and community levels.

SDG 5: Improving gender equality generally enables the achievement of better health. Mothers make most health decisions for their children, so their empowerment leads to improved child health outcomes.

SDG 6: Improving water quality and access leads to improved health– without clean water and adequate sanitation it is difficult to achieve health gains. The latter are immediate in terms of decreased water-borne infections and improved nutrition; improving water quality and sanitation also leads to long-term developmental gains.

SDG 8: Increased health/well-being supports people to enter the workforce and contributes to economic growth and employment.

SDG 11: Improving access to adequate housing reduces crowding and hence exposure to communicable disease.

SDG 14: The health of marine systems is directly connected to human health in coastal areas and where populations depend on marine food sources. Marine pollution and collapse of fish stocks from overfishing can have direct impacts on nutrition.

SDG 3 has several strong and reinforcing synergies with other SDGs. Figure 3.3 highlights those goals that have the strongest reinforcing synergies with SDG 3.

QUESTIONS FOR DISCUSSION AND ACTIVITIES

1. Describe key threats to health and well-being.

2. What are trends around health in various world regions?

3. What are major obstacles to achieving good health in the Global North? In the Global South?

4. Explain the connection between poverty and health.

5. Describe the leading causes of death in the highest-income countries versus the lowest-income countries and explain the reasons for these differences.

6. Which solutions for improving health would you prioritize, where, and why?

7. How might the impact of the COVID-19 pandemic change health services and health access?

8. Select a disease and explain how it is connected to at least two other SDGs. What solution might advance targets in both SDG 3 and the other SDG?

TERMS

access to health services
accessibility of health services
burden of disease
communicable diseases
degenerative disease (see noncommunicable disease)
diarrheal diseases
endemic diseases
epidemic diseases
equity in health
health
Health in All Policies
health service
health system
human immunodeficiency virus (HIV/AIDS)
immunity, herd
infectious diseases (see communicable disease)
malaria

maternal mortality
morbidity
mortality rate
mortality rate, child
mortality rate, infant
mortality rate, neonatal
Neglected Tropical Diseases
noncommunicable disease
pathogenic
pandemic
people-centered care
primary health care
syndemic
tuberculosis (TB)
Universal Health Coverage (UHC)
vector (disease vector)
vector-borne disease
vector control
zoonotic disease

RESOURCES USED AND SUGGESTED READINGS

Access to Medicine Index. (2021). https://accesstomedicinefoundation. org/access-to-medicine-index

Bambra, C., Riordan, R., Ford, J., et al. "The COVID-19 Pandemic and Health Inequalities." *Journal of Epidemiology and Community Health* 74: 964–68 (2020). https://doi.org/hrttps://doi.orgHhhh10.1136/jech-2020-214401

Benatar, S., and Brock, G. (eds.). *Global Health: Ethical Challenges.* 2nd ed. New York: Cambridge University Press (2021).

Buckingham, R. W. *A Primer of International Health.* Boston: Allyn and Bacon (2001).

Gostin, L., and Meier, B. (eds.). *Foundations of Global Health and Human Rights.* New York: Oxford University Press (2020).

Guterres, A. *"Video Message to the World Health Assembly."* United Nations (2021, May 24). https://news.un.org/en/story/2021/05/1092592

Hira, A. "The Political Economy of the Global Pharmaceutical Industry: Why the Poor Lack Access to Medicine and What Might Be Done About It." *International Journal of Development* 8: 84–101 (2009, October).

Lemery, J., Knowlton, K., and Sorensen, C. (eds.). *Climate Change and Human Health: From Science to Practice.* 2nd ed. Hoboken, NJ: Jossey-Bass (2021).

Merson, M., Black, R., and Mills, A. *Global Health: Diseases, Programs, Systems, and Policies.* 4th ed. Burlington, MA: Jones & Bartlett Learning (2020).

Peoples Health Movement. *Global Health Watch 5: An Alternative World Health Report.* New York: Blackwell (2017).

Rogers, D. *Partnering for Health Early Warning Systems.* World Meteorological Organization 60 (1) (2011). https://public.wmo.int/en/ bulletin/partnering-health-early-warning-systems

Short, J. R. *Stress Testing the USA: Public Policy and Reaction to Disaster Events.* 2nd ed. Cham, Switzerland: Palgrave McMillan (2021).

Stop TB Partnership. *"The Paradigm Shift: Global Plan to End TB*: 2018–2022." Geneva, Switzerland (2019). http://www.stoptb.org/assets/ documents/global/plan/GPR_2018-2022_Digital.pdf

Sustainable Development Solutions Network. *Health in the Framework of Sustainable Development. Technical Report for the Post-2015 Development Agenda* (2014). https://irp-cdn.multiscreensite.com/ be6d1d56/files/uploaded/Health-For-All-Report.pdf

Sustainable Development Solutions Network. "Achieving SDG 3: Policy Brief Series 2018" (2018). https://resources.unsdsn.org/achieving-sdg-3-policy-brief-series—2018

United Nations International Children's Emergency Fund. "Progress towards Ending Preventable Maternal Mortality" (2015). https:// data.unicef.org/resources/strategies-toward-ending-preventable-maternal-mortality/

United Nations International Children's Emergency Fund. "Tracking the Situation of Children during COVID-19" (2020, August). https:// data.unicef.org/resources/tracking-the-situation-of-children-during-covid-19-august-2020/

United Nations Water. "WHO and UNICEF Launch Updated Estimates for Water, Sanitation and Hygiene" (2019). https://www.unwater.org/news/who-and-unicef-launch-updated-estimates-water-sanitation-and-hygiene#:~:text=Some%202.2%20billion%20people%20around,World%20Health%20Organization%20(WHO)

Vision Zero Network. "What Is Vision Zero?" (2021). https://visionzeronetwork.org/about/what-is-vision-zero/

World Health Organization. "Early Warning, Alert and Response Systems" (n.d.). https://www.who.int/emergencies/surveillance/early-warning-alert-and-response-system-ewars

World Health Organization. "Sustainable Development Goals" (n.d.). https://www.who.int/health-topics/sustainable-development-goals#tab=tab_1

World Health Organization. "The 8th Global Conference on Health Promotion, Helsinki, Finland, 10–14 June 2013" (2013, June). http://www.who.int/healthpromotion/conferences/8gchp/statement_2013/en.

World Health Organization. *Global Status Report on Road Safety 2018...* Geneva, Switzerland: World Health Organization (2018), 30. License: CC BYNC-SA 3.0 IGO. https://www.who.int/publications/i/item/9789241565684

World Health Organization. *Ending the Neglect to Attain the Sustainable Development Goals: A Road Map for Neglected Tropical Diseases 2021–2030.* Geneva, Switzerland: World Health Organization (2020). https://www.who.int/publications/i/item/9789240010352

World Health Organization. *Stronger Collaboration for an Equitable and Resilient Recovery towards the Health-Related Sustainable Development Goals: 2021 Progress Report on the Global Action Plan For Healthy Lives and Well-Being for All.* Geneva, Switzerland: World Health Organization (2021). https://www.who.int/initiatives/sdg3-global-action-plan/progress-reports/2021

Education

LEARNING OBJECTIVES

After reading this chapter, you should be able to:

- Describe trends and patterns in educational attainment at various levels
- Explain the reasons for educational inequalities
- Identify and explain the various targets and goals associated with SDG 4
- Describe some of the strategies and progress toward SDG 4
- Explain how the three Es are present in SDG 4
- Explain how SDG 4 strategies connect to other SDGs

Malala Yousafzai was shot at age 15 by the Pakistani Taliban for her outspoken campaign over girls' right to an education. Here is an excerpt of her 2014 speech when she was awarded the Nobel Peace Price for her advocacy:

"I am those 66 million girls who are deprived of education. And today I am not raising my voice, it is the voice of those 66 million girls.

We see many people becoming refugees in Syria, Gaza and Iraq. In Afghanistan, we see families being killed in suicide attacks and bomb blasts.

Many children in Africa do not have access to education because of poverty. And ... we still see girls who have no freedom to go to school in the north of Nigeria.

Many children in countries like Pakistan and India are deprived of their right to education because of social taboos, or they have been forced into child marriage or into child labor....

Leaders must seize this opportunity to guarantee a free, quality, primary and secondary education for every child.

Some will say this is impractical, or too expensive, or too hard. Or maybe even impossible. But it is time the world thinks bigger....

Why is it that countries which we call 'strong' are so powerful in creating wars but are so weak in bringing peace? Why is it that giving guns is so easy but giving books is so hard? Why is it that making tanks is so easy, but building schools is so hard?

Let us become the first generation that decides to be the last that sees empty classrooms, lost childhoods, and wasted potentials.

Let this be the last time that a girl or a boy spends their childhood in a factory.

Let this be the last time that a girl is forced into early child marriage.

Let this be the last time that a child loses life in war. Let this be the last time that we see a child out of school.

Let this end with us.

Let's begin this ending together, today, right here, right now...."*

(See Photo 4.1.)

---

*"Malala Yousafzai: Nobel Peace Prize Acceptance Speech." *The Malala Fund* (2014, December 10). https://www.malala.org/newsroom/archive/malala-nobel-speech

PHOTO 4.1 Malala Yousafzai and Kailash Satyarthi at the Nobel Peace Prize Ceremony in Oslo, Norway in 2014
Malala's speech focused on inequalities in education and called on young people to insist on the right to education.

EDUCATION: UNDERSTANDING THE ISSUE

"Everyone has the right to education," according to Article 26 of the 1948 Universal Declaration of Human Rights. Education enables upward socioeconomic mobility and is a key to escaping poverty. It is also a crucial driver of progress in health, fertility, politics, and social empowerment, and creates effective and productive members of society who are essential for sustainable growth and development. Today, we have the most educated and informed generation in history.

The international community recognized educational attainment as a development priority in the 2000 Millennium Development Goals (MDGs). The education-related MDG targets were able to make tremendous progress in expanding primary education. SDG 4 is more ambitious than the MDG targets and shifts from the narrow focus on universal primary education to expanding opportunities across all phases of education—preprimary, primary, secondary, vocational, higher, and adult education. The targets include outcomes in literacy, numeracy, and education in topics such as global citizenship, sustainability, and gender equality. SDG 4 aims to "ensure inclusive and equitable quality education and promote lifelong learning opportunities for all."

The world has made impressive progress on education over the last two decades and many experts believe that it is possible to achieve universal primary education by 2030. Since 2010, access to primary education and school enrollment rates have risen at all levels, particularly for girls. Beyond primary education, there has been progress in secondary education, with rates of schooling rising steadily since 1970. One contributing factor to this recent increase is that some countries have made the full twelve years of primary and secondary education compulsory. Figure 4.1 shows regional trends in primary and secondary education between 1990 and 2018. Between 1995 and 2018, the percentage of countries with gender parity in education rose from 56 percent to 65 percent in primary education and from 13 percent to 24 percent in secondary education.

Despite the progress, about 260 million children were out of school in 2018—nearly one-fifth of the global population in that age group. Further, more than 58 percent —617 million—of children and adolescents worldwide do not meet minimum proficiency standards in reading and mathematics. Experts also argue that there has been less attention and investment in adult education and higher education.

To achieve quality education for all, there must be equality of opportunity and universal access; but many obstacles remain. For example, children with disabilities are often denied education because they are unable to access educational facilities. Gender inequality persists; young pregnant mothers are often expelled from schools. In some instances, students and teachers have been incarcerated and even killed for demanding their right to education or better working conditions, respectively. There are also stark disparities in education along the lines of race, ethnicity, and urban-rural locations.

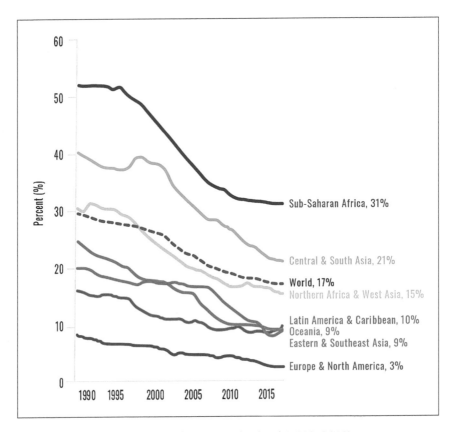

FIGURE 4.1 Percent of Children Out of School (1990–2018)
The percent of children out of school has steadily declined in all regions of the world between 1990 and 2018. However, sub-Saharan Africa still has 31 percent of its children out of primary and secondary school.

PHOTO 4.2 School Children in Nagar, Pakistan
The lack of basic infrastructure in this school including no electric
lights, no proper desks, and lack of learning materials makes it
challenging for young learners.

Primary Education (Ages 5–11)

Primary education involves children from ages five through eleven or twelve.
The objective of primary education is to provide basic education in reading,
writing, and mathematics and establish an elementary understanding of other
subjects. The most progress has been made in primary education, largely
because of the efforts that began with the MDGs in 2000. Not only have most
regions seen improvements, but primary education has almost reached gender parity (Figure 4.2 shows trends since 1970 and Map 4.1 shows primary
school enrollment rates for 2019.)

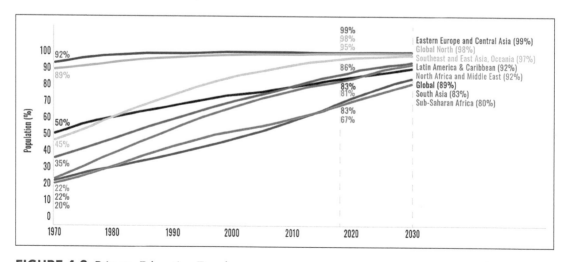

FIGURE 4.2 Primary Education Trends
Primary education enrollment has improved for many countries in the Global South, in part due
to the attention and investment in the MDGs that stressed keeping girls in school. However,
there is still room for improvement in regions such as sub-Saharan Africa and South Asia.

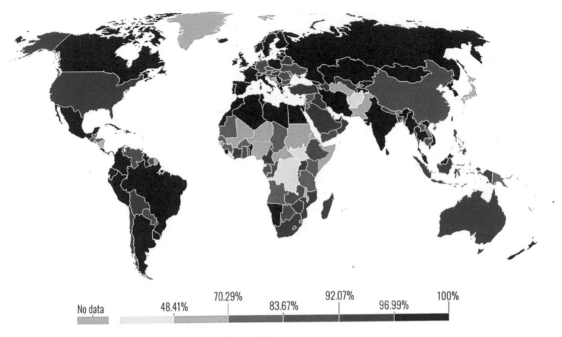

| No data | | 48.41% | 70.29% | 83.67% | 92.07% | 96.99% | 100% |

MAP 4.1 Enrollment in Primary School (2019)
While there is good primary school enrollment in many countries, both in the Global North and Global South, there are some countries where enrollment levels are concerning. This includes Afghanistan, Honduras, El Salvador, and numerous countries in Africa.

South Asia has made the greatest strides in reducing the gender gap in education, where a girl can now expect to receive twelve years of education compared to just six years in 1990. Boys in the region have similar school life expectancy rates.

However, wide disparities along gender lines are still found across countries mainly in Southeast Asia and sub-Saharan Africa. For example, in Eritrea, Djibouti, the Central African Republic, and Niger, girls receive less than six years of schooling on average. Other challenges include a lack of basic infrastructure in the schools (see Photo 4.2). For example, only 35 percent of primary schools in the Global South have electricity and less than 40 percent are equipped with basic handwashing facilities.

Finally, data highlight the urgent need to improve the quality of education for many. According to some estimates, 55 percent of children and adolescents of primary school age are not achieving minimum proficiency levels in reading and 60 percent are not reaching minimum proficiency levels in mathematics.

Secondary Education (Ages 12–18)

Secondary education involves children from ages twelve to eighteen. It is often characterized by the transition from the single-class teacher, who delivers all content to a class, to one where content is delivered by a series of subject specialists. The educational aim is to build on basic skills and to begin some advancement of skills in certain subjects such as history, geography, or chemistry. Together, primary education and secondary education form **basic education**.

SDG 4 calls for youth to complete secondary education, placing secondary education attainment as a goal on par with primary education attainment and representing a recent trend toward establishing secondary schooling as a right. Historically, secondary schooling was not designed to be inclusive of all children. On the contrary, its initial function in many countries was to select and train only those who would be joining the country's professional and ruling social classes. For example, in classical and medieval times, secondary education was provided by the church to the sons of nobility and to boys preparing for universities and the priesthood. The purpose, content, and curricular focus of secondary schools varies worldwide; in some countries, vocational and technical programs run parallel to upper secondary education.

Recent studies have highlighted that secondary schooling is important for increasing employment, accelerating economic development, and improving individual well-being. Over the last decade, many countries have made free universal secondary education a priority, some by making secondary education compulsory until the age of sixteen. Ghana, Malawi, Sierra Leone, and Tanzania have promised to substantially expand access to secondary school. However, despite these pledges, secondary enrollment in sub-Saharan Africa lags way behind other regions. Figure 4.3 shows trends in secondary enrollment from 1970 to 2018, with forecasts into 2030.

Higher Education

Higher education refers to various forms of postsecondary education, including trade, technical and vocational education, colleges, academies, institutes of technology, and universities that award professional or academic diplomas and degrees. Higher education is considered the driver of new knowledge and technologies, and helps develop scientists, technicians, medical practitioners, teachers, civil service members, business leaders, and many other types of professionals.

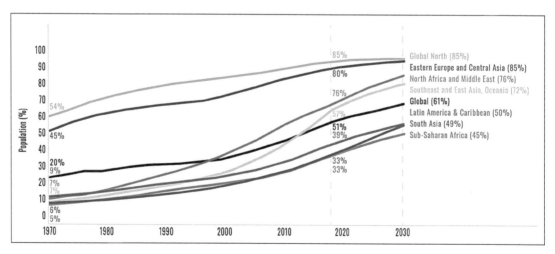

FIGURE 4.3 Secondary Education Trends
Achieving secondary schooling has lagged behind the gains in primary school achievements. In 2020 Sub-Saharan Africa and South Asia reported only more than one-third of students complete secondary school, compared to 85 percent in the Global North.

Higher educational attainment rates are highest in the high-income countries of the Global North. Over the past few decades, there has been increasing demand for higher education in low-income and middle-income countries. This is partly a response to the awareness that many higher-paying or desirable jobs around the globe require postsecondary education and degrees. UNESCO reports that globally, women's enrollment in higher education tripled, from 38 million to 116 million, between 1995 and 2018, accounting for 54 percent of the total increase in enrollment. Figure 4.4 shows regional trends in higher educational attainment.

Half the population in high-income countries in the Global North and countries in East Asia are projected to complete tertiary education by 2030. There has also been progress in tertiary education completion in North Africa, the Middle East, Southeast Asia, and Oceania. Tunisia had among the highest participation rates as recently as 2010 but has since stagnated at around 35 percent. Saudi Arabia's enrollment rate more than doubled, jumping from 32 to 70 percent, between 2009 and 2017. China has also been experiencing rapid growth in higher education. In 1998, the Chinese government announced their intention to expand higher education. From 1999 to 2009, they increased tertiary enrollment sixfold, from one million to six million students. China also raised the share of the urban labor force with a college degree by 60 percent. By 2018, gross enrollment in tertiary education in China was more than 50 percent with women enrolling at higher rates than men.

In many low-income countries, however, the expansion of higher education has not kept pace with investment and improvements in primary and secondary education. This has created an imbalanced educational system where many countries have a shortage of higher education infrastructure, facilities, and qualified academic faculty. For example, many African governments decided to invest and prioritize primary and secondary education, sometimes at the expense of higher education, casting higher education as a luxury.

While the SDG Goal 4, Target 4.3, calls for ensuring "equal access for all women and men to affordable and quality technical, vocational and tertiary education, including university," the SDGs lack targets that would make this a reality in many countries where higher education requires reform and rebuilding.

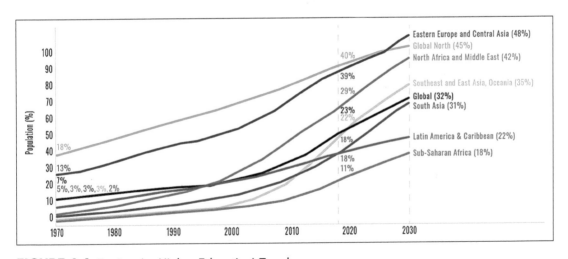

FIGURE 4.4 Tertiary (or Higher Education) Trends
Trends in tertiary schooling remain unacceptably low in many countries in the Global South. The global average is around 32 percent of all adults have achieved higher education.

Researchers Savo Heleta and Tohiera Bagus criticize the SDGs for neglecting higher education, pointing out that while higher education is receiving more development aid than before, a large majority of the funding goes to middle-income countries instead of low-income countries. Heleta and Bagus also draw attention to the fact that much of the aid for higher education goes toward scholarships for individuals to study at universities in the Global North instead of toward strengthening the origin country's higher educational system. They worry that many low-income countries will lack the capacity to develop higher educational systems and people from those countries will be left behind unless there is a change.

Adult Education

Globally, at least 750 million adults—two-thirds of them women—remain illiterate. Many adults are left without basic skills to succeed in life when they are denied access to education or forced to leave school. Half the global illiterate population live in South Asia and a quarter live in sub-Saharan Africa. Other regions that have illiterate adults include southeastern Asia and northern Africa with a smaller percent in Latin America and the Caribbean. Countries such as Haiti, Mali, Iraq, Niger, and Afghanistan have adult literacy rates below 50 percent.

Significant gender gaps in adult literacy are the result of many countries' policies from a generation ago not prioritizing education for young women. Northern Africa, Western Asia, South Asia, and sub-Saharan Africa still report higher rates of illiteracy among women. According to Pakistan's recent national survey, the adult Pakistani literacy rate is 67 percent for males and 42 percent for females.

Teaching literacy to adults is different from teaching children with school textbooks. A major challenge is that adult literacy educators receive much less policy attention than schoolteachers. They are among the lowest paid teachers and receive the least training. Many are either volunteers, part-time, or not officially qualified teachers. A 2020 UNESCO report noted that despite their importance, youth and adult literacy educators find themselves in the most precarious position of any group of educators. Youth and adult literacy educators in rural areas typically juggle part-time teaching jobs with other jobs, receive inadequate pay, and often lack meeting places and teaching materials. Many youth and adult literacy educators often work independently, without support from any well-established organization or professional group, and feel neglected by their national education systems.

Another challenge is that today's literacy requirements go beyond a set of reading, writing, and counting skills; they include understanding, interpretation, and communication in an increasingly digital and information-rich world. A lack of access to technology hinders literacy anywhere, but it is particularly a problem in rural areas.

Governance of Educational Systems

The weak governance of education systems is among the biggest challenges in education. Dr. Koumbou Boly Barry from Burkina Faso is the UN Special Rapporteur on the right to education. Her view is that when the education sector is mapped out and planned in a country, it is important to have all stakeholders present so that they participate and contribute to the success of education in their own communities and to build a holistic vision giving more attention to early childhood, vocational training, and research. Barry also says that the budgeting and decision making for education must be decentralized because it must be tailored to the needs of specific localities and specific vulnerable groups such as nomads, refugees, families in poverty, and people with disabilities. Communities and schools must be able to make decisions that fit their needs. Local actors, such as parents' associations and nongovernmental organizations (NGOs) need to be involved.

INEQUALITIES IN EDUCATION

Inequality can affect many different facets of education: resources, access, participation, and attainment, to name a few. Educational disparities can also be found along dimensions of gender, wealth, ethnicity, race, ability, and location (especially in terms of urban vs. rural areas).

Gender Inequality

Progress has been made in reversing educational disparities for women globally, and in all regions of the world. A World Economic Forum 2018 report observed women had, in fact, nearly closed the gender gap in higher education. Yet, gender inequality remains an issue which is why target 4.5 calls for the elmination of all gender disparities in education. In addition, SDG 5 draws attention to gender and education by broadly calling for gender equity in many dimensions.

Gender inequalities are not limited to girls. In Latin America and the Caribbean, disengagement from education is strongly linked to disaffection with school, underlying gender stereotypes, and social norms. In these regions, as poverty rises, boys' participation in education declines.

Large gender gaps exist in education access, achievement, and continuation in many settings, most often at the expense of girls. Globally, there are still more girls than boys that remain out of school. Estimates are that more than 130 million girls are regularly truant and/or eventually drop out of school and around sixteen million girls will never set foot in a classroom. Gender disparities are more acute in sub-Saharan Africa, North Africa, and the Middle East. Gender gaps are projected to persist for girls in much of the Global South and widen for boys in a subset of countries in Latin America, North America, and Europe.

There is also a significant gender gap in digital education and literacy. Even in high-income countries such as the United States, women account for less than 1 percent of the Silicon Valley applicant pool for technical jobs in artificial intelligence and data science. For example, in 2019 women filled only 21 percent of technical roles and accounted for only 10 percent of employees working on machine intelligence at Google. Women in lower-income countries are 25 percent less likely than men to know how to leverage Information and Communication Technology (ICT) for basic purposes, such as using a spreadsheet. Globally, 327 million fewer women than men have a smartphone, shutting them out of gaining and using ICT skills.

Even in the present day, many obstacles stand in the way of women and girls fully exercising their right to participate in, complete, and benefit from education. These obstacles include but are not limited to poverty, geographical isolation, minority status, disability, early marriage and pregnancy, gender-based violence, and traditional attitudes about the status and role of women. As recently as 2015, Sierra Leone banned pregnant girls from school. The good news is that activism and accountability mechanisms can help protect pregnant girls' right to go to school—in 2019, the ban was ruled discriminatory by the Court of Justice of the Economic Community of West African States and was lifted.

Gender inequality in education can be magnified by other disadvantages. For example, in at least twenty countries, poor, rural young women rarely complete secondary school. Women from the poorest households are four times more likely to be illiterate than those from the richest households.

Gender inequality in education has rippling effects throughout society. For many girls, school is crucial for a better future—staying in school just one more year can increase a girl's earnings as an adult by up to 20 percent. In sub-Saharan Africa and South Asia, education beyond primary school reduces adolescent pregnancies by as much as 10 percent. Finally, some countries may lose more than $1 billion a year by failing to educate girls to the same level as boys.

Some data shows that gender inequality in primary education may have reached a peak globally in 2017 and is projected to decrease steadily until 2030. Gender equality in secondary and tertiary education is more mixed: Women are overtaking men in most regions of the world, whereas large-magnitude disparities seen in sub-Saharan Africa, North Africa, and the Middle East are projected to persist. There needs to be continued investment and attention on closing the gender gap to meet the targets for all education levels.

Other Forms of Inequalities in Education

Although gender inequity is significant, it only captures one dimension of inequality in education. Other inequities in educational attainment include ethnicity, race, socioeconomic status, ability, and other identities. Poverty is a main determinant of educational attainment. In all but the high-income countries in Europe and North America, only 18 of the poorest youth complete secondary school for every 100 of the richest youth. In at least twenty countries, mostly in sub-Saharan Africa, few poor rural young women complete secondary school.

A 2018 UNESCO report found that Indigenous peoples are more likely to be unable to access education that upholds their cultural heritage, traditional knowledge, or language. Indigenous peoples account for about 370 million people, living across more than ninety countries. Among the many inequalities they face, the lack of access to quality education is particularly problematic. In many countries, Indigenous peoples encounter more barriers to completing primary school and are less likely to obtain a diploma, certificate, or degree than non-Indigenous people. Moving on to higher levels of education remains a challenge for many Indigenous peoples, especially for girls.

The lack of access to quality education is mainly due to the marginalized status of Indigenous peoples, which is rooted in a complex mix of histories, policies, and perceptions, including decades of assimilation policies that did not value Indigenous cultural heritage, knowledge, traditional cultural expressions, and languages. In some countries, formal education was often seen as a way to assimilate Indigenous populations, separating children from their families, cultural practices, and languages. As a result, resentment and suspicion of a country's education system remain high among Indigenous communities.

People with disabilities continue to struggle to access education. According to another UNESCO report in 2020, despite specific SDG 4 targets to provide adapted infrastructure and materials for students with disabilities, many countries have done little to work toward these targets. For example, Burundi and Niger reported that no primary or lower secondary school in their territory met these criteria. In Niger, only 10 of the 162 teachers working in special needs and inclusive schools are trained to work with children with disabilities. Even though Barbados and Panama have laws on inclusion, children with disabilities are often still segregated in special schools.

Many learning platforms and a great deal of digital content are not accessible to blind or deaf students, even those with access to assistive technology. Blind students struggle with information shared in images their software cannot read and end up left behind by frequent changes of online platforms.

The right to an education free of discrimination goes back to the 1960 Convention Against Discrimination in Education, yet the world has struggled to make that aspiration a reality (see *SDGs and the Law* 4.1). The right to inclusive education is enshrined in the 2006 UN Convention of the Rights of Persons with

SDGs AND THE LAW 4.1

The 1960 Convention against Discrimination in Education

In 1960 United Nations Educational, Scientific and Cultural Organization (UNESCO) sponsored the Convention against Discrimination in Education, which expresses the fundamental principles of nondiscrimination and equality in the field of education, while it also takes into account the diversity of the national educational systems.

The convention has some level of accountability because of the existence of a monitoring mechanism. Under Article VIII of the UNESCO Constitution, member states are required to submit a report on the legislative and administrative provisions they have adopted and on other measures taken to implement the conventions and recommendations.

Article 1 defines a "discrimination" as "any distinction, exclusion, limitation or preference which, being based on race, color, sex, language, religion, political or other opinion, national or social origin, economic condition or birth, has the purpose or effect of nullifying or impairing equality of treatment in education" and "education" as "all types and levels of education, including access to education, the standard and quality of education, and the conditions under which it is given."

Article 3 takes measures to eliminate and prevent discrimination in education (e.g., between nationals and foreigners).

Article 4 enshrines several principles to formulate and apply a national policy that promotes equality in education such as making primary education free and compulsory; making secondary education in its different forms generally available and accessible to all; and ensuring higher education is equally accessible to all on the basis of individual capacity.

Article 5: notes that education is a human right and fundamental freedom, while also acknowledging the liberty of parents to choose for their children's education in conformity with their moral and religious convictions and the right of members of national minorities to carry on their own educational activities.

It is the first international instrument that covers the right to education extensively and has a binding force in international law.

Source: United Nations Educational, Scientific and Cultural Organization. Convention on Discrimination in Education (1960, December). https://en.unesco.org/about-us/legal-affairs/convention-against-discrimination-education

Disabilities. In 2016, General Comment 4 by the UN Committee on the Rights of Persons with Disabilities broadened the concept of inclusion, stating that among the core features of inclusive education must be respect for the diversity of all learners, irrespective not only of disability but also of other characteristics, such as sex. Still, many governments have yet to establish this principle in their laws, policies, and practices. Only 50 percent of countries have issued policies or established laws to promote more equality and inclusion in education.

Education in Emergencies

Conflicts, disasters caused by natural hazards, and health crises, such as pandemics, can keep millions of children out of school. In crisis-affected countries, school-age children are more than twice as likely to be out of school as their peers in other countries. The trauma of conflict, disaster, or displacement can affect both access to education and educational performance.

For many children, education becomes a crisis within a crisis. Children living in refugee camps or who are internally displaced often have difficulty accessing education. In 2018, an estimated four million refugee children were out of school. In Kenya, which hosts many refugees in schools in or near refugee settlements, there are problems with understaffing of teachers and overcrowding of children. In addition, many refugee children face linguistic barriers, and some schools require documents such as birth certificates or school transcripts, which poses additional administrative barriers.

Providing education in emergency and conflict situations is made even more difficult by the fact that humanitarian aid rarely addresses education during such crises. In 2016, an Overseas Development Institute (ODI) paper proposed a new fund dedicated to education in crisis situations. The ODI paper estimated that $8.5 billion is needed each year to close the education gap for children in crisis. To date, pledges of just $500 million have been made and still only around 2 percent of humanitarian aid is spent on education.

It is difficult to predict how many people will be affected by climate displacement in the future. In 2019, almost twenty-four million people were newly displaced by weather-related disasters across 140 countries, almost three times the number of displacements caused by conflict and violence.

Yet educational needs have been almost invisible in climate change discussions and the participation of education systems in these conversations has been marginal.

Children affected by climate migration are likely to lack access to quality education, further intensifying inequalities. Crisis-affected populations, including refugees and displaced persons, will have vulnerabilities that multiply and increasingly intersect in the context of climate change. Education may have already been severely disrupted and educational resources, teachers, and infrastructure are likely to be lacking. Social safety nets may be inadequate or be at risk of disruption, adding to the precariousness of displaced learners. These crises are all happening against the backdrop of hardening of migration policies impacting the ability of learners to access and continue learning.

The Digital Divide

Nearly half the world is offline. The **digital divide** is the gap that exists between people in the world who have access to computers, mobile devices, and the internet, and those who do not. In some areas, internet access is either limited, unavailable, or unaffordable. While 87 percent of households in the United States currently have access to a computer, smartphone, tablet, or other internet-enabled device and 73 percent have access to the internet, a digital divide remains. The issue lies mostly with access to high-speed broadband, which is required to make use of much of what is available on the internet. Marginalized communities including people of color, low-income individuals, English-language learners, people with disabilities, and populations experiencing homelessness are among those most likely to lack access to high-speed internet. While there are millions in the Global North without access to reliable internet in their homes, the differences in connectivity rates between the richest and the poorest countries is most pronounced.

The digital divide has become more apparent during the COVID-19 pandemic, when the lack of reliable internet access hindered education—even in the Global North. As education pivoted to virtual instruction, the digital divide affected millions. Students from low-income families often did not have devices like laptops or tablets, which are often necessary tools. Completing assignments, taking supplementary online classes, or even access to virtual tutoring becomes a daily challenge. Even internships, mentorship programs, and networking are harder to manage when students face the digital divide.

Ultimately, the digital divide is a symptom that points to a much deeper problem of inequalities in our economic systems and economic development. To address the digital divide, we must address the underlying social and cultural challenges that have created it. Targets in SDG 9 emphasize investments in ICT, particularly in the Global South.

SOLUTIONS FOR IMPROVING EDUCATION

Table 4.1 lists the targets for SDG 4. According to the United Nations, the following strategies are priorities for achieving these targets:

- Establish policies for inclusion at all levels of education;
- Increase funding for education and target funding for those most vulnerable or those furthest behind;
- Integrate literacy and nonformal education into a national educational plan to secure budgets;
- Improve and increase educational infrastructure (schools, equipment, desks, etc.);
- Improve education quality;
- Enhance the status and training of literacy educators; and
- Engage and coordinate key actors such as government authorities, schools, NGOs, parents, community, and civil society.

Without the support of parents and community, it is often difficult to prevent families from taking their children out of (formal or nonformal) education programs so that they can help with herding and other subsistence activities. Similarly, governmental and NGO support are important to ensure sustainable financial and technical support; however, these sources may not be sufficient or may have strings attached, which complicates financial and technical support.

Joseph Friedman and colleagues suggest that inequalities could be reduced with relatively simple policies. For example, policies that eliminate school fees, improve local access to schools, increase the number of years of compulsory schooling, and provide food, stipends, and other resources to children at school can increase participation among the most disadvantaged children.

Other solutions that may advance more inclusive education include improving sanitation facilities in schools, paying greater attention to school-related

TABLE 4.1 SDG 4 Targets and Indicators to Achieve by 2030

Target	Indicators
4.1 Ensure that all girls and boys complete free, equitable and quality primary and secondary education leading to relevant and effective learning outcomes	**4.1.1** Proportion of children and young people: (a) in grades 2–3; (b) at the end of primary; and (c) at the end of lower secondary achieving at least a minimum proficiency level in (i) reading and (ii) mathematics, by sex
4.2 Ensure that all girls and boys have access to quality early childhood development, care and preprimary education so that they are ready for primary education	**4.2.1** Proportion of children under five years of age who are developmentally on track in health, learning and psychosocial well-being, by sex **4.2.2** Participation rate in organized learning (one year before the official primary entry age), by sex
4.3 Ensure equal access for all women and men to affordable and quality technical, vocational and tertiary education, including university	**4.3.1** Participation rate of youth and adults in formal and nonformal education and training in the previous twelve months, by sex
4.4 Substantially increase the number of youth and adults who have relevant skills, including technical and vocational skills, for employment, decent jobs, and entrepreneurship	**4.4.1** Proportion of youth and adults with information and communications technology (ICT) skills, by type of skill
4.5 Eliminate gender disparities in education and ensure equal access to all levels of education and vocational training for the vulnerable, including persons with disabilities, Indigenous peoples, and children in vulnerable situations	**4.5.1** Parity indices (female/male, rural/urban, bottom/top wealth quintile and others such as disability status, Indigenous peoples, and conflict-affected, as data become available) for all education indicators on this list that can be disaggregated
4.6 Ensure that all youth and a substantial proportion of adults, both men and women, achieve literacy and numeracy	**4.6.1** Percentage of population in a given age group achieving at least a fixed level of proficiency in functional (a) literacy and (b) numeracy skills, by sex
4.7 Ensure that all learners acquire the knowledge and skills needed to promote sustainable development, including, among others, through education for sustainable development and sustainable lifestyles, human rights, gender equality, promotion of a culture of peace and nonviolence, global citizenship and appreciation of cultural diversity and of culture's contribution to sustainable development	**4.7.1** Extent to which (i) global citizenship education and (ii) education for sustainable development, including gender equality and human rights, are mainstreamed at all levels in (a) national education policies, (b) curricula, (c) teacher education, and (d) student assessment
4.A Build and upgrade education facilities that are child, disability, and gender sensitive and provide safe, nonviolent, inclusive, and effective learning environments for all	**4.A.1** Proportion of schools with access to (a) electricity; (b) the internet for pedagogical purposes; (c) computers for pedagogical purposes; (d) adapted infrastructure and materials for students with disabilities; (e) basic drinking water; (f) single-sex basic sanitation facilities; and (g) basic handwashing facilities (as per the WASH indicator definitions)

(continued)

Target	Indicators
4.B Substantially expand globally the number of scholarships available to developing countries, in particular least developed countries, small island developing states and African countries, for enrollment in higher education, including vocational training and ICT, technical, engineering and scientific programs, in developed countries and other developing countries	**4.B.1** Volume of official development assistance flows for scholarships by sector and type of study
4.C Substantially increase the supply of qualified teachers, including through international cooperation for teacher training in developing countries, especially least developed countries and small island developing states	**4.C.1** Proportion of teachers in (a) preprimary; (b) primary; (c) lower secondary; and (d) upper secondary education who have received at least the minimum organized teacher training (e.g., pedagogical training) preservice or in-service required for teaching at the relevant level in a given country

Source: United Nations Sustainable Development Goals. https://sustainabledevelopment.un.org/sdg4

gender-based violence, in-person and online, and encouraging pregnant girls to go back to school. The *Solutions* 4.1 box highlights a program in Argentina aimed at adolescent pregnant girls.

 SOLUTIONS 4.1

Flexible Learning Programs in Argentina

Flexible learning programs in Buenos Aires have allowed pregnant girls and young parents to return to school.

Between 2001 and 2002, Argentina went through a severe financial crisis, with devastating social consequences. The proportion of people living below the poverty line increased from 24 percent in 1998 to 57 percent in 2002. In response to the crisis, the city of Buenos Aires created the *Puentes Escolares* (Education Bridges) program, which provided homeless and vulnerable children and adolescents with access to education and prepared them for reentry into education through workshops run by NGOs. The program guides individual future students from school selection and registration to graduation, acting as a bridge between adolescents and formal or nonformal education. Although the program is not specifically targeted to pregnant adolescents or young parents, many participate in the workshops. It also helped create a nursery at one education center where young parents attending school themselves could leave their children.

The city also created an alternative secondary education program, *Bachilleratos Populares* (Popular Baccalaureates), in vulnerable neighborhoods, with students coming predominantly from low-income families. Most female students are young and adult mothers who are excluded from mainstream schools due to early motherhood. The program's key advantages are flexibility and close contact between teachers and students. It lasts three years and students attend four hours a day, five times a week, usually in the evening to accommodate work and childcare needs. As of 2016, there were 2,293 graduates and 93 *Bachilleratos*, with 20 percent offering play areas for children.

Source: United Nations Educational, Scientific and Cultural Organization. *Global Education Monitoring Report 2020: Gender Report.* Paris: United Nations Educational, Scientific and Cultural Organization (2020). https://unesdoc.unesco.org/ark:/48223/pf0000374514?posInSet=6&queryId=N-cae3db85-bc97-4f3f-bebf-1f4c3b1b5c7f

While these are relatively obvious solutions, some education experts are critical of these targets, noting that they do not account for the larger, systemic changes that may be needed (see *Critical Perspectives* 4.1).

Reduce Inequalities: Pay Attention to Indigenous Communities

Importantly, some countries have made efforts to correct educational inequalities among Indigenous people. In Australia, improving school attendance rates among Aboriginal and Torres Strait Islander students is a key priority. The government is investing $287 million over seven years (2015 to 2022) to improve student outcomes. In New Zealand, there are special provisions available for Māori parents who wish to have their child educated in the Māori language. Section 155 of the Education Act provides for the establishment of *kura kaupapa Māori* (a kura is a school, and "kaupapa Māori" is a Māori philosophy and approach). These are state schools that use *te reo Māori* (the Māori language) as the medium of instruction. Other legislation also

---

 CRITICAL PERSPECTIVES 4.1

Equity Is a Process, Not a Number

Dr. Elaine Unterhalter, Professor of Education and International Development at University College London Institute of Education, has studied education for many years. She criticizes SDG 4 for overemphasizing metrics that do not include clearer goals for equity and the broader systemic changes needed.

Unterhalter notes that equity is often portrayed as a numerical relationship (parity or equivalence), instead of an undoing of structural inequalities, such as those associated with charging fees for schooling or challenging racial or gender-based violence. This narrow notion of inequity is concerned with addressing only the question of *how much* participation certain groups have in various stages or forms of education.

She argues ensuring "equal access" is not simply a matter of enrolling "marginalized" or underserved groups in schools or universities, which might be sites of discrimination and violence. Building equitable provision entails looking critically at the ways in which education might reproduce inequalities and working in multiple ways to address this at interconnected levels from the classroom up to the administration and policy formulation.

She also notes that the broader meanings of inclusion, quality, and equalities are not well

developed in SDG 4 and that the indicators are formulated as either measures of inputs or outcomes. While numbers and metrics are important, she believes we must be cautious because sometimes measurement and the power of numbers in education has been a form of domination and imposing particular hierarchies.

Unterhalter says we need to ask how people who experience the injustice of education exclusion, locally, nationally, and internationally, view the process of developing metrics. In what ways can they participate in reviewing metrics and indicators? She argues if we are to engage in measuring the targets for SDG 4 not in a spirit of imposing particular frameworks of evaluation, but of consulting how to establish these to express quality and equality, we need greater insight into a range of normative, epistemological, conceptual, empirical, and numerical resources to undertake this; a different kind of power with numbers.

Source: Unterhalter, E. "The Many Meanings of Quality Education: Politics of Targets and Indicators in SDG4." *Global Policy. Special Issue: Knowledge and Politics in Setting and Measuring SDGs* 10 (51): 39–51 (2019). https://doi.org/10.1111/1758-5899.12591

establishes policies that assist groups underrepresented in education, such as scholarships for Māori and Pacific Islanders.

Finally, Ecuador has to overcome the unique challenge of helping more than thirteen different Indigenous peoples who have their own languages. These include the Awa, Épera, Chachi, Tsa'chi, Kichwa, A'i (Cofán), Pai (Seco-ya), Bai (Siona), Wao, Achuar, Shiwiar, Shuar, Sapara, and Andwa. Ecuador has focused on bilingual and intercultural education and established a program called Community-Family Infant Education. Community-Family Infant Education provides training to mothers and fathers and to the community on how to educate children from the moment that they are conceived. They aim to strengthen the growth of abstract thinking, literacy, skill development, personality, construction of identity, and self-esteem of children from indigenous groups and nationalities.

Reform the School-to-Prison Pipeline

In the United States, school discipline and the criminal justice system have been intertwined, in a phenomenon called the **school-to-prison pipeline**. The school-to-prison pipeline is the disproportionate tendency of minors and young adults from disadvantaged backgrounds to become incarcerated because of increasingly harsh school and municipal policies. According to the ACLU, many of these children have learning disabilities or histories of poverty, abuse, or neglect, and would benefit from additional educational and counseling services. Instead, they are isolated, punished, and pushed out. Videos have circulated of police "take-downs" of students at school.

Reasons for this phenomenon include "zero tolerance" policies that criminalize minor infractions of school rules, which have increased school suspensions. Yet Black students are three times more likely to be suspended or expelled than White students. In addition to zero tolerance, many schools started relying on actual police officers stationed in schools. Allowing police officers to arrest a student may increase the chances for a student to get a juvenile record, even if the infraction is a disorderly conduct or talking back to a teacher.

Solutions to this inequality will require financial investments to increase workers and mental health professionals in the schools, and to reduce class size. Other solutions include reducing the number of suspensions and eliminating school zero tolerance policies. Some schools are exploring restorative justice programs, which achieve accountability through the development of caring, supportive relationships and through strategies that allow students to reflect on their behavior and make amends when needed to preserve the health of the community.

Invest in Education and Educational Systems

Long-term investment in education requires predictable and sustained financing. This is a pretty obvious solution, but it remains a challenge in a world that is overly driven by a focus on economic returns because investing in education does not bring short-term, quick wins. Rather, education is an investment over the long term. To achieve the targets in SDG 4, many countries will need to increase public funding for education at all levels, but particularly for secondary, adult, and higher education. This may require widening the tax base and increasing the share of the national budget allocated to education. Education experts David Archer and Tanvir Muntasim propose countries should spend about 20 percent of aid on basic education;

presently, only about 4 percent of global aid goes to basic education. They also recommend that more money be spent on lower-income countries and point out that it is not just a matter of increasing funding, but making sure existing money is better spent.

The biggest single costs for education are recurring costs, such as teacher salaries. To achieve the first target of SDG 4—universalizing access to primary and secondary education—millions more trained teachers will need to be employed and that requires a long-term, steady source of financing. It also requires proper training for teachers, which is lacking in many regions. In one-third of all countries, fewer than 75 percent have been trained according to national standards; UNESCO estimates indicate that presently around half of teachers in Africa are undertrained. Untrained teachers can be paid lower salaries, so there is a disincentive to train teachers. Yet many studies have demonstrated that teachers—and their level of knowledge about their subject—are the most important determinant of education quality.

Use Information Communication Technologies in Education

Some experts regard ICT as highly promising for accelerating progress toward the literacy target. It may help reduce the digital divide that many experience. These technologies include the use of radio (particularly in remote areas), TV, mobile phones, tablets, and computers. ICT may help motivate learners, promote quality and effective learning, and to deliver services more efficiently. However, ICTs also have their share of limitations. Some people of older generations find it difficult to catch up with ICT skills and are at risk of being left behind in education, the job market, and the larger economy. In addition, the most impoverished often lack literacy skills, and are unable to access and efficiently use these technologies. Meanwhile, despite growing use of mobile phones and personal computers, access to the internet is lacking in many parts of the world. For example, in Kenya, about 72 percent of the population own a mobile phone, but only around 32 percent are internet users. The *Solutions* 4.2 box highlights a program targeting the use of ICT in rural areas in Mongolia.

MOOCs

Bridging the digital divide may provide more with opportunities for **MOOCs** (Massive Open Online Courses). These are free online courses for anyone. MOOCs were first developed in 2008 and became widespread by 2012. Many universities offer MOOCs, and there are also nonprofit organizations such as Coursera and EdX that specialize in producing MOOCs. Millions of people have taken advantage of free MOOCs for career development, supplemental learning, higher education prep, and lifelong learning. MOOCs are diverse and include courses for banking and bookkeeping, regional studies, art history, chemistry, and computer programming.

MOOCs may offer an alternative form of higher education for low-income individuals and communities and could contribute to the democratization of higher education, particularly in the Global South. However, a drawback to MOOCs is that despite their potential to support learning and education, only a very small portion of the enrolled learners complete the course. Some of the impediments include poor internet connection, language and translation barriers, or accessibility barriers for users who are differently abled.

SOLUTIONS 4.2

Addressing Rural Illiteracy

Target 4.5: Eliminate gender disparities in education and ensure equal access to all levels of education and vocational training for the vulnerable, including persons with disabilities, Indigenous peoples, and children in vulnerable situations.

Target 4.6: Ensure that all youth and a substantial proportion of adults, both men and women, achieve literacy and numeracy.

Mobile technologies may prove valuable for accelerating progress toward the literacy targets. When available, they can allow learners to practice and progress at their own pace, at their convenience, and in different locations, which can be especially important in rural and more isolated communities.

Mongolia is a country reimagining how to reach rural populations with ICT. Mongolia has made improvements in its education system, including new laws that guarantee free and compulsory education for all children under the age of sixteen. The government has increased funding for development of schools and adult nonformal education programs. Overall, youth and adult literacy rates are high at 98 percent.

However, education remains largely inaccessible for rural and predominately nomadic people in remote areas. More than 15,000 children are out of school and many adult rural Mongolians are illiterate or semiilliterate. In some rural communities, the nomadic herders do not see the direct benefit of education to their incomes or livelihood, and as a result many boys drop out of school to assist their parents.

Mongolia has responded with their Literacy Through Distance Learning Program. The program employs an intergenerational approach to literacy skills training and learning and focuses on the learning needs of entire families. The approach is furthermore designed to foster a positive attitude toward learning and to enable parents and their children to assist one another in the learning process. The program encompasses a range of themes including but not limited to:

- Health (e.g., preventive measures against HIV/AIDS or nutrition and hygiene);
- Literacy for economic self-sufficiency and community/rural development; and
- ICT skills training.

Since its inception in 2004, the program has succeeded in improving literacy among marginalized Mongolians by providing literacy skills training to 3,500 illiterate and 4,500 semiilliterate people each year. Family-based literacy training has fostered the development of positive social relationships and communication patterns between parents and their children. This has also motivated parents to ensure that their children attend and stay in school.

Sources: United Nations Educational, Scientific and Cultural Organization. "Effective Literacy and Numeracy Practices Database—LitBase" (2021). http://www.unesco.org/uil/litbase United Nations Educational, Scientific and Cultural Organization. *Harnessing the Potential of ICTs Literacy and Numeracy Programmes using Radio, TV, Mobile Phones, Tablets and Computers.* Hamburg, Germany: UNESCO (2016).

SUMMARY AND PROGRESS

While national governments are mainly responsible for managing educational systems and advancing more inclusionary education, SDG 4 will require global and regional cooperation. However, despite some of the challenges, the current generation is the most educated in history.

The coronavirus outbreak is one of the most unprecedented disruptions in the history of education and will likely have far-reaching consequences that may jeopardize the good progress made in improving global education. Starting in the spring of 2020, most governments around the world temporarily closed educational institutions in an attempt to contain the spread of the virus, disrupting

school for 1.3 billion students in 190 countries (see Map 4.2). UNESCO estimates that two-thirds of an academic year were lost due to COVID-19 school closures. By 2021, some schools were open, but many remained closed throughout the school year. Experts worry that as many as twenty-four million children may not return to the classroom due to the pandemic's economic impacts. Even more concerning, the pandemic is aggravating chronic difficulties that the most vulnerable—especially girls—already struggle to resolve. Girls are more likely to experience gender discrimination in the allocation of household chores, gender-based violence, early and forced marriage, and early and unintended pregnancy. School closures have led to increased childcare and chore responsibilities at home, which are likely to disadvantage girls even more. The shift to online distance learning could also disadvantage girls in low- and middle-income countries where they are 8 percent less likely than boys to have a mobile phone and 20 percent less likely to use the internet. It is possible that eleven million young girls may not ever return to school at all, even after the health crisis is over.

Making progress toward achieving SDG 4 requires a systemic approach both to education reform and education financing. Efforts need to be coordinated by national education systems working closely with the global community and global financing mechanisms to strengthen educational systems over the medium and long term. While the current pandemic is disrupting educational systems, some see this as a window of opportunity to "build back equal" by transforming education systems through more inclusive measures, but this remains to be realized.

Finally, there are many ways education connects to other SDGs. Those goals that have the strongest reinforced synergies with SDG 4 are highlighted in Figure 4.5. The goals of gender equality, climate change, and reduced inequalities have particularly strong synergies with education. Some of the other SDGs will require higher levels of educational attainment; on the flip side, achieving goals around gender equality and reduced inequalities should open up educational opportunities for those communities that have been behind.

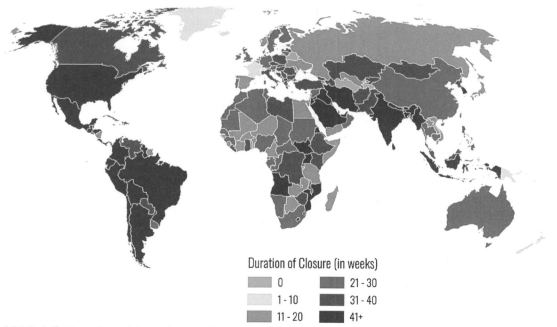

Duration of Closure (in weeks)

0	21 - 30
1 - 10	31 - 40
11 - 20	41+

MAP 4.2 Duration of Complete and Partial Closures Due to COVID-19 (2020)
School closures for 2020 affected millions of children around the world.

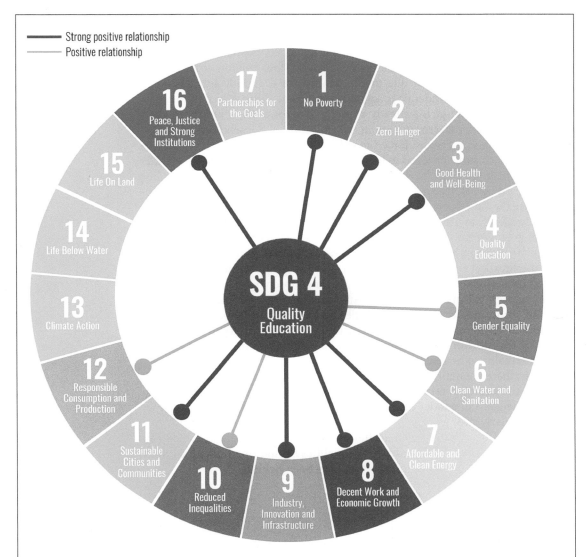

SDG 1: Higher levels of educational attainment often open up more economic opportunities for higher income.

SDG 3: Access to high-quality education is associated with better health, at both individual and community levels.

SDG 5: Given that the educational gaps are highest among girls and women, eliminating discrimination against women and girls may increase educational attainment, and preventing early and forced marriage will mean girls can stay in school longer.

SDG 6: Achieving access to adequate and equitable sanitation and hygiene may mean that women and girls spend less time traveling for water, reducing vulnerabilities while doing so, and free up time for education.

SDG 8: Working on SDG 4 targets may result in increasing women's economic opportunities for decent work and labor rights protections.

SDG 9: To achieve many of the SDG 9 targets around scientific research and expanding technological capabilities will require a more educated work force.

SDG 10: By working on targets in SDG 10 such as eliminating discriminatory laws, policies and practices will expand educational opportunities for many of the most vulnerable, such as refugees, migrants, indigenous peoples.

FIGURE 4.5 Interconnections in SDG 4 and the Other Goals
Education connects to all of the SDGs in some way; this figure highlights several of the stronger connections.

QUESTIONS FOR DISCUSSION AND ACTIVITIES

1. Explain the reasons for different types of educational inequalities.

2. Compare progress toward better educational attainment in the Global North to the Global South.

3. Explain how inequalities impact education in the Global South. Are these the same inequalities that impact education in the Global North?

4. Is higher education necessary for a sustainable future?

5. Do you think MOOCs are an effective solution to expand access to higher education?

6. Which of the solutions for achieving SDG 4 targets would you prioritize and why?

7. Why are girls less represented in STEM and especially IT? What could be done to encourage more gender parity in these areas?

8. What was your experience learning during the COVID-19 pandemic?

TERMS

basic education
digital divide
higher education
MOOCs

primary education
school-to-prison pipeline
secondary education

RESOURCES USED AND SUGGESTED READINGS

Archer, D., and Muntasim, T. "Chapter 8 Financing SDG 4: Context, Challenges, and Solutions." In A. Wulff (ed.). *Grading Goal Four: Tensions, Threats, and Opportunities in the Sustainable Development Goal on Quality Education (pp. 170–93).* Leiden: Koninklijke Brill (April 2020). https://doi.org/10.1163/9789004430365_008.

Filho, W. L., Azul, A. M., Brandli, L., Salvia, A. M., and Wall, T. *Gender Equality.* Cham, Switzerland: Springer (2021).

Friedman, J., York, H., Graetz, N., Woyczynski, L., Whisnant, J., Hay, S. I., and Gakidou, E. "Measuring and Forecasting Progress towards the Education-Related SDG Targets." *Nature* 580: 636–39 (2020).

Heleta, S., and Bagus, T. "Sustainable Development Goals and Higher Education: Leaving Many Behind." *Higher Education* 81: 163–77 (2020). https://doi.org/10.1007/s10734-020-00573-8

Internal Displacement Monitoring Center. "Global Report on Internal Displacement 2020" (2020). https://www.internal-displacement.org/global-report/grid2020/#:~:text=Previous%20Editions-,Global%20Report%20on%20Internal%20Displacement%202020,and%20disasters%20across%20145%20countries

Islam, M., and Shamsuddoha, M. "Socioeconomic Consequences of Climate Induced Human Displacement and Migration in Bangladesh." *International Sociology* 32(3): 277–98 (2017, February).

Milana, M., Webb, S., Holford, J., Waller, R., and Jarvis, P. (eds.). *The Palgrave International Handbook on Adult and Lifelong Education and Learning.* London: Palgrave Macmillan (2018).

Nhamo, G., and Mjimba, V. (eds.). *Sustainable Development Goals and Institutions of Higher Education.* New York, NYSpringer Nature (2020).

Shabliy, E. V., Kurochkin, D., and Ayee, G. *Global Perspectives on Women's Leadership and Gender (In)Equality.* London: Palgrave Macmillan (2020).

Smith, A. (ed.). *Gender Equality in Changing Times: Multidisciplinary Reflections on Struggles and Progress.* London: Palgrave Macmillan (2020).

Spring, J. *American Education.* 18th ed. New York: Routledge (2018).

United Nations Educational, Scientific and Cultural Organization. *Education 2030: The Incheon Declaration and Framework for Action for the Implementation of Sustainable Development Goal 4.* Paris: United Nations Educational, Scientific and Cultural Organization (2016). http://uis.unesco.org/sites/default/files/documents/education-2030-incheon-framework-for-action-implementation-of-sdg4-2016-en_2.pdf

United Nations Educational, Scientific and Cultural Organization. *Global Education Monitoring Report. Background Paper Prepared for the 2019 Global Education Monitoring Report: Migration, Displacement and Education: Building Bridges, Not Walls: Inclusion of Refugees in National Education Systems.* Paris: United Nations Educational, Scientific and Cultural Organization (2018).

United Nations Educational, Scientific and Cultural Organization. "Indigenous Peoples' Right to Education." Paris: United Nations Educational, Scientific and Cultural Organization (2019). https://unesdoc.unesco.org/ark:/48223/pf0000369698

United Nations Educational, Scientific and Cultural Organization. "COVID-19 Education Response" (2020). https://en.unesco.org/covid19/educationresponse/

United Nations Educational, Scientific and Cultural Organization. *Global Education Monitoring Report Summary 2020: Inclusion and Education: All Means All.* Paris: United Nations Educational, Scientific and Cultural Organization (2020).

United Nations Educational Scientific and Cultural Organization. "Q&A with the UN Special Rapporteur on the Right to Education, Dr. Koumbou Boly Barry" (2020, December 7). https://en.unesco.org/news/qa-special-rapporteur-right-education

United Nations Educational, Scientific and Cultural Organization. "World Teacher Day 2020: Teachers Leading in Crisis, Reimagining the Future." Paris: United Nations Educational, Scientific and Cultural Organization (2020). https://unesdoc.unesco.org/ark:/48223/pf0000374450

United Nations International Children's Emergency Fund. "Education Uprooted: For Every Migrant, Refugee and Displaced Child, Education." New York: United Nations International Children's Emergency Fund (2017).

Wade, S. *Inclusive Education: A Casebook and Readings for Prospective and Practicing Teachers.* Mahwah, NJ: L. Erlbaum Associates (2000).

World Economic Forum. "Global Gender Gap Report 2018" (2018). https://reports.weforum.org/global-gender-gap-report-2018/key-findings/?doing_wp_cron=1665262814.9456059932708740234375

Gender Equality

LEARNING OBJECTIVES

After reading this chapter, you should be able to:

- Identify and understand the geography of gender inequality and how it is measured

- Identify and explain the impacts of gender inequality

- Explain the causes of gender inequality

- Identify and explain the various targets and goals associated with SDG 5

- Explain how the three Es are present in SDG 5

- Explain how SDG 5 strategies connect to other SDGs

In 2013, the Rana Plaza building, which housed five garment factories in Dhaka, Bangladesh, collapsed, killing at least 1,132 people, and injuring more than 2,500. It remains one of the worst industrial accidents on record and helped spotlight the poor labor conditions in which women garment workers work.

Rana Plaza is not an outlier. Since this tragedy, more than 100 other accidents have occurred where women workers were injured or died. In Bangladesh, women are about 90 percent of the workforce in the garment industry. Most women come from low-income families and are paid far less than men mainly due to their lack of education. On average, these workers make only 3,000 taka a month (about $35 a month), which is far below the Bangladeshi living wage of 5,000 taka per month.

*In addition to this wage disparity, garment workers in Bangladesh deal with harsh working conditions, often finishing their day at 3 AM after starting at 7:30 AM. Female workers are regularly exposed to numerous hazards including long working hours, absence of leave facilities (such as lack of restrooms), congested and overcrowded working conditions, absence of health facilities and safety measures, absence of amenities, and lack of safe drinking water. Many women endure physical hazards such as exposure to toxic agents and repetitive motion injuries. Many women are vulnerable to sexual, verbal, and psychological harassment and violence at these factories.**

---

*Bratton, H. "3 Organizations Helping Garment Workers in Bangladesh." *Borgen Magazine*, November 19, 2020. https://www.borgenmagazine.com/3-organizations-helping-garment-workers-in-bangladesh/

GENDER INEQUALITY: UNDERSTANDING THE ISSUE

Gender inequality has been a historic and persistent challenge that not a single country has eliminated yet.

Half the people on this planet—almost four billion people—are female. Although gender inequality takes many different forms, it is usually women who are disadvantaged relatively to similarly situated men. Gender equality is said to be achieved when women and men enjoy the same rights and opportunities across all aspects of life, including social interactions, economic participation, and decision making, and when the different behaviors, aspirations, and needs of women and men are equally valued and favored. A World Economic Forum 2020 report concluded that "at the present rate of change, it will take nearly a century to achieve parity, a timeline we simply cannot accept in today's globalized world, especially among younger generations who hold increasingly progressive views of gender equality."

The United Nations has set gender equality as a goal instead of gender equity. The terms **gender equality** and **gender equity** are related but have different meanings. Equity is the fairness of treatment for both men and women, according to their respective needs. If equality is the end goal, equity is the means to get there.

The United Nations Educational, Scientific and Cultural Organization (UNESCO) explains that gender equality, equality between men and women, "does not mean that women and men have to become the same, but that their rights, responsibilities and opportunities will not depend on whether they were born male or female. Gender equity means fairness of treatment for men and women according to their respective needs. This may include equal treatment or treatment that is different but which is considered equivalent in terms of rights, benefits, obligations, and opportunities."

Countries in the Global North have paid more attention to women's right to work and their rights in the workplace, the issues of unpaid care and domestic work, and efforts to change negative social norms and gender stereotypes. In contrast, Global South countries tend to prioritize access to health care, education, the elimination of violence against women and girls, and increased political participation and representation of women.

Measuring Gender Equality

Experts identify four principal components of gender equality that are often used to measure and quantify inequality: (1) education, (2) health, (3) economic and labor participation, and (4) political empowerment. Various organizations have developed different composite indices for quantitatively understanding gender inequality. For example, the Economist Intelligence Unit has an index for women's economic opportunity that tracks labor policies and practices, access to finance, education and training, legal and social status, and participation in the general business environment. Their index focuses primarily on economic statistics, which represents just one dimension of gender inequality.

There are other regional and international indicators. The European Institute for Gender Equality assesses gender equality among EU member states. Similarly, the African Gender and Development Index (AGDI) is a tool that assesses gender inequality in Africa. Three of the most widely used gender indices are the UN **Gender Inequality Index (GII)**, the Organization for Economic Cooperation and Development's (OECD) Social Institutions and Gender Index, and the World Economic Forum's Global Gender Gap Index.

The GII measures inequalities across three dimensions:

1. Reproductive health (based on maternal mortality ratio and adolescent birth rates);

2. Empowerment (based on proportion of parliamentary seats occupied by females and proportion of adult females aged twenty-five years and older with at least some secondary education), and;

3. Economic status (based on labor market participation rates of female and male populations aged fifteen years and older).

Map 5.1 shows the GII for the most recent year. The GII is a value of 0 to 1, where 1 indicates complete inequality.

The **Social Institutions and Gender Index (SIGI)** is compiled by the OECD, and measures discrimination against women in social institutions. It attempts to measure how gender gaps in social institutions translate into gender gaps in development outcomes such as labor force, poverty levels, marginalization, education, vulnerability to violence, and public leadership positions. The SIGI is comprised of four subindices:

1. Discriminatory family code (such as laws on child marriage or divorce);

2. Restricted physical integrity (such as laws on sexual harassment or violence);

3. Restricted resources and assets (such as women not being allowed to have their own bank accounts); and

4. Restricted civil liberties (such as laws on citizenship or freedom of movement).

Each subindex includes several subcategories so the index scores countries on fourteen indicators in total. The SIGI ranges from 0 percent for no

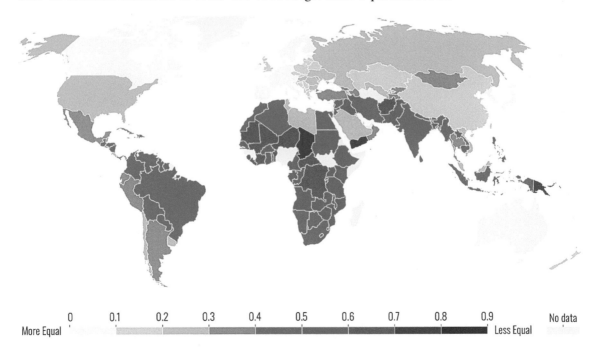

MAP 5.1 Gender Inequality Index (2019)

Countries in the Global North tend to have higher levels of gender equality than countries in the Global South. The United States has less gender equality than many EU countries, Australia, New Zealand, and Canada. Regionally, countries in Africa, the Middle East, and South Asia have the highest levels of gender inequality.

discrimination to 100 percent for very high discrimination and ranks inequality in five categories: very low, low, medium, high, and very high.

The **Global Gender Gap Index (GGI)**, created by the World Economic Forum in 2006, measures broader gender-based disparities. The GGI is based on the premise that gender inequality is the combined result of various socioeconomic, policy, and cultural variables. Unlike other indices, the GGI measures "gaps" rather than levels and quantifies the relative gaps between women and men across four key areas:

1. Health (such as differences in life expectancy);

2. Education (differences between men and women in primary, secondary, and tertiary education);

3. Economy (proportion of women among skilled professionals, wage equality, overall income disparities), and;

4. Politics (women's representation in parliament/ministry positions, etc.).

The state of the gender gap in each of these four areas varies: Generally, the world has made better progress on gender equality in educational attainment and health than in economic opportunity and political empowerment. The overall rankings for the 2021 index are shown in Table 5.1.

TABLE 5.1   Selected Global Gender Gap Index 2021 Rankings

The World Economic Forum's Gender Gap Index is a compilation of fourteen different measurements across four key areas. It is reported on a 0–1 scale with "1" being full gender parity.

Rank	Highest Ranked Countries	Score
1	Iceland	0.892
2	Finland	0.861
3	Norway	0.849
4	New Zealand	0.840
5	Sweden	0.823
6	Namibia	0.809
7	Rwanda	0.805
8	Lithuania	0.804
9	Ireland	0.800
10	Switzerland	0.798
	Lowest Ranked Countries	
147	Saudi Arabia	0.603
148	Chad	0.593
149	Mali	0.591
150	Iran	0.582
151	Democratic Republic of Congo	0.576
152	Syria	0.568
153	Pakistan	0.556
154	Iraq	0.535
155	Yemen	0.492
156	Afghanistan	0.444

Source: World Economic Forum. *Global Gender Gap Report, 2021* (2021): 10. https://www.weforum.org/reports/global-gender-gap-report-2021

Although these three widely used gender indices may use different data and methodologies, they yield remarkably similar results and rankings. All three indices rank sub-Saharan Africa, the Middle East, and South Asia as regions that experience the most gender inequality. Some of the countries with the highest gender inequality include Yemen, Pakistan, Chad, Iran, Iraq, Jordan, Lebanon, Cote d'Ivoire, and the Democratic Republic of Congo. Saudi Arabia, which is notorious for its repression of women, has made some recent reforms, but women's equality remains a distant promise (see *Understanding the Issue* 5.1).

 ## UNDERSTANDING THE ISSUE 5.1

Repression and Reform in Saudi Arabia

In Saudi Arabia, women and girls continue to face systematic discrimination in law and in practice in other areas such as marriage, divorce, and inheritance. The Saudi government has approved international and domestic declarations regarding women's rights, but it is often cited as among the worst countries for gender equality. Saudi Arabia has a male guardianship system: A man controls a Saudi woman's life from her birth until her death. Every Saudi woman must have a male guardian, normally a father or husband, but in some cases a brother or even a son has the power to make a range of critical decisions on her behalf. The Saudi state essentially treats women as permanent legal minors. This male guardianship system is not concretized in law per se but remains entrenched in cultural practices. For many years, the Saudi government placed the following restrictions on Saudi women:

- They could not apply for a passport or travel outside the country without their male guardian's approval;
- They could not enter into marriage without permission of the male guardian;
- The government did not penalize or prevent companies who required a male guardian's consent for women to work or restrict jobs to men;
- Women were prevented from some professions, like judges and drivers;
- Women were not able to vote in elections or be elected to any political office;
- Girls were not allowed to participate in sports (including the Olympic Games); and
- Women were not allowed to drive cars.

There have been some recent reforms to counter the repression. In recent years, Saudi Arabia has increased employment opportunities for women in areas previously closed to them. In 2013, the Saudi government sanctioned sports for girls in private schools for the first time and, in 2016, four Saudi women were allowed to participate in the Olympic Games in Rio de Janeiro. In 2015, women were allowed to vote in elections and, in 2018, women were allowed to drive cars. In 2019, there were major reforms to the discriminatory male guardianship system. Among other things, women aged over twenty-one are now allowed to apply for and obtain a passport and travel, register a divorce or a marriage, and apply for official documents without male guardian permission.

While the reforms recognized women's rights in certain areas and eased major restrictions on women's freedom of movement, they did not abolish the guardianship system. For example, women still cannot marry without the permission of a guardian. Other forms of discrimination also continue: There is no law defining the minimum age for marriage in Saudi Arabia, harassment of women persists, and women's rights activists are often arrested and jailed.

Despite the reforms, the discrimination and repression that women still experience in Saudi Arabia is real: It is estimated that every year, 1,000 women try to flee Saudi Arabia.

Source: Al Mofawez, M. *Oppression of Women in the Islamic World and Gender Inequality in Saudi Arabia* (2016). *LMU/LLS Theses and Dissertations*: 347. https://digitalcommons.lmu.edu/etd/347

Europe is the best-performing region. Several countries (Belgium, Denmark, Finland, Norway, and Switzerland) all appear in the top ten of at least two of the three indices. The region is home to the four most gender-equal countries in the world. While the indices agree that Europe has the highest gender equality, the indices vary at the country scale. For example, for many years, the GGI has ranked Iceland as the most gender-equal country in the world, while SIGI and GII have consistently ranked Switzerland or Norway first. Other countries ranked in the top ten by at least one index include France, Slovenia, New Zealand, Ireland, Spain, and Germany. Some indices also highly rank countries in the Global South such as Rwanda, Nicaragua, and Colombia. Notably, the United States is not highly ranked on any of the three indices. It is ranked 53rd in the world in the GGI, and 51st in the GII (far behind neighbor Canada, which is ranked 16).

THE CONSEQUENCES OF GENDER INEQUALITY

There are many ways women may experience gender inequality in their lifetimes in the twenty-first century. Some are more visible than others (see Photo 5.1). These include discrimination in the family (such as the burden of domestic work or inability to divorce); restricted access to financial sources; restricted civil liberties (such as lack of voting rights); restricted physical integrity (such as lack of freedom of movement or travel or inability to access family planning); higher adolescent birth rates; lower educational attainment; and reduced access to health services. This section briefly considers several of these.

Women and Education

Education for girls is particularly important, as it breaks the cycle of poverty and transforms lives for many. The previous chapter provides an in-depth discussion on the importance of education and the educational disparities

PHOTO 5.1 The Tehran Metro
In Tehran, Iran, the Tehran Metro has separate trains for women and men.
Is this an example of separate but equal? Or separate and unequal?

disadvantaging many young girls and women. This chapter offers a brief reminder of how educational setbacks translate into lack of access to skills and limited opportunities in the labor market. Today, 10 percent of girls between the ages of fifteen and twenty-four in the world are illiterate, with a high concentration of these girls living in the lowest-income countries.

Many countries have made good progress; thirty-five countries have already achieved full educational parity. Some of these countries include Australia, Austria, Jamaica, Botswana, Uruguay, the United States, and Namibia. Another 120 countries have closed the gap to at least 95 percent and include countries such as the Dominican Republic, Spain, Ireland, Fiji, Serbia, Italy, and Poland. Educational attainment gaps are relatively small on average, but there are still countries where investment in women's education is weak. Countries that have not made as much progress include Togo, Angola, Mali, Benin, Yemen, and the Democratic Republic of Congo.

The increase in girls' school enrolment has been one of the most remarkable achievements of the past decades. Each additional year of postprimary education for girls has important multiplier effects, improving women's employment outcomes, decreasing the chance of early marriage, and improving their health and well-being. Interestingly, some of the strongest progress has been in improving parity at higher levels of education, particularly college and university.

Women and Health

There have also been improvements in the gap between men and women's access to health services due in part to investments started during the Millennium Development Goals. Parity has been essentially achieved in all countries in terms of life expectancy as women tend to live longer in all countries. The 2020 Global Gender Gap Report states that 95.7 percent of the global health gap has been closed, and that forty-eight countries have achieved near-parity and seventy-one others have closed at least 97 percent of the gap. Countries that trail this global figure include Pakistan (94.6 percent), India (94.4 percent), Vietnam (94.2 percent), and China (92.6 percent). However, given the significant population sizes of India, China, and Pakistan, this means millions of women in these countries are not yet granted the same access to health services as men.

Despite progress on access to health services, health equity remains an issue. Maternal mortality is alarmingly high in countries affected by conflict and crisis and there is inequity at the rural-urban scale. Women in rural areas are still much less likely to have access to skilled health personnel when they give birth than their urban counterparts. In some countries, females in rural areas have less access to health care or proper nutrition, leading to a higher mortality rate. In addition, social conditions such as early child marriage continues to be a persistent problem. Recent estimates are that globally, 750 million women and girls were married before the age of eighteen.

Access to health services for women also includes access to reproductive health. The proportion of women with unmet needs for family planning has stagnated at 10 percent globally since 2000. In 2019, 190 million women of reproductive age worldwide who wanted to avoid pregnancy did not use any contraceptive method. A 2020 report noted that ending preventable maternal deaths, covering all unmet needs for family planning, and eliminating gender-based violence by 2030 will require some $264 billion over the next decade. However, only a fraction of that, $42 billion, is expected to be spent on those areas between now and 2030.

🔍 UNDERSTANDING THE ISSUE 5.2

The #MeToo Movement

There remains much to do in the United States regarding gender equality, notably around sexual harassment and sexual abuse. Tarana Burke is the founder of the "Me Too" movement. Her hashtag "#MeToo" has been used more than nineteen million times on Twitter alone.

In 1996, Burke was working as the director of a youth camp when she met a young woman who asked to speak with her privately about experiencing sexual abuse. Although Burke recalls she was unable to respond thoroughly in the moment, this encounter would become the foundation for the Me Too movement she created a decade later. She continued to focus on young women of color and cofounded an African-centered Rites of Passage program for girls called Jendayi Aza. Jendayi Aza evolved into her nonprofit JustBe, Inc. which was founded in 2006 to empower and encourage young Black girls in their process of self-discovery through unique programming and workshops. JustBe, Inc. had a tremendous impact on the Selma, Alabama, community and the program was adopted by every public school in the city. Shortly after, the "Me Too" movement was born as a way for young women of color to share their stories of surviving sexual abuse and sexual harassment. Burke began using the phrase "me too" to promote the idea of "empowerment through empathy." Her campaign was not only designed to facilitate healing but also to train survivors to work in communities of color.

Burke's work returned to the spotlight. During the 2017 Harvey Weinstein sexual abuse scandal, Burke's hashtag "#metoo" went viral. People all over the world began posting the phrase on their social media accounts to align with the movement.

Following this surge of support, Burke became a recognized global leader and helped get a larger conversation started around sexual violence. In 2017, *Time* magazine named Burke and other "Silence Breakers" as Person of the Year. In 2019, she earned the Sydney Peace Prize in Sydney, Australia. Burke currently serves as the Senior Director of Girls for Gender Equity in Brooklyn, New York.

Source: Alexander, K. L. "Tarana Burke." National Women's History Museum (2020). www.womenshistory.org/education-resources/biographies/tarana-burke

Harmful Practices That Affect Girls and Women

Issues such as harm and violence continue to disproportionately affect girls and women. A 2021 World Health Organization report says approximately 852 million women worldwide have been subjected to physical and/or sexual violence, although most surveys tend to underestimate the true prevalence of violence against women as there will always be women who do not disclose these experiences. Around forty-nine countries lack laws that protect women from domestic violence. The United Nations has been focused on reducing four harmful practices: childhood sexual violence, female genital mutilation, child marriage, and son preference.

The United Nations Children's Fund (UNICEF) estimates that, globally, 35 percent of women between fifteen and forty-nine years of age have experienced physical and/or sexual intimate partner violence or nonpartner sexual violence. Acts of sexual violence can range from direct physical contact with the use of force or restraint to less direct forms such as unwanted exposure to sexual language and images. While both boys and girls can be the target of sexual violence, data suggest that girls are generally at a heightened risk, particularly after puberty. A 2017 UNICEF report indicated that nine million girls aged fifteen to nineteen were forced into sexual intercourse or other sexual acts between 2016 and 2017. In thirty-eight low- and middle-income countries, close to seventeen million adult women reported having experienced forced sex in childhood. Sub-Saharan Africa has among the highest reports of childhood experiences of forced sex. In Cameroon, one in six young women report childhood experiences of forced sex. However, even the highest-income countries report sexual violence against women: In the United Kingdom, Luxembourg, France, the Netherlands, and Spain, more than one in ten young women report experience of contact and noncontact forms of sexual violence before the age of fifteen. In twenty-eight countries in Europe, around 2.5 million young women report experiences of contact and noncontact forms of sexual violence before age fifteen. *Understanding the Issue* 5.2 profiles the #MeToo movement that began in the United States.

Female Genital Mutilation/Cutting

Another harmful and violent practice is female genital mutilation/cutting or female circumcision (FGM/C). The World Health Organization (WHO) defines FGM/C as "any procedure that involves partial or total removal of the external female genitalia, or other injury to the female genital organs for non-medical reasons." UNCIEF estimates at least 200 million girls and women in thirty countries have been subjected to the practice. This practice is geographically concentrated in a swath of countries from Western Africa to the Horn of Africa with Mali, Sudan, Somalia, and Egypt reporting the highest levels. It is also concentrated in the Middle East in Iraq and Yemen, and in some countries outside those regions such as Indonesia. Variations of this practice have also been reported in India, Malaysia, Oman, Saudi Arabia, and the United Arab Emirates (Map 5.2). The practice has also been found

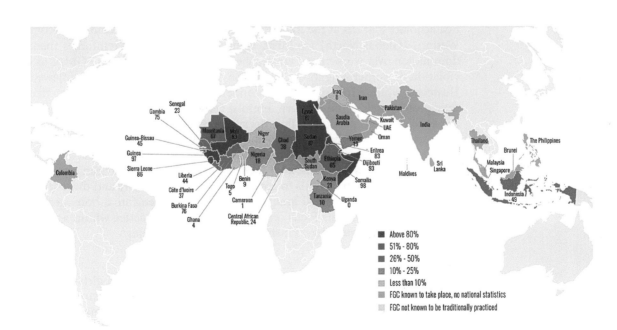

MAP 5.2 Female Genital Cutting, Prevalence among Women (Age 15–49)
FGM/C is geographically concentrated in a swath of countries from Western Africa to the Horn of Africa, with Mali, Sudan, Somalia, and Egypt reporting the highest levels. It is also found in Iraq and Yemen, and in some countries outside those regions such as Indonesia. Variations of this practice have also been reported in India, Malaysia, Oman, Saudi Arabia, and the United Arab Emirates.

in countries in Europe and North America, which have been destinations for migrants from countries where the practice still occurs.

FGM/C is often carried out by "traditional practitioners," such as birth attendants, who may have little or no medical training or knowledge of the potentially dangerous medical and psychological impacts of FGM/C on the girls they cut. It often takes place in unsterile, domestic environments, using unsafe tools. Females that have experienced FGM/C may have a high risk of prolonged bleeding, infection (including HIV), childbirth complications, infertility, and death.

The elimination of FGM/C has become a global concern. In 2012, the UN General Assembly adopted a milestone resolution calling on the international community to intensify efforts to end the practice. As a result, there has been some decline in the prevalence of FGM/C over the last three decades. For example, the percentage of girls aged fifteen to nineteen who have undergone FGM/C in Burkina Faso declined from 89 percent in 1980 to 58 percent today; in Kenya it declined from 41 percent in 1984 to 11 percent today. But the pace of progress has been uneven; in Egypt the percentage of girls aged fifteen to nineteen who have undergone FGM/C declined only somewhat from 97 percent in 1985 to 70 percent today. UNICEF says that by 2030, more than one in three girls worldwide will be born in the thirty-one countries where FGM/C is most prevalent, and this could put sixty-eight million girls at risk.

According to many feminist scholars, FGM/C reflects deep-rooted inequality between the sexes and constitutes an extreme form of discrimination against women. FGM/C is a social norm that is held in place by an entire community; men and women alike hold a traditional belief in the need to control a girl's sexuality and ensure her virginity until marriage, or to prepare her for marriage.

However, there is a more complex and nuanced conversation among feminist scholars that sits at the intersection of feminism and postcolonial theory. Western (Global North) feminists argue that the practice reinforces men's domination over women and creates inequalities. They condemn FGM/C as a violation of women's rights and claim the cultural practice must be outlawed. However, while African feminists condemn FGM/C as a cultural practice, they protest against the Western feminist framing of FGM/C, which they see as neocolonial. In particular, they decry the type of vocabulary Western feminists use such as barbaric, brutal, and torture to indicate African inferiority. They also reject descriptions of African women as cruel, ignorant, or helpless and say this is an example of Western ethnocentrism. African feminists claim Western feminism fails to consider women's autonomy and diversity in the African context, instead lumping all African women into a homogenous group. African feminists focus on understanding the practice within a wider context of ethnic groupings, kinship, and extended families.

Early Child Marriage

Early **child marriage** is also an issue. Each year, twelve million girls are married before the age of eighteen. According to UNICEF, many factors place a child at risk of marriage, including poverty, the perception that marriage will provide "protection," family honor, social norms, customary or religious laws that condone the practice, an inadequate legislative framework, and the state of a country's civil registration system.

Regionally, levels of child marriage are highest in sub-Saharan Africa, where 35 percent of young women are married before age eighteen, followed by South

Asia, where nearly 30 percent are married before age eighteen. Lower levels of child marriage are found in Latin America and the Caribbean (24 percent), the Middle East and North Africa (17 percent), and Eastern Europe and Central Asia (12 percent).

One in three of the world's child brides live in India. Indian girls are often deprived of the right to move freely and make choices about their work, education, marriage, and social relationships. Many get married and become mothers when they are still children. Girls who live in rural areas or come from households in poverty are at greater risk of child marriage and a higher proportion of child brides are those with little or no education.

Child marriage may result in early pregnancy and social isolation, interrupt education, and limit opportunities for career and vocational advancement. Child marriage may also cause maternal and infant deaths and encourage child labor.

Beyond this, however, early child marriage is a fundamental violation of her human right to choose her own destiny. This harmful practice can be imposed on girls by family members, community members, or society at large, regardless of whether the victim provides, or is able to provide, full, free, and informed consent. Child marriages are often transactional—girls are sold to a spouse to ease the financial burden of continuing to raise them or because the family receives money. Although these practices have been banned, it is difficult to enforce as they occur in areas of crisis or displacement, or tend to be concentrated in the poorest, rural areas. According to the UN Population Foundation, approximately 650 million women alive today were married as children.

Preference for Sons

Finally, the preference for a son over a daughter continues to be the social norm in many societies. When boys are valued more highly than girls, pressure to have a son is intense. The preference for sons over daughters may be so pronounced that couples will go to great lengths to avoid giving birth to a girl or will fail to care for the health and well-being of a daughter they already have. Data, such as differences in mortality rates and excess deaths among female infants and young girls, confirms that this is still occurring.

The preference for sons has also led to imbalances in the number of men and women, distorting the sex-ratio balance of countries' populations. Women, on average, live longer than men. This means that all else being equal, females should account for slightly more than half the total population. In 2020, the global sex ratio is 101 males to 100 females, but in some countries the sex ratio is skewed beyond the expected sex ratio. China's male to female ratio in 2021 was 114 males per 100 females, which translates into nearly thirty-four million more males than females. India, another country with a deeply held preference for sons and male heirs, reported a ratio of 108 males for 100 females, an excess of thirty-seven million males. Vietnam, Pakistan, and Azerbaijan also have skewed sex ratios. The term "missing women" refers to the shortfall in the number of women relative to the expected number of women in a country. It is estimated that there are more than 130 million missing women in the world as a result of selective abortion and excess female deaths.

The sex-ratio imbalances can have far-reaching consequences: They entrench gender inequality within a society, distort labor markets, and exacerbate human trafficking. In countries where there is a strong son preference, girls' mortality rates are much higher than would be expected. Sex-ratio imbalances also have consequences for males. Large numbers of men in India and China may be unable to find partners and have children.

Women's Economic Participation and Opportunity

Since the second half of the twentieth century, women's labor force participation has grown significantly. Despite this progress, women around the world are less present in the labor market than men and men tend to earn more than women. This contributes to an inequality in economic participation and opportunity. The World Economic Forum reports that globally, only 55 percent of adult women are in the labor market versus 78 percent of men. Low female participation is especially acute higher up on the career ladder. Even though there has been an increase in the number of women as managers, in 2019, women only held 28 percent of managerial positions worldwide. In many low-income countries in the Global South, women hold most of the jobs in agriculture (Photo 5.2).

In many countries, women are significantly disadvantaged in accessing credit, land, or financial products, which prevents them from starting or improving a business. There are still seventy-two countries where at least some women from specific social groups do not have the right to open a bank account or obtain credit, and twenty-five countries where not all women have full inheritance rights.

These economic inequalities have consequences on economic growth, social cohesion, and social mobility. For example, in agriculture, gender differences in access to inputs, including land and credit, can lead to gaps in earnings through lower productivity. An inability to own or inherit land or assets, or to open a bank account, can become a critical problem in case of divorce or the husband's death. Even when women earn an income, they may not be involved in decisions about how it is spent. In many countries in sub-Saharan Africa, the Middle East, and South Asia, a large percentage of women are not involved in household decisions about spending their personal earned income. This pattern is stronger among low-income households within low-income countries.

PHOTO 5.2 Woman Laborer
Kandy, Sri Lanka. Women are the major source of labor, picking tea leaves for tea production. In many countries in the Global South, women hold a majority of jobs in agriculture, although these are typically the lowest paid jobs in an economy.

Unpaid Domestic Work

Another factor that contributes to financial disparities and overall economic participation and opportunity gaps between women and men is the disproportionate burden of household and care responsibilities placed on women. In every country in the world, men spend less time on unpaid work (such as domestic and volunteer work) than women; in many countries, women spend significantly more time than men on these activities. Even in countries where this ratio is lowest—for example, Norway or the United States—women spend almost twice as much time as men on unpaid domestic work. In Japan, another high-income country, women spend an average of more than four times as much time than men. According to the International Labour Organization, women's unpaid work amounts to twelve billion hours per day. These vital unpaid contributions are typically ignored in official statistics, GDP estimates, and economic policies in general.

The disparities in unpaid domestic work suggest that ongoing cultural and social transformations take a long time to occur. Some experts suggest that policies offering cost- and time-effective solutions to house care needs such as day care within a company's office building or changing incentives for men to take on more household and care duties (e.g., paternity leave) may be effective in improving women's career opportunities.

The Gender Wage Gap

In addition to a difference in labor force participation, significant wage gaps between men and women persist—particularly for women of color. The **gender wage gap** refers to the difference in earnings between women and men. It is commonly calculated by dividing women's wages by men's wages and is often expressed as either a percent or in dollar terms. This tells us how much a woman is paid for each dollar paid to a man. On average, there is a 40 percent wage gap (the ratio of the wage of a woman to that of a man in a similar position) and a 50 percent income gap (the ratio of the total wage and nonwage income of women to that of men).

The gender wage gap links to legal, social, and economic inequality and captures a concept that is broader than the concept of equal pay for equal work. **Equal pay** is the idea that individuals in the same workplace and in the same job be given equal pay and benefits, including basic pay, bonuses, and allowances. Many countries have equal pay laws.

While the gender wage gap measures inequality, it may not necessarily measure discrimination. This is because discrimination can still exist in the absence of pay gaps. There may be large pay gaps in the absences of discrimination in hiring practices—for example, women may get fair treatment but tend to apply for lower-paid jobs.

Experts have calculated the gender wage gap in a multitude of ways, which can create a misconception that data on the gender wage gap are unreliable. However, the data on the gender wage gap point to a consensus: No matter how you measure it, there is a gap. Women everywhere consistently earn less than men. This is true at every wage level, regardless of whether the woman is in a low- or high-earning position. Women also earn less than men at every education level, suggesting that women cannot "educate" themselves out of the wage gap. The wage gap is even wider for most women of color.

The most common way of discussing the wage gap is in terms of dollars and cents, but this makes the issue more abstract and may unintentionally obscure the real impact on working women and their families. More concretely, in 2018, a woman in the United States working full time over the year earned on average $10,194 less than her male counterpart (Map 5.3 highlights the variations in this wage gap by state). If this wage gap were to remain unchanged, she would earn about $407,760 less than a man over the course of a forty-year career. Perhaps it is more powerful to think of the wage gap in the United States as the "half-million dollar" gap.

Gender parity in pay is proving hard to achieve. Pay differentials between men and women are a persistent form of gender inequality in the workplace and the World Economic Forum's Global Gender Gap Index 2020 found that the progress toward closing the gender gap on this aspect has stalled. No country has yet achieved gender parity in wages.

Part of the challenge in closing the wage gap is that it is difficult to assess the extent of gender inequality in the workplace. It is possible to overestimate or underestimate wage gaps. For example, women tend to hold a higher share of part-time work, and so wage gaps may be a result of working fewer hours. Women are more likely to take time away from careers to care for children, which affects their overall earnings. Men are concentrated in professions where salaries are higher, while a relatively higher share of women are working in occupations that are less well-paid than men. Finally, men are concentrated in senior roles that are higher paying than other positions, and because there are fewer women in senior roles, this can lead to an overestimation of wage gaps.

A positive trend is that wage gap among the highest-income countries has been closing, though at a very slow rate. The bad news is that it appears that the wage gap is worsening throughout the rest of the world, although experts cannot yet explain why. As a result, the overall global progress toward closing the wage gap has stalled. There are many calls to remove barriers that prevent women from attaining the same economic opportunities as men, especially in the lowest-income countries.

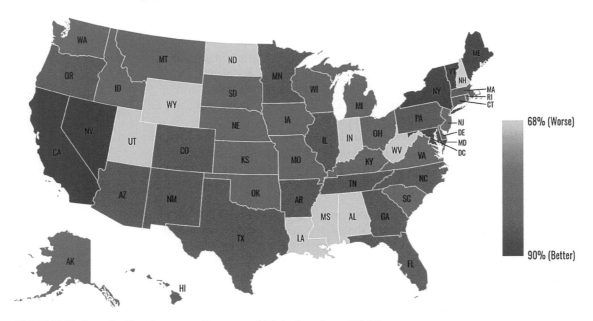

MAP 5.3 Female Earnings as a Percent of Male Earnings (2018)
This map highlights the gender wage gap, and shows there are variations by state, with the lowest wage gap in states such as New York, Vermont, Maryland, Nevada, and California, and the highest wage gaps in states such as North Dakota, Alabama, Wyoming, and Utah.

Women and Political Representation

Women continue to be underrepresented at all levels of political leadership. Few women head their national governments or occupy positions in state or national legislatures. Currently, women represent only 8 percent of all national leaders, and only 2 percent of all presidential posts. In the United States only 25 percent of mayors were female (407 out of 1,621 mayors of US cities with populations more than 30,000) as of 2021. Yet, women's political representation—at all levels—local and national—has a positive impact on public spending patterns and service provision.

The 2020 World Economic Forum Report found that the political empowerment gap is and has been the largest area of gender disparity. To date, only 25 percent of the 35,127 global seats in parliament are occupied by women; in some countries, women are not represented at all. Over the past fifty years, eighty-five countries—more than half the countries in the world—have never had a female head of state, including the United States, Italy, Japan, Mexico, the Netherlands, Russian Federation, South Africa, Spain, and Sweden.

However, there has been some progress. Women's representation in the political arena is higher than ever before (just not in proportion to their population). Many countries have improved in this area with a significant increase in the number of women in parliamentary positions compared to several years ago. Notably, in some countries such as Latvia, Spain, and Thailand the number of women in parliament has increased substantially. In forty-six countries, women now hold more than 30 percent of seats in national parliament in at least one chamber. Rwanda is an interesting case study where more than 60 percent of seats are held by women, far more than any other country.

When examining higher-level institutional roles, the presence of women grows even thinner. Only 21 percent of the 3,343 ministers worldwide are women and there are thirty-two countries where women represent less than 10 percent of ministers in office today. In Azerbaijan, Belize, Brunei Darussalam, Iraq, Lithuania, Saudi Arabia, Vanuatu, Papua New Guinea, and Thailand there are no women ministers at all.

Iceland—which has closed approximately 70 percent of its political empowerment gap—has the most widespread presence of women across parliament, ministries, and heads of states. Iceland's score is 10 percentage points higher than the second-ranked Norway and is almost four times higher than the global average.

SOLVING GENDER INEQUALITY

Table 5.2 outlines the specific targets for SDG 5, which aims to eliminate gender inequality. Because there are likely to be multiple types of gender inequality occurring simultaneously within a country, there is no one set of policy prescriptions for a country to follow. Different countries will need to target or prioritize elements of health, or economic participation, and so forth, depending on the local circumstances.

Solutions can be implemented at various scales. At the global scale, there have been numerous treaties and other agreements that unequivocally hold that discrimination against women is a violation of human rights and that governments, communities, and individuals have a duty to end them. The Beijing Declaration and Platform for Action was adopted at the Fourth World

TABLE 5.2 SDG 5 Targets and Indicators to Achieve by 2030

Target	Indicators
5.1 End all forms of discrimination against all women and girls everywhere	**5.1.1** Whether or not legal frameworks are in place to promote, enforce, and monitor equality and nondiscrimination on the basis of sex
5.2 Eliminate all forms of violence against all women and girls in the public and private spheres, including trafficking and sexual and other types of exploitation	**5.2.1** Proportion of ever-partnered women and girls aged fifteen years and older subjected to physical, sexual or psychological violence by a current or former intimate partner in the previous twelve months, by form of violence and by age **5.2.2** Proportion of women and girls aged fifteen years and older subjected to sexual violence by persons other than an intimate partner in the previous twelve months, by age and place of occurrence
5.3 Eliminate all harmful practices, such as child, early and forced marriage, and female genital mutilation	**5.3.1** Proportion of women aged 20–24 years who were married or in a union before age fifteen and before age eighteen **5.3.2** Proportion of girls and women aged 15–49 years who have undergone female genital mutilation/cutting, by age
5.4 Recognize and value unpaid care and domestic work through the provision of public services, infrastructure, and social protection policies and the promotion of shared responsibility within the household and the family as nationally appropriate	**5.4.1** Proportion of time spent on unpaid domestic and care work, by sex, age, and location
5.5 Ensure women's full and effective participation and equal opportunities for leadership at all levels of decision making in political, economic, and public life	**5.5.1** Proportion of seats held by women in national parliaments and local governments **5.5.2** Proportion of women in managerial positions
5.6 Ensure universal access to sexual and reproductive health and reproductive rights as agreed in accordance with the Program of Action of the International Conference on Population and Development and the Beijing Platform for Action and the outcome documents of their review conferences	**5.6.1** Proportion of women aged 15–49 years who make their own informed decisions regarding sexual relations, contraceptive use, and reproductive health care **5.6.2** Number of countries with laws and regulations that guarantee women aged 15–49 years access to sexual and reproductive health care, information, and education
5.A Undertake reforms to give women equal rights to economic resources, as well as access to ownership and control over land and other forms of property, financial services, inheritance, and natural resources, in accordance with national laws	**5.A.1** (a) Proportion of total agricultural population with ownership or secure rights over agricultural land, by sex; and (b) share of women among owners or rights-bearers of agricultural land, by type of tenure **5.A.2** Proportion of countries where the legal framework (including customary law) guarantees women's equal rights to land ownership and/or control

5.B	5.B.1
Enhance the use of enabling technology, in particular information and communications technology, to promote the empowerment of women	Proportion of individuals who own a mobile telephone, by sex
5.C	5.C.1
Adopt and strengthen sound policies and enforceable legislation for the promotion of gender equality and the empowerment of all women and girls at all levels	Proportion of countries with systems to track and make public allocations for gender equality and women's empowerment

Source: United Nations Sustainable Development Goals. https://sustainabledevelopment.un.org/sdg5

Conference on Women in 1995 as a comprehensive and visionary agenda for achieving gender equality, the empowerment of women, and the realization of human rights for women and girls. It calls for the removal of systematic and structural barriers that prevent women and girls from enjoying their human rights across social, economic, political, and environmental domains, as well as policy actions to achieve that vision. There are other treaties and agreements that member states of the United Nations have agreed to implement, yet gender inequality is persistent. *SDGs and the Law* 5.1 details the Convention of the Elimination of All Forms of Discrimination against Women (CEDAW).

 ## SDGs AND THE LAW 5.1

The Convention on the Elimination of All Forms of Discrimination against Women

Equality of rights for women is a basic principle of the United Nations, expressed in Article 1 of the United Nations Charter. Subsequent international legal instruments have strengthened and emphasized the human rights of women. For example, the Universal Declaration of Human Rights proclaims equality for women. For many years it was assumed that women's rights were best protected and promoted by the general human rights treaties. However, the 1960s saw a new awareness of the patterns of discrimination against women and some criticized existing treaties for failing to deal with discrimination against women in a comprehensive way. In addition, there was concern that the general human rights regime was not, in fact, working as well as it could to protect and promote the rights of women.

In 1979, the UN General Assembly adopted the *Convention on the Elimination of All Forms of Discrimination against Women (CEDAW)*.

The convention defines discrimination against women as:

> any distinction, exclusion or restriction made on the basis of sex which has the effect or purpose of impairing or nullifying the recognition, enjoyment or exercise by women, irrespective of their marital status, on a basis of equality of men and women, of human rights and fundamental freedoms in the political, economic, social, cultural, civil or any other field.

By accepting the convention, UN member states commit to undertake a series of measures to end discrimination against women in all forms, including:

- To incorporate the principle of equality of men and women in their legal system, abolish all discriminatory laws, and adopt appropriate ones prohibiting discrimination against women;

- To establish tribunals and other public institutions to ensure the effective protection of women against discrimination; and
- To ensure elimination of all acts of discrimination against women by persons, organizations, or enterprises.

Countries that have ratified or acceded to the convention are legally bound to put its provisions into practice. They are also committed to submit national reports, at least every four years, on measures they have taken to comply with their treaty obligations. For example, shortly after ratifying CEDAW:

- Honduras created policies to make agricultural training and loans available to women farmers;
- Austria amended policies for maternity protection and paternity leave;
- Cambodia created a women's ministry;
- Uganda created and funded programs to reduce domestic violence; and
- Israel allocated funding to mammograms.

Some countries, such as Spain, adopted recommendations made by the CEDAW into their own domestic laws.

Legal scholar Frances Raday writes that CEDAW is radical in that it embraces all aspects of women's lives—political, public, and diplomatic; economic, employment, and rural; educational; health; marriage and family; and protection against violence, including domestic violence. In countries that have ratified the treaty, she notes that CEDAW has proved invaluable in opposing the effects of discrimination, which include violence, poverty, and lack of legal protections, along with the denial of inheritance, property rights, and access to credit. CEDAW has also brought to light the gendered impacts of environmental disasters such as climate change.

Although there is considerable evidence that CEDAW has contributed to increasing women's right to equality in many countries, this impact varies geographically. Not all countries—even those who adopted the convention—have implemented all the actions they agreed to, and there is little that can be done to enforce this. It is clear that an enormous amount remains to be done before women enjoy full equality with men in all countries.

Sources: Raday, Frances. "Gender and Democratic Citizenship: The Impact of CEDAW." *International Journal of Constitutional Law* 10 (2): 512–30 (2012, March). https://doi.org/10.1093/icon/mor068
United Nations Women. "Convention of the Elimination of All Forms of Discrimination against Women" (1979). https://www.un.org/womenwatch/daw/cedaw/cedaw.htm

Create National Action Plans

At the national scale, national action plans can bring together communities, local and religious leaders, and service providers, ensuring broad support and buy-in. Governments can also create policies that provide talent development, integration, and deployment opportunities for all genders, diversify the leadership pool, and provide support to families and caregivers, in both youthful and aging societies alike.

A good example is Rwanda. Rwanda experienced the 1994 genocide that killed nearly a tenth of its population. However, Rwanda has also become known for progressive strides in gender empowerment and political representation. In 2003, Rwanda adopted a new constitution that committed to "a state governed by the rule of law, a pluralistic democratic government, equality of all Rwandans and between women and men reflected by ensuring that *women are granted at least 30 percent* of posts in decision making organs." In that year's election, women won 48.8 percent of seats in its lower house of parliament, placing Rwanda first among all nations in terms of women's political representation (in contrast, women in the United States comprise only 23 percent of the House and 25 percent of Senate). These dramatic gains for Rwandan women are the result of specific mechanisms used to increase women's political participation, among them a constitutional guarantee, quota

MAKING PROGRESS 5.1

Improving Women's Access to Health Care in Rwanda

Target 5.6: Ensure universal access to sexual and reproductive health and reproductive rights as agreed in accordance with the Program of Action of the International Conference on Population and Development and the Beijing Platform for Action and the outcome documents of their review conferences.

In a region where weak health systems and high birth rates are common, Rwanda has made impressive progress in rapidly increasing access to family planning and reproductive health services, with measurable benefits for women, families, and the broader society. In 1970, modern contraception only met less than 1 percent of demand for family planning in Rwanda. By 2020, it reached 71 percent for all married women, well above the average for sub-Saharan Africa of 62 percent.

Over the past four decades, Rwanda's total fertility rate more than halved from 8.3 to 3.8 live births per woman, while the maternal mortality ratio declined from 1,300 deaths per 100,000 live births in 1990 to 248 in 2017. This progress is the result of government action that prioritized strengthening the health system. Rwanda established nursing and midwifery schools to increase the number of trained nurses, midwives, and doctors. Today, community health-care workers also play an important role in service delivery, distributing condoms, and contraceptives. Community-based health insurance was introduced in 1999 and, by 2018, coverage rates had reached 83 percent of Rwandans.

Effective mobilization of both domestic and external finance has been key. Rwanda has insisted that all donor support and official development assistance is channeled to existing government priorities. Health-care expenditure has increased from $7.91 per capita in 2002 to $58 per capita in 2018, exceeding the World Health Organization's (WHO) recommended minimum of $44 per capita.

A number of challenges remain. Discriminatory attitudes and beliefs remain entrenched in some contexts. Many women report that they cannot access contraception without their husband's approval. Unmarried adolescent girls' access to family planning is often limited for many reasons, including disapproving attitudes among some community health workers. Youth corners in health centers or stand-alone youth centers help provide services and privacy for young women, but these practices are not yet widespread.

Sources: Family Planning. "Rwanda: Commitment Maker since 2012" (2020). https://www.familyplanning2020.org/rwanda
Population Reference Bureau. "Rwanda's Success in Improving Maternal Health" (2015). https://www.prb.org/resources/rwandas-success-in-improving-maternal-health/
Rwanda Ministry of Health et al. *Success Factors for Women's and Children's Health: Rwanda.* Geneva, Switzerland: Partnership for Maternal, Newborn, & Child Health and World Health Organization (2014).

system, and innovative electoral structures. They have also made tremendous progress in reducing other indicators of gender inequality such as maternal mortality (see *Making Progress* 5.1). These mechanisms could serve as a model for other countries to adopt.

Policies on women's economic empowerment should focus not only on increasing their participation to boost economic growth but also on expanding public investment to redistribute the burden of unpaid care and domestic work. The *Solutions* 5.1 box details how Tunisia has made progress in improving gender equity with the introduction of new laws that were coordinated among governments, community organizations, and religious leaders.

SOLUTIONS 5.1

Making Laws to Empower Women in Tunisia

Target 5.5: Ensure women's full and effective participation and equal opportunities for leadership at all levels of decision making in political, economic, and public life.

Target 5.5 is to "ensure women's full and effective participation and equal opportunities for leadership at all levels of decision making in political, economic and public life." Tunisia has been a leader in the Arab world on gender-equitable laws since 1956, when its Code of Personal Status established guidelines for marriage based on mutual consent and equality for women in divorce proceedings.

In 2014, the Tunisian government enacted a new constitution that granted women far-reaching new rights. Women's civil society organizations played a critical role in this achievement. They worked across party lines and historical divisions between Islamist and secular women's rights groups to establish the National Dialogue for Women, which developed an inclusive platform for their demands regarding the new constitution. They used social media to generate awareness of, and opposition to, a draft clause that would have positioned women as "complementary" to men. As a result, the draft was amended to provide that "all citizens, male and female, have equal rights and duties, and are equal before the law without any discrimination" (Article 21).

The new constitution provided a firm foundation for significant legislative changes in 2017, including passage of the Law on Eliminating Violence against Women, repeal of the penal code provision that had allowed a rapist to escape punishment if he married his victim, and changes to laws that prevented Muslim women from marrying non-Muslims.

In 2018, Tunisia took steps to become the first country in the region to legislate on equal inheritance rights. Islamic feminists in the region argue that inheritance laws require reform not only on the basis of equality and justice but also to keep pace with changes in the structure and dynamics of family life. Women's lesser access to inheritance has historically been justified because of men's roles as providers for women and children. Now that a sizeable proportion of households in some countries in the region are maintained by women alone, Islamic feminists argue that the case for change is irrefutable and urgent.

Source: UN Women. *Progress of the World's Women 2019–2020: Families in a Changing World.* New York: United Nations (2019). https://www.unwomen.org/-/media/headquarters/attachments/sections/library/publications/2019/progress-of-the-worlds-women-2019-2020-executive-summary-en.pdf?la=en&vs=3513

Engage Locally

In many instances, national policies that engage or involve community organizations such as health-care centers and community advocates such as nonprofit organizations, schools, or community educators are best poised for success. The practice of child marriage in India has declined over the last thirty years and progress has been accelerating in the last decade as there have been coordinated efforts to engage the local community. In 1970, more than 70 percent of girls were married before the age of eighteen. Today, that figure is only 27 percent.

One example of a national social program in India that takes action at the local level is an initiative called *Beti Bachao, Beti Padhao* ("Save the Daughter, Educate the Daughter") that promotes keeping girls in school. The initiative has reached 1.9 million parents. A *precheta* (a community educator and/or women's advocate) works with *sathins* (grassroots-level women's advocates), *anganwadis* (village-level health centers), auxiliary nurse midwives, health activists, and teachers to lead community meetings and teach about domestic violence, child marriage, physical and sexual exploitation, and other issues affecting women. *Sathins* and *prechetas* are part of the local government's funded development program for women.

The private sector has also been advancing gender equality and human rights principles through voluntary codes and agreements, with a focus on providing women with decent working conditions, meeting environmental and labor standards, and paying a fair share of taxes. There is still more businesses can do to treat people with dignity and respect, offer equal opportunities to all members of the society, and leverage gender diversity by investing in all their talent.

Change Social Norms

Despite decades of treaties and national action plans, gender discrimination remains entrenched in many societies. There needs to be a key strategy at all levels to implement programs to change social norms that perpetuate violence and eliminate harmful practices. These programs should address broader issues such as the subordinate position of women and girls, their human rights, and how to elevate their status and access to opportunities. More specifically, programs should address systemic societal beliefs and attitudes that perpetuate violence against children, in any setting, including the home, school, community, or online. Meaningful change requires shifting deeply ingrained social and cultural norms and behaviors toward equality. We must change the idea that some forms of violence are not only normal but even justifiable and thus tolerated.

The UN Commission on the Status of Women has identified eight priorities for achieving progress in gender equality, female empowerment, and the realization of human rights of women and girls:

1. Remove all discriminatory laws and prioritizing gender-responsive implementation and institutional frameworks.

2. Break silos and building integrated approaches to implementation based on human rights standards and principles.

3. Reach the most marginalized groups of women and girls and ensuring that no one is left behind.

4. Provide adequate funding to meet gender equality commitments.

5. Accelerate the growth in women's participation in all aspects of decision making and creating enabling environments for women's rights organizations.

6. Transform social norms to create cultures of nonviolence, respect, and equality.

7. Harness technology for gender equality.

8. Close data and evidence gaps to effectively monitor progress.

Some of these strategies are obvious, such as removing discriminatory laws. Some are more challenging, such as transforming social norms.

Consider the strategy to provide and increase investment in policies and programs that align with targets in SDG 5 and many of the other SDG goals where targets focus on women. In many countries, essential services on which millions of women and girls depend—water and sanitation, early childhood education and care, and services for survivors of gender-based violence such as shelters, legal services, and health services—are chronically underfunded, of poor quality, or simply unavailable. Yet investment in such services generally pays off. One report noted that in South Africa, a gross annual investment of 3.2 percent of GDP in early childhood education would not only result in universal coverage for all children ages 0 to 5 but also create 2.3 million new jobs and increase female employment by 10 percentage points. These new jobs could generate new tax and social security revenue of $3.8 billion.

Another priority is improving access to and collection of data to better monitor progress and accountability. Despite increasing attention to gender statistics in recent decades, there are problems with data. Eighty percent of the indicators for gender equality across the SDGs are lacking data. For example, only 41 percent of countries regularly produce data on violence against women; only 15 percent of countries have legislation that mandates specialized gender-based surveys; and just 13 percent of countries have a dedicated gender statistics budget. Part of the problem is that there is an absence of internationally agreed-upon standards for data collection. Compounding the problem is the uneven availability of gender statistics across countries and over time.

Finally, an emerging priority is the gender digital divide, which has received the least attention globally. Women face challenges in accessing information and communication technologies (ICT), which affects their educational and employment opportunities. In low-income countries, women's internet use is estimated to be 30 percent lower than men.

A Critical Perspective on Solving Gender Inequality: Feminism

Feminism is a social movement whose basic goal is equality between men and women. There have been many waves of modern feminism, starting in the nineteenth century. Many of the early efforts included the right to vote and establishing equal legal rights for women. More recent waves of feminism works see women's inequality as a social creation that is reproduced by a patriarchal social and political system.

A main theoretical contribution in feminism since the 1990s is to make visible the structure, practices, and inequalities of the gendered social order. As feminist scholar Judith Lorber notes, this takes gender beyond individual attributes and identifies and shows that gender, like social class and racial categories, is imposed on, rather than developed from, individuals. A gendered social order is resistant to individual challenge.

Feminists work to end gender inequality in a variety of ways and there is no unified approach. Some focus on reforms to change language, knowledge, and history to reflect women's previously invisible experiences and contributions. They try to make visible discriminatory practices. Other feminists have confronted the gendered social order by resisting the idea that the gender order can be made equal because they say men's dominance is overwhelming. They have focused on making visible the ways that organizations, institutions, and even daily practices allow men to control women's lives. They argue a new perspective is needed and it should be women's voices and perspectives that reshape the gendered social order. Still others are not content with reforming or resisting patriarchy and misogyny and rebel against deep-seated assumptions about what is male/female, noting that there are multiple gender identities, such as transgender or nonbinary. They argue the gendered social order pits men against men as well as men against women, so a binary approach to gender fails to account for the ways in which everyone is affected by the gendered social order. This radical framework forces us to look at how we know or understand "gender" and how in the process we have contributed to and maintained an unequal social order. For many radical feminists, the UN targets for SDG 5 are

PHOTO 5.3 Women's March in DC in 2017.
Hundreds of thousands of women gathered on the National Mall in Washington, DC, and in many other cities throughout the United States and abroad to support gender equality, civil rights, and other issues that were expected to face challenges under the then newly inaugurated President Trump.

unlikely to genuinely change the gendered social order; they fail to address the root of gender inequality in a transformative way.

There are competing viewpoints about the causes and solutions to gender inequality. Regardless of which viewpoint you find most compelling, the fact is that the major players in international development—governments, the UN system, and so forth—are focusing on strategies and solutions that will make progress toward SDG 5 targets.

SUMMARY AND PROGRESS

The year 2020 marked the 25th anniversary of the Beijing Declaration and Platform for Action. A report by the UN Commission on the Status of Women noted that many countries have introduced new legislation and policies and programs to advance gender equality since 2015.

There are indications that women are increasingly able to exercise agency and voice within their families and in the workplace. These include an increase in the age of marriage to sixteen or eighteen years of age; greater social and legal recognition of a diversity of partnership forms; declines in birth rates as women are better able to choose whether and when to have children, and how many; and women's increased economic autonomy. These areas of progress are both causes and consequences of large-scale demographic changes, dramatic shifts in women's and girls' access to education and employment, legal reform, and, hopefully, indicative of changing social norms.

Around the world, young women are leading movements for change on issues ranging from democracy, education, migrants' rights, and the rights of lesbian, gay, bisexual, transgender, and intersex people, on the understanding that only when the human rights of women and girls are fulfilled can other forms of inequality, exclusion, and injustice be ended (Photo 5.3). Young women are also demanding an end to violence against women and girls and that their sexual and reproductive health rights be upheld. Men, too, are increasingly using their positions of power to challenge gender inequality and advocate change.

Despite these improvements, in some areas, progress has stalled or, even worse, been reversed. Women continue to be significantly underrepresented in leadership roles across all sectors. The gender gap in labor force participation has remained unchanged over the last twenty years. Few countries have strengthened women's participation in environmental sustainability, or crafted gender-responsive disaster risk reduction and resilience programs. Persistent areas of concern include violence against and disproportionate burdens of unpaid care and domestic work on women and girls. Figure 5.1 not only shows the gender gap that has been closed to date but also shows how many more years will be needed to achieve gender equality.

The 2020 coronavirus outbreak has exacerbated existing areas of inequality for many women and girls, from health and the economy to security and social protection. Women play a disproportionate role in responding to the virus, including as frontline health-care workers and caring for those at home. Women's unpaid care work has increased significantly as a result of school closures and the increased needs of older family members. Women are also harder hit by the economic impacts of COVID-19, as they disproportionately work in insecure labor markets.

The pandemic has also led to an increase in violence against women and girls. With lockdown measures in place, reports of violence against women, especially domestic violence, have increased as security, health, and money worries create tensions and strains exacerbated by more confined living conditions. Many women have been unable to access services that are suffering from cuts and restrictions. In France, reports of domestic violence increased 30 percent in 2020; in Cyprus and Singapore, helplines registered a 30 percent increase in calls. The pandemic is likely to have a negative impact on some of the targets in SDG 5.

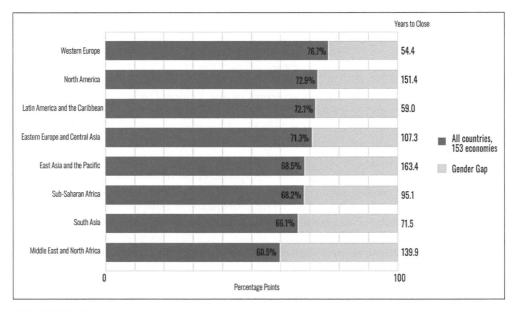

FIGURE 5.1 Gender Gap Closed to Date by Region (2020)
This graph depicts how many years it will take to close the gender gap if there are no interventions. For example, it may take 151 years to close the gender gap in the United States without any change in policy, practice, or values.

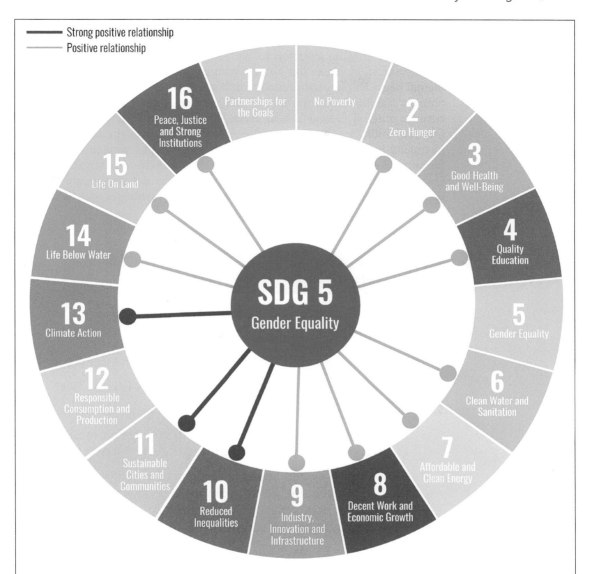

FIGURE 5.2 Interconnections in SDG 5 and the Other Goals
Gender inequality connects to all the SDGs in some way; this figure highlights several of the stronger connections.
Which of the connections do you believe has the strongest synergy to realize improvements in targets?

Finally, women's rights advocates fought hard to achieve both a stand-alone goal on gender equality as well as integrating gender equality across other goals and targets, drawing attention to the gender dimensions of poverty, hunger, health, education, water and sanitation, employment, climate change, environmental degradation, urbanization, conflict and peace, and financing for development. Taking action on SDG 5 will provide numerous benefits and help advance other SDG targets. Within the SDGs, there are thirty-nine targets and fifty-four indicators that are gender related, pointing to the fact that gender inequality is found in many dimensions. For example, women and girls are far more likely to face disadvantages in education: Achieving targets in SDG 4 Quality Education will also reinforce gains in gender equality. Figure 5.2 highlights those goals that have the strongest reinforcing synergies with SDG 5.

QUESTIONS FOR DISCUSSION AND ACTIVITIES

1. Describe the geography of gender inequality.

2. Explain the causes of gender inequality.

3. Why has gender discrimination persisted, even in high-income countries?

4. Why is the United States not ranked more highly in most GIIs?

5. Why does gender inequality often result in violence against women?

6. Select one of the SDGs that has a strong connection to SDG 5 and find a real-world example of progress that positively impacts both SDGs.

7. If you live in the United States, investigate the gender wage gap for the state in which you live. What might explain why the gap is more or less in your state than others? Go to https://ww3.aauw.org/resource/gender-pay-gap-by-state-and-congressional-district/

8. Read *Still I Rise* by Maya Angelou and discuss the types of hardships she overcame.

9. Which one or two strategies for reducing gender inequality do you feel should be prioritized?

10. Describe the impacts of gender inequality. Have you experienced or know anyone who has experienced gender inequality? What was your response?

TERMS

child marriage
equal pay
gender equality
gender equity
Gender Inequality Index (GII)

gender wage gap
Global Gender Gap Index (GGI)
missing women
Social Institutions and Gender Index (SIGI)

RESOURCES USED AND SUGGESTED READINGS

Adichie, C. *We Should All Be Feminists*. New York: Random House (2014).

Barant, N., MacFeely, S., and Peltola, A. "Comparing Global Gender Inequality Indices: How Well Do They Measure the Economic Dimension?" *Journal of Sustainability Research* 1: e190016 (2019). https://doi.org/10.20900/jsr20190016

Beard, M. *Women and Power: A Manifesto*. New York: W. W. Norton & Company (2017).

Castañeda Camey, I., Sabater, L., Owren, C., and Boyer, A. E. *Gender-Based Violence and Environment Linkages: The Violence of Inequality*. Gland, Switzerland: IUCN (2020). https://doi.org/10.2305/IUCN.CH.2020.03.en

Chant, S. (ed.). *The International Handbook of Gender and Poverty: Concepts, Research, Policy*. Northhampton, MA: Edward Elgar Publishing (2010).

De Beauvoir, S. *The Second Sex*. New York: Vintage (2011).

Filho, W. L., Azul, A. M., Brandli, L., Salvia, A. M., and Wall, T. (eds.). *Gender Equality*. Cham, Switzerland: Springer (2021).

Friedan, B. *The Feminine Mystique*. New York: W. W. Norton & Company (2013) (originally published 1963).

Lorber, J. *Gender Inequality: Feminist Theories and Politics*. 4th ed. New York: Oxford University Press (2010).

Shabliy, E. V., Kurochkin, D., and Ayee, G. (eds.). *Global Perspectives on Women's Leadership and Gender (In)Equality*. London: Palgrave Macmillan (2020).

Smith, A. (ed.). *Gender Equality in Changing Times: Multidisciplinary Reflections on Struggles and Progress*. London: Palgrave Macmillan (2020).

United Nations Economic and Social Council, Commission on the Status of Women. *Review and Appraisal of the Implementation for the Beijing Declaration and Platform for Action*. New York: CSW (2020). https://undocs.org/E/CN.6/2020/3

United Nations International Children's Emergency Fund. "Ending Child Marriage: A Profile of Progress in India" (2019). https://data.unicef.org/resources/ending-child-marriage-a-profile-of-progress-in-india/

United Nations International Children's Emergency Fund. "UNICEF'S Data Work on FGM/C" (2020). https://data.unicef.org/topic/child-protection/female-genital-mutilation/

United Nations Population Fund. "Frequently Asked Questions about Gender Equality" (2005). http://www.unfpa.org/resources/frequently-asked-questions-about-gender-equality

United Nations Women. "Revisiting Rwanda Five Years after Record Breaking Parliamentary Elections" (2018, August 13). https://www.unwomen.org/en/news/stories/2018/8/feature-rwanda-women-in-parliament

World Economic Forum. *Global Gender Gap Report 2021*. Geneva, Switzerland: World Economic Forum (2021). http://www3.weforum.org/docs/WEF_GGGR_2021.pdf

World Health Organization. "Female Genital Mutilation" (2020). https://www.who.int/news-room/fact-sheets/detail/female-genital-mutilation

World Health Organization. *Violence against Women Prevalence Estimates, 2018: Global, Regional and National Prevalence Estimates for Intimate Partner Violence against Women and Global and Regional Prevalence Estimates for Non-partner Sexual Violence against Women*. Geneva, Switzerland: World Health Organization (2021). License: CC BY-NC-SA 3.0 IGO.

Water and Sanitation

LEARNING OBJECTIVES

After reading this chapter, you should be able to:

- Identify and describe the current trends and challenges associated with water supply and water sanitation

- Describe the process of water treatment

- Identify and explain the consequences of unsafe drinking water

- Identify and explain the consequences of water pollution

- Explain the challenges in providing safe water and sanitation to all

- Identify and explain the various targets and goals associated with SDG 6

- Explain how the three Es are present in SDG 6

- Explain how SDG 6 strategies connect to other SDGs

The water crisis in Flint, Michigan began in 2015 after residents began to complain about the water. It looked brown. It smelled and tasted bad. State official assured residents the water was safe to drink. But it wasn't. The Detroit Free Press reported children were developing rashes and suffering from mysterious illnesses. A few months later, a research team from Virginia Tech University tested the water of hundreds of homes and issued a report indicating that 40 percent of Flint homes had elevated lead levels. Thousands of children were affected.

Young children under the age of five are the most vulnerable to the effects of lead because their body, brain, and metabolism are still developing. The health impacts of lead in children have a wide range of effects on development and behavior, including increased behavioral effects, delayed puberty, and decreases in hearing, cognitive performance, and growth or height. Lead also affects adults and causes cardiovascular effects, nerve disorders, decreased kidney function, and fertility problems.

Flint, located seventy miles north of Detroit, is a city of 98,000, where 41 percent of residents live below the poverty line and nearly 57 percent of residents are Black. Civil rights advocates wondered: If Flint were wealthy and White, would Michigan's state government have responded more quickly and aggressively to complaints about its lead-polluted water? Would it have paid for the needed equipment upgrades?

It may be difficult to prove whether or not race and class were factors in the state's slow response, but the result was the same: Thousands of Flint's poorest residents, Black and White, have been exposed to lead in their drinking water. And the long-term health effects of that poisoning may not be fully understood for years.*

---

*Michigan Civil Rights Commission. "The Flint Water Crisis: Systemic Racism through the Lens of Flint" (2017). http://www.michigan.gov/documents/mdcr/VFlintCrisisRep-F-Edited3-13-17_554317_7.pdf

WATER AND SANITATION: UNDERSTANDING THE ISSUE

A fresh and dependable supply of water is critical to sustaining life and supporting healthy communities, economies, and environments. Every day, we use water for bathing, cooking, drinking, irrigating crops, and washing clothes.

Currently, only 70 percent of the global population has access to safe drinking water. While substantial progress has been made in increasing access to clean drinking water and sanitation, billions of people—mostly in rural areas—still lack these basic services. An estimated 2.2 billion people lack safely managed drinking water and about 4.2 billion lack safely managed sanitation (Maps 6.1 and 6.2). Three billion people worldwide lack basic handwashing facilities at home. More than 892 million people still practice open defecation, a practice that has significant consequences, as discussed in *Understanding the Issue* 6.1.

There are two main and often related water challenges: (1) supplying sufficient quantities of high-quality fresh water to residents and consumers and (2) assuring that water quality in nearby water bodies, like rivers and lakes, are not overly impacted by pollution. Solutions to both water quality and water supply issues rely on **water infrastructure** such as aqueducts, the underground network of pipes, treatment plants, and other facilities (Figure 6.1). This chapter will discuss both types of water issues.

 ## UNDERSTANDING THE ISSUE 6.1

Why Toilets Are Important

Open defecation is when people defecate in the open—for example, in fields, forests, bushes, lakes, and rivers—rather than using a toilet. UN agencies report that of the 673 million people practicing open defecation, 91 percent live in rural areas.

Globally, the practice is decreasing steadily, however its elimination by 2030 (SDG 6 Target 6.2) requires a substantial acceleration in toilet use particularly in Central and Southern Asia, Eastern and Southeast Asia, and sub-Saharan Africa.

Open defecation is an affront to one's dignity, health, and well-being, especially for girls and women. For example, hundreds of millions of girls and women around the world lack privacy when they are menstruating. Open defecation also poses risks like public health dangers, endangerment of personal safety, and exposure to increased sexual exploitation.

One gram of feces can contain ten million viruses, one million bacteria, and one thousand parasite cysts. Poor sanitation and hygiene practices (e.g., not handwashing with soap and water after defecation and before eating) contribute to more than 800,000 deaths from diarrhea annually, according to the World Health Organization.

Open defecation has been practiced for centuries; it is an ingrained cultural norm in some societies. Stopping it requires a sustained shift in the behavior of whole communities so that a new norm, toilet use by all, is created and accepted. Ending open defecation requires an ongoing investment in the construction, maintenance, and use of latrines, and other basic services.

The World Health Organization estimates that every $1 invested in water and toilets returns an average of $4 in saved medical costs, averted deaths, and increased productivity. Hygiene promotion is also ranked as one of the most cost-effective public health interventions. Conversely, a lack of sanitation holds back economic growth.

Source: United Nations News. "'Transformational Benefits' of Ending Outdoor Defecation: Why Toilets Matter." United Nations (2019, November 18). https://news.un.org/en/story/2019/11/1051561

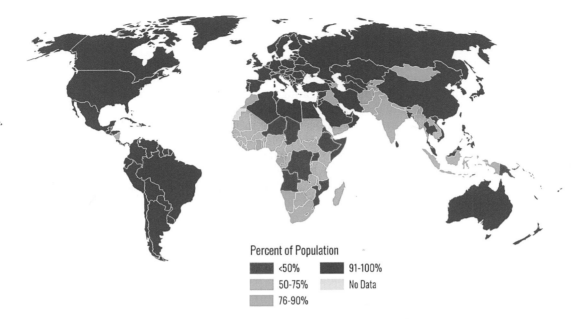

MAP 6.1 Proportion of the Population with Access to Safely Managed Water (2015)
Although there has been substantial progress in increasing access to clean drinking water, an estimated 2.2 billion people lack safely managed drinking water; a majority of these live in Africa.

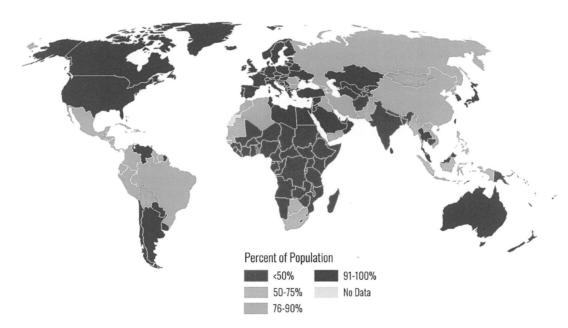

MAP 6.2 Proportion of the Population Using Basic Sanitation Services (2015)
About 4.2 billion people lack safely managed sanitation. Of note are India and Bangladesh and many countries in sub-Saharan Africa, where more than half the population lacks basic sanitation services.

Fresh Water Supply

Some 71 percent of the Earth's surface is covered with water and the oceans hold nearly 96 percent of all Earth's water. Water also exists in the air as water vapor, in rivers and lakes, and in icecaps and glaciers (see Figure 6.2). Water on the planet is part of a large and complex **hydrologic cycle**, as shown in Figure 6.3. Water is moving around constantly, from liquid, to vapor to ice and back again. In this chapter, we are concerned with fresh water, not saline water that is found in the oceans and is the focus of SDG 14 Life Below Water.

Fresh water is essential to human life: drinking, cooking, and irrigation all depend on fresh water. Humans take fresh water from lakes, rivers, streams, and groundwater (aquifers). **Aquifers** can be either unconfined or exposed aquifers (which may contain pollutants) or confined aquifers that are filtered and cleaner. In many countries and communities, fresh water from these sources first goes into a water treatment facility where it is purified to certain set standards. An underground network of pipes delivers drinking water to homes and businesses served by a public water system. In some low-income countries, this infrastructure is not well developed, or may not exist.

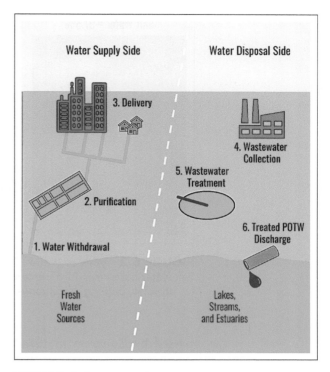

FIGURE 6.1 Water Infrastructure

It involves two systems: water supply and water sanitation. To deliver clean water, it is first withdrawn from a source, purified, and delivered to homes and businesses. On the other end, dirty water is moved through an extensive system of pipes and underground infrastructure that delivers the water to a wastewater treatment plant, where it is filtered and cleaned before being discharged back into lakes, streams, and estuaries.

The Global North is characterized by having extensive and well-developed water infrastructure and water treatment facilities. Water suppliers use a variety of treatment processes to remove contaminants from drinking water. These processes include **coagulation**, **filtration**, and **disinfection**. Some water systems also use ion exchange and absorption. Water utilities select the treatment combination most appropriate to treat the contaminants found in the source water of that particular system. Water utilities must test their water frequently for specified contaminants and report these results to the states. If a water system does not meet the minimum standards, it is the water supplier's responsibility to notify its customers. Many water suppliers now are also required to prepare annual reports for their customers, and the use of the web has allowed the public to stay better informed.

In the Global North, there are extensive public water systems that serve a diverse range of populations from small rural areas to cities greater than several million. For example, each day water utilities in the United States supply nearly thirty-four billion gallons of water.

Many countries in the Global North have established laws that provide for safe water. In the United States, the Safe Drinking Water Act was

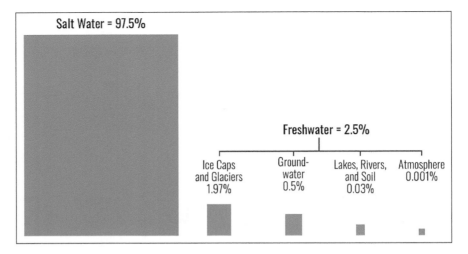

FIGURE 6.2 Types of Water
Salt water comprises 97.5 percent of all water on the planet. Freshwater—used for drinking, bathing, irrigation—accounts for only 2.5 percent.

passed by Congress in 1974 to protect public health by regulating the nation's public drinking water supply. The US Environmental Protection Agency (EPA) has also set standards for some ninety chemicals, microbes, and other physical contaminants in drinking water. The law requires cities and states to protect drinking water and its sources: rivers, lakes, reservoirs, springs, and ground water wells. As a result of this act, drinking water quality in the United States remains among the safest in the world, and the EPA guidelines are used internationally by the World Health Organization to set global standards.

While most populations in the Global North take safe drinking water for granted, there remain several threats to drinking water as crises in both Flint, Michigan, and Jackson, Mississippi, reminded us. Possible contaminants include lead, arsenic, and chromium. Improperly disposed-of chemicals, animal and human wastes, wastes injected underground, and naturally occurring substances also have the potential to contaminate drinking water. Drinking water that is not properly treated or disinfected, or that travels through an improperly maintained distribution system, may also pose a health risk. Drinking-water utilities also face new responsibilities due to concerns over water system security and threats of infrastructure terrorism.

The United States uses thirty-nine billion gallons of water each day to support daily life from cooking and bathing in homes to use in factories and offices across the country (down from forty-two billion gallons in 2017). Water consumption is not just the water we use for domestic consumption but also for the products we use (clothes, food, etc.) (Table 6.1). This is called **virtual water**. The average American uses about 100 to 175 gallons of water each day in the home (compare this to the average Britain who uses fifty-two gallons and the average Rwandan who uses five gallons). Map 6.3 shows that of the countries in the Global North, the United States is one of the largest total water withdrawals. Other countries with high rates of water consumption per capita include China, Chile, and Argentina, and arid countries Iraq

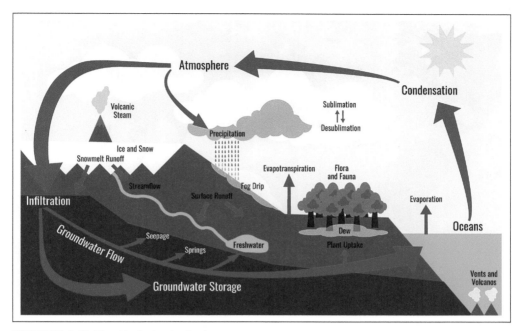

FIGURE 6.3 The Hydrologic Cycle
Water on the planet is part of a large and complex hydrologic cycle. Water is moving around constantly, from liquid, to vapor to ice and back again.

and Turkmenistan. In Iraq, irrigation of agriculture is partly the issue; similarly in Turkmenistan water withdrawn from the Aral Sea for agriculture has also led to chronic water shortages.

In the United States, it costs about $2 to provide 1,000 gallons of clean water, which equates to approximately $300 per household per year. While this is a small cost, compared with fueling our automobiles, the cost of making water safe continues to rise. In many other high-income countries much of the existing drinking water infrastructure was built during the Sanitation Revolution of the late nineteenth century. This infrastructure is now more than 100 years old, and in many cases cannot handle the volume of demand due to population growth, or the pipes may be leaking or deteriorating. There are an estimated 240,000 water main breaks per year in the United States, wasting more than two trillion gallons of treated drinking water. According to the American Water Works Association, an estimated $1 trillion is necessary to maintain and expand service to meet demands over the next twenty-five years and to ensure the continued source development, storage, treatment, and distribution of safe drinking water. In most high-income countries, clean and safe drinking water is a given, but there are still challenges ahead from protecting the supply and the need to upgrade or expand infrastructure.

TABLE 6.1 Virtual Water	
Below are Measures of Virtual Water Content of Some Selected Products.	
Product	**Virtual water content (gallons)**
1 glass of beer	20
1 glass of milk	53
1 cup of coffee	37
1 potato	6.6
1 bag of potato chips	49
1 hamburger	634
1 cotton t-shirt	1,083
1 pair of leather shoes	2,114

Source: World Water Council. www.worldwatercouncil.org

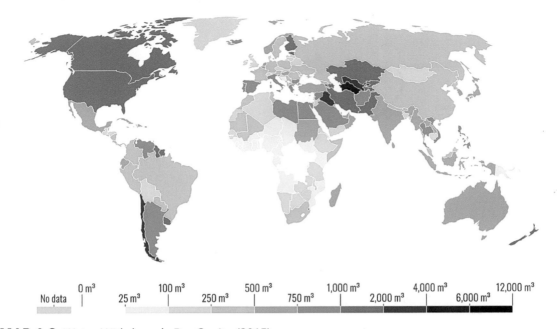

| No data | 0 m³ | 25 m³ | 100 m³ | 250 m³ | 500 m³ | 750 m³ | 1,000 m³ | 2,000 m³ | 4,000 m³ | 6,000 m³ | 12,000 m³ |

MAP 6.3 Water Withdrawals Per Capita (2015)
Of the countries in the Global North, the United States and Canada are among the largest total water withdrawals. Other countries with high rates of water consumption per capita include Chile and Argentina, and arid countries Iraq and Turkmenistan.

Drinking water supply issues are often more pronounced in the Global South where the building of water supply systems has failed to keep pace with continued rapid population growth, particularly in large cities. Many governments were not only unprepared to manage high population growth rates in cities but also did not have the necessary financial, technical, or managerial capacities to do so. These places fell behind in constructing and properly managing water supply and wastewater treatment systems. The right to safe water and adequate sanitation remains an unfulfilled promise.

According to the World Health Organization, unsafe drinking water continues to be responsible for more than 80 percent of diseases and 30 percent of deaths in the lowest-income countries. At any given time, close to half the people in the Global South (almost three billion) are suffering from one or more of the main water-linked disease such as diarrhea, cholera, enteric fevers, guinea worm, and trachoma.

Every day in communities across the Global South, residents depend on a variety of informal, and often illegal, techniques and practices to access water and sanitation. Although most cities have water supply infrastructure, it is often unreliable, inconsistent, and incomplete. Water runs only intermittently, or under irregular or insufficient pressure; it may be contaminated and the system can suddenly break down, causing water to cease flowing for days or weeks at a time. This particularly impacts the city's poor and slum dwellers, many of whom have no legal access to piped water supply and are especially vulnerable to limited water access. As a result, those without water must spend

PHOTO 6.1 A Girl Carries Water Containers
In Ethiopia, this girl walks carrying water containers. Sometimes, girls will
be kept out of school to help with procuring water; it is not uncommon for
women and girls to travel many hours to procure water.

considerable time each day to access water and sanitation facilities. This can
include missing work to procure water, walking miles in search of sanitation
or clean water, and buying water from illegal and informal sources. In addition, some will tap illegally into the water supply.

It is estimated that women and girls are responsible for water collection in eight out of ten households because it falls under their overarching
responsibility of household management (Photo 6.1). Water access in the
Global South is more than just about water: The time spent trying to get
water curtails other opportunities, such as generating income or getting
an education. It can also be dangerous. For example, women and girls
who travel to procure water may be at risk of physical or sexual assault. In
addition, pregnant women who carry heavy vessels of water put substantial
strain on the body.

Water Scarcity

Water scarcity affects every continent. According to United Nations Water,
1.2 billion people, or almost one-fifth of the world's population, live in areas
of physical water scarcity, and another 500 million people are approaching
this level. Map 6.4 shows that an increasing number of regions are chronically short of water. Physical water scarcity occurs when demand outstrips
water supply, leading to issues of access to water. Economic water scarcity is
due to a lack of water infrastructure. In many cases, there is sufficient water
to meet human and environmental needs, but access is limited by lack of
infrastructure. Water scarcity is both a natural and a human-made phenomenon. There may be enough freshwater on the planet for eight billion people,
but it is distributed unevenly and too much of it is wasted, polluted, and
unsustainably managed.

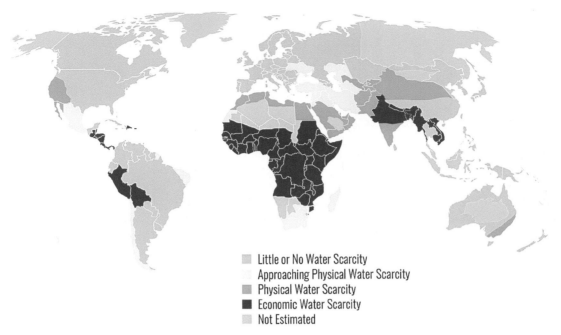

Little or No Water Scarcity
Approaching Physical Water Scarcity
Physical Water Scarcity
Economic Water Scarcity
Not Estimated

MAP 6.4 Global Physical and Economic Water Scarcity
This map shows many regions facing physical and economic water security. Physical water scarcity is prevalent in the western part of the United States, and countries in Northern Africa and the Middle East. Physical water scarcity occurs when demand outstrips water supply, leading to issues of access to water. Many countries in Sub-Saharan Africa, South Asia, and Central America are experiencing economic water scarcity. Economic water scarcity is due to a lack of water infrastructure. In many cases, there is sufficient water to meet human and environmental needs, but access is limited by lack of infrastructure. Water scarcity is both a natural and human-made problem.

Many countries are confronting recurring droughts. Climate change may add even more stress in water scarce areas. Climate change is projected to increase the number of water-stressed regions and exacerbate shortages in already water-stressed regions. Climate change will significantly impact precipitation patterns and water supply. Higher temperatures and more extreme, less predictable weather conditions are projected to affect availability and distribution of rainfall, snowmelt, river flows, and groundwater, and also further deteriorate water quality. Some areas will see more floods, some areas may see an increase in severe droughts, and some areas may see both floods and droughts at different times of the year.

Water availability is becoming less predictable in many places and increased incidences of flooding threaten to destroy water points and sanitation facilities and contaminate water sources. By 2050, the number of people at risk of floods will increase from its current level of 1.2 billion to 1.6 billion. And there may be 3.2 billion people living in potential severely water-scarce areas.

In the Global North, Australia is the driest continent and on its way to becoming drier. Nearly all Australia's major cities have applied water restrictions and efficiency measures while they deal with a projected decrease in available supply. In the United States climate change is likely to exacerbate existing water infrastructure deficiencies. Models predict that much of the Midwest and Eastern United States will experience more frequent and severe storm events. Existing antiquated stormwater infrastructure was not designed to manage these larger events. Coastal areas in particular are vulnerable to flooding due to a combination of rising sea levels and more extreme storm events. Other areas, particularly in the Southwest, can expect increased drought associated with altered precipitation patterns, increased temperatures, and evaporation, which will stress water resources at the same time that continued population growth is expected. The greatest risk of water shortages will concentrate in thirteen states: Arizona, Arkansas, California, Colorado, Florida, Idaho, Kansas, Mississippi, Montana, Nevada, New Mexico, Oklahoma, and Texas.

In the Global South, North Africa and the Middle East are the most water stressed. These regions are hot and dry, so water supply is low to begin with, and growing demands have pushed countries further into extreme stress. According to the World Resources Institute, seventeen countries are experiencing extremely high levels of water stress, and include Algeria, Libya, Egypt, Tunisia, Qatar, and Jordan. Beyond North Africa and the Middle East, countries such as India and Namibia also face extremely high-water stress. Landlocked countries, such as Bolivia, Paraguay, and Mongolia, are at risk of water stress because many are in dry regions where semiarid and arid conditions prevail. In those places, water challenges are further exacerbated by the effects of climate change, such as desertification, drought, and land degradation.

Water scarcity affects people and places at different scales. Cities have been dealing with water scarcity for some time. In 2018, the city of Cape Town, South Africa confronted water scarcity brought on by three consecutive years of poor rainfall. Local officials created a plan for "Day Zero," a shorthand reference for the day when the water level for the major dams supplying the city's water would fall below 13 percent and signal the largest drought-induced municipal water failure in modern history. Day Zero would shift the city into the most stringent of water restrictions: Municipal water supplies would be switched off and residents would have to line up for their daily ration of water.

Since water scarcity became an ongoing issue starting in the early 2000s, the city has enacted various **water conservation** measures to reduce the demand for water. It successfully reduced water use by more than 50 percent during the drought from 2015 to 2018 using a series of water restrictions. Residents were requested not to flush the toilet after urinating, to flush using rainwater or graywater after defecating, and to reduce the length and frequency of showers. Using municipal water to top up pools, irrigate lawns, or hose down surfaces was forbidden. The city created an online map with green dots showing which houses were doing a good job saving water and which were not; and they published the names of excessive water users. City officials even drove through neighborhoods that were using too much water with a bullhorn calling them out. It is estimated that around 50 percent of households adhered to water restrictions.

In June 2018, the region saw average rainfall for the first time in four years. With the rain, dam levels rose and Cape Town narrowly avoided Day Zero. It is unlikely that this will be the only time the city will confront a countdown to Day Zero.

A Critical Perspective on Water Scarcity

Water scarcity is not always a result of natural processes, changing climate, or droughts. Issues of water scarcity can result from inadequate or poorly planned water management policies. The work of urban political ecologists has uncovered how water is connected to social power in the city. Unequal water distribution is often the outcome of discrimination, poor policies, and inequality.

Geographer and political ecologist Erik Swyngedouw's work on the city of Guayaquil, Ecuador shows that the consequences of water practices shape power, rights, and citizenship in the city. He notes there are processes that produce water inequality in the city that are often linked to city-wide structures of water governance, in other words, water politics. His research has shown that the poor pay far more for water than middle class or wealthy residents in Guayaquil. Similarly, Matthew Gandy's work on water in Mumbai reveals that the city has been dominated by middle-class interests to "modernize" by building elevated highways and new information technologies, while the urban poor and their need for clean, safe water has been ignored. Similar problems exist in Lima, Peru, where discriminatory practices against low-income residents have exacerbated water scarcity. Here, improvements in water services have tended to concentrate in higher-income areas. Peru embraced water privatization, but, on average, poor families are spending 48 percent of their income on food and drinking water.

Political ecologists welcome water conservation measures and policies to reduce water demand, but also argue that city leaders and decision makers need to understand how past policies and practices have led to water inequality. They would be skeptical of plans to privatize water because this is likely to punish the poor.

WATER TREATMENT INFRASTRUCTURE

Water is treated for three main outcomes: drinking water (which must meet the highest standards of purity), irrigation or industrial supply (nonpotable), and safe return to water bodies such as lakes, rivers, and streams (also nonpotable). Water treatment removes contaminants and other undesirable components in the water.

Treatment for drinking water is the removal of contaminations from water to produce water that is pure enough for human consumption without any short-term or long-term risk to health. Wastewater (or sewerage) treatment is the removal of contaminations from water that has been used by households and industry, with the purpose of returning it safe enough for release to lakes, rivers, or streams. Both forms of treatment rely on physical treatment (such as filters) as well as chemical (such as the use of chlorination to kill bacteria).

Treatment for drinking water has the highest standards of purity because of its use by humans in cooking, drinking, and other household uses. Wastewater or sewerage treatment is designed to remove primarily biological contamination (bacteria) from the water, such as feces, or other organic solids. Table 6.2 outlines the typical treatment process.

TABLE 6.2 Steps in the Water Treatment Process

Step		Process
1	Catchment	A series of pumps and pipes that connect a water source to the treatment plant
2	Screening	Physically separate biological solids from the water. Screens or filters capture large debris such as plants, tree limbs, trash, and fish, channeling them to sludge ponds where the organic materials are broken down by bacteria. Often sludge that has dried is shipped to landfills.
3	Coagulation and Flocculation	Chemicals called coagulants are added to the water. The positive charge of these chemicals binds to the dirt particles and together they form a large gelatinous particle called floc.
4	Sedimentation and Filtration	Floc is separated from the water and is pumped to a settlement pond. During sedimentation, floc settles to the bottom of the water supply, due to its weight. This settling process is called sedimentation. Once the floc has settled to the bottom of the water supply, the clear water on top will pass through filters of varying compositions (sand, gravel, and charcoal) and pore sizes, to remove dissolved particles, such as dust, parasites, bacteria, viruses, and chemicals.
5	Ozonation	Ozone, a highly reactive gas, is added to the water to kill bacteria, viruses, and protozoans and reduces the concentration of certain minerals like iron, magnesium, and sulfur that helps to correct bad odors and tastes.
6	Disinfection	A disinfectant (e.g., chlorine, chloramine) may be added to kill any remaining parasites, bacteria, and viruses. • If treatment is for drinking water, fluoride may be added to the water for dental health.
7	Distribution and/or discharge	• If the treatment is processing drinking water, that drinking water is then pumped from drinking water treatment facilities into a raised tower or a tank and gravity allows it to flow. • If the treatment is processing wastewater, the treatment plant will discharge the clean water into the nearby lake, river or stream.

Source: By author compiled from several sources.

In the Global North, most cities and communities have extensive drinking water treatment facilities and wastewater treatment facilities, as shown in Figure 6.4. This is expensive infrastructure that requires continual maintenance and investment. It is estimated that the United States spent $340 billion on drinking water improvements and $300 billion on wastewater treatment between 1980 and 2020. However, even this considerable investment has not kept up with maintenance and upgrading needs. The American Society of

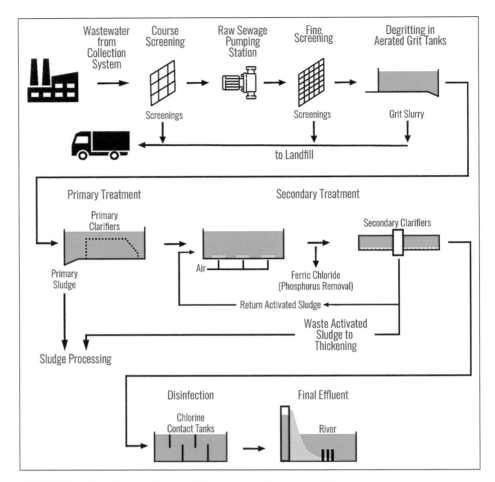

FIGURE 6.4 A Typical Urban Wastewater Treatment Plant
Contaminated water goes through a series of low-technology and high-technology processes before it is discharged as clean. Wastewater treatment plants are an expensive and complex infrastructure that requires continual maintenance and investment. It can cost between $200–500 million for a treatment plant serving a small city.

Civil Engineers rates the US drinking water infrastructure a grade of "C–" and wastewater infrastructure at a "D+." Their 2021 report notes that upgrading existing water systems and meeting the drinking water infrastructure needs of a growing US population will require at least $1 trillion investment.

In the Global South, many low-income countries have lacked the capital for such expensive infrastructure or have not prioritized such investment. Water and sanitation systems require a vast infrastructure that connects households and businesses to treatment facilities. The United Nations estimates 1.6 billion people, or almost one-quarter of the world's population, live in countries that lack the necessary infrastructure to take and treat water from rivers and aquifers. The lack of water treatment also helps to explain why thousands of children die each day due to preventable water- and sanitation-related diarrheal diseases.

There has been far better progress on drinking water treatment than wastewater treatment. Between 1990 and 2015, the proportion of the global population using an improved drinking water source increased from 76 percent to 90 percent. While three in ten people still lack access to safe drinking water, some six in ten people lack access to safely managed sanitation facilities.

WATER-QUALITY TRENDS

Improperly disposed of chemicals and pharmaceuticals, animal and human wastes, wastes injected underground, and naturally occurring substances all have the potential to contaminate water sources and impact water quality. This section will focus on water-quality problems that result from inadequate wastewater systems.

There are two main sources of water pollution, point and nonpoint. **Point sources** are those where there is a clear discharge mechanism such as effluent pipes or outfalls. These stationary devices can be measured for the amount of pollution discharged. The main point sources of pollution in cities are industrial and municipal facilities. Point sources of water pollution are relatively easy to monitor and control. As a result, point source pollution in the United States has declined significantly since the implementation of clean water legislation. However, in some emerging industrial economies in the Global South, point source pollution is on the rise.

A **nonpoint source** is any source from which pollution is discharged that does not have a single or clearly obvious source. Nonpoint sources are often intermittent and diffuse, making it hard to quantify individual contributions—and therefore harder to regulate and reduce.

There are three main types of nonpoint sources of water pollution. The first type is agricultural runoff, where water from irrigation, rain, or melted snow flows from farm fields into the ground or bodies of water. Agricultural runoff can contain chemicals from pesticides and fertilizers as well as soil and organic debris. Often excess nitrogen and phosphorus are washed from farm fields and these can cause eutrophication of water bodies. The second type is urban runoff, a term that refers to the various pollutants that accumulate in soil and on roadways that are washed into the sewage systems during floods or rains. As rainwater journeys over roads, parking lots, and other urban structures, it picks up a variety of pollutants. In the Global North urban runoff is now the largest source of pollution in estuaries and beaches. The third type of nonpoint sources includes wastes and sewage from residential areas. These wastes are collected in the larger sewer system and it is impossible to tell their origin. Today, nonpoint sources of pollution are the main challenge in the Global North, and an increasing challenge in the Global South.

Wastewater Infrastructure

The Global North constructed vast sewer systems to collect wastewater and wastewater treatment facilities back in the nineteenth and early twentieth centuries. However, population growth over the last hundred years has meant that, today, the volume of sewage and stormwater exceeds the processing ability of most treatment plants. This is noticeable during heavy rains. Many cities in the United States, for example, have **Combined Sewer Systems** (CSSs) (Figure 6.5). Unlike separate sewers, in which sewage and stormwater are carried in separate systems, combined sewers carry raw sewage and stormwater together in one pipe to a sewage treatment plant. During dry weather, CSSs transport wastewater directly to the sewage treatment plant. When it rains, CSS facilities cannot handle the sudden increase in the volume of water, and as a result, the excess volume of sewage, clean water, and stormwater is discharged untreated into rivers, lakes, tributaries, and oceans. In some places, an overflow can be triggered by as little as a tenth of an inch of rain. These overflows, called **Combined Sewage Overflows** (CSOs), release untreated raw sewage and stormwater into rivers or lakes.

CSOs contain raw sewage from homes, businesses, and industries, as well as stormwater runoff and all the debris and chemicals that wash off the street or are poured in storm drains. This toxic brew is unappealing and quite dangerous. CSOs also contain untreated human waste, oxygen-demanding substances, ammonia, pesticides (such as malathion sprayed on the city to fight West Nile Virus), nutrients, petroleum products (from sources such as gas stations, auto repair shops, and garages), and other potential toxins and pathogenic microorganisms associated with human disease and fecal pollution.

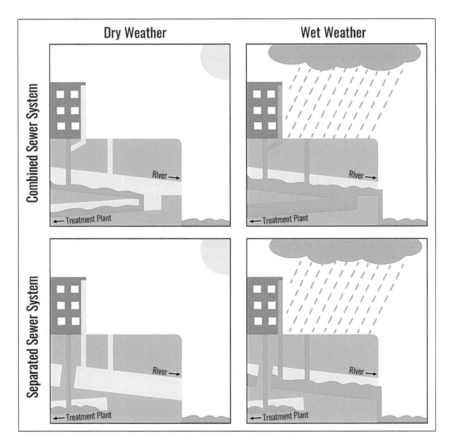

FIGURE 6.5 Combined and Separated Sewer Systems
Unlike separate sewer systems, in which sewage and stormwater are carried in separate systems, combined sewer systems carry raw sewage and stormwater together in one pipe to a sewage treatment plant. However, when it rains, combined sewer system facilities cannot handle the sudden increase in the volume of water, and as a result, the excess volume of sewage, clean water, and stormwater is discharged untreated into rivers, lakes, tributaries, and oceans.

In the Global North, CSOs are among the major sources responsible for beach closings, shellfish restrictions, and drinking water contamination. Residents are often warned to avoid contact with river water or beach water for several days after periods of heavy rainfall.

Increased Nonpoint Source Pollution

Today, **nonpoint source pollution**—runoff from farms and from cities—is the leading cause of water pollution in the Global North. The diffuse nature of these sources makes them more difficult to control, so they remain the largest source of water-quality issues.

One type of nonpoint pollution, **urban stormwater runoff**, impacts water quality in several ways. First, urban stormwater picks up pesticides, fertilizers, and pet waste from lawns and gardens as well as salts, antifreeze, oil, and heavy metals (from brakes) off road surfaces. These chemicals can be released directly and entirely untreated into a nearby waterbody.

The situation is better in Europe, where they have an integrated water framework that applies to all countries in the European Union. Nearly all European countries collect and treat sewage at the tertiary level for most of their populations. Tertiary treatment removes nitrogen and phosphorus that can cause nutrient pollution (such as algae blooms and eutrophication—a topic covered in more detail in Chapter 14 The Ocean).

Yet, even European countries with effective infrastructure systems continue to deal with overflows. In 2019, there were more than 200,000 discharges of untreated sewage into United Kingdom rivers and almost 2,000 discharges into UK coastal waters during the May to September vacation season alone. Although the United Kingdom has some 22,000 combined sewer outfalls, the country has done a better job than the United States at investing in infrastructure upgrades.

A 2021 European Environment Agency report noted that 95 percent of coastal waters, rivers, and lakes met the minimum water-quality standards set by the bathing water directives, improving significantly over the last several decades. The exception was Poland, where only 22 percent of their coastal waters met the water-quality standards.

The discharge of untreated or partially treated wastewater has significantly contaminated water sources in and around the Global South, ranging from India, Nigeria, and Egypt to Mexico. Water expert Asit Biswas argues that the real water crisis the world is likely to face in the coming decades may come from widespread water contamination due to inadequate water sanitation systems.

Communities in the Global South have much more serious nonpoint sources of water pollution than those in the Global North because large sections of their population are not served by sewers, drains, or solid-waste collection. Many cities in the Global South also suffer from significant point pollution and, in some cases, environmental regulations on point source pollution, such as those from industrial manufacturing, are not always enforced.

Water quality provided remains a serious, critical, and somewhat neglected issue compared to water supply.

SOLUTIONS FOR WATER AND SANITATION

Table 6.3 outlines the targets for SDG 6. As with other SDGs, there are some targets that are more pertinent in the Global South, and some targets that challenge all countries. In addition, many targets are interconnected with targets for other SDGs.

TABLE 6.3 SDG 6 Targets and Indicators to Achieve by 2030

Target	Indicators
6.1 Achieve universal and equitable access to safe and affordable drinking water for all	**6.1.1** Proportion of population using safely managed drinking water services
6.2 Achieve access to adequate and equitable sanitation and hygiene for all and end open defecation, paying special attention to the needs of women and girls and those in vulnerable situations	**6.2.1** Proportion of population using safely managed sanitation services, including a handwashing facility with soap and water
6.3 Improve water quality by reducing pollution, eliminating dumping, and minimizing release of hazardous chemicals and materials, halving the proportion of untreated wastewater and substantially increasing recycling and safe reuse globally	**6.3.1** Proportion of wastewater safely treated **6.3.2** Proportion of bodies of water with good ambient water quality
6.4 Substantially increase water-use efficiency across all sectors and ensure sustainable withdrawals and supply of freshwater to address water scarcity and substantially reduce the number of people suffering from water scarcity	**6.4.1** Change in water-use efficiency over time **6.4.2** Level of water stress: freshwater withdrawal as a proportion of available freshwater resources
6.5 Implement integrated water resources management at all levels, including through transboundary cooperation as appropriate	**6.5.1** Degree of integrated water resources management implementation (0–100) **6.5.2** Proportion of transboundary basin area with an operational arrangement for water cooperation
6.6 Protect and restore water-related ecosystems, including mountains, forests, wetlands, rivers, aquifers, and lakes	**6.6.1** Change in the extent of water-related ecosystems over time
6.A Expand international cooperation and capacity-building support to developing countries in water- and sanitation-related activities and programs, including water harvesting, desalination, water efficiency, wastewater treatment, recycling, and reuse technologies	**6.A.1** Amount of water- and sanitation-related official development assistance that is part of a government-coordinated spending plan
6.B Support and strengthen the participation of local communities in improving water and sanitation management	**6.B.1** Proportion of local administrative units with established and operational policies and procedures for participation of local communities in water and sanitation management

Source: United Nations Sustainable Development Goals. https://sustainabledevelopment.un.org/sdg6

A safe drinking water supply and protecting water quality are framed by local needs, regional watershed features, and national regulations. However, water issues are also complicated by fragmented responsibility at local, regional, and national levels and across national scales. Other challenges include climate change, aging infrastructure, and, in some places, increased use from population growth. This means that countries, cities, and communities will have different issues and there will be no one-size-fits-all solution. However, there are many best practices that can be adapted around the world to move toward more sustainable water practices.

Improve Water, Sanitation, and Hygiene

Targets 6.1 and 6.2 focus on improving drinking water and sanitation systems. In contrast to unimproved facilities such as unprotected wells, or pit latrines, improved drinking water facilities include tap water, public standposts, water storage tanks, or protected wells and springs. Improved sanitation systems include flush toilets connected to sewers or septic tanks, composting toilets, or, at minimum, pit latrines set on concrete slabs.

A key strategy is called **WASH**—water, sanitation, and hygiene. WASH is a comprehensive approach that involves coordinating improvements in drinking water treatment, latrine design and construction, and hygiene promotion. Water experts say WASH projects cannot be met by relying solely on official development assistance; it is incumbent upon national governments to dramatically increase the amounts of public funding made available for the expansion of WASH services.

Many countries have not invested in water and sanitation to the same degree as they have health, energy, transportation, and education. One of the greatest barriers to achieving these targets is inadequate financing and financing gaps (the shortfall in what it costs to provide water services compared to what is generated through fees). To meet Targets 6.1 and 6.2, capital financing would need to triple to $114 billion per year. This is approximately three times the historic spending on extending services to the underserved. Massive investments are needed—far above what has currently been committed. Part of the problem is that public utilities find it difficult to recoup the high costs of installation. As a result, financing gaps lead to deferred maintenance, deterioration of assets, and increased failure rates. There is also great variation in financing water infrastructure from country to country. Some countries rely on major contributions from household fees (such as Brazil, Costa Rica, Serbia, and Uruguay), others rely more on external aid (such as Kenya, Lesotho, and Tajikistan), and a few countries rely on national financing to support the majority of water infrastructure expenditures (including Bhutan, Fiji, Pakistan, and Peru).

Official development assistance has played a vital role in helping support water supply and water sanitation efforts. Development agencies such as USAID and UKAid have financed projects in many slums and rural areas. Population density can greatly influence the costs of both water supply and sanitation systems in low-income urban or rural areas. Sizeable high-density urban communities provide opportunities for large-scale centralized infrastructure and facilities though resource sharing and economies of scale.

In lower-density urban areas and rural areas, the implementation of **decentralized wastewater treatment systems** (such as the communal water pump shown in Photo 6.2) offers a cost-effective alternative. These can include piped supplies (into households, yards, or public standposts) or nonpipe supplies (such as protected wells and springs, rainwater, and packaged or delivered water). For people in low-density rural areas, shared facilities can offer a more affordable alternative to household-level services. Once population density (and the economic ability of consumers to pay) reaches a critical mass, these households may be added to a centralized system.

PHOTO 6.2 Communal Water Standpipe in Cameroon
This is a communal water standpipe where young girls and women collect water for the household. This low technology is a relatively inexpensive solution providing clean drinking water for the village. If the village grows in population and density, there may be opportunities to build a centralized system for clean water.

PHOTO 6.3 A Household Rainwater Harvesting Tank
A rainwater tank is an inexpensive way to improve access to clean drinking water.

In addition to official development assistance, other forms of repayable finance can be used, including commercial bank loans, bonds, and microfinance. **Microfinance** may provide a promising solution. Microfinance is the offering of relatively small loans for shorter periods to communities and households. Microfinance programs are alternative financing for groups unable to obtain credit through other sources. An example of water microfinance is Water.org. Their program, WaterCredit, uses microcredit to fund sanitation projects. Millions of people pay high prices for water from vendors or collect water from unsafe natural sources. WaterCredit provides small, affordable loans to families to purchase long-term safe water and sanitation solutions that solve their immediate need and over time will cost less than continuing to pay for temporary fixes. WaterCredit also works with rural utilities to build their capacity through technical assistance and staff training to help them upgrade infrastructure such as water filtration and treatment systems.

One of the countries Water.org works with is Cambodia. More than three million people in Cambodia lack access to safe water, and five million lack access to improved sanitation. With approximately 77 percent of Cambodians living in rural areas, lack of access to safe water and sanitation disproportionately affects its rural communities. Water.org offers small loans for households to construct drinking water systems that include a rain storage tank, rain catchment systems, and a toilet on their own property (similar to the one shown in Photo 6.3). The impacts can be transformative: A family can draw filtered, clean rainwater from a storage barrel, saving hours of time they once spent traveling for water. Having a toilet not only improves hygiene but also provides safety to millions of women and children who once had to use outdoor latrines, where open defecation poses risks such as snake and insect bites. The *Solutions* 6.1 box discusses an example of improvements made in a school in Tanzania.

⊜ SOLUTIONS 6.1

Harvesting Rainwater in Bagamoyo, Tanzania

Target 6.1: Achieve universal and equitable access to safe and affordable drinking water for all.

Sometimes simple solutions can have a profound impact. Journalist Jessica Troni investigated the impact of installing a "low-tech" rainwater harvest system in a schoolyard in a Tanzanian village where water is scarce and bottled water can be expensive. She noted the lack of water affected children at school, impeding their education:

"The students at Kingani school in the Tanzanian town of Bagamoyo used to have two choices for drinking water at school: get sick or remain thirsty.

Rising sea levels, increased drought and reduced or erratic rainfall made the drinking wells so salty it would cause headaches, stomach aches and ulcers. To make matters worse, the water that students would spend time fetching from watering holes was so dirty that it spread disease. Ismat Hassan, who came from Tanzania's capital Dar es Salaam to study and board at Kingani, got stomach ulcers from drinking the well water, typhoid from the water collected from watering holes, and pain and exhaustion when she chose to drink neither.

The school administration tried to address the problem by having new wells dug, but they quickly turned saline from seawater intrusion. It then started paying for water to be trucked in, but this quickly became unaffordable, as did the students'

spending on bottled water from nearby shops.

Kingani's 650 pupils would also suffer from dizziness, fatigue and constipation due to dehydration. In addition to the absenteeism, students were spending time to look for and fetch water from holes instead of studying, even though it kept making them sick. During the dry season, the well water sometimes became so salty that it could not even be used for laundry, as it would bleach clothes.

That changed after the construction of a rainwater harvesting system, involving rooftop guttering and a series of large tanks for storing water that students can use for drinking, washing and cooking. The concrete and plastic tanks they set up there can contain an estimated 147,000 litres of water which will help them store water on rainy days so that they can use that water when there is no rain. Nowadays, the playground is dotted with big water tanks that students drink freely from and without the fear of health repercussions. There has been a huge improvement in student attendance, motivation and well-being. The new system is a sustainable solution that has eased concerns and expenditures at all levels."

Source: Troni, J. "Drink Salty Water or Go Thirsty—Climate Change Hits Tanzanian School Children," United Nations News, February 15, 2019. https://www.unenvironment.org/news-and-stories/story/drink-salty-water-or-go-thirsty-climate-change-hits-tanzanian-school

Focus on Water Efficiency, Security, and Conservation

One way to stretch existing water supplies is to focus on water efficiency. As mentioned, water losses from aging infrastructure are a significant and costly challenge in the Global North. For example, the city of London, which has hundreds of miles of Victorian-era water pipes, loses 132 million gallons each day to leaking pipes. The Thames Water company, which supplies water to London, stepped up leak detection operations in 2018, incorporating the use of drones and satellite observation to help identify leaking pipes and more efficiently repair them.

In addition to addressing water wastage, cities and states are also encouraging the use of water efficient fixtures and appliances in their jurisdictions. This can include setting water conservation standards and adopting newer technologies that track water use. Other common options include rebates, incentive programs, and recognition for water conservation efforts. Even Los Angeles, California, known for images of people watering lawns and washing their cars in the midst of drought, has seen some improvement in water conservation (see *Making Progress* 6.1).

MAKING PROGRESS 6.1

Water Conservation in a Thirsty Los Angeles

Target 6.4: Substantially increase water-use efficiency across all sectors and ensure sustainable withdrawals and supply of freshwater to address water scarcity and substantially reduce the number of people suffering from water scarcity.

Los Angeles' water supply challenges are legendary enough to be the central plot line in Hollywood movies (e.g., the 1974 film noir classic *Chinatown*). Recent studies show that climate change is further threatening water supplies. Currently, 85 percent of the metro area's water supply comes from imported sources, whose supplies are depleting.

The mayor established a Water Cabinet to coordinate water policy and management. The city is taking a collaborative, integrated approach to managing all the city's water resources and integrating them within a large sustainability agenda. The result will be a series of plans and actions, organized under the umbrella *One Water Plan LA 2040*. In this way, the city will integrate efforts in wastewater and stormwater management, water recycling, and conservation.

Between 2015 and 2017, Los Angeles successfully reduced per capita potable water use from 131 to 104 gallons per capita per day, a 20 percent decrease, showing that even in thirsty Los Angeles, change can happen. Los Angeles

has increased its commitment to water conservation measures and forbidden hosing down sidewalks as well as outdoor watering within two days after a rain. Residents have also been encouraged to shower for only five minutes, install high-efficiency toilets, and turn off the water while brushing their teeth.

Other projects include the Purple Pipe Recycled Water Project, which enables residents to use recycled water for irrigation and industrial purposes. It also features public engagement activities like free landscaping classes that feature drought-tolerant "California Friendly® Plants" and incentives to "Cash in Your Lawn," a program that reimburses homeowners who remove their water-thirsty lawns and replace them with drought-resistant plantings (xeriscaping) or artificial turf.

Sources: "Cash for Grass Rebate Program." Los Angeles County Waterworks District (2018). https://dpw.lacounty.gov/wwd/web/Conservation/CashforGrass.aspx

Mann, M. E., and Gleick, P. H. "Climate Change and California Drought in the Twenty-First Century." *Proceedings of the National Academy of Sciences of the United States* 112 (13): 3858–59 (2015). https://doi.org/10.1073/pnas.1503667112

"One Water LA." City of Los Angeles (2018). https://www.lacitysan.org/san/faces/home/portal/s-lsh-es/s-lsh-es-owla?_adf.ctrl-state=2qhj32cth_5&_afrLoop=419478517667671#

Desalination

As countries confront water scarcity, **desalination**—removing salts, minerals, and other impurities from ocean water—is a potential option for places facing extreme and persistent water shortages.

Saudi Arabia is a leader in desalination and is the largest producer of desalination water in the world. Between 1975 and 2000, the country invested a total of more than $100 billion in water supply and sanitation and may spend another $130 billion by 2030. There are currently more than a dozen desalination plants in the country.

A drawback is that desalination is extremely energy intensive and, as a result, desalinated water is at least four times the cost of water that comes from other traditional infrastructure. Lower-income countries are less likely to adopt desalination because of the expense. Desalination is not an appropriate solution for places that are deep in the interior of a continent, or at high elevation—which unfortunately is true for many of the places with the biggest water problems.

Another drawback is the possible impacts that concentrated salt water (brine) discharged from the desalination plant could have on aquatic ecosystems. Brine tends to sink to the bottom and suffocate marine life. There is also the possibility that water intake may suck fish and shellfish or their eggs into the filter system if the process is not careful.

Despite the drawbacks, the increased pressure to find freshwater has driven technology investment in the desalination industry. It is an alternative that may help as part of a broader range of efforts to cut water use and water supplies, and as its technology grow more energy-efficient (and costs come down) it may be more widely used.

Water Reuse Systems

One form of water conservation is to reuse water (water recycling). Water reuse generally refers to projects that use technology to treat water according to the water-quality requirements of its planned reuse. Wastewater produced by a household includes greywater from sources such as baths, showers, bathroom sinks, washing machines, and dishwashers and blackwater from sources such as toilets, dishwashers, and kitchen sinks.

In an on-site system (such as house or office building) graywater can be reused without treatment for purposes other than human contact or consumption, such as landscape irrigation, toilet flushing, or industrial processes. In contrast, blackwater must be treated before it can be used and should only be reused outside in gardens or lawns. The reuse of blackwater can be more expensive than graywater.

In addition to saving water, water recycling systems also save energy through reductions in water treatment and distribution. Golf courses, large parks, and green areas are increasingly being irrigated with reclaimed water. Recycling water not only conserves a vital resource but it can also provide financial savings.

A more intensive process is large-scale water recycling, which centrally captures and reclaims previously used water, sometimes redistributing it for large-scale irrigation, using it to replenish a ground water basin, or even fully treating it to potable standards.

The term **Direct Potable Reuse** means that wastewater is treated and then returned directly into the drinking water distribution system or into pipelines to a drinking water treatment plant. Such systems use a variety of technologies including microfiltration, reverse osmosis, and ultraviolet disinfection to prepare water.

While some cities such as Singapore have done this successfully, Direct Potable Reuse has been less common in the Global North as concern about public perception has delayed these efforts. While some critics deride this as "toilet to tap," Direct Potable Reuse systems already exist in some US and Australian cities. There is potential to expand Direct Potable Reuse.

Reduce Pollution to Improve Water Quality

In the Global North, point source pollution has been dramatically reduced and controlled and the challenge comes from nonpoint pollution, particularly from urban stormwater runoff. In the Global South, countries are confronted by pollution from both point source and nonpoint sources.

One of the most important ways to address water quality is to have strong federal or national regulations. In the United States, the US Clean Water Act (examined in *SDGs and the Law* 6.1) continues to guide efforts to regulate water pollution and serves as an example for the creation of many other countries' water regulations. However, in some countries, there may be a lack of regulations, or it may be ineffectively enforced.

Strategies to advance sustainable sanitation can be simple: Begin with a toilet that effectively captures human waste in a safe, accessible, and dignified setting. The waste then gets stored in a tank, which can be emptied later by a collection service, or transported away by pipework. The next stage is treatment and safe disposal. However, in reality, sanitation that covers hundreds, or hundreds of thousands of people, will involve an expansive series of connected infrastructure that must be constructed, coordinated, and then managed.

Nature-Based Solutions: Green Infrastructure

Over the last century, the world has turned primarily to human-built, or "gray," infrastructure to improve water management. In so doing, it has often ignored traditional and Indigenous knowledge that embraces greener approaches. Although many countries in the Global South have been encouraged to construct or expand conventional wastewater infrastructure that follow the western models, some experts see the potential for returning to traditional solutions to wastewater. These are called **Nature-Based Solutions (NBS)**. NBS approaches can include more ecosystem-friendly forms of water storage, such as natural wetlands, improvements in soil moisture, and more efficient recharge of groundwater, all of which could be more sustainable and cost-effective than traditional gray infrastructure such as dams. In addition, forests, wetlands, grasslands, soils, and crops, when managed properly, play important roles in regulating water quality by reducing sediment loadings, capturing and retaining pollutants, and recycling nutrients. Where water becomes polluted, both constructed and natural ecosystems can help improve water quality. Nonpoint source pollution from agriculture, notably nutrients, remains a critical problem in both the Global North and Global South. It is also the most amenable to NBS, which can rehabilitate ecosystem services that enable soils to improve nutrient management, and thus lower fertilizer demand and reduce nutrient runoff and/or infiltration

SDGs AND THE LAW 6.1

The US Clean Water Act

Most environmental experts point to the 1972 Federal Water Pollution Control Act Amendments (commonly called the 1972 Clean Water Act) as setting the framework for the last fifty years of water pollution policy in the United States. It also influenced many other countries that have passed similar laws.

The Clean Water Act was remarkable for several reasons. It is the principal law governing pollution of the nation's surface waters. For the first time, a single federal agency was placed in charge of water pollution, taking some control from the states and establishing basic federal standards, to which all states had to comply. The Clean Water Act also provided technical tools and financial assistance to address the many causes of water pollution. Another revolutionary aspect to the legislation was that it required reporting of discharge information and making it publicly available. This gave citizens a strong role to play in protecting water resources and the tools to help them do so. In today's information economy, discharge release and toxic release inventory are available to citizens using the internet, and citizens have access to even more information such as permit documents, pollution management plans, and inspection reports. The provision for public information also allows citizens to bring action (or sue) to enforce the Clean Water Act. This has proved instrumental in making progress in cleaning up water pollution.

The Clean Water Act was highly ambitious: Its major goals were to eliminate the discharge of water pollutants, restore water to "fishable and swimmable" levels, and completely eliminate all toxic pollutants. Today these three aspirational goals of the Clean Water Act of 1972 have not been achieved, yet they continue to provide the framework for how the federal government (and municipalities) deal with water pollution.

For its time, the Clean Water Act was genuinely revolutionary. The act made it illegal for any point source—that is a specific source of pollution—to discharge any pollutant into the waters unless specifically authorized by permit. This is one reason the United States has seen reduced point source pollution.

In the almost fifty years since the Clean Water Act, there have been successes and failures. We now have a more complete picture of the volume and types of pollution, but the results are mixed. Point source pollution from industrial sources has been reduced. In many rivers and lakes, oxygen levels have recovered due to the filtering out of organic wastes. Some pollutants have declined, but others are on the rise. In 1972, two-thirds of the lakes were too polluted for swimming or fishing; today two-thirds of the lakes, rivers, and waterways are safe for swimming and fishing. In 1970, more than 70 percent of industrial discharge was not treated at all; today 99 percent is.

However, much remains to be done. Almost half of US waters are still impaired—too polluted to serve as sources of drinking water or to support good fish and wildlife. Wetlands continue to be lost to pollution and development. Today, nonpoint source pollution—runoff from farms and from cities—is the leading cause of water pollution, but many experts say this is still inadequately addressed by the Clean Water Act.

to groundwater. *Critical Perspectives* 6.1 explores nontraditional water management in the Ganges.

Vegetation can also improve the environment and provide aesthetic amenities when "decentralized" and distributed throughout the city or community. **Green infrastructure** elements include street trees and other vegetation, green roofs, green façades, permeable pavements, rain gardens, and stormwater treatment swales. These are called decentralized efforts

CRITICAL PERSPECTIVES 6.1

Is Western Technology the Best Way to Treat Wastewater?

The Ganges River begins as pristine, clear waters in the Himalayas. But pollution, untreated sewage and use by hundreds of millions of people transform the river into a toxic sludge by the time it reaches the Bay of Bengal, some 1,500 miles later. The Ganges winds through twenty-nine cities, each with populations of more than 100,000.

Known as *Ganga Ma* (Mother Ganges), the river is revered as a goddess whose purity cleanses the sins of the faithful and helps the dead on their path toward heaven. Hindus believe that if the ashes of their dead are deposited in the river, they will be ensured a smooth transition to the next life or be freed from the cycle of death and rebirth. It is said that a single drop of Ganges water can cleanse a lifetime of sins. In cities along the Ganges, daily dips are an important ritual among the faithful. Many cities are pilgrimage sites.

Despite its spiritual importance, the physical purity of the river has deteriorated dramatically. A significant amount of the Ganges' pollution is organic waste: sewage, trash, food, human waste, and animal remains. Today, half a billion people live in the basin of the Ganges and more than 100 cities dump their raw sewage directly into the river. Coliform bacteria measurements have consistently been at levels too high to be safe for agricultural use, drinking and bathing. Waterborne illnesses are common killers, accounting for the deaths of some 1.5 million Indian children each year.

The Ganges presents an interesting case study. Because the river is holy, it attracts tens of thousands of pilgrims each day for ritual bathing, exposing large numbers of people to untreated, contaminated water (see Photo 6.4). It is estimated that 40 percent of the people who take a dip in the river regularly have skin or stomach ailments.

Over the last three decades, several initiatives have been undertaken to clean the river but have failed to deliver. In 1985, the Indian government initiated the Ganga Action Plan to clean up the river in selected areas by installing sewage treatment plants. Several more plans have followed. In 2015, the Indian government announced it would invest $3 billion to ensure that by 2020, no untreated municipal sewage or industrial runoff enters the river. However, by 2020, the practice of discharging untreated sewage remained widespread and much of the money allocated for the project had not yet been spent. The United Nations estimates that as much as 80 percent of sewage discharged into the two major tributaries that feed the river is still untreated. Approximately 1.5 billion gallons (6 billion liters) of untreated sewage is dumped into the Ganges *each day*.

Strategies for cleaning up the river focus on building new or upgrading existing wastewater treatment plants. The Indian government spent close to $500 million between 1985 and 2000 on treatment plants, and almost that much between 2000 and 2020 with mixed results. Many of the treatment plants malfunctioned, were designed improperly, or could not handle the bacterial load. Corruption and ineffective monitoring contributed to the problems. Some have become skeptical of the adoption of expensive, multimillion-dollar Western-style treatment plants.

There is growing criticism over the adoption of Western-style technology to solve issues in other parts of the world. Western-style technology tends to be very expensive, relies on highly trained engineers and workers to maintain the technology, and requires a stable and consistent supply of electricity. Further, Western-style waste treatment plants are engineered for use in countries where there are no monsoon rains and where the population does not drink directly from the water source. Few considered the radically different ways that people use rivers in India.

As one alternative to the high-technology treatment plants, engineers have developed

PHOTO 6.4 Bathing in Ganges
Devotees gather to take a holy dip at Sangam, located at the confluence of the
Ganga and Yamuna rivers in Allahabad in India. In cities along the Ganges, daily dips
are an important ritual among the faithful. However, pollution in the Ganges makes
this ritual more problematic by increasing the likelihood of contracting waterborne
diseases.

different types of nonmechanized, low-technology sewage treatment plants. This plant is more compatible with India's climate and replaces the high-technology solution with a wastewater oxidation pond system that stores sewage in a series of ponds and uses bacteria and algae to break down waste and purify the water. The ponds allow waste to decompose naturally in water. Bacteria grow on the sewage and decompose it; the algae feed on the nutrients released by the bacteria and produce oxygen for the water. This alternative treatment does not require electricity and relies on sunshine to speed up the decomposition. The pond system is much less expensive than mechanical treatment plants.

The debate between Western-style technology and lower-cost alternatives in the case of Ganges highlights general themes applicable to many of the solutions to pollution in the Global South. While there have been large-scale investments by governments in high-technology solutions, the lack of results

has caused many to seek alternatives that are locally sensitive and economically affordable. Experts contend that solutions should respond to local demands and should be as simple, sturdy, and inexpensive as possible. Low-cost, low-technology solutions such as the pond system or pour-flush latrines or even improved pit latrines have been successful. An important element in pollution control is the role of public participation. The involvement of the local community and households is a crucial component to success.

Sources: Hamner, S. Pyke, D., Walker, M., Pandy, G., Mishra, R. K., Mishra, V. B., Porter, C., and Ford, T. "Sewage Pollution of the River Ganga: An Ongoing Case Study in Varanasi, India." *River Systems* 20 (3–4): 157–67 (2013). https://doi.org/10.1127/1868-5749/2013/0058
Natarajan, P. M., Kallolikar, S., and Ganesh, S. "Transforming Ganges to Be a Living River through Waste Water Management." *World Academy of Science, Engineering and Technology International Journal of Environmental and Ecological Engineering* 10 (2):251–60 (2016).

because they create a system of numerous, but individual, types of vegetation. Advocates of green infrastructure often point to the multiple benefits this provides, including energy savings, mitigation of the urban heat island effect, reduction air pollution, wildlife habitat provision, and provision of an aesthetically pleasing amenity.

To keep stormwater out of combined systems, municipalities have started to examine the efficacy and cost-effectiveness of "gray-green hybrid" systems that integrate decentralized green infrastructure throughout, and areas serviced by existing gray infrastructure to manage runoff. In the United States, the EPA has supported efforts to expand green infrastructure and it has begun to promote applications of decentralized green infrastructure as a more cost-effective, sustainable path to clean water.

Yet green infrastructure techniques require a substantially different set of implementation, operation, and maintenance approaches than traditional gray infrastructure. Green infrastructure fundamentally involves land-use changes, multiple stakeholders, community outreach and buy-in, frequent decentralized maintenance regimes, and a different set of skills than is associated with traditional gray stormwater management systems.

Around the world, architects and planners are rethinking building design. Green roofs, for example, serve multiple purposes: stormwater catchment, reduced temperatures, and increased building energy efficiency. The city of Chicago has implemented a number of fascinating programs, including incentive programs for **green roofs** and a municipal "green alley" program. Green alleys are made of highly reflective recycled materials that reduce the urban heat island effect and the pavement used is permeable to allow stormwater to filter through and into the soil below, instead of draining to the sewer system.

SUMMARY AND PROGRESS

Since 2000, millions of people have gained access to drinking water, sanitation, and hygiene services, but many countries still have a long way to go to fully realize the targets of SDG 6. Currently 138 countries have safely managed drinking water, yet nearly half the global population (3.6 billion) lack safe sanitation. The positive news is that 70 percent of the global population has a basic handwashing facility with soap and water available at home. A 2021 UNICEF report notes that the world is not on track to achieve SDG Targets 6.1 and 6.2, and that achieving universal coverage by 2030 will require a quadrupling of current rates of progress in safely managed drinking water services, safely managed sanitation services, and basic hygiene.

Discrimination, exclusion, marginalization, entrenched power asymmetries, and material inequalities are among the main obstacles to achieving the human rights to safe drinking water and sanitation for all and realizing the water-related SDGs of the 2030 agenda. Poorly designed and inadequately implemented policies, inefficient and improper use of financial resources, and policy gaps fuel the persistence of inequalities in access to safe drinking water and sanitation. Unless exclusion and inequality are explicitly addressed in both policy and practice, water interventions will continue to fail to reach those most in need and who are likely to benefit most.

The lack of adequate financial resources constrains progress in many countries. More than 80 percent of countries report insufficient finance for both urban and rural areas in meeting national targets for drinking water and sanitation, as well as those for water quality, a major component of SDG 6. Increasing donor commitments to the water sector remains crucial in sustaining progress toward Goal 6.

With regard to water quality, some places have seen dramatic decreases in certain types of pollutants, while other pollutants have increased. Most countries have successfully implemented pollution laws and regulations, but may lack the resources to adequately enforce such regulations. The pressures of population growth and increasing consumption of resources have, in some cases, offset new laws and regulations designed to decrease or prevent pollution.

The COVID-19 pandemic has underscored the importance of water, sanitation, and hygiene for protecting human health. The 2020 UN SDG Progress Report noted that handwashing is one of the cheapest, easiest, and most effective ways to prevent the spread of the coronavirus. However, only 70 percent of people had a basic handwashing facility with soap and water at home. In the lowest-income countries, that share was only 28 percent. According to the report, an estimated three billion people worldwide lacked the ability to safely wash their hands at home. The regional disparities are stark: In sub-Saharan Africa, 75 percent of the population (767 million people) lacked basic handwashing facilities, followed by Central and Southern Asia at 42 percent (807 million people), and Northern Africa and Western Asia at 23 percent (116 million people). Although many countries are steadily improving access to handwashing facilities, some are not on pace to meet the targets.

Many of the targets in SDG 6 overlap with several others as highlighted in Figure 6.6, but perhaps most significantly with SDG 3 Good Health and Well-Being. According to UN Water, investing in WASH can have a beneficial impact across a number of issues covered by the SDGs, including health and education. For example, it has been estimated that 26 percent of childhood deaths and 25 percent of the total disease burden in children under five could be prevented through the reduction of environmental risks, including by reducing unsafe water and sanitation and inadequate hygiene. Specifically, diarrheal diseases are among the main contributors to global child mortality, causing about 10 percent of all deaths in children under five years. Improvements in WASH can also lead to improved nutrition.

In health-care facilities, improvements in WASH will lead to a reduction in maternal mortality, as well as increased use of health centers and facilities. Investing in projects to achieve Targets 6.1 and 6.2 provides cobenefits that expand beyond the water and sanitation sector and improves health outcomes as well.

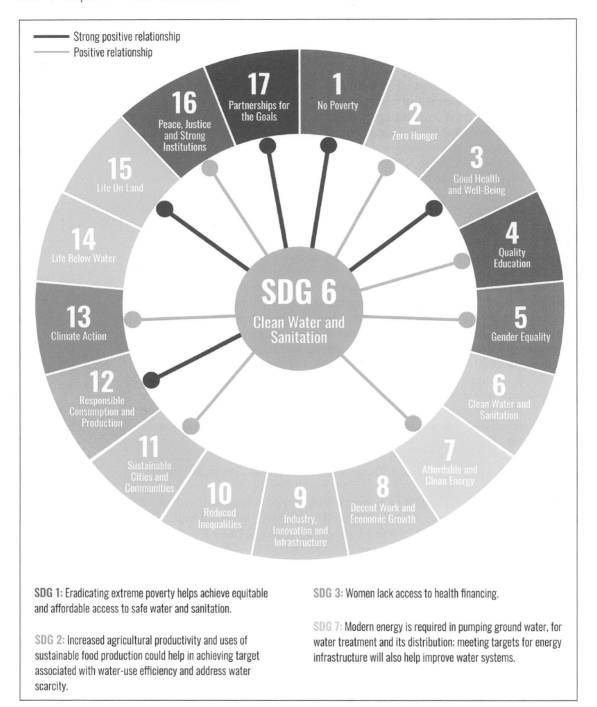

SDG 1: Eradicating extreme poverty helps achieve equitable and affordable access to safe water and sanitation.

SDG 2: Increased agricultural productivity and uses of sustainable food production could help in achieving target associated with water-use efficiency and address water scarcity.

SDG 3: Women lack access to health financing.

SDG 7: Modern energy is required in pumping ground water, for water treatment and its distribution; meeting targets for energy infrastructure will also help improve water systems.

FIGURE 6.6 Interconnections in SDG 6 and the Other Goals
Water and sanitation connect to all the SDGs in some way; this figure highlights several of the stronger connections.

QUESTIONS FOR DISCUSSION AND ACTIVITIES

1. Where in the world is water sanitation the most pressing issue?

2. Describe the process of water treatment.

3. Explain the consequences of unsafe drinking water and water pollution on human health.

4. Is water supply or water quality the more important issue where you live?

5. What water conservation measures are in place where you live? Why do you think water conservation programs or policies in the United States have been weak or absent?

6. Which one or two strategies for improving water quality do you feel should be prioritized in lower-income countries?

7. Explain how achieving targets in SDG 6 could improve gender inequality in the Global South.

8. Go to the USGS Water Quality Trends Interactive Map (https://nawqa-trends.wim.usgs.gov/swtrends/). Describe the patterns and discuss why some riversheds are seeing improvement in water quality and some are declining.

9. View this short video about Singapore's New Water (https://www.youtube.com/watch?v=zFGy2t1AGZ8). Would you have any qualms about a Direct Potable Reuse water system in your community? How would you convince others to support it?

TERMS

aquifer
coagulation
Combined Sewage Overflows
Combined Sewer systems
decentralized wastewater
    treatment systems
desalination
Direct Potable Reuse
disinfection
filtration
green infrastructure
green roof

hydrologic cycle
microfinance
Nature-Based Solutions (NBS)
nonpoint source pollution
point source pollution
urban stormwater runoff
virtual water
water conservation
water infrastructure
water, sanitation, and hygiene
    (WASH)

RESOURCES USED AND SUGGESTED READINGS

Amnesty International/WASH United. "Recognition of the Human Rights to Water and Sanitation by UN Member States at the International Level: An Overview of Resolutions and Declarations That Recognise the Human Rights to Water and Sanitation" (2015). www.amnesty.org/download/Documents/IOR4013802015english.pdf

Asian Development Bank. "Asian Water Development Outlook 2016: Strengthening Water Security in Asia and the Pacific." Manila: ADB (2016). www.adb.org/sites/default/files/publication/189411/awdo-2016.pdf

Bakker, K. *Privatizing Water: Governance Failure and the World's Urban Water Crisis*. Ithaca, NY: Cornell University Press (2010).

Bhattacharya, S., and Banerjee, A. "Water Privatization in Developing Countries: Principles, Implementations and Socio-economic Consequences." *World Scientific News* 4: 17–31 (2015).

Biswas, A. K., and Tortajada, C. "Water Quality Management: A Globally Neglected Issue." *International Journal of Water Resources Development* 35 (6): 913–16 (2019). https://doi.org/10.1080/07900627.2019.1670506

"Combined Sewage Overflows." *Riverkeeper*, video (2010). http://www.riverkeeper.org/campaigns/stop-polluters/sewage-contamination/cso/

Engel, K., Jokiel, D., Kraljevic, A., Geiger, M., and Smith, K. *Big Cities, Big Water, Big Challenges: Water in an Urbanizing World*. World Wildlife Fund (2011).

European Environment Agency. "European Bathing Water Quality in 2021" (2021). https://www.eea.europa.eu/publications/bathing-water-quality-in-2021/european-bathing-water-quality-in-2021#:~:-text=In%202021%2C%2088.0%25%2C%20of,compared%20to%20 2%25%20in%202013

Gandy, M. "Landscapes of Disaster: Water, Modernity, and Urban Fragmentation in Mumbai." *Environment and Planning A* 40: 108–30 (2008).

Gumprecht, B. *The Los Angeles River: Its Life, Death and Possible Rebirth*. Baltimore: The Johns Hopkins University Press (2001).

International Energy Agency. "Water Energy Nexus: Excerpt from the World Energy Outlook 2016." Paris: IEA Publications (2016). https://www.iea.org/reports/water-energy-nexus

Jones, J. A. A. *Water Sustainability: A Global Perspective*. New York: Routledge (2010).

Karvonen, A. *Politics of Urban Runoff: Nature, Technology, and the Sustainable City*. Cambridge, MA: MIT Press (2011).

Lewin, T. *Sacred River: The Ganges of India*. Boston: Houghton Mifflin/Clarion Books (2003).

Melosi, M. V. *Precious Commodity: Providing Water for America's Cities*. Pittsburgh: University of Pittsburgh Press (2011).

Pielou, E. C. *Fresh Water*. Chicago: University of Chicago (1998).

Sedlak, D. *Water 4.0: The Past, Present, and Future of the World's Most Vital Resource*. New Haven, CT: Yale University Press (2014).

Swyngedouw, E. *Social Power and the Urbanization of Water: Flows of Power*. Oxford: Oxford University Press (2004).

UN General Assembly. Seventy-first Session, Official Records. *Human Rights to Safe Drinking Water and Sanitation.* Note by the Secretary-General. A/71/302 (2016). undocs.org/A/71/302

UN General Assembly. Seventy-second Session, Official Records. *Human Rights to Safe Drinking Water and Sanitation.* Note by the Secretary-General. A/72/127 (2017). undocs.org/A/72/127

United Nations. *Sustainable Development Goal 6: Synthesis Report 2018 on Water and Sanitation.* New York: United Nations (2018). https://www.unwater.org/publication_categories/sdg-6-synthesis-report-2018-on-water-and-sanitation/

United Nations Water. "Eliminating Discrimination and Inequalities in Access to Water and Sanitation" (2015). www.unwater.org/publications/eliminating-discrimination-inequalities-access-water-sanitation/

United Nations Water. "Summary Progress Update 2021: SDG 6-Water and Sanitation for All" (2021). https://www.unwater.org/publications/summary-progress-update-2021-sdg-6-water-and-sanitation-all

United Nations Water. "Water and Climate Change" (2021). https://www.unwater.org/water-facts/climate-change/

United Nations Water. "Water Scarcity" (2021). https://www.unwater.org/water-facts/water-scarcity

United Nations Water. Decade Programme on Advocacy and Communications/Water Supply and Sanitation Collaborative Council. "The Human Right to Water and Sanitation" (n.d.). www.un.org/waterforlifedecade/pdf/human_right_to_water_and_sanitation_media_brief.pdf

Water.org. "WaterCredit Initiative" (2022). https://water.org/solutions/watercredit/

World Health Organization and UNICEF. *Progress on Household Drinking Water, Sanitation and Hygiene 2000–2020: Five Years into the SDGs.* Geneva, Switzerland: World Health Organization and the United Nations Children's Fund (2021). https://data.unicef.org/resources/progress-on-household-drinking-water-sanitation-and-hygiene-2000-2020/?utm_source=newsletter&utm_medium=email&utm_campaign=JMP%202021

World Health Organization and UNICEF. "Progress on Sanitation and Drinking-Water." Geneva, Switzerland: WHO and UNICEF (2017). https://www.unicef.org/reports/progress-on-drinking-water-sanitation-and-hygiene-2019#:~:text=The%20report%20shows%20that%20in,cent%20to%2045%20per%20cent

World Health Organization and United Nations Water. *Global Analysis and Assessment of Sanitation and Drinking-Water: Financing Universal Water, Sanitation and Hygiene under the Sustainable Development Goals.* Geneva, Switzerland: WHO and UN-Water (2017). https://www.who.int/water_sanitation_health/publications/glaas-report-2017/en/

Energy

LEARNING OBJECTIVES

After reading this chapter, you should be able to:

- Describe the types of renewable and nonrenewable energy resources

- Describe the concept of an energy portfolio

- Identify and explain the impacts of fossil fuel energy on air pollution, climate change, and human health

- Explain the challenges to transitioning to renewable energy

- Identify and explain the various targets and goals associated with SDG 7

- Explain how the three Es are present in SDG 7

- Explain how SDG 7 strategies connect to other SDGs

The automobile has shaped cities in the United States for over a century, dominating urban design. Vast amounts of land have been set aside for roads, highways, and parking garages (Photo 7.1). There are television shows, movies, and songs about cars. For decades, getting a driver's license has been a benchmark of teenage independence.

There are broad environmental impacts of our car dependency. Almost 30 percent of the greenhouse gasses (GHGs) emitted in the United States are from transportation and the largest sector of these emissions is from cars and light trucks. In addition to being a major contributor to climate change, traditional fuel-burning cars contribute to local air and water pollution problems. Oils, gas, and heavy metals from brakes collect on paved surfaces and are then washed into nearby streams and rivers when it rains.

Equity and public health impacts arise from our car culture. The health impacts of air pollution—for example, asthma and lung cancer—are not trivial. Some 53,000 Americans die prematurely every year from vehicle pollution, losing ten years of life on average compared to their potential lifespans in the absence of tailpipe emissions. These losses are not equally borne. Many major urban highways are located in low-income communities or communities of color, making these populations disproportionately exposed to high levels of air pollution.

Although Americans know the negatives that come from our dependency on cars and fossil fuels, we just can't seem to quit the habit. Or can we?

PHOTO 7.1 Los Angeles Freeways
Infrastructure for automobiles takes up considerable space; this stretch of the
Route 405 in Los Angeles is sixteen lanes across. Highways are impermeable
surfaces, which contribute to stormwater runoff into waterways.

ENERGY: UNDERSTANDING THE ISSUE

Energy resources impact our everyday lives in many ways. We need **energy**
to turn on a light or cook our meals, to power the tractors and trucks that
deliver food, and to commute to jobs. In their book *Energy Resources*, Wes and
Colin Reisser note energy resources are complex and interconnected; energy
is a crucial factor in climate change, rising prices for commodities, political
instability in the Middle East, and environmental degradation in many parts
of the world. The increasing global demand for energy continues to be one of
the most significant and disruptive forces impacting the global environment.

Creating a sustainable global energy system will require substantial retool-
ing of energy resources and a rethinking of both production and consumption
of energy. It is a monumental challenge.

Inequalities in Energy Consumption

Energy is used for two broad purposes: transportation (including cars, trucks,
planes, and trains) and electricity generation, which powers industry, homes,
and commercial buildings. Nonrenewable sources are the predominant energy
source of both transportation and electricity. In the United States, energy is
measured in **British thermal units (Btus)**; other countries use the metric-based
system joules (Btu measures the amount of heat required to raise the temperature of 1 pound of water 1 degree Fahrenheit).

There are large inequalities in energy consumption between countries in
the Global North and the Global South (see Map 7.1). The average American
consumes more than ten times the energy of the average Indian, four to five
times that of the average Brazilian, and three times more than the average Chinese. The gulf between higher-income and very low-income countries is even
greater—several low-income nations consume less than 1 Btu equivalent per

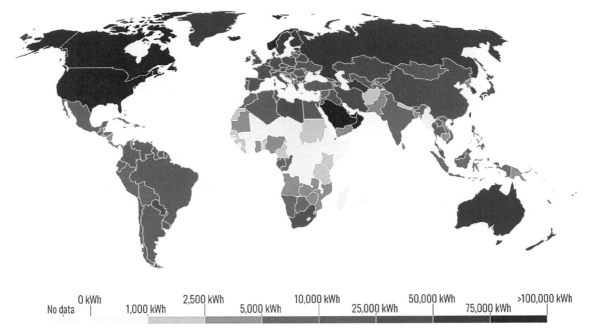

0 kWh   2,500 kWh   10,000 kWh   50,000 kWh   >100,000 kWh

No data   1,000 kWh   5,000 kWh   25,000 kWh   75,000 kWh

MAP 7.1 Energy Use Per Person (2019)

Energy use per capita is highest in the United States, Canada, Norway, and oil-producing countries such as Saudi Arabia, the UAE, and Oman. It is lowest in countries in sub-Saharan Africa.

person. The starkest contrast in the energy portfolio can be seen by comparing energy consumption in the United States versus sub-Saharan Africa. The world consumed 593 quadrillion Btus in 2017; the United States consumed 97.6 quadrillion Btus; the entire region of sub-Saharan Africa consumed 19.8 quadrillion Btus; and the country of Kenya consumed 0.323 quadrillion Btus.

Global average per capita energy consumption has been consistently increasing over the last thirty years. Growth in per capita energy consumption varies among countries and regions. Although countries in the Global North still consume the most per capita in the world, many have seen a decline in overall energy consumed as a result of adoption of energy efficiency technology. Most of the recent growth in per capita energy consumption has been driven by increased consumption in transitioning middle-income countries (and to a lesser extent, low-income countries).

Energy and Electricity

Electricity is one of the most important sources of secondary energy. It is secondary because other energy resources are used to create electricity, which is then transmitted to users such as industry, businesses, and individual residences. Electricity can be produced from both renewable and nonrenewable resources. In many countries, steam turbines produce most of the electricity. An energy input (e.g., coal) is used to boil water in a boiler, which produces steam that is then pumped through a turbine system that spins to produce electrical current. This electrical current is transmitted to its end uses through a series of transmission lines and transformers.

Most countries in the Global North, such as the United States, use a significant amount of energy to generate electricity. Today's US electricity system is a complex network of power plants, transmission and distribution wires, and end users of electricity. Most Americans receive their electricity from centralized power plants that use a variety of energy resources to produce electricity, including coal, natural gas, nuclear energy, or renewable resources such as water, wind, or solar energy. Since 2015, the majority of new energy generation capacity in the United States has come from renewable energy sources, which have become more economically competitive. Yet approximately 80–85 percent of electricity in the United States is still derived from fossil fuels.

Because fossil fuels comprise a majority of electricity in the Global North, energy is a major sustainability challenge. Although the United States produces a good deal of its own energy, it must rely on imported fossil fuels to meet domestic demand, notably oil for transportation. Producing energy from fossil fuels releases pollutants and carbon dioxide into the air, and dependence on these fuels leaves the economy vulnerable to global volatilities and price fluctuations.

While countries in the Global North enjoy affordable and widely available electricity, electricity remains inaccessible for many of those in poverty in the Global South. A lack of energy access is strongly tied to having a low income. The United Nations estimates that one billion people living in sub-Saharan Africa and South Asia do not have access to electricity and hundreds of millions more live with unreliable or expensive power, which poses a barrier to economic development. Map 7.2 highlights populations without electricity in Africa.

Energy inequalities also occur within many countries in the Global South. Those without service are either remote (rural), or poor, or both. Rural areas lack electrical infrastructure, leaving many to rely on other sources for heating and lighting, such as wood, kerosene, or animal waste. In urban areas, impoverished communities remain unserved because they may lack the permanent infrastructure necessary to receive electricity from the formal grid. In total, more than three billion people still rely on wood, coal, charcoal, or animal waste for cooking and heating. A lack of access to electricity has a large impact on a wide range of development indicators, including health, education, food security, gender equality, livelihoods, and poverty reduction.

There are many different reasons for lack of access: lack of sufficient power-generation capacity, poor transmission and distribution infrastructure, high costs of supply to remote areas, or simply a lack of affordable electricity. These reasons, however, all connect to a broader underlying reason for the lack of access: inequitable global structure and an enduring legacy of the conditions created by uneven development. The lack of energy power in and within many countries in the Global South is the result of the imbalance of political and economic power.

Access to electricity in low-income countries has been growing, energy efficiency continues to improve, and renewable energy is making impressive gains in the electricity sector. The World Bank reports that since 2010, the number of people gaining access to electricity has been increasing to around 118 million each year, but these efforts will need to accelerate if the world is going to meet SDG 7 by 2030.

NONRENEWABLE ENERGY RESOURCES

Energy resources can be categorized into two broad groups: nonrenewable and renewable. Today, about 80 percent of the world's energy needs are met using **nonrenewable energy resources**. Nonrenewable energy resources are energy resources that, once converted for use by people, are gone forever. These include coal, oil, natural gas, and nuclear energy. In contrast, **renewable energy resources** derive energy from ongoing natural processes that are constantly being replenished. These include **biomass**, **hydroelectricity** wind, **geothermal energy**, and solar energy.

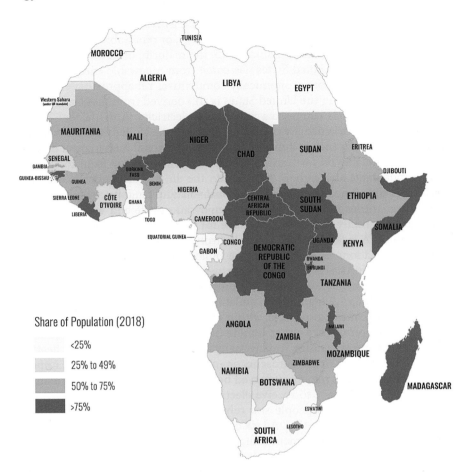

MAP 7.2 **Population without Access to Electricity by Country in Africa (2018)**
Central African countries such as Uganda, Democratic Republic of the
Congo, Malawi, Chad, Niger, and South Sudan have more than 75 percent of
their population without access to electricity.

The story of modernization and industrialization in the Global North
(and after the 1960s the Global South) is largely one of harnessing fossil
fuel energy resources: first coal, then oil, and more recently natural gas.
The availability of inexpensive oil supplies has allowed 150 years of economic growth and supported a worldwide industrial revolution. It has
also enabled a culture of consumerism and a dependence on automobiles.
The economies of Global North grew wealthy on carbon-based energy, but
there was little thought given to how these practices impacted health and
the environment.

Oil, coal, and natural gas are called **fossil fuels** because all three are derived
from plant materials and living organisms that are converted into hydrocarbons
over millions of years. The type of fossil fuel formed depends on the nature of
the organic material buried, the specific environmental conditions (types of
sediment), and the time it takes for conversion. Natural gas is trapped methane,
crude oil is the residual sludge of marine phytoplankton, and coal is highly

compressed terrestrial organic matter—mostly plants—that decomposed less than the others. Today, the three major fossil fuels—**petroleum**, natural gas, and coal—account for most energy production.

Extracting fossil fuels can cause ecosystem degradation and destruction, land subsidence, fires, and explosions of oil wells. In addition, workers in extraction can be injured or killed, and there is the impact on wildlife due to oil spills and pipeline leaks. Combusting fossil fuels also has impacts. Because fossil fuels contain high levels of carbon, they release carbon dioxide into the atmosphere when they are combusted. They also release particulate matter, nitrogen oxides, sulfur dioxide, mercury, and heavy metals.

The story of fossil fuels is a complex one. On the one hand, fossil fuels have served us well, allowing economic development of the Global North. On the other hand, this prosperity and wealth has come at a steep cost everywhere: It has exacerbated the concentration of wealth, accelerated resource exploitation, and generated air and water pollution that have led to respiratory diseases. Further, fossil fuels are driving climate change, creating disasters from leaks and spills, encouraging mountaintop removal, and subjecting economies to market volatility, and the list goes on. In 2020, an estimated 6.6 million people died prematurely from exposure to air pollution, most of this generated by the combustion of fossil fuels.

Coal

Coal: Coal emerged as the leading fuel of the Industrial Revolution and remains a main source of energy. Coal is fossilized organic plant material created when plants decay, compress, and heat over millions of years. The process occurs primarily in wetland regions where woody plants accumulate and are buried by sediments. The repeated advance and retreat of seas 200 to 360 million years ago transformed the plants into peat layers, and harder coals are formed over long periods and with more pressure added by overburden and soil. Coal is classified based on its hardness and the relative amount of carbon within the rock: peat (the softest and youngest), lignite, sub-bituminous, bituminous, and anthracite (the hardest and oldest). The harder and higher carbon such as anthracite burns at higher temperatures and burns cleaner than lower-rank coals such as lignite and sub-bituminous. In the United States, 90 percent of the coal is sub-bituminous or bituminous. In contrast, in China, much of the coal is lignite and bituminous.

Coal is cheap, and widely distributed around the world. However, it is concentrated in four regions: North America, Eastern Europe, China, and Central Asia. Of the three fossil fuels, combusting coal emits the highest percentage of carbon dioxide, followed by petroleum, and natural gas in third. In addition to emitting GHGs and other air pollutants, coal extraction has other severe environmental impacts (surface mining more visibly damages ecosystems, but underground mining can create land subsidence and underground fires) and social impacts (mining work is hazardous and long-term health consequences are common).

China, the world's largest producer and consumer of coal, relies on coal for three-quarters of its electricity. The large amount of lignite and bituminous coal used for energy is the main reason Chinese cities are heavily polluted with sulfur and smog. The particulates released by coal burning have led to major increases in pulmonary and respiratory diseases. Coal-fired power plants have come under increasing public scrutiny, in part due to the emissions of mercury, a toxic pollutant. Smog levels in cities such as Beijing and Chongqing can reach such unhealthy levels that the government orders most residents to stay inside. On the production side, the coal-mining industry in China has a poor safety record. News stories highlight the periodic coal mine disasters that kill hundreds of Chinese coal miners annually.

China's most recent Five-Year Plan features targets to increase renewable energy sources such as hydropower, wind, solar, and nuclear. Currently, China has thirty-eight operational nuclear power reactors and another nineteen under construction. Despite the increase in renewables, a recent study by the US Energy Information Agency suggests that by 2040, fossil fuels are still expected to make up most of China's electricity generation mix.

Oil

Oil: Although coal started the Industrial Revolution, oil emerged as the primary source of energy in the twentieth century and has powered economic growth and development since. The United States has less than 5 percent of the world's population but consumes 20 percent of its oil and is the top-consuming country for oil, followed by China, India, and Japan. The United States consumes 19.6 million barrels of oil per day and depends on imported oil for 25 percent of its petroleum demand. In the United States, oil has become central for all forms of industrial production and transport. Many say that without cheap oil, the US economy would grind to a halt. While public debate has focused on the need to reduce the use of coal, there has been less discussion about reducing the use of petroleum. Crude oil is used to make the petroleum products that fuel transportation vehicles (e.g., airplanes, cars, and trucks), to heat homes, and to make products such as medicines and plastics. Very few have suggested phasing out of petroleum by-products such as plastic chairs, cosmetics, and synthetic carpets. Crude oil, petroleum, and crude oil by-products are integrated within the global economy. As mentioned at the beginning of the chapter, US dependence on both the automobile and oil present a challenge to a more sustainable energy future.

Natural Gas

Natural gas: Natural gas is largely methane and is called a "cleaner" fossil fuel because when burned it emits less CO_2 and other air pollutants than other fossil fuels. Natural gas has become the major source of energy production in some countries including the United States, the European Union, and Russia. Recently, natural gas has replaced coal as a leading source of energy. The industrial sector uses natural gas as a fuel for process heating, while the residential sector uses natural gas to heat buildings and water, cook, and dry clothes. The commercial sector uses natural gas to heat buildings and water and to operate refrigeration and cooling equipment, and the transportation sector uses natural gas as a fuel and most of the vehicles that use natural gas as a fuel are in government vehicle fleets.

The Impacts of Fossil Fuels: Photochemical Smog

Energy consumption is directly responsible for the worst air pollution. The term "**smog**" (a contraction of the words "smoke" and "fog") was coined in the early twentieth century to refer to visible air pollution. Smog can be natural (from volcanic activity or forest fires) or man-made (from the combustion of fossil fuels). Smog also occurs when volatile organic compounds (VOCs) are released from burning fuel (gasoline, oil, wood, coal, natural gas). VOCs include chemicals such as benzene, toluene, methylene chloride, and methyl chloroform. When VOCs are emitted, they can react with nitrogen oxides and oxygen in the presence of heat and sunlight and form ground-level ozone, or smog. Photochemical smog is not emitted directly into the air but is created through a series of chemical reactions.

A majority of smog is produced by transportation sources—cars, trucks, and trains. Smog is found primarily in urban areas and is often worse in the summer months when heat and sunshine are more plentiful.

For many cities, smog represents the single most challenging air pollution problem. For example, in the United States, 140 million Americans live in urban areas that exceed air quality standards. Los Angeles, Sacramento, San Diego, Phoenix, and Denver experience many days each year where smog exceeds safe levels. In the Global South cities such as Bangkok, Beijing, Kolkata, Delhi, and Tehran experience similarly unsafe levels of air pollution.

The health effects are serious and widespread. Exposure to smog can cause eye and nose irritation, a decrease in the lungs' working capacity, shortness of breath, and wheezing and coughing. It can also exacerbate any preexisting respiratory issues such as asthma. Studies have shown that there is a link between long-term exposure and premature death. In extreme cases, it can cause death. In a 2021 study, Karn Vohra and colleagues reported that pollution from fossil fuels caused four million deaths in China and India combined, and a million each in Bangladesh, Indonesia, Japan, and the United States.

Beijing in China is one of the world's worst cities for smog. Delhi, India also suffers poor air quality. One study estimated that smog in Delhi causes the death of about 10,500 people every year and another study reported that 2.2 million children in Delhi have irreversible lung damage due to the poor quality of the air. Children are more vulnerable to the negative effects of air pollution as they are growing and developing, which means that they breathe a higher rate of air per kilogram of their body weight. They also spend more time outside and are thus more exposed to pollutants.

There is no doubt that fossil fuels fail to meet the metric of sustainability: The consumption of nonrenewable fossil fuels not only makes them unusable by later generations but also undermines human health, weakens the environment, and contributes to climate change. For decades, progress in addressing these impacts has been hampered by the assumption that economic growth is dependent on current energy sources. Early in the Industrial Revolution, smokestacks belching smoke were a sign of success and economic vibrancy. While conservative economists are concerned that efforts to reduce fossil fuels, conserve energy, or switch to renewables could negatively impact economic growth, the good news is that cost-effective sustainable options—renewable energy—are available, cost competitive, and being used in many countries.

RENEWABLE ENERGY

A renewable-energy future is absolutely essential, yet the inertia of "business as usual" has stalled the development of renewable energy until relatively recently. Renewable energy such as solar, wind, and geothermal have become cost competitive with fossil fuels and offer tremendous potential to creating a nonpolluting, more sustainable energy system. Many countries are making efforts to increase the use of renewable energy sources and have set ambitious targets for the proportion of electricity emanating from renewable sources. For example, China, the United States, Germany, and India have set targets to dramatically increase wind energy capacity over the next decade. Since 2010, more than half of all new wind power was added outside the traditional markets of Europe and North America as China and India experienced tremendous growth in wind and solar capacity.

Solar

Solar energy, once slow to develop due to of a lack of consistent government support and investment in many countries, has seen tremendous gains in the last ten years. Markets for solar energy are maturing rapidly and solar electricity is increasingly economically competitive with nonrenewable energy sources. Both the costs of solar materials and the costs of installation have decreased considerably. China's investment in solar energy has helped reduce the cost of manufacturing solar panels, making this option more affordable than before. There are several types of solar energy systems, including photovoltaic (which converts sunlight directly), parabolic troughs (which concentrate energy to drive a generator), and solar water heaters. Solar power is more affordable, accessible, and prevalent than ever before. Solar installations have grown tremendously in the last ten years, and the costs of installing photovoltaic panels have dropped by more than 60 percent, making solar installations affordable to many. Homes in many countries in the world have solar panels and solar water heaters; large-scale solar farms provide electricity to hundreds of thousands of customers. Moreover, the solar industry is a proven incubator for job growth. For example, in the United States, the number of solar jobs increased significantly over the past ten years and there are now approximately 250,000 solar workers. Job growth in solar has far outpaced job growth in conventional energy such as coal and oil. While solar power is still a small part of total energy, its growth puts it on target to be one of the larger sources of renewable energy by 2040.

Wind

The industry of generating electricity from wind power has been growing exceptionally. However, 80 percent of the total installed wind power capacity of the world is located in only eight countries. China leads the world in using **wind power**, followed by the United States, Germany, and India. When considering the percentage of wind power as part of an overall energy portfolio, however, many European countries rank highest. Wind power provides 48 percent of Denmark's electricity, 33 percent in Ireland, and 27 percent in Portugal and Germany. In comparison, wind power energy only provides 7 percent of electricity in the United States.

Wind power energy offers many advantages, primarily that it is abundant in many areas, inexhaustible, and relatively cost-effective. In many countries, the wind industry already employs more people than coal. However, wind is not without its drawbacks. Wind turbine blades can cause noise and visually impact the landscape. Birds and bats have been killed by flying into spinning turbine blades. Finally, land suitable for wind-turbine installation competes with alternative uses for the land (such as residential development), which may be more highly valued than electricity generation, making expansion of wind power a challenge.

Hydropower

The force of falling water can be used to drive turbines, which in turn supplies electricity. While waterpower is a largely nonpolluting, renewable energy source, harnessing it with hydroelectric dams still involves ecological, social, and cultural trade-offs. Constructing dams can drown habitats and farmland. Dams can also impede or prevent the migration of fish, even when fish ladders are provided. Ecological consequences extend throughout river systems.

Biofuels

Biofuels are fuels made from solid, liquid, or gas biomass derived from living matter. The most touted biofuel is ethanol made from corn. Recently farmers have benefited from the boost that ethanol has given to corn prices, but this trend is unlikely to be sustainable in the long term. Using corn for energy production contributes a relatively large carbon footprint, increases food prices, and is an inefficient use of prime agricultural land. Seeking a nonfood crop that can be grown on marginal land (so agricultural lands can be used more efficiently) may be a more responsible approach to increase available biomass energy.

Trends in Renewable Energy Production and Consumption

As prices for renewable energy continue to fall, they are increasingly becoming an economical energy choice for homeowners and businesses, primarily in the Global North. The most significant hurdle to affordable solar and wind energy remains the soft costs—permitting, zoning, and hooking a system up to the power grid. Part of the problem has been that renewable energy has been chronically underfunded, especially in the United States, while fossil fuels have been subsidized for decades. The positive news is that, in the last ten years, there have been increases in subsidies for renewables. Energy subsidies keep prices for customers below market levels or reduce costs for customers and suppliers. Recently total subsidies for all renewable energy in the United States have been higher than for fossil fuels. Programs and policies that help overcome the costs of renewable energy will help level the playing field and speed our uptake of renewable energy sources.

Another important barrier to an increased reliance on renewable energy is the periodicity of many of these energy sources. Solar panels only produce energy when the sun is shining; wind turbines turn at the whim of winds. With renewable energy, power generation does not necessarily align with times of power demand. Traditional fossil-fuel-burning power plants are unable to turn on and off quickly to "fill in" in response to this periodicity. For these reasons, developing advanced electricity storage capacity and updating the electrical grid will be important steps in phasing in renewable energy.

THE ENERGY PORTFOLIO

The global energy economy relies on a diverse set of energy resources that typically includes both nonrenewable and renewable sources. Some resources are used close to where they are extracted, while others travel thousands of miles. Energy resources are part of a complex global supply chain that moves energy from one geographic location to another.

The **energy portfolio** or energy mix varies at different geographic scales—by city, province, country or region. Currently, the global energy portfolio is dominated by fossil fuels, which account for approximately 85 percent of global energy consumption.

Figure 7.1 shows the most recent energy portfolio of the United States. A majority of energy in the United States—63 percent—comes from nonrenewables such as natural gas, petroleum, and coal. The United States consumes large amounts of every energy resource, with natural gas comprising the single largest provider of energy, followed by petroleum, then coal. Only 11 percent of energy consumed comes from renewable resources. The sources of energy used by each sector vary widely. For example, petroleum provides about 91 percent of the transportation sector's energy consumption, but less than 1 percent of the electric power sector's primary energy use. The mix of energy sources in an energy portfolio changes over time. There has been a decrease in the use of coal and an increase in the use of renewables over the last ten years. In fact, use of renewables is currently at an all-time high.

Figure 7.2 shows the distribution of energy within the various sectors. Not all resources in the US energy portfolio go to the same sectors, nor do all sectors consume the same amount of energy from each source. Generation of electricity is by far the largest single use of energy resources in the United States. Furthermore, not all energy is easily substituted. For example, electricity cannot replace petroleum for transportation without a shift to electric vehicles.

Energy portfolios can vary greatly among countries. Compare the US energy portfolio to that of Germany where renewables contribute 38 percent of energy for electricity generation, with more than 24 percent coming from wind power (coal still remains an important source at 30 percent)

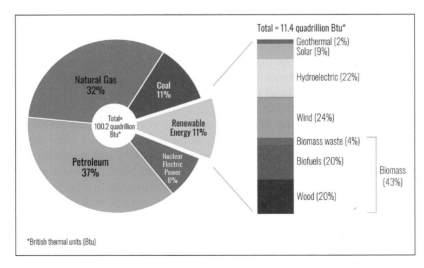

FIGURE 7.1 US Primary Energy Consumption by Energy Source (2019)
This figure shows the most recent energy portfolio of the United States.
A majority of energy in the United States—63 percent—comes from
nonrenewables such as natural gas, petroleum, and coal. Only 11 percent
of energy consumed comes from renewable resources (although this has
been increasing over the last decade).

(Figure 7.3). **Nuclear power** is also an important source in Germany, providing 13 percent of electricity produced. Germany is on track to make renewable energy sources deliver more than 65 percent of its consumed electricity by 2030. Tanzania's energy portfolio provides an interesting contrast to the Global North. Almost 55 percent of its electricity comes from natural gas, followed by hydroelectrical power at 36 percent. Oil represents only 5 percent and is mostly used for backup generators. Despite the impressive numbers in hydroelectrical power, there is very little other renewable energy (solar, wind, or geothermal) being used.

These comparisons of these three countries remind us that a country's energy portfolio is influenced by the geographic disparities in energy sources, production, and total consumption. Energy resources are distributed unevenly around the globe, giving some geographic locations natural advantages and disadvantages. The rich oil deposits in the Middle East, an abundance of coal in the United States and China, and uranium deposits in Australia are not just key providers of domestic energy; in some cases, they have allowed these countries to become exporters of high-demand commodities.

Renewables are also unevenly distributed. Wind and solar radiation occur at varying intensities and durations in different latitudes and climates. Precipitation (in the form of snowfall or rain) can be seasonal and concentrated in specific areas.

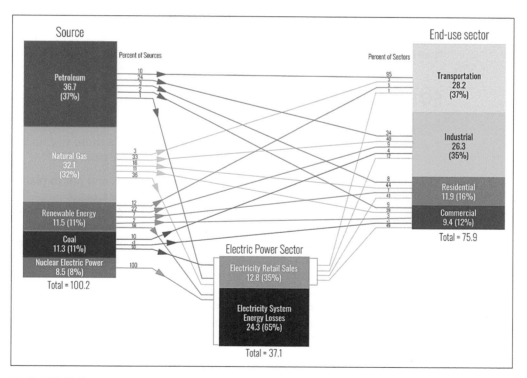

FIGURE 7.2 Energy Distribution in the United States
The distribution of energy within the various sectors in the United States occurs through a series of complex and interconnected energy infrastructure. Not all resources in the US energy portfolio go to the same sectors, nor do all sectors consume the same amount of energy from each source. Generation of electricity is by far the largest single use of energy resources in the United States. In the United States, electricity is generated using primarily nonrenewables such as coal, petroleum, and natural gas. Renewable energy also figures in, but less than fossil fuels.

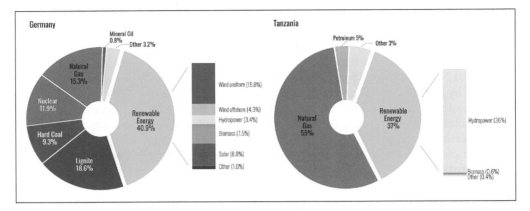

FIGURE 7.3 Germany and Tanzania's Energy Portfolio
Comparing the United States', Germany's, and Tanzania's energy portfolios underscores how energy resources can vary greatly among countries.

Further, a country's energy portfolio is set within a political economy, that includes policies that subsidize or encourage certain types of energy use, and issues of access such as affordability. High rates of poverty in Tanzania are one reason only 38 percent of all households are connected to electricity.

At any scale—global, country, or local—the energy portfolio will shift in response to changing prices for certain resources, policies (subsidies or incentives), and regulations (e.g., on air pollution) or technological advances.

SOLUTIONS TO INCREASING SUSTAINABLE ENERGY

There are many actions that can be taken to work toward a more sustainable energy future. Experts suggest that the following should be priorities:

- Make clean-cooking solutions a top political priority, and put in place specific policies, cross-sectoral plans, and public investments.
- Double the financing for SDG 7 globally, from the current annual level of about US$500 billion to US$1 to $1.2 trillion per year until 2030.
- Close the electricity access gap by establishing detailed plans of action nationally, regionally, and globally to "leave no one behind."
- Accelerate the pace of transition toward renewable energy, especially in end-use sectors such as transport, buildings, and industry.
- Harness the potential of decentralized renewable energy solutions. **Decentralized energy systems** are characterized by locating energy production facilities closer to the site of energy consumption and allows for more optimal use of renewable energy.
- Scale up investments in energy efficiency across all sectors of the economy.
- Invest in data collection systems and data analysis to build institutional capacities at the national level and ensure effective monitoring of the SDG 7 targets.

While there are many actions a country or locale can take, generally there are two broad strategies to achieve many of the targets: focusing on increasing electricity in the poorest countries where access is missing or unreliable and increasing renewable energy share in a country's energy portfolio. Table 7.1 outlines the targets for SDG 7.

TABLE 7.1 SDG 7 Targets and Indicators to Achieve by 2030	
Target	**Indicators**
7.1 Ensure universal access to affordable, reliable, and modern energy services	**7.1.1** Proportion of population with access to electricity **7.1.2** Proportion of population with primary reliance on clean fuels and technology
7.2 Increase substantially the share of renewable energy in the global energy mix	**7.2.1** Renewable energy share in the total final energy consumption
7.3 Double the global rate of improvement in energy efficiency	**7.3.1** Energy intensity measured in terms of primary energy and GDP
7.A Enhance international cooperation to facilitate access to clean energy research and technology, including renewable energy, energy efficiency, and advanced and cleaner fossil-fuel technology, and promote investment in energy infrastructure and clean energy technology	**7.A.1** Mobilized amount of US dollars per year starting in 2020 accountable toward the $100 billion commitment
7.B Expand infrastructure and upgrade technology for supplying modern and sustainable energy services for all in developing countries, in particular least developed countries, small island developing states, and land-locked developing countries, in accordance with their respective programs of support	**7.B.1** Investments in energy efficiency as a percentage of GDP and the amount of foreign direct investment in financial transfer for infrastructure and technology to sustainable development services

Source: United Nations Sustainable Development Goals. https://sustainabledevelopment.un.org/sdg7

The good news is that the global electrification rate rose from 83 percent in 2010 to 90 percent by 2018. Access to clean cooking fuels and technologies was at 63 percent in 2018, although this varies by region as shown in Figure 7.4. And the renewable energy share of total energy consumption is steadily increasing. International financial flows to countries in support of clean and renewable energy (such as hydropower and solar) has reached $21.4 billion currently, twice what it was in 2010.

Power Africa

Target 7.1 aims to provide clean electricity to the almost one billion people who currently live without it. Despite being home to almost a fifth of the world's population, Africa accounts for little more than 3 percent of global electricity demand. Nearly half of Africans (about 590 million people) did not have access to electricity in 2020, while around 80 percent of sub-Saharan African companies suffered frequent electricity disruptions leading to economic losses. In addition, more than 70 percent of the population of Africa, around 900 million people, lack access to clean cooking. Sub-Saharan Africa's current electrification rate of 45 percent is very low compared to other parts of the world. To scale up the electricity sector, this region will need substantial investments of up to $120 billion a year through 2040.

Women and children bear the greatest burden of energy poverty. Lack of access to clean fuels is one of the most significant contributors in low-income

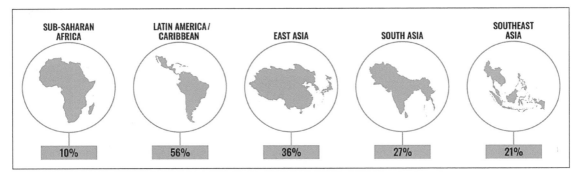

FIGURE 7.4 Population with Access to Modern Energy Cooking Services by Region
Although most regions in the Global South are home to some people without access to modern cooking services, it is extremely low in sub-Saharan Africa, where only 10 percent of people have access to clean cooking services such as gas stoves.

countries to women's workloads and poses a barrier to the economic advancement of women. The *Solutions* 7.1 box profiles initiatives that target women. Beyond households, gaining access to modern energy services is also essential for businesses, farmers, and community buildings. A priority for policy makers is to address the persistent lack of access to electricity and clean cooking and the unreliability of electricity supply. These have acted as brakes on development.

It is possible that Africa could become the first major region to develop its economy primarily by using energy efficiency, renewables, and natural gas—all of which offer huge untapped potential and economic benefits. This is an exciting prospect, provided investment and loans do not come with strings attached. Currently, renewables account for three-quarters of new electrical generation in the region. With an average of 300 days of sunshine each year, many areas in Africa have excellent solar resources, but currently have less than 1 percent of the global installed capacity. Solar could overtake hydropower and natural gas in Africa. There is also potential to develop wind power in Ethiopia, Kenya, Senegal, and South Africa. Solutions to providing electricity in this region will likely involve a combination of expanding the electrical grid; creating mini-electrical grids and **stand-alone power systems**; using Liquified Petroleum Gas (LPG) for clean cooking in urban areas, particularly slums; investing in improved cookstoves for rural areas, where the majority of people without access to electricity live; and using biofuels, solar, hydro, and other renewables.

Promote Renewable Energy in Africa

Africa is home to abundant renewable energy resources and its renewable energy power potential is substantially larger than the current and projected power consumption of the continent. Growth has been constrained to date by limited access to financing, underdeveloped grids, and infrastructure, and, in some countries, by an uncertain policy environment. Despite this, recent advances in renewable energy technologies and accompanying cost reductions enable a large-scale deployment of renewable energy, giving Africa a cost-effective path to sustainable and equitable growth (Table 7.2).

There have been some promising improvements. The number of people gaining access to electricity in Africa has increased steadily, outpacing population growth. Kenya, Ethiopia, and Tanzania have led recent progress and account for more than 50 percent of the African population gaining electricity access.

SOLUTIONS 7.1

Investing in Women's Entrepreneurship in Clean Energy

Target 7.1: Ensure universal access to afford-able, reliable and modern energy services.

The **off-the-grid energy** industry is gaining traction in the Global South as a key strategy. Many small companies offer "solar kits" comprising of a solar panel, battery, and accompanying appliances such as light bulbs, a flashlight or lantern, a mobile phone charger, and a radio—some even include a digital television.

Initiatives such as Solar Sister, Envirofit, Barefoot College, Kopernik, and Grameen Shakti have reached millions of low-income people in African and Asian countries with these products and many of these focus their outreach on women.

Solar Sister, for example, recruits, trains, and supports female clean energy entrepreneurs. The women entrepreneurs sell a range of low-cost clean energy products, including small portable lights and lanterns, solar radios, and clean cookstoves. One of their main products is a solar home system. This home system includes a roof-top solar panel that has a phone charger, three lamps (one is portable), and up to fifteen hours of light. A system like this may cost around $30, with daily installments of $.50 paid for a period of twelve months, after which customers own the system outright. To date, Solar Sisters has supported 7,400 entrepreneurs who have sold more than 400,000 clean energy products to some 3.4 million people.

Increasingly, these small-scale off-grid energy initiatives use the pay-as-you-go purchase model, which is a flexible payment plan. Pay-as-you-go models have become increasingly attractive in many markets and one of the biggest advantages of this system is that people can pay in installments, often with mobile money. Many households are able to pay back the low-cost solar lights and lanterns in a few months and realize an overall cost savings as they no longer need to

PHOTO 7.2 Solar Power in a Home in Kenya
This home in Kenya has a small solar panel on the roof and a solar light for indoor lighting. The NGO Wema has supplied abut 100 solar lights to the village. This solar panel does not require significant infrastructure and can improve access to energy. A panel of this modest size can charge a lightbulb or two inside the house and recharge a mobile phone.

buy kerosene, candles, or batteries. This allows the household to use that savings to send children to school or save up for additional solar infrastructure.

Families benefit from better health and economic stability by using solar light. Children have reliable, bright lighting to study by at night. Women in particular benefit from time savings thanks to increased lighting after dark and those who use solar lights report increased productivity after the sun sets. Solar lights replace traditionally used kerosene lamps, which can emit a thick black smoke (carbon dioxide). Kerosene can also be expensive; some families may spend 10 to 15 percent of their household income on the fuel. Kerosene lighting causes a serious fire hazard because the lamps are easily knocked over and the fuel is highly explosive. Schools and homes have been burnt down and people killed because of fires caused by kerosene lamps.

Initiatives like Solar Sister was founded to create benefits for individual women and their households and communities by enhancing income and autonomy, teaching business skills, and improving household health, child education, and community safety (see Photo 7.2).

Sources: "Africa Energy Outlook, 2019." International Energy Agency (2019, November). https://www.iea.org/reports/africa-energy-outlook-2019. "Home." Solar Sister (2022). https://solarsister.org/

East Africa, particularly Kenya, Rwanda, and Ethiopia, stands out as a beacon of progress. The region has steadily increased its electrification rate every year for the last twenty years. A combination of factors can be attributed to progress in Kenya. In 2018, the government launched the Kenya National Electrification Strategy aiming to provide electricity to all citizens by 2022. This included a program called the Last Mile Connectivity Project that supported both grid and off-grid solutions. Although the program has not brought electricity to all, currently Kenya has the highest electricity rate in East Africa with 75 percent. Despite a shift to modern and more efficient energy sources over this period, Africa's current policy settings are not enough to put it on track to meet its development needs and provide reliable and modern energy services for all.

Solid biomass—including fuelwood, charcoal, and dung—is the most widely used fuel across the region. Several governments, including Ghana, Cameroon, and

TABLE 7.2 Selected Energy Infrastructure Projects in Africa

Project	Countries Involved	Proposed Construction Start	Cost Estimate (USD)
Batoka George Hydropower Project	Zambia and Zimbabwe	Construction contract awarded, project not yet started	$4.5 billion
Grand Ethiopian Renaissance Dam	Ethiopia	Under construction (completion date of 2025–27)	$4.8 billion
Ruzizi III Hydropower Project	Burundi, DR Congo, and Rwanda	No start date yet	$625 million
Trans-Saharan Gas Pipeline	Nigeria, Niger, and Algeria	No start date yet	$10–13 billion
Zambia-Tanzania-Kenya Transmission Line	Kenya, Tanzania, Zambia	Project funded but construction not yet started	$500 million
East African Crude Oil Pipeline	Uganda, Tanzania	Under construction	$3.5 billion

Source: By author compiled from several sources.

Kenya, are promoting Liquefied Petroleum Gas (LPG) as a better alternative. Ghana has been promoting LPG and currently 24 percent of the population rely on LPG. In addition, the Ghanaian government recently distributed LPG cookstoves to 150,000 households in 108 districts under the LPG Promotion Programme launched in 2017. It intends to have 50 percent of households using LPG in Ghana by 2030. In other countries such as Nigeria, LPG uptake primarily displaces kerosene. Clean cooking has only increased by 0.7 percent since 2013 in rural sub-Saharan Africa, in part because supply chains for cleaner fuels lack the necessary scale to reach many rural communities (Map 7.3). The importance of clean cooking is detailed in *Solutions* 7.2, which highlights strategies to increase access to clean cookstoves.

MAP 7.3 Population without Access to Clean Cooking by Country in Africa (2018)
Many countries in Africa have more than 50 percent of the population without access to clean cooking methods. Women (and their children) without clean cooking stoves are exposed to higher levels of indoor pollution, which can cause both short- and long-term respiratory issues.

SOLUTIONS 7.2

Clean Cookstoves

Target 7A: Enhance international cooperation to facilitate access to clean energy research and technology, including renewable energy, energy efficiency, and advanced and cleaner fossil-fuel technology, and promote investment in energy infrastructure and clean energy technology.

Target 7B: Expand infrastructure and upgrade technology for supplying modern and sustainable energy services for all in developing countries, in particular least developed countries, small island developing states, and land-locked developing countries, in accordance with their respective programs of support.

A 2020 World Bank report finds that four billion people around the world still lack access to clean, efficient, convenient, safe, reliable, and affordable cooking energy. The report also notes that the lack of clean cooking is costing the world more than \$2.4 trillion each year, driven by adverse impacts on health, climate, and gender equality. For example, household air pollution, mostly from cooking smoke, is linked to around four million premature deaths annually, with half those deaths being children under the age of five. Women and children are disproportionately affected by household air pollution because they are most exposed as they often spend a significant part of their day collecting the fuel—firewood, for instance—needed to cook a meal. Smoke and particulate matter are emitted when cooking over fire.

Families who switch to clean cookstoves, like those shown in Photo 7.3, significantly reduce time spent collecting wood, money spent on solid fuels, and smoke output, significantly improving the health of women and children. In many rural areas and urban slums, clean cookstoves can

PHOTO 7.3 Cookstoves
This photo shows three different models of clean cookstoves. The Mimi-Moto runs on wood or pellets, but at higher thermal efficiency that means a far cleaner burn. The two-burner stove uses liquefied petroleum gas (LPG). The Nuwave hotplate runs on electricity. All are relatively small in size but can have a significant impact on reducing exposure to indoor air pollutants.

have positive impacts. A charcoal stove can save a family around $200, which can pay for a whole year of schooling for one child. It may not seem like a great deal, but this is a huge cost saving for the average family.

While clean cooking fuels and technologies are now more available, consumer awareness, accessibility, and affordability remain significant challenges. The provision of clean cooking solutions does not guarantee that rural and urban communities will stop using traditional cooking methods. One study found that in Kenya, while only 26 percent of households said that charcoal was their primary cooking fuel, almost 70 percent were using it some of the time. Although many of the proposed cookstoves allowed faster and more efficient cooking, they were much less flexible in adapting to women's needs than traditional three-stone fireplaces. Many recipients of clean cookstove programs thus continue to use traditional fuels and solutions for sociocultural, economic, and pragmatic reasons. Programs to replace traditional but unsustainable fuel use will likely only succeed if they prioritize user preferences and local cooking contexts.

Unfortunately, success stories are few and far between in the area of clean cooking. In sub-Saharan Africa, gains in clean cooking have been marginal. The United Nations reports that the need for the rapid deployment of clean cooking fuels and technologies has not received the political attention it deserves. High entry costs for many clean cooking solutions, a lack of consumer awareness of their benefits, financing gaps for producers seeking to enter the market, slow progress in the innovation of clean cookstoves, and lack of infrastructure for fuel production and distribution have together kept widespread solutions to this challenge out of reach.

Experts suggest countries formalize cooking energy demand in national energy planning, and dramatically scale up public and private financing. To achieve universal access to clean energy cooking by 2030, the World Bank estimates that approximately $150 billion is needed per year, including $39 billion from the public sector, $11 billion from the private sector, and the remaining $100 billion from household purchases of stoves and fuels. To that end, the World Bank's Energy Sector Management Assistance Program has established a Clean Cooking Fund to support public and private investments in the clean cooking sector.

Sources: "The State of Access to Modern Energy Cooking Services." The World Bank (2020, September 24). https://www.worldbank.org/en/topic/energy/publication/the-state-of-access-to-modern-energy-cooking-services

"SDG7: Data and Projections." US International Energy (2019, November). https://www.iea.org/reports/sdg7-data-and-projections/access-to-electricity

However, it should be noted that sometimes these clean cookstoves still release emissions, just fewer. In addition, some clean cookstoves have performed well in the lab, but not in the field, where some have cracked or broken under constant heat.

Beyond countries in Africa, several other countries in the Global South have made clear progress on expanding electricity access in recent years. They share common factors that include sustained political commitment and financing as well as policies and incentives that help support energy suppliers while keeping consumer prices affordable. Other common factors are that they use targeted public funding for renewable investment and they have an effective balance of grid and off-grid systems.

Improve Energy Conservation and Efficiency

In the Global North (and to a lesser extent in the Global South), buildings account for approximately 40 percent of all energy use, 38 percent of carbon dioxide emissions, and 12 percent of total water consumption. Where buildings are located can also have far-reaching impacts, such as its influence on transportation behavior, the landscape, and its connections to the electric, gas, and water grids.

For these reasons, investing in sustainable building practice is a critical way to move toward a more sustainable use of resources, including energy. Many countries and local governments have policies and programs in place to encourage energy conservation.

Renovation, retrofit, and refurbishment of existing buildings are opportunities to upgrade energy performance and decrease energy demand. This can involve measures ranging from superficial, low-cost measures and modifications to full overhauls or replacement of major systems. Retrofitting existing buildings saves on embodied energy (the energy required to extract, manufacture, and deliver building materials and construct the building) and avoids the waste of building demolition. Energy efficiency retrofits can reduce the operational costs, particularly in older buildings.

For example, Washington, DC, has recently spent $81 million on energy efficiency and renewable energy services for low-income residents in single-family homes and multifamily buildings. In a strategic move that likely increased program visibility among target communities, the city paid for Ben's Chili Bowl—an iconic restaurant for the city's African American community—to install energy-efficient lighting. The long-lasting lighting will reduce energy use by approximately 50 percent and save the business about $1,200 in yearly energy costs. Washington, DC, has also installed 158 solar photovoltaic systems on roofs of low-income residents' homes.

Build Green Buildings

In addition to retrofitting buildings, the rise of green building standards has revolutionized the construction industry. "Green building" is a broad term with multiple meanings. One definition is "structures designed to increase efficiency of resource use, including energy, water, indoor environmental quality, siting, infrastructure and pollution." Alternative definitions prioritize occupant health, reducing the impacts of the built environment on the natural environment, or focus on considerations over a building's full life cycle including design, construction operation, maintenance, and removal.

Green buildings are an important shift in building and construction activities, for several reasons. First, cutting-edge buildings utilize new energy and resource-efficient technologies and demonstrate their efficacy. Builders who work on green buildings gain practical experience in installing these systems and can bring them into the mainstream. Second, countries and cities have begun to change their building codes to encourage green buildings. This is necessary because, in some cases, building codes prohibit or otherwise complicate use of these new technologies.

A number of green building certification systems now exist, including the US Green Building Council's Leadership in Energy and Environmental Design (LEED), which is the world's most widely used green building rating system (although there are many other versions of green building councils around the world and some have different rating tools). Many countries are making significant strides in sustainable building design and construction. Countries that have been recognized for their growing commitment for LEED-certified green buildings include China, India, Brazil, and South Korea. The green building materials market is now a trillion-dollar industry. Currently, there are more than 82,000 commercial projects participating in LEED worldwide, totaling more than 1.4 billion gross square meters of space worldwide.

Switch to Renewable and Alternative Energy

In addition to promoting renewable energy as a solution in Africa, switching to renewable and alternative energy everywhere is a key solution. Around the world, many homes and businesses are installing solar energy and in many of the low-income countries in the Global South, solar is making a significant difference to the quality of life. *SDGs and the Law* 7.1 discusses the role of the International Renewable Energy Agency in facilitating a shift to renewable energy.

There may be solutions to promoting adoption of small-scale alternative energy systems to achieve alternative energy goals, even in the Global North. Chicago may be known as the Windy City, but its sustainability plan notes that the city is also well-suited to produce solar energy throughout the year. To promote more adoption of solar energy, the city plans to reduce the time and paperwork involved in the solar permit approval process for small-scale solar installations.

Wind power has also dramatically increased in the past ten years. There are new small-scale wind electric systems for homes that have wind turbines that can be mounted on roofs, lower electricity bills by 50 to 90 percent, and are a source of power during utility outages. Most of the small-scale wind turbines are vertical turbines that have clean, aerodynamic lines, and are both quiet and affordable. For example, the company Aleko sells a wind turbine for use in homes, boats, and telecommunications towers for a modest $159.

 ## SDGs AND THE LAW 7.1

The Statute of the International Renewable Energy Agency

The proposal for an international agency dedicated to renewable energy was made in 1981. As global interest in renewable energy steadily increased, world leaders convened in several settings to focus on renewable energy policies, financing, and technology. Finally, in 2009, the International Renewable Energy Agency became the first intergovernmental organization dedicated to promoting renewable energy.

The International Renewable Energy Agency supports countries in their transition to a sustainable energy future and serves as the principal platform for international cooperation, a repository of policy, technology, resource, and financial knowledge on renewable energy. It promotes the widespread adoption and sustainable use of all forms of renewable energy, including bioenergy, geothermal, hydropower, ocean, solar, and wind energy, in the pursuit of sustainable development, energy access, energy security, and low-carbon economic growth and prosperity.

The International Renewable Energy Agency's role is to seek out, establish, and develop new synergies, facilitate dialogue, share best practices, promote enabling policies, build capacity, and foster cooperation at the global, regional, and national levels. It encourages investment flows and works to strengthen technology and innovation, with diverse stakeholders contributing to these shared goals. It serves as the convening platform to advance the widespread adoption and use of renewable energy, with the ultimate goal of safeguarding a sustainable future.

Source: International Renewable Energy Agency "Vision and Mission" (2021). https://www.irena.org/statutevisionmission

There has also been promising new technology development with regard to aviation as a number of firms are developing plans and prototypes for low and zero-emission airplanes. Rolls-Royce's first all-electric aircraft completed its maiden flight in 2021. Currently the company is working with another airline company to develop an all-electric passenger aircraft. The *Making Progress* 7.1 box reports on the increase of renewable energy in the United States.

Set Carbon Neutrality Goals

As part of efforts to switch to renewables, an emerging trend among countries, cities, and businesses is pledging to become carbon neutral. The concept **carbon neutral** or having a net zero carbon footprint refers to achieving net zero carbon emissions by offsetting any carbon released with an equivalent amount of sequestered or offset carbon.

There are three main ways to achieve carbon zero. The first is *reducing* or limiting energy usage and emissions from transportation (by walking, using bicycles or public transport, avoiding flying, using low-energy vehicles), as well as from buildings, equipment, animals, and processes.

The second way is to obtain electricity and other energy from a *renewable energy* source. This can be done either by generating energy directly from the renewable source (e.g., installing solar panels on the roof) or by selecting an approved green energy provider and by using low-carbon alternative fuels such as sustainable biofuels.

The third way involves buying carbon offsets. Examples of sequestering carbon includes planting trees, funding carbon projects that would prevent

---

MAKING PROGRESS 7.1

Renewable Energy in the United States

Target 7.2: Increase substantially the share of renewable energy in the global energy mix.

In 2019, US annual energy consumption from renewable sources exceeded coal consumption for the first time in 130 years, according to the US Energy Information Administration. This outcome mainly reflects the continued decline in the amount of coal used for electricity generation over the past decade and the growth in renewable energy, mostly from wind and solar.

Although coal was once commonly used in the industrial, transportation, residential, and commercial sectors, today, coal is mostly used in the United States to generate electricity. About 90 percent of US coal consumption is in the electric power sector and the other 10 percent in the industrial sector. Electricity generation from coal has declined significantly over the past decade while natural gas consumption in the electric power sector has significantly increased in recent years. Natural gas has also displaced much of the electricity generation from retired coal plants.

Total renewable energy consumption in the United States grew for the fourth year in a row to a record-high 11.5 quadrillion Btu in 2019. Since 2015, the growth in US renewable energy is almost entirely attributable to the use of wind and solar in the electric power sector. In 2019, electricity generation from wind surpassed hydro for the first time and is now the most-used source of renewable energy for electricity generation in the United States on an annual basis.

Source: "US Renewable Energy Consumption Surpasses Coal for the First Time in over 130 Years." US Energy Information Administration (2020, May 28). https://www.eia.gov/todayinenergy/detail.php?id=43895

future greenhouse gas emissions, and buying carbon credits. An example of a carbon credit could be investing in a sustainable forestry project in the Pacific Northwest or a biogas project from farms in the region. **Carbon offsets** enable organizations to reduce their environmental impact by supporting projects that reduce, absorb, or prevent carbon and other emissions from entering the atmosphere. A carbon offset is created when one ton of greenhouse gas is captured, avoided, or destroyed to compensate for an equivalent emission made.

Carbon offsets have been criticized, particularly for the case of forest-based offsets. For example, several years of wildfires in the western United States have added carbon back into the atmosphere, negating gains in reforestation, making this less viable as a genuine option. Janet Fisher and colleagues found that although offsetting potentially generates new revenue streams and opportunities for poor communities in the Global South, it often comes with trade-offs, injustices, unintended consequences, or obstacles to implementation that tend to be insufficiently acknowledged by proponents. In some instances, buying an offset is a privilege of the wealthy, but does little to address the more complex work of changing behavior or values around fossil fuels.

A growing number of countries, businesses, cities, and states have pledged to reach carbon neutrality by the year 2050 or sooner. Being carbon neutral is increasingly seen as good corporate or state social responsibility. Two countries, Bhutan and Suriname, have achieved or surpassed carbon neutrality (although it should be noted that their emissions are very low to begin with, so it is not likely to make a significant impact). Currently some twenty countries have set carbon neutrality targets including Austria, the Bahamas, Belgium, Canada, France, the European Union, Japan, South Korea, and the United Kingdom, with most setting 2040 or 2050 as their target years. In 2019, Costa Rica became the first country in Latin America and the Caribbean to announce a comprehensive plan to become a zero-emissions economy by 2050. Their National Decarbonization Plan covers ten sectoral areas including transportation, industry, waste management, and agriculture (including livestock). With some exceptions, most of the countries that pledge carbon neutrality are not big emitters (such as the United States, India, Russia, and China). Several initiatives like Climate Neutral Network, Carbon Neutrality Coalition, and Pathzero assist countries, businesses, and individuals in reducing their carbon footprint or achieving climate neutrality.

Achieving nationwide carbon neutrality will be a challenge for many countries, yet there have been encouraging results at smaller scales such as corporations and cities. Still, critical scholars see carbon neutrality as potentially "greenwashing" if such efforts do not require the wealthiest countries to abandon fossil fuels any time soon.

Dell, Google, and PepsiCo are Fortune 500 corporations that have pledged carbon-neutral initiatives. In 2006, Google began installing thousands of solar panels on its Bay Area corporate campus, enough to provide approximately 30 percent of the Bay Area campus's energy needs. The company has pursued strategies to achieve carbon neutrality like investing in wind farms in several locations, including a wind farm in Iowa and two wind power plants in South Dakota. By 2017, Google announced that it reached 100 percent renewable energy. In September 2019, Google's chief executive announced plans for a $2 billion wind and solar investment, the biggest renewable energy deal in corporate history, making Google the largest corporate buyer of renewable power. Businesses that invest in renewable energy can save money, show commitment to sustainability, and leverage their capital and brand recognition to further investment and interest in renewable energy.

A small but growing number of cities (most of them small-sized) have already switched to 100 percent renewable energy with zero net carbon emissions. More than fifty cities have pledged to become carbon neutral over the next several decades and together have formed the Carbon Neutral Cities Alliance. Cities achieve carbon neutrality by changing operating practices of the buildings and utilities they use or by leveraging influence as a major consumer of electricity. Many cities are able to find considerable energy and cost efficiencies by managing their water utilities, replacing streetlights with high-efficiency fixtures, and updating municipal purchasing policies to prioritize purchase of energy-efficient appliances and other materials. Cities also influence citizens' and businesses' behavioral choices by educating the community about weatherization and sustainable energy options. Finally, cities can help finance projects in the private sector by offering incentives or low-interest loans to enable sustainable energy projects that offer savings in the long run but have high up-front costs.

Cities also support alternative energy through municipal aggregation programs. Cleveland coordinates a community aggregation program that allows residential and small commercial customers of the Cleveland Electric Illuminating Company to leverage their power as a group to purchase green electricity at a price lower than otherwise possible. Customers can then choose the percentage of green energy (0, 50, or 100 percent) they would like to purchase. Cleveland believes that this program provides their residents with price protections while they support renewable energy and help reduce the city's carbon footprint. Finally, many cities are trying to improve energy efficiency by introducing a **Smart Grid** (see *Key Terms and Concepts* 7.1).

Ensuring Equity in Renewables

In the Global North, renewable energy tends to be adopted by wealthier households or businesses, but it is important that low-income residents not be left out of the opportunities for renewable energy. Too often, options for going green are reserved for the wealthy. This is because switching to smart heating systems, solar power, or wind power can have expensive start-up costs, even in the Global North. Furthermore, cost savings are often unavailable to renters because installing alternative energy systems is usually a low priority for building owners, even if some subsidy is available. As we move toward renewable energy, we must ensure equity of these systems and their benefits as well.

In 2019 California launched an innovative new $1 billion program, California's Solar on Multifamily Affordable Housing (SOMAH), which is funded by the statewide greenhouse gas cap-and-trade program (https://calsomah.org/). The program will provide up to $100 million annually for ten years to incentivize installations of solar on multifamily low-income housing properties across California. By 2029, the program is expected to reach renters in more than 200,000 units. The program will cover the costs of outfitting low-income properties with solar so savings can be passed along to renters living there. The law notably allows tenants to receive credits for electricity produced by the systems, thereby allowing them to directly benefit from the solar installation.

Similarly in 2018, Washington, DC, launched the initiative Solar for All, to bring the benefits of **solar energy** to 100,000 low-to-moderate income families in the District of Columbia. The district is using the concept of equity or "equal access" to local, clean energy to reduce residents' utility burdens and protect underserved communities in the face of the changing climate.

KEY TERMS AND CONCEPTS 7.1

The Smart Grid

Going carbon neutral will require fundamental changes in our relationship to energy and the emergence of a Smart Grid may be an essential part of this transformation. A Smart Grid is an electrical grid that includes a variety of operational and energy measures like smart meters, smart appliances, renewable energy resources, and energy efficient resources that relay information to users and energy producers such as power companies. According to the US Department of Energy, a Smart Grid is characterized by the:

- Increased use of digital information and dynamic optimization to improve reliability, security, and efficiency of the electric grid;
- Better integration of distributed power generation and renewable resources;
- Development and incorporation of demand response, demand-side resources, and energy-efficiency resources;
- Deployment of "smart" technologies (real-time, automated, and interactive technologies that optimize the physical operation of appliances and consumer devices) for metering, communications concerning grid operations and status, and distribution automation; and
- Usage of electricity storage systems including plug-in electric and hybrid electric vehicles.

A first step in this direction is "smart meters" that provide homes and businesses with detailed information about their energy use, which helps consumers identify ways of conserving energy and reaping financial savings. In one version of a Smart Grid, information flows between the utility and its customers, enabling real-time management of the home's electricity consumption. The idea is that if customers know that energy is cheaper in the middle of the night when demand is low, they will shift some energy uses to that time. This is called "demand side management," which either reduces the consumption of electricity or can help synchronize demand and supply by shifting noncritical usage of electricity from peak to nonpeak periods.

Already, electricity transmitters have developed programs that compensate large customers for temporarily reducing their electricity usage during periods of high demand, or other system challenges. Smart Grids could greatly expand this. Power would be flexibly priced according to availability (e.g., systems relying on solar power would have lower prices during the day). This information could be communicated to appliances, set, for instance, to automatically run a load of dishes when a low price is reached.

Providing solar energy to low-income residents may provide multiple benefits. It can reduce emissions of GHGs, reduce energy costs for low-income families, and provide energy during disruptions to the electrical grid.

A Critical Perspective: The Political Economy of Fossil Fuels

Moving toward a renewable energy future is more complex than just ending our dependence on fossil fuels. Fossil fuels are integrated in so many systems that it is likely an abrupt switch away from fossil fuels would cause economic collapse and an increase in food insecurity. Currently, some sectors of energy production have insufficient renewable sources. Coal is still critical for electricity production in many countries; a substitute for a coal-fired power plant must include energy storage systems that can store excess power for when wind or solar is not available. Similarly, there will need to be new technology to allow electric vehicles to become more widespread. Of all the fossil fuels, oil may be the most difficult to transition from because of its primary use in transportation and there are no existing equivalent alternatives yet.

Unfortunately, some countries seem to be backtracking on reducing fossil fuel use. Some are reducing coal production but increasing gas production. In China, coal mining increased in 2021.

The interdisciplinary field of political ecology highlights how social power affects access to energy services, participation in energy decision making, and allocation of energy's environmental and social costs. Energy is bound in a flow of raw materials that create economic and political power at a range of scales, from the geopolitics of international oil trade to relations of responsibility, autonomy, and identity associated with energy consumption and citizenship. As we attempt to move toward more sustainable energy, it is vital to understand these contexts and possible consequences. Gavin Bridge and colleagues show that energy transitions in the Global North have been achieved through a global process of unequal exchange. The partial decarbonization of northern European economies today is due to deindustrialization that led to the relocation of carbon-intensive production elsewhere (in other words, we have just moved the problem to a new location). Their work suggests it is also possible that pursuing "clean energy" may result in land and water grabs, perpetuating further inequalities. Finally they point out that Indigenous communities that challenge conventional energy projects—such as the KeyStone Pipeline—have faced police violence and had their concerns dismissed. These examples show how deeply embedded conventional energy is within the global political economy. And how much deep work remains to be done to change power relations, attitudes, and social norms.

SUMMARY AND PROGRESS

There are many ways individuals can work toward energy sustainability. Individuals can switch from incandescent light bulbs to light-emitting diode (LED) lights or reduce time spent driving personal cars and use mass transit or bicycles instead. Wealthier individuals may be able to install renewable energy for their homes.

However, significant advances will depend on changes in government policy and new technology. This is because the production, distribution, and access to energy involves complex infrastructure that is determined by a national energy framework. For example, an individual may want to use renewable energy within their home, but without direct access to affordable renewable services, electricity will come from the conventional established power grid, which is more likely to use nonrenewable sources. Similarly, an individual may want to use wind power, but accessing this will require the development of a large-scale grid storage system or permits that allow such installation particularly if the home or structure is protected by historic preservation laws.

Another challenge is that there is no universal solution to energy sustainability. Many energy strategies must be localized as an area's geographies determine their energy needs, optimal efficiency strategies, and opportunities for alternative energy production. The energy demand profiles in cold versus hot climates is very different. Significant variation in energy use exists at multiple scales—within a country, within a city, and even between neighborhoods in the same city. Diversifying energy sources may also reduce the impact of price volatility many consumers experience.

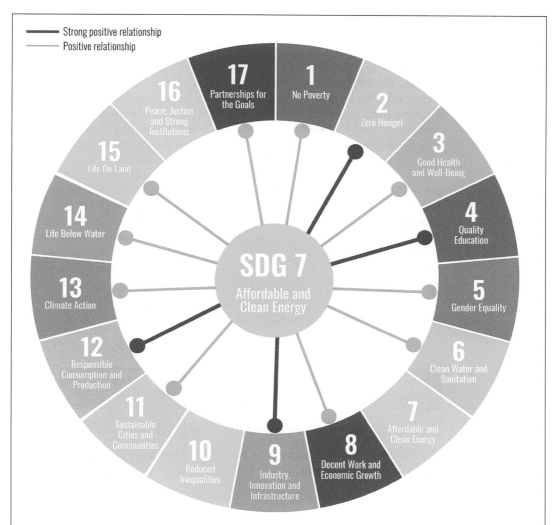

SDG 3: Access to clean fuels and technologies has the potential to save millions of lives each year. Household air pollution resulting from the inefficient use of clean fuels and technologies for cooking alone is responsible for some 4 million deaths annually.

SDG 4: Globally, over 291 million children go to primary schools without any electricity. Well-lit, well-heated, and well-cooled schools and households are essential for creating effective educational spaces for children and adults.

SDG 6: Energy and water are closely interlinked and interdependent. If we continue with business as usual, it will be impossible to meet the simultaneous huge increases in water and energy demands in the next decades. The inextricable linkages between these two critical resources requires a suitably integrated approach.

SDG 8: The clean energy transition generates jobs and the renewable energy sector worldwide employed 9.8 million people worldwide in 2016. Energy efficiency and renewable energy are creating more jobs than the fossil fuel industry, enabling net employment gains.

SDG 11: Cities globally consume up to 75 percent of energy and are responsible for 70 percent of greenhouse gas emissions. Promoting sustainable cities requires coordinated multi-sectoral investments and integrated policies.

SDG 13: Effective action towards a low-carbon and climate-resilient energy system is essential for achieving the objectives of the Paris Agreement and the 2030 Agenda. The energy sector accounts for roughly two-thirds of all anthropogenic greenhouse-gas emissions.

SDG 14: Increasing the share of renewable energy in the global energy mix and improving energy efficiency, reliability and affordability will enhance sustainability and help reduce ocean acidification through reduced carbon dioxide emissions.

SDG 15: Ensuring that the world's poor have access to modern energy services would reinforce the objective of halting deforestation, since firewood taken from forests is a commonly used energy resource among the poor.

FIGURE 7.5 Interconnections in SDG 7 and the Other Goals
Energy connects to all of the SDGs in some way; this figure highlights several of the stronger connections.

An additional challenge is there are many who resist the economic and political costs of transitioning to renewable energy, arguing that the costs seem too high to make action worthwhile. Some believe that ending the use of fossil fuels and transitioning to renewables requires us to sacrifice economic growth. Further, lower-income countries object that countries in the Global North used fossil fuels without restraint to realize economic growth and to improve their standards of living and that it is unfair for the international community to hold these countries to a different standard. For these reasons and more, it has been difficult to convince both policy makers and the public that it is imperative to move to cleaner, renewable energy.

The COVID-19 pandemic underscored that energy services are key to preventing diseases and fighting pandemics. Energy is needed to power health-care facilities, supply clean water for essential hygiene, and enable communications and ICT services that connect people while maintaining social distancing. Places and communities that lack access to energy may have more difficulty containing COVID-19 and providing needed health services. For example, in sub-Saharan Africa, only 28 percent of health-care facilities have reliable electricity.

Advancements in SDG 7 have the potential to spur progress on other SDGs and these reinforcing synergies are highlighted in Figure 7.5.

QUESTIONS FOR DISCUSSION AND ACTIVITIES

1. Describe the health and environmental impacts of using fossil fuels.

2. Explain why reducing and/or eliminating use of fossil fuels will require a shift in cultural values.

3. Describe two benefits and two challenges in transitioning to renewable energy.

4. Which strategies for bringing power to low-income countries do you think should be prioritized for investment?

5. Why is improving energy efficiency an important strategy? What community-wide energy conservation programs are in place where you live? If there are none, why do you think that is so? What actions do you take to conserve energy in your residence?

6. Explain two ways that increasing access to energy may positively impact gender equality in the Global South.

7. Explain how energy use is directly implicated in climate change.

8. Why is a target on clean cooking important?

9. Investigate whether there are any LEED buildings on your university campus or community by looking up LEED by state or region (https://www.usgbc.org/organizations/region).

10. View this short video on "Solar for All." Describe three benefits in renewable energy for low-income residents and communities. Explain some of the obstacles that exist in providing renewable energy for low-income residents (Shink, J. M., and Skinner, M. "Solar for All." Documentary Film (2020). https://dceff.org/film/solar-for-all/

TERMS

biofuel
biomass
British thermal unit (Btu)
carbon neutrality
carbon offset
coal
decentralized energy system
energy
energy portfolio
fossil fuels
geothermal energy
hydroelectricity

natural gas
nonrenewable energy resource
nuclear power
off-the-grid energy
oil
petroleum
renewable energy resource
smart grid
smog
solar energy
stand-alone power system
wind power

RESOURCES USED AND SUGGESTED READINGS

Bridge, G., Barca, S., Özkaynak, B., Turhan, E., and Wyeth, R. "Towards a Political Ecology of EU Energy Policy," in C. Foulds and R. Robison (eds.). *Advancing Energy Policy* (pp. 163–65). Cham, Switzerland: Palgrave Pivot (2018). https://doi.org/10.1007/978-3-319-99097-2_11

Caldwell, N. "California to Invest $1 Billion in Solar for Low-Income Housing Units." *Greenmatters*, November 2, 2017. http://www.greenmatters.com/living/2017/11/02/ZSWXp8/california-low-income-solar

"Carbon Neutral Cities Alliance Members." *Carbon Neutral Cities Alliance* (2021). https://carbonneutralcities.org/cities/

Fisher, J., Cavanagh, C. J., Sikor, T., and Mwayafu, D. "Linking Notions of Justice and Project Outcomes in Carbon Offset Forestry Projects: Insights from a Comparative Study in Uganda." *Land Use Policy* 73: 259–68 (2018). https://doi.org/10.1016/j.landusepol.2017.12.055

Funk, M. *Windfall: The Booming Business of Global Warming*. New York: Penguin Press (2014).

"Home." US Energy Information Administration (2021). www.eia.gov

Huber, M. "Energy and Social Power—From Political Ecology to the Ecology of Politics," in T. Perreault, G. Bridge, and T. McCarthy (eds.). *The Routledge Handbook of Political Ecology* (pp. 481–92). London: Routledge (2015).

Jacobson, M. *100% Clean, Renewable Energy and Storage for Everything*. New York: Cambridge University Press (2021).

Lelieveld, J., Klingmüller, K., Pozzer, A., Burnett, R.T., Haines, A., and Ramanathan, V. "Effects of Fossil Fuel and Total Anthropogenic Emission Removal on Public Health and Climate." *Proceedings of the National Academy of Sciences of the United States of America* 116 (15): 7192–97 (2019). https://doi.org/10.1073/pnas.1819989116

Lovins, A. *Reinventing Fire: Bold Business Solutions for the New Energy Era*. White River Junction, VT: Chelsea Green Publishing (2011).

"Power Africa." US Agency for International Development (2020). https://www.usaid.gov/powerafrica

Reisser, W., and Reisser, C. *Energy Resources: From Science to Society*. New York: Oxford University Press (2019).

Smil, V. *Energy and Civilization: A History*. Cambridge, MA: MIT Press (2017).

Van de Graaf, T., and Sovacool, B. K. *Global Energy Politics*. Medford, MA: Polity Press (2020).

Vohra, K., Vodonos, A., Schwartz, J., Marais, E., Sulprizio, M., and Mickley, L. "Global Mortality from Outdoor Fine Particle Pollution Generated by Fossil Fuel Combustion: Results from GEOS-Chem." *Environmental Research* 195: n.p. (2021). https://doi.org/10.1016/j.envres.2021.110754

World Bank. "Kenya Launches Ambitious Plan to Provide Electricity to All Citizens by 2022" (2018). https://www.worldbank.org/en/news/press-release/2018/12/06/kenya-launches-ambitious-plan-to-provide-electricity-to-all-citizens-by-2022

Yergin, D. *The Quest: Energy, Security, and the Remaking of the Modern World*. New York: Penguin Books (2012).

Zehner, O. *Green Illusions: The Dirty Secrets of Clean Energy and the Future of Environmentalism*. Lincoln: University of Nebraska (2012).

Decent Work

LEARNING OBJECTIVES

After reading this chapter, you should be able to:

- Describe the concept of decent work
- Explain the challenges and obstacles that hinder decent work including slavery and child labor
- Identify and explain the various targets and goals associated with SDG 8
- Explain strategies to achieving decent work and economic growth
- Explain how the three Es are present in SDG 8
- Explain how SDG 8 strategies connect to other SDGs

Chances are your chocolate bar was made possible by child labor. In Western Africa, cocoa is a commodity crop grown primarily for export; cocoa is the Ivory Coast's primary export and makes up about half of the country's agricultural exports in volume. As the chocolate industry has grown over the years, so has the demand for cheap cocoa.

Most cocoa farmers earn less than $1 per day, an income below the extreme poverty line. As a result, they often resort to the use of child labor to keep their prices competitive. Approximately 2.1 million children in the Ivory Coast and Ghana work on cocoa farms, most of whom are likely exposed to the worst forms of child labor.

The children of Western Africa are surrounded by poverty, and many begin working at a young age to help support their families. Some children end up on the cocoa farms because they need work and traffickers tell them that the job pays well. Other children are sold to traffickers or farm owners by their own relatives, who are unaware of the dangerous work environment and the lack of any provisions for an education. Often, traffickers abduct the young children from small villages in neighboring African countries, such as Burkina Faso and Mali. In one village in Burkina Faso, almost every mother in the village has had a child trafficked onto cocoa farms. Once they have been taken to the cocoa farms, the children may not see their families for years.*

---

*Child Labor and Slavery in the Chocolate Industry. *Food Empowerment Project* (n.d.). https://foodispower.org/human-labor-slavery/slavery-chocolate/

WHAT IS DECENT WORK?

There is a global consensus that not all work is decent and respects the fundamental rights of the person. **Decent work** is defined as jobs that deliver a fair income, provide security in the workplace and social protection for families, and help realize better prospects for personal development and social integration. Child labor, slavery, and unpaid labor are still realities for many around the world.

To achieve sustainability and the ability to provide for needs, all people must be productively employed. Paid work is the main source of income for the vast majority of households worldwide, and yet unsafe working conditions, underemployment, and unemployment are the main challenges to economic sustainability.

It is true that economic growth has delivered a better quality of life and brought prosperity and security to billions of people globally. Yet, growth as it has been practiced has come at an ecological cost, undermining the foundations of human well-being and generating tremendous inequalities. SDG 8 "Decent Work and Economic Growth" is conceptually challenging because there is a fundamental question as to whether economic growth is compatible with sustainability.

The icon for SDG 8 consists of three pillars. The first pillar represents employment. The second pillar stands for international standards of work and includes fair incomes, fundamental rights at work (such as limits to the hours of work required), work that meets a minimum wage, and treating workers fairly. The third pillar highlights social protections, which include social security, unemployment benefits, food stamps, and investments in education and job training for workers.

Over the past twenty years, the number of workers living in extreme poverty has declined dramatically. In many countries in the Global South, the middle class now makes up more than 34 percent of total employment—a number that nearly tripled between 1991 and 2015. Despite economic growth, there are deficits in opportunities for decent work. According to a 2019 **International Labour Organization (ILO)** report, a majority of the 3.3 billion people employed globally experienced a lack of material well-being, economic security, equal opportunities, or scope for human development. Access to paid work is no guarantee of decent work. Many workers find themselves having to take up jobs that are informal or are characterized by low pay and little or no access to social protection and rights at work. Another report noted that in 2019, more than 630 million workers worldwide—almost one in five, or 19 percent, of all those employed—did not earn enough to lift themselves and their families out of extreme or moderate poverty, defined as earning less than $3.20 per day in purchasing power parity terms. In the lowest-income countries, employment may not enable people to escape poverty. Even in middle-income countries, access to social security systems, job security, collective bargaining, and compliance with labor standards and health and safety standards are elusive to varying degrees.

A person's geographic location can determine opportunities for decent work. People in low-income countries have the most difficulty finding paid work that is of good quality, as many vulnerable workers feel forced to take up any job, regardless.

In addition, as population in some low-income countries continues to increase, and significant numbers of young people reach adulthood, it is estimated that 470 million new jobs will be needed by 2030. However, the number of jobs being created in some regions in the Global South is not keeping pace with this need.

Overall, progress toward decent work is uneven—and slowing—in many parts of the world, as measured by employment levels, gender wage gaps, fairness, and respect for workers' rights.

CHALLENGES TO DECENT WORK

There are numerous challenges and obstacles to achieving decent work including unemployment, lack of social protection, discrimination in the workforce, child labor, and modern slavery.

Unemployment

An estimated 172 million people worldwide were unemployed in 2019, which corresponds to an unemployment rate of 5 percent (this grew to about 220 million or 6.3 percent during the height of the COVID-19 pandemic in 2021). Apart from the unemployed, there is a less visible measurement: An additional 140 million people were in the "potential labor force" in 2018, which means that they are classified as underutilized labor. Underutilized labor are people who are willing and available to work more than they are employed to do. The statistics on underutilized labor include far more women (eighty-five million) than men (fifty-five million) globally. Women are much more likely to work part time although a significant proportion say they would prefer more hours of employment. Furthermore, women, young people (ages 15–24), and persons with disabilities continue to be much less likely to be in formal employment. Equally worrisome is the fact that more than one in five young people are not in employment, education, or training.

While progress reducing unemployment in the Global South has been mixed, the Global North has seen a steady decline in unemployment (pre-COVID-19). Among the major regions of the world, the unemployment rate is highest in North Africa (12 percent) and Central and Western Asia (9 percent), while the lowest rates are observed in Southeast Asia and the Pacific (3 percent) and North America (4 percent).

There are also significant geographical disparities within countries. At the global level, the unemployment rate of the working-age population living in rural areas tends to be higher than that in urban areas. The urban-rural disparities often facilitate rural to urban migration as workers search for better job opportunities. Unequal access to decent work can exacerbate income inequalities.

The Lack of Social Protection

According to the United Nations, social protection is a human right. **Social protection** is the set of policies and programs designed to reduce and prevent poverty and vulnerability. It includes benefits for children and families, maternity and paternity leave, unemployment, employment injury, sickness, old age, disability, and health protection. Social protection systems provide assistance through social insurance, tax-funded social benefits, social assistance services, public works programs, fee waivers (for basic health and education), subsidies (for food and fuel), and other schemes that guarantee basic income security. Social protection benefits can either be in cash or in-kind (such as reduced or free school meals). One type of social protection is social security, which contributes cash for those who have been part of the formal sector of employment.

Currently, a majority of the global population—four billion people worldwide—are not covered by at least one social benefit. A 2019 ILO report on social protection shows that only 29 percent of the global population enjoys access to comprehensive social security. The report found the following:

- Only 35 percent of children worldwide have access to social protection. Almost two-thirds of children globally—1.3 billion children—are not covered, most of them living in Africa and Asia.
- On average, just 1.1 percent of GDP is spent on child and family benefits for children aged zero to fourteen, pointing to significant underinvestment in children.
- Only 41.1 percent of mothers with newborns receive a maternity benefit and eighty-three million new mothers remain uncovered.
- Only 21.8 percent of unemployed workers are covered by unemployment benefits, while 152 million unemployed workers remain without coverage.

Table 8.1 highlights regional trends in social protection and Map 8.1 displays the variation in government spending on social protection varies around the world. Regions that need to increase public expenditure on social protection include North Africa, sub-Saharan Africa, South Asia, the Middle East, and Southeast Asia. ILO Director-General Guy Ryder explained the importance of social protection:

> The lack of social protection leaves people vulnerable to ill-health, poverty, inequality and social exclusion throughout their lifecycle. Denying this human right to 4 billion people worldwide is a significant obstacle to economic and social development. While many countries have come a long way in strengthening their social protection systems, major efforts are still necessary to ensure that the right to social protection becomes a reality for all.

TABLE 8.1 Social Protection Varies Regionally

Region	Percent of the population that receives at least one social protection	Notable achievements
Sub-Saharan Africa	17.8%	Botswana, Lesothos, Namibia, and Mauritius have reached universal pension coverage
South America	67.6%	Argentina, Brazil, and Chile have achieved universal coverage of children
North America	70%	Universal coverage of persons with disability and older persons
Middle East	Lack of data	28% have old-age pension; unemployment protection in Bahrain, Kuwait and Saudi Arabia
Europe	84%	80% achievement for child and family benefits, maternity benefits, disability benefits and old-age pension. Several countries have reached universal coverage
South Asia	39%	Extended maternity protection in Bangladesh and India

Source: International Labour Organization. *World Social Protection Report, 2017–2019*. Geneva, Switzerland: ILO (2017). ilo.org/wcmsp5/groups/public/---dgreports/---dcomm/---publ/documents/publication/wcms_605078.pdf

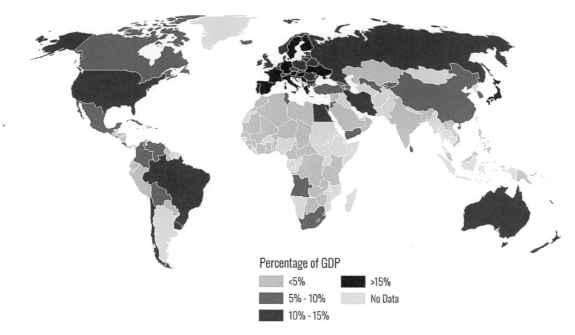

MAP 8.1 Public Social Protection Expenditure, Excluding Health (2019)
The map shows countries in Europe invest the highest in social protections such as unemployment, housing, disability, and family leave. Regions that need to increase public expenditure on social protection include North Africa, sub-Saharan Africa, South Asia, the Middle East, and Southeast Asia.

Discrimination in Employment and Occupation

Hundreds of millions of people suffer from discrimination in the workplace. It is a persistent and multifaceted problem. Discrimination stifles opportunities, wastes the human talent needed for economic progress, and accentuates social tensions and inequalities. Combating discrimination is seen as an essential part of decent work. Discrimination in the workplace may occur based on religion, age, caste, HIV status, gender, nationality, disabilities, or lifestyle (e.g., body weight or smoking).

Consider the challenges confronting many migrants where they may be an ethnic, racial, or religious minority in their host country. Research shows that migrants consistently face widespread discrimination when attempting to access employment, and many experience discrimination while employed (Photo 8.1). Migrant workers have also been particularly affected by economic crises, including reduced employment or migration opportunities, increased xenophobia, deteriorating working conditions, and even violence.

Migrants often face unfair working conditions. Some countries exclude migrant workers from social insurance programs. Others only allow migrants access to short-term programs such as health care but deny them long-term benefits such as old-age pensions. Certain countries may allow access to long-term benefits but do not permit portability between countries, which in turn discourages return migration. More recently, discriminatory tendencies have recently been aggravated by hostile political discourse and there is a risk that this may lead to exclusion, rejection, and expulsion of migrant workers. Social tensions and hardening attitudes toward migrants—as with any social group—can result in systematic and widespread discrimination.

PHOTO 8.1 Migrant Laborers in Dubai
Migrant workers at a construction site in Dubai. Over two-thirds of the Dubai population is migrant labor with 1.1 million working in construction. Dubai and the UAE have been the subject of complaints of mistreatment of workers. Migrant workers say they often face brutal work conditions, shifts that exceed twelve hours, and companies that withhold paychecks or workers' passports, thereby restricting mobility.

Populist policies can foster greater xenophobia and discrimination directed toward migrants.

On the positive side, there have been advances in antidiscrimination legislation and policies and, in general, a growing awareness of the need to overcome discrimination at work. However, having laws and institutions to prevent discrimination at work and offer remedies is not always enough. There can be a lack of political will and a lack of enforcement or inadequate institutional capacity to identify and address issues of discrimination.

Progress has been made in recent decades in advancing gender equality in the work force, but challenges remain such as sexual harassment, a lack of maternity and paternity protections, and the gender wage gap, a topic also explored in Chapter 5. Recent statistics show that men continue to earn 12 percent more than women in most of the countries with this data. Women's wages are on average 70 to 90 percent of men's, and without action, it may take many decades to achieve equal pay. Women's labor force participation rate is 63 percent while that of men is 94 percent. However, women do twice the unpaid care and domestic work that men do. Reducing gender discrimination in the labor market is critical because gender equality in employment gives women more decision-making power and enhances family well-being: Women will typically invest more of their income than men in the health, nutrition, and education of their children.

Statistics on gender wage gaps do not typically include women who receive no direct pay for their involvement in family work (as this is part of the informal economy). If these women, and the many others engaged in other forms of informal work, were included in the statistics, women's relative disadvantage in the labor market would be even larger.

Employment in the **formal economy** involves the sale of labor in the marketplace, which is formally recorded in government and official statistics. The **informal economy** is the unrecorded sector where few, if any, taxes are paid. There is a diverse range of informal sector jobs, including retail distribution, small-scale transport, and personal services such as child care, security services, and recycling enterprises (Photo 8.2). Even babysitters are part of the informal economy.

Workers in the informal sector can be paid a wage, but because this is typically not recorded, these services are not "counted" in a country's formal GDP. As a result, a country's GDP does not always accurately reflect the level of productivity and economic activity occurring in the economy.

The informal sector is a response to the lack of formal employment opportunities. When the formal sector provides sufficient employment opportunities, the informal sector becomes less important. However, if the formal sector provides few jobs or only jobs with low wages, then the informal sector becomes a coping strategy as people try to make a living.

Half the global workforce—1.6 billion people—support themselves and their families through insecure and often unsafe jobs in the informal economy. In 2016, informal employment was much more widespread in the agricultural sector (94 percent) than in the nonagricultural sector (51 percent). Reliance on informal workers was also more prevalent in certain regions, including sub-Saharan Africa (89 percent) and Central and Southern Asia (86 percent). The informal economy is often highly gendered: More than 80 percent of women in nonagricultural jobs are in informal employment in South Asia, 74 percent in sub-Saharan Africa, and 54 percent in Latin America and the Caribbean.

PHOTO 8.2 Street Market in Cusco, Peru
People buying and selling fruits in a street market. Street vendors are an integral part of urban economies, and many offer goods and services in public spaces such as this street market.

Child Labor

Not all work done by children is classified as child labor (Figure 8.1). In some cases, children's or adolescent's participation in work can positively affect their health and personal development. Whether or not particular forms of "work" can be called "child labor" depends on the child's age, the type and hours of work performed, the conditions under which it is performed, and the objectives pursued by individual countries. The answer varies from country to country, as well as among sectors within countries.

However, millions of children are classified as child labor. A 2021 UNICEF report noted that 160 million children—63 million girls and 97 million boys—were in child labor, or one in ten children worldwide (Map 8.2). Nearly half those children in labor were in hazardous work directly endangering their health, safety, and moral development.

According to the ILO, the term **child labor** is defined as work that deprives children of their childhood, their potential, and their dignity, and is harmful to physical and mental development. It refers to work that is mentally, physically, socially, or morally dangerous and harmful to children and/or interferes with their schooling by depriving them of the opportunity to attend school; forcing them to leave school prematurely; or requiring them to attempt to combine school attendance with excessively long and heavy work.

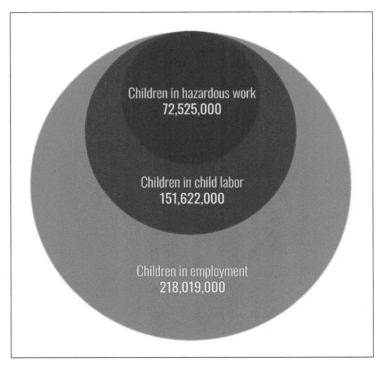

FIGURE 8.1 Global Estimate of Children in Hazardous Work, Child Labor, and Employment
Not all work done by children is classified as child labor. Estimates place 73 million children in hazardous work, the worst form of child labor.

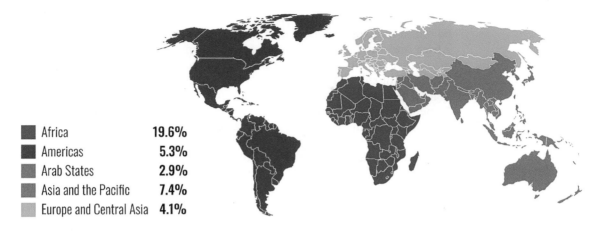

■ Africa	**19.6%**
■ Americas	**5.3%**
■ Arab States	**2.9%**
■ Asia and the Pacific	**7.4%**
■ Europe and Central Asia	**4.1%**

MAP 8.2 Regional Prevalence of Child Labor (2017)
Africa is home to the highest percentage and the most numbers of children in child labor.

The **worst forms of child labor** as defined by Article 3 of the ILO Convention No 182 are:

- All forms of slavery or practices similar to slavery, such as the sale and trafficking of children, debt bondage and serfdom, and forced or compulsory labor, including forced or compulsory recruitment of children for use in armed conflict;
- The use, procuring, or offering of a child for prostitution, for the production of pornography, or for pornographic performances;
- The use, procuring, or offering of a child for illicit activities, in particular for the production and trafficking of drugs as defined in the relevant international treaties; and
- Work that, by its nature or the circumstances in which it is carried out, is likely to harm the health, safety, or morals of children.

Among the worst forms of child labor is **hazardous child labor**. Hazardous child labor is work in dangerous or unhealthy conditions that could result in a child being killed, injured, sickened, permanently disabled, or psychologically harmed because of poor safety and health standards and working arrangements. An estimated seventy-three million children aged five to seventeen work in dangerous conditions in a wide range of sectors, including agriculture, mining, construction, and manufacturing, as well as in hotels, bars, restaurants, markets, and domestic services.

It is alarming that child labor continues to be a problem around the world, as evidenced by the fact that 2021 was declared the International Year for the Elimination of Child Labor. Although several international agreements exist, many children are still subject to child labor or exploitation. The good news is that the struggle against child labor has gained momentum since 2000. Between 2000 and 2016 alone, there was a 38 percent decrease in child labor globally. Latin America, the Caribbean, and Southeast Asia saw decreases in child labor. The bad news is that progress in sub-Saharan Africa has become elusive and the region has seen an increase in both the number and percentage of children in child labor. UNICEF predicts that the number of children in child labor could rise to 169 million by the end of 2022.

Child labor is most prevalent in the lowest income countries, but it can also be found in many countries with higher incomes. The ILO notes that North Africa and sub-Saharan Africa are the regions most affected by situations of conflict and disaster, which in turn heighten the risk of child labor. The incidence of child labor in countries affected by armed conflict is 77 percent higher than the global average, while the incidence of hazardous work is 50 percent higher in countries affected by armed conflict. Nearly 70 percent of all children in child labor are in agriculture, underscoring the need for improving rural development.

Child labor is a violation of children's rights and of the right of everyone to live in a world free from child labor. It is reflected in Target 8.7, which calls for immediate and effective measures to eliminate the worst forms of child labor, including hazardous child labor, and to end child labor in all its forms by 2025.

Modern Slavery

According to the ILO, the term **modern slavery** is used as an umbrella term for two situations: forced labor and forced marriage. Currently, an estimated forty million people are in modern slavery, including 25 million in forced labor and 15.5 million in forced marriage. Modern slavery affects all population groups—young and old, male and female. But some groups are more vulnerable than others. Women and girls account for 71 percent of modern slavery victims, and 99 percent of victims of forced labor in the commercial sex industry. Map 8.3 shows the prevalence in modern slavery.

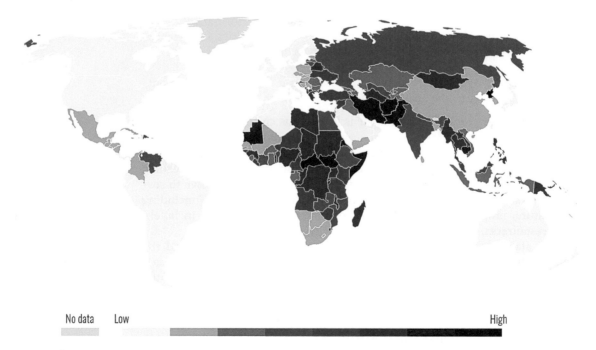

No data　Low　　　　　　　　　　　　　　　　　　　　　　　　　High

MAP 8.3 Prevalence of Modern Slavery (2019)
Countries in Central Africa and Central Asia have the highest rates of modern slavery. Other notable countries include North Korea and Cambodia.

Forced labor is defined as being forced to work under threat or coercion as domestic workers, on construction sites, in sweatshops, on farms and fishing boats, or in other sectors. The term implies the use of deception or coercion, either by the state and public agencies or by private individuals and enterprises, to force people to enter work or service against their will to work in conditions they did not accept or understand. Those in forced labor are often prevented from leaving. They may have their wages withheld or they may be threatened with violence or violence against their family.

One of the most common techniques used to entrap laborers is through false debts. An owner lures a person into slavery by offering a small advance payment for their labor. The owner then ensures it is impossible for those enslaved to ever repay by inflating the debt owed with exorbitant interest charges, not paying the victim the promised wages, and prohibiting him or her from working anywhere else.

In many cases, the products made by forced labor and the services they provided end up in seemingly legitimate commercial channels. Forced laborers produced some of the food we eat and the clothes we wear, and they have cleaned the buildings in which many of us live or work.

Many forced labor victims are in situations of debt bondage. **Debt bondage** is defined as being forced to work to repay a debt and not being able to leave or being forced to work and not being able to leave because of a debt. Most of those in forced labor are domestic workers, followed by construction, and manufacturing. The *Understanding the Issue* 8.1 box highlights the problem of forced labor in the fisheries industry.

An estimated 15.5 million people are living in a **forced marriage**. This term refers to situations in which persons have been forced to marry without their consent. A marriage can be forced through a range of different mechanisms, including physical, emotional, or financial duress; deception by family members, the spouse, or others; or the use of force or threats or severe pressure. While men and boys can also be victims of forced marriage, most victims (88 percent) were women and girls, with more than a third (37 percent) of victims under eighteen years of age at the time of the marriage.

Ending modern slavery will require multiple approaches; responses need to be adapted to the diverse places and cultures in which modern slavery still occurs. The vast majority of forced labor exists in the private sector, underscoring the need for partnering with the business community along with civil society organizations.

Research has shown that much of modern slavery today occurs in contexts of state fragility, conflict, and crisis. Providing support for those countries in precarious situations can be effective at reducing modern slavery.

SOLUTIONS FOR ACHIEVING DECENT WORK

Table 8.2 lists the targets and indicators for SDG 8. It was once thought that there is an inherent tension or contradiction between economic growth and jobs and sustainability. However, achieving sustainability can create jobs. Environmental laws, regulations, and policies that include labor protections

UNDERSTANDING THE ISSUE 8.1

Forced Labor in Fisheries

Recent reports by the International Labour Organization indicates that forced labor and human trafficking in the fisheries sector are a severe problem. These reports suggest that fishers, many of them migrant workers, are vulnerable to human rights abuse aboard fishing vessels. Migrant workers in particular may be deceived and coerced by brokers and recruitment agencies and forced to work on board vessels under the threat of force or by means of debt bondage.

Victims describe illness, physical injury, psychological and sexual abuse, deaths of crewmates, and being stranded on board vessels in remote locations of the sea for months and years at a time. Fishers are forced to work for long hours at very low pay and the work is intense, hazardous, and difficult. Capture fisheries have one of the highest occupational fatality rates in the world.

Recent trends within the fisheries sector, such as overfishing, illegal fishing, and a shift in sourcing the workforce from high-income to middle- and low-income countries have increased the use of relatively low-cost migrant workers. However, lack of training, inadequate language skills, and lack of enforcement of safety and labor standards make these fishers particularly vulnerable to forced labor and human trafficking. A substantial proportion of these people are trapped on board fishing vessels or in the wider fish and seafood industry, though it has been difficult to obtain an accurate estimate of the true extent of the problem. Trafficking is, by its very nature, criminal and hidden, which makes it very difficult to research and estimate. The recent repatriation of thousands of Cambodian, Thai, and Burmese fishers from vessels that had been fishing in Indonesia shows that this issue is widespread and affects many. Too few resources globally are spent to help victims to reintegrate. Trafficked persons may need any number of services—medical care, counseling, job placement, legal assistance, education, training, or retraining—and these services need to be available to them as part of a protection response.

The ILO has adopted two instruments in the last decade that are central to addressing these matters: the 2007 Work in Fishing Convention and the 2014 Protocol to the Forced Labour Convention. Together, they provide a comprehensive framework for regulating work in fishing and preventing fishers from becoming victims of forced labor and without access to decent working conditions. Yet the problem persists.

Addressing forced labor and trafficking requires the commitment and engagement of governments, nongovernmental organizations (NGOs), trade unions, researchers, the private sector, the media, and consumers. The work of NGOs such as Anti-Slavery International or Amnesty International combat slavery by rescuing victims, bringing criminals to justice, creating individualized care plans for survivors, working to strengthen justice systems, and educating the public about these issues.

_____

Source: International Labour Organization. "Forced Labour and Human Trafficking in the Fisheries" (n.d.). https://www.ilo.org/global/topics/forced-labour/policy-areas/fisheries/lang--en/index.htm

offer a powerful means of integrating decent work with environmental objectives. There are many strategies for advancing SDG 8, some of which are detailed in Table 8.2.

Invest in and Expand Social Protection

Social protection is a key strategy to reducing poverty; it is also a key strategy to decent work. There are many barriers to decent work. There are structural barriers that affect individuals such as lack of education and skills, work inexperience, and information constraints. There are also structural barriers that affect the broader labor market such as social and economic inequalities

TABLE 8.2 SDG 8 Targets and Indicators to Achieve by 2030 or Earlier

Target	Indicators
8.1 Sustain per capita economic growth in accordance with national circumstances and, in particular, at least 7 percent GDP growth per annum in the least developed countries	**8.1.1** Annual growth rate of real GDP per capita
8.2 Achieve higher levels of economic productivity through diversification, technological upgrading, and innovation, including through a focus on high-value added and labor-intensive sectors	**8.2.1** Annual growth rate of real GDP per employed person
8.3 Promote development-oriented policies that support productive activities, decent job creation, entrepreneurship, creativity and innovation, and encourage the formalization and growth of micro-, small-, and medium-sized enterprises, including through access to financial services	**8.3.1** Proportion of informal employment in nonagriculture employment, by sex
8.4 Improve progressively global resource efficiency in consumption and production and endeavor to decouple economic growth from environmental degradation, in accordance with the ten-year framework of programs on sustainable consumption and production, with developed countries taking the lead	**8.4.1** Material footprint, material footprint per capita, and material footprint per GDP **8.4.2** Domestic material consumption, domestic material consumption per capita, and domestic material consumption per GDP
8.5 By 2030, achieve full and productive employment and decent work for all women and men, including for young people and persons with disabilities, and equal pay for work of equal value	**8.5.1** Average hourly earnings of female and male employees, by occupation, age, and persons with disabilities **8.5.2** Unemployment rate, by sex, age, and persons with disabilities
8.6 By 2020, substantially reduce the proportion of youth not in employment, education, or training	**8.6.1** Proportion of youth (aged 15–24 years) not in education, employment, or training
8.7 Take immediate and effective measures to eradicate forced labor, end modern slavery and human trafficking, and secure the prohibition and elimination of the worst forms of child labor, including recruitment and use of child soldiers, and by 2025 end child labor in all its forms	**8.7.1** Proportion and number of children aged 5–17 years engaged in child labor, by sex and age
8.8 Protect labor rights and promote safe and secure working environments for all workers, including migrant workers, in particular women migrants, and those in precarious employment	**8.8.1** Frequency rates of fatal and nonfatal occupational injuries, by sex and migrant status **8.8.2** Increase in national compliance of labor rights (freedom of association and collective bargaining) based on ILO textual sources and national legislation, by sex and migrant status
8.9 By 2030, devise and implement policies to promote sustainable tourism that creates jobs and promotes local culture and products	**8.9.1** Tourism direct GDP as a proportion of total GDP and in growth rate **8.9.2** Number of jobs in tourism industries as a proportion of total jobs and growth rate of jobs, by sex

(*continued*)

Target	Indicators
8.10 Strengthen the capacity of domestic financial institutions to encourage and expand access to banking, insurance, and financial services for all	**8.10.1** Number of commercial bank branches and automated teller machines (ATMs) per 100,000 adults **8.10.2** Proportion of adults (fifteen years and older) with an account at a bank or other financial institution or with a mobile-money-service provider
8.A Increase Aid for Trade support for developing countries, in particular least developed countries, including through the Enhanced Integrated Framework for Trade-Related Technical Assistance to Least Developed Countries	**8.A.1** Aid for Trade commitments and disbursements
8.B By 2020, develop and operationalize a global strategy for youth employment and implement the Global Jobs Pact of the ILO	**8.B.1** Total government spending in social protection and employment programs as a proportion of the national budgets and GDP

Source: United Nations Sustainable Development Goals. https://sustainabledevelopment.un.org/sdg8

or a lack of labor rights. Governments that prioritize investment in social protection policies and programs can reduce these barriers. Governments that combine income support during periods of joblessness with active labor market policies (ALMPS) can prevent people from being forced by necessity to accept any new job that comes their way, regardless of its quality. Ideally, this would also be supplemented with programs to teach people the necessary skills to gain better jobs and find them new employment opportunities. ALMPs can include workforce development or job training programs and employment subsidies that help individuals to improve their level of education and skills and gain work experience. The *Solutions* 8.1 box highlights a successful workforce development program in Detroit, Michigan.

Social protection systems support the economy by stabilizing household incomes. Four policy areas are particularly noteworthy:

1. Unemployment protection,
2. Cash transfer programs,
3. Public employment programs (PEP), and
4. Payments for ecosystem services (PES).

Unemployment protection schemes and cash transfer programs play an essential role in supporting workers facing job loss related either to the transition to environmental sustainability or to a natural disaster. They facilitate the transition to new jobs, particularly when combined with skills development and job placement or relocation measures. In addition, access to safe and regular labor migration opportunities can foster economic diversification and increase adaptive capacity through remittances and skills transfer.

Cash transfer programs contribute to preventing poverty and reducing the vulnerability of households and communities. Public employment (or government jobs) programs are used to increase demand for labor in markets where there is insufficient availability in the formal sector. Often public employment

SOLUTIONS 8.1

Doing Economic Development Differently in Detroit

Target 8.5: By 2030, achieve full and productive employment and decent work for all women and men, including for young people and persons with disabilities, and equal pay for work of equal value.

Deindustrialization—the loss of manufacturing employment starting in the 1980s—marked a critical shift in the economies of many US cities. Pittsburgh, Syracuse, Buffalo, Cleveland, and Detroit watched as companies fired or relocated workers, closed factories, or left the region or country entirely. These cities were transformed from "industrial" to "Rustbelt." Many of these cities, once vibrant manufacturing centers, have struggled to reinvent themselves and these impacts reverberate even today.

Federal and state dollars have been used to rebuild these economies. Millions of public-sector dollars have been invested in Detroit—in housing, transportation, infrastructure, and so forth—with little input or direct benefit to local community residents.

In response, a broad community and labor coalition formed to ensure that investments in public infrastructure and housing in Detroit create economic opportunities for residents and local businesses in communities impacted by public project investments. This coalition—Doing Development Differently in Metro Detroit (D4)—is working to create community benefit agreements in several key areas, including residential housing development projects. For example, they worked with Better Buildings for Michigan, a state-sponsored and federally funded effort to promote mainstream adoption of home energy upgrades, to incorporate language in their neighborhood competitive bid process that encourages bidding building contractors to indicate whether and how they plan to hire local residents. D4 consistently connects infrastructure projects and local community economic development opportunity. This is an example of how integrating sustainability invites a rethinking of economic development.

Source: DiRamio, M., and Coxen, T. "Working toward a Sustainable Detroit: Investing in Sustainable Industry and 'Green Collar' Careers for Residents in Detroit." Detroit, MI: Detroit Regional Workforce Fund and United Way for Southeastern Michigan (2012). http://skilledwork.org/wp-content/uploads/2014/01/Working_Towards_a_Sustainable_Detroit.pdf

programs provide the construction or expansion of public or social goods such as infrastructure. The work experience and skills acquired through these programs can improve a person's employability. There are, however, challenges to public employment programs. Governments at any scale must have effective capacity for planning, supervision, and implementation. The costs of infrastructure can be high and these types of projects can be short term. These programs combine the goals of job creation, income security, poverty reduction, and the provision of public goods and services. Ethiopia's Productive Safety Net Program is an example of a PEP. It contributes to improving food security through land restoration and reforestation and has become Africa's largest climate resilience program. Public works provided through the program have boosted food output by increasing land productivity and have enhanced community resilience. The program mitigates climate change by promoting land use practices that increase carbon sequestration and increase yields by reducing soil erosion and sediment loss.

In the Global North, government programs or those funded by the government and run by nonprofit organizations are linking sustainability goals with job-training programs that can help people develop skills for the workforce. For example, Earth Conservation Corp in Washington, DC, trains unemployed, disadvantaged youth to work on a variety of environmental projects. The program helps the youth build skills in environmental management such as wetland restoration, raptor rehabilitation, and tree planting while teaching leadership and life skills. Explicitly connecting green spaces and jobs helps to break down the dichotomy of sustainability and environmental protection versus economic growth.

Finally, PES are payments to farmers or landowners who have agreed to take certain actions to manage their land or watersheds to provide an ecological service such as clean water, habitat for animals, or carbon storage in forests. The Socio Bosque program in Ecuador offers the lowest-income communities in private and communal forests yearly payments in return for maintaining forest cover. The program also seeks to improve the socioeconomic situation of beneficiaries by asking them to submit a plan on how the payments will be spent and encouraging them to invest. Similarly, in Brazil, the PES program, Bolsa Floresta, tries to generate employment and income while encouraging the conservation of forests. It was started in 2007 and pays a modest incentive of about $30 a month for households to conserve their forest. The program rewards Indigenous peoples for their conservation work in tropical forests, provides training and support for sustainable production, and strengthens community associations. With more than 8,500 participating families in fifteen conservation units covering ten million hectares, it is one of the largest PES programs in the world. Another example is in South Africa where the Working for Water Program provides income support to participants in exchange for protecting the natural environment of rivers.

Many advocates claim that these PES programs have the potential to reduce the tension between conservationists and local communities. Recent research, however, has found that evaluating the effectiveness of PES is proving very difficult to do. It is not clear that these programs are reducing tensions, nor achieving conservation objectives.

Ensure Fair and Effective Migration Governance

A high percentage of victims of modern slavery are exploited outside their country of origin, underscoring the link between migration and modern slavery. The unique vulnerabilities faced by migrants should be addressed by reforms to migration governance designed to maximize the benefits and minimize the risks and social costs of migration. For example, ensuring security *en route* is critical as this is a time of high vulnerability, especially for children. The compliance of host governments with international conventions governing the rights of migrants, especially child migrants, is necessary for protecting migrants at their destination.

Countries and regions experiencing conflict and disaster need assistance from the global community in handling political, economic, and natural shocks. These fragile situations—characterized by unemployment or reduced income, a breakdown in formal and family social support networks, displacement, and disruptions in basic services provision—create an elevated risk of child labor and modern slavery. This underscores the need for the international community to identify the early warning signs within these conflict areas and to provide support to enhance responses to child labor and modern slavery among emergency-affected populations. Humanitarian organizations could be facilitators of such responses, but experts suggest that other tools are needed to rapidly assess risks of child labor and modern slavery to guide responses.

Embrace Green Growth

Sustainability challenges us to move past the conceptual framework that prioritizes growth and progress while discounting negative environmental or social impacts. Instead, the economy is seen as dependent on the environment for all production inputs and as a sink for all waste outputs. Work is intrinsically linked to the natural environment. Jobs in agriculture, fisheries, forestry, tourism, and other industries including pharmaceuticals, textiles, and food depend on a healthy environment. The environment also impacts the quality of work and, in turn, the worker. For example, temperature rises like those expected due to climate change will increase the number of days that are too hot to work, putting workers' health at risk and reducing productivity.

We must resist the common misconception that economic development and sustainability are mutually exclusive activities. Conservative, neoclassical economists argue that regulations limiting carbon emissions curb innovation and limit employment opportunities in the private sector. On the contrary, a sustainable economy can create jobs. First, green businesses—those that use environmentally benign processes or manufacture goods that are environmentally benign—employ workers directly, and often locally. Second, sustainable practices reduce material and energy consumption, which results in cost savings that can be reinvested in new job-creating activities or that make companies more financially secure. Third, sustainable practices improve worker health, productivity, and security, as workplaces become healthier and less toxic. Finally, all these practices increase competitiveness, which leads to sustained job growth.

The concept of **green growth** has emerged as an approach to reframe the conventional growth model and to reassess energy, agriculture, water needs, and the resource demands of economic growth. The intellectual origins of green growth date to the 1980s and 1990s, when it started as an alternative to the idea of unrestrained, unguided economic growth. The concept of green growth accepts that economic growth in some form will continue to occur, but that it must incorporate new behaviors, technologies, and investments. And it must be predicated on the notion of ecological limits.

The OECD defines green growth as a means to foster economic growth and development while ensuring that natural assets continue to provide the resources and environmental services on which our well-being relies. Green growth is a combination of both economic policy and sustainable development policy. It tackles two key imperatives together: the continued inclusive economic growth needed to reduce poverty and improve well-being and the improved environmental management needed to tackle resource scarcities and climate change. Today, the terms "green growth" and "sustainable growth" are used similarly. Table 8.3 summarizes the benefits of green growth. Two key strategies around green growth are ecotourism and green jobs.

TABLE 8.3	Benefits of Green Growth
The following may be the sustainable outcomes of green growth.	
Economic	• Increased and more equitably distributed GDP—production of conventional goods and services
	• Increased production of unpriced ecosystem services (or their reduction prevented)
	• Economic diversification, i.e., improved management of economic risks
	• Innovation, access, and uptake of green technologies, i.e., improved market confidence
Environmental	• Increased productivity and efficiency of natural resource use
	• Natural capital used within ecological limits
	• Other types of capital increased through use of nonrenewable natural capital
	• Reduced adverse environmental impact and improved natural hazard/risk management
Social	• Increased livelihood opportunities, income and/or quality of life, notably of the poor
	• Decent jobs that benefit poor people created and sustained
	• Enhanced social, human, and knowledge capital
	• Reduced inequality

Source: OECD. *Green Growth and Developing Countries: A Summary for Policy Makers* (2021, June). http://www.oecd.org/greengrowth/green-development/50526354.pdf

Promote Ecotourism

One trend on the rise has been ecotourism. **Ecotourism** is defined as "responsible travel to natural areas that conserves the environment, sustains the well-being of the local people, and involves interpretation and education." In theory, ecotourism can provide economic benefits to economically weaker communities living around protected areas and inspire them to protect the biodiversity in their own interest. Ecotourism ventures may be large scale, such as the overall planning and development of a scenic area or a village, or small scale such as the operation of an eco-inn or an ecotour route. The International Ecotourism Society has listed the following as the principles of ecotourism:

- Minimize physical, social, behavioral, and psychological impacts;
- Build environmental and cultural awareness and respect;
- Provide positive experiences for both visitors and hosts;
- Provide direct financial benefits for conservation;
- Generate financial benefits for both local people and private industry;
- Deliver memorable interpretative experiences to visitors that help raise sensitivity to host countries' political, environmental, and social climates;
- Design, construct, and operate low-impact facilities; and
- Recognize the rights and spiritual beliefs of the Indigenous people in the community and work in partnership with them to create empowerment.

Costa Rica has long been considered an exemplary ecotourism country. It is almost one-fourth rainforest—a major draw—and boasts volcanoes and beaches and rich biodiversity. Over the past two decades, the ecotourism industry in Costa Rica has expanded extensively, making Costa Rica one of the world's best ecotourism destinations (see Photo 8.3).

The development of ecotourism in Costa Rica resulted from a forward-thinking 1998 Biodiversity Law. The law helped establish entrepreneurship training programs tailored to the needs of each community. The training programs teach business development with a focus on environmental and social responsibility and is hosted by various environmental organizations including the Nature Conservancy and Conservation International.

One study by Carter Hunt and colleagues on the social impact of ecotourism on local communities focused on the Osa Peninsula, a region in southwest Costa Rica that has been heavily dependent on the influx of international tourists and foreign investment. The study found that the tourism industry tends to hire more local people than other sectors in the Costa Rican economy. Ecotourism also provides jobs with higher salaries and better opportunities for advancement than other jobs in the region, especially for young people, who often have lower skills and less experience than the labor force as a whole, and women with children, who need a more flexible working schedule to balance childcare. Moreover, workers employed in ecotourism are less likely to engage in illegal logging or the extraction of nontimber products, further reducing deforestation. This highlights that effective ecotourism can have multiple positive benefits for a community and its surrounding environment.

However, ecotourism is not always successful or sustainable. Revenues from ecotourism do not always reach local communities; and there have been reported negative impacts from ecotourism such as increased pollution, solid waste generation, degradation of forest, trail erosion, and disturbance to plants and animals. Political ecology scholars have found that ecotourism can replicate power relationships that turn locally owned and operated ecotourism resources into state-run territories that further marginalize local communities.

Priyanka and Aditya Gosh examined a case study on Sundarban Biosphere Reserve, home of highly endangered Royal Bengal Tiger in one of the largest mangrove forests in India. This project was touted as catering to both biodiversity conservation and socioeconomic development of local communities living around the protected area. They found the Sundarban ecotourism project failed to offer any benefits at all to the poorest and most marginal communities that surround the reserve. On the contrary, it offered disproportionately larger returns to the ecotourism companies and hotels who control lodging, food, and transportation and are not owned locally, thus defeating the principle behind the mechanism. Additionally, locals blamed tourists for increasing pollution and harming the health of the ecosystem. Their study

PHOTO 8.3 Selvatura Treetop Hanging Bridge in Costa Rica

This hanging bridge is in the Monteverde cloud forest reserve. In addition to having one of the best ziplines in the area, guided tours also showcase the wildlife and nature in the reserve. Costa Rica has a reputation for ecotourism, in part because it has one of the largest percentages (26 percent) of protected land in the world.

finds ecotourism as a concept cannot have a one-size-fits-all approach or cannot be considered a magic bullet for biodiversity conservation and simultaneous local socioeconomic development.

Generate Green Jobs

The ILO defines **green jobs** as decent jobs that directly contribute to environmental sustainability, either by producing environmental goods or making more efficient use of natural resources. A green economy is seen as reconciling two competing agendas of environment and economy and offers a pragmatic middle ground. Green jobs reduce the consumption of energy and raw materials, limit greenhouse gas emissions, minimize waste and pollution, protect and restore ecosystems, and enable enterprises and communities to adapt to climate change. Table 8.4 lists examples of some of the fastest-growing green jobs.

Many green jobs are in the environmental goods and services sector and directly benefit the environment or conserve natural resources. Examples include jobs in waste and wastewater management and treatment, energy and water-saving activities, conservation and protection, environmental goods (such as the installation of renewable energy production technologies), or adapted goods that have been modified to be cleaner or more resource efficient (such as buses with lower emissions). However, a green job does not necessarily have to be limited to jobs in and around the environment. A job that improves the environmental impact of production processes in enterprises in a car manufacturing facility is an example.

We must be cautious as not all green jobs are decent: An individual can be employed in the green sector, but not have social protection or good wages.

The green economy will be a major source of job growth in the future of work. For example, transitioning away from fossil fuels to renewable could create around twenty-four million jobs, largely offsetting any job losses in the traditional energy sectors such as oil and coal extraction. In the United States, the clean energy sector is now creating new jobs twelve times faster than almost any other sector in the American economy. In Latin America and the Caribbean, the clean energy sector has the potential to generate fifteen million new jobs in the region by 2030. However, it is important to note that many of the jobs in energy are traditionally dominated by men; efforts will be needed to ensure these green energy jobs are available to women.

Developing a strong green industry sector is also a smart business strategy. Investors are increasingly recognizing the business case for "going green," and

TABLE 8.4 Examples of the Fastest-Growing Green Jobs
• Clean car engineers
• Water-quality technicians
• Urban growers for sustainable food and restaurant industry
• Recyclers
• Green builders and green design professionals
• Wave energy producers
• Biofuels jobs
• Ecotourism
• Socially responsible investing
• Developing clean technology
• Developing the next generation of electric vehicles
• Constructing energy efficient buildings
• Energy efficiency and building retrofit industry (weatherization, energy auditing and retrofits, building operations, etc.)
• Landscaping and forestry
• Cleaning up brownfields
• Installing solar and wind energy systems
• Deconstructing, salvaging, and reselling building materials
• Electricians building wind turbines
• Pollution control technicians and analysts

Source: By author, compiled from several green job sites.

many see sustainability as an indicator of effective management. A Brookings Institute report noted that jobs in the "clean tech" sector experienced robust growth rates over the last decade. The highest job growth was in wave/ocean power, solar thermal, and wind. A recent study of some 90 percent of the world's leading CEOs suggests that sustainability will become a market and business driver over the next decade. Being seen as sustainable is not only an economic advantage but is also increasingly the new normal for businesses.

There are several strategies governments—at various scales from local to national—can do to support the growth of green jobs (see *SDGs and the Law* 8.1). Table 8.5 lists examples of direct public-sector investments and incentives to encourage green jobs in the private sector. Although an increasing number of legal and policy frameworks have been adopted on the "green economy" or "green growth," this is not yet a widespread practice in all countries and regions.

A Critical Perspective on Growth and Green Growth

Is economic growth sustainable? Is green growth genuinely different from growth imperatives of the past, or is this merely greenwashing?

SDG 8 Targets 8.1 and 8.2 focus on increased economic growth, which will be measured by increasing GDP rates. Development scholars Shirin Rai, Benjamin Down, and Kanchana Ruwanpura are concerned about the possible

 SDGs AND THE LAW 8.1

Laws and Policies for Greening Jobs in Sub-Saharan Africa

Sub-Saharan Africa is one of the fastest-developing regions in the world. However, due to factors such as widespread poverty, recurrent droughts, and dependence on rain-fed agriculture, it is more vulnerable to environmental degradation and its impacts. This intensifies the vulnerability of the region's economies, which are dependent on natural resources.

Increasingly, countries in this region have passed legislation to defend the right to work and live in a clean and healthy environment, and most promote environmental impact assessments. Further, some have laws that emphasize the role of workers in respecting and protecting the environment. This is the case for the Mining Codes in Benin (2006) and Burkina Faso (2003), the Forest Code in the Central African Republic (2008), the Oil Code in the Comoros (2012), and agriculture laws in the Democratic Republic of the Congo (2011).

Legislative and policy measures also try to combine the goal of job creation with the preservation of the environment, including training for workers and the introduction of environmental concerns into educational programs. Examples of development plans or national strategies on climate change that include labor issues exist in Chad, Burkina Faso, Burundi, Mali, Niger, and Senegal. Niger's National Policy on Climate Change (2012) promotes the creation of "green jobs" and the adoption of tax incentives for employers that create them. Benin, Burkina Faso, Burundi, Djibouti, and Mali similarly established tax incentives to support jobs in energy and waste management.

Finally, some countries have adopted green jobs programs. In Senegal, a joint program with the UN Development Programme seeks to promote and develop new sectors with green jobs, build the capacities of certain groups (e.g., women), and provide training on the creation of green jobs. At least 1,000 green jobs have been created in Senegal since 2015, and 10,000 more are projected by 2020.

Source: International Labour Organization. *World Employment and Social Outlook 2018: Greening with Jobs*. Geneva, Switzerland: ILO (2018).

negative impacts of SDG 8 targets on gender equality. They note that the SDG 8 Targets 8.1 and 8.2's focus on increasing annual growth rates of GDP could result in economic growth but may simultaneously exacerbate gender inequality, as women take low-paying jobs that do not allow them to escape poverty.

They also argue that the emphasis on increasing GDP and per capita growth neglects the value and costs of unpaid social reproductive work. Social reproductive work are the tasks necessary for everyday life: cooking, types of care (caring for children, the sick, the elderly, the disabled), cleaning, and biological reproduction. The work of feminist economists has shown that GDP measurement is deeply flawed because it does not include unpaid domestic work, a component in the informal economy. There is a long history of undervaluing or ignoring women's unpaid labor not only by neoclassical economics but also the modern labor movement. Their analysis concludes that unless SDG 8 accounts for unpaid work that continues to be largely performed by women, it cannot address the decent work agenda in a comprehensive and gendered way. They also argue that it is important to challenge a growth-led approach to development and there is a need for a gendered approach to measuring development. Focusing on increasing the GDP growth rate continues business as usual.

Rai and colleagues are not alone in their criticism of targets that emphasize economic growth using traditional measurements such as GDP. Critics argue that economic expansion *inevitably* causes ecological harm and that the relationship between economic growth and ecological protection is more complex than green growth assumes. Tim Jackson's book *Prosperity without Growth* challenges the assumption that a larger, constantly growing economy

TABLE 8.5 Policy Options to Drive Demand for Green-Collar Jobs and Businesses

Public sector investments in

- Energy efficiency retrofits of public buildings
- Solar or other renewable energy systems on public buildings financed with capital budgets, bonds, or performance contracting
- New public buildings constructed to green standards
- Public transit infrastructure
- Green products and services from local providers

Incentives or requirements to drive private-sector investment

- Tax incentives, rebates, reduced fees, or streamlined permitting for private building owners that invest in energy efficiency, renewable energy, or green building
- Technical assistance or innovative financing for private investment in renewal energy, efficiency, green building, alternative vehicles, or green space
- Green building codes, energy conservation ordinances, or other requirements for new green buildings or retrofits of existing buildings
- Land use and infrastructure policies to support green manufacturing companies

Source: Apollo Alliance and Green for All, 2008. *Green-Collar Jobs in America's Cities* (2018, January). https://www.cows.org/_data/documents/1165.pdf

is a better one. Political economists and political ecologists among others argue that political moderates embrace green jobs and green growth because it allows everyone to enjoy the benefits of continued growth while avoiding the necessary structural changes that must take place. They are particularly critical of liberal free-market reforms that believe government can create an inclusive, green economy by setting standards, spurring innovation, and realigning investments, and see economic incentives, payment for ecosystem services, and ecotourism as "having your cake and eating it too."

Daniel Fiorino has noted that green growth is, at best, a misleading and, at worst, cynical effort to justify continued, rapid, and ultimately socially and ecologically destructive patterns of economic growth. The concepts of the green economy, green growth, and green jobs perpetuate the capitalist, neoliberal values and institutions that are at the source of our unsustainability. Critics say that green growth is a new way to legitimize capitalism and to continue to make the case for the necessity of economic growth, but fails to deal substantively with poverty, inequality, and empowerment.

An alternative to the green economy or green growth is **degrowth**. The idea behind degrowth is a reduction in production and consumption in a way that increases human well-being and enhances ecological conditions. Advocates want to replace GDP as the main indicator of prosperity and replace it with other measures of social well-being. Degrowth would fundamentally restructure the global economy within the context of the Global South, where growth is directed only within lower-income countries, not wealthy ones. Specific actions that could advance a degrowth society include:

- Promote local currencies and end national currencies;
- Transition to nonprofit and small-scale companies;
- Reduce working hours;
- Facilitate volunteer work;
- Reuse empty housing;
- Introduce a basic income and income ceiling; and
- Transition from a car-based system to a more local, biking- and walking-based one.

Degrowth requires a transformation of not just the economic system, but all the systems on which it relies (political and social). For now, degrowth is intriguing, but remains a radical theory.

SUMMARY AND PROGRESS

According to the most recent UN SDG Progress Report, between 2010 and 2019, the world witnessed rising labor productivity and improved unemployment rates, despite large disparities across regions. Economic growth has taken place in conjunction with improvements in decent work. However, significant challenges remain. Many of the lowest-income countries experienced annual GDP growth of less than 5 percent over the past five years, falling short of the SDG 8 target of at least 7 percent growth per annum. However, global labor productivity had steadily increased over the last twenty years.

The economic impacts of the COVID-19 pandemic crisis are sobering: The world is now facing its worst recession in generations. Even the highest-income countries are struggling to cope with the health, social, and economic fallout of the pandemic, but the poorest and most disadvantaged countries will inevitably be hit the hardest. Estimates are that world trade will decline by 13 to 32 percent, foreign direct investment will decline by up to 40 percent, and remittances to low- and middle-income countries will fall by 20 percent in 2020. Tourism, for example, is still reduced significantly due to the closure of borders, travel bans, and lockdown measures. Many small nations, particularly islands, rely on the revenue earned from guests on visiting ships and the related tourism activities including restaurants, attractions, and tours. The industry is said to contribute $2 billion a year to the Caribbean; the shutdown of tourism in 2020 has dramatically affected countries in this region.

More than 1.6 billion workers in the informal economy risk losing their livelihoods. Many poorer countries are already experiencing acute food insecurity. The crisis poses a serious threat to the occupational safety and health of workers and may increase the risk of child labor as a result of rising poverty driven by the pandemic.

Among the world's poor, the COVID-19 pandemic, together with job losses, fragile health systems, insufficient basic services, and low coverage of social protection systems, has aggravated their vulnerabilities. The United Nations says that urgent policy measures are needed to support businesses, boost labor demand, and preserve existing jobs to achieve full and productive employment and decent work for all women and men. Even prior to the coronavirus pandemic, there were challenges to achieving the targets set in SDG 8, and some targets were not making sufficient progress.

Figure 8.2 highlights the numerous connections SDG 8 has with other SDGs. For example, Target 8.7 calls for effective measures to end forced labor, modern slavery, human trafficking, and all forms of child labor. These include eliminating all forms of violence against all women and girls in public and private spheres, including trafficking and sexual and other types of exploitation (SDG 5.2 Gender Equality); eliminating all harmful practices, such as child, early, and forced marriage and female genital mutilations (SDG 5.3 Gender Equality); ending abuse, exploitation, and trafficking of children (SDG 16.2 Peace, Justice and Strong Institutions); and facilitating orderly, safe, and responsible migration and mobility of people, including through implementation of planned and well-managed migration policies (SDG 10.7 Reduced Inequalities).

Another interconnection is around climate change. Although climate change mitigation measures may result in short-term employment losses, their negative impact on GDP growth, employment, and inequality can be reduced through appropriate policies. Climate change mitigation could bring down the share of women in total employment unless action is taken to reduce occupational segregation, as employment gains associated with the 2°C scenario are likely to create jobs in currently male-dominated industries (renewables, manufacturing, and construction). Coordination between social partners can reduce inequality and promote efficiency gains, while coordination at the international level is necessary to achieve meaningful cuts in emissions. Certain mitigation policies (such as limiting the increase in temperature by promoting renewable energy) may act as an incentive for enterprises to develop and adopt more efficient technology, thereby boosting employment in key occupations, as well as productivity. Adaptation policies

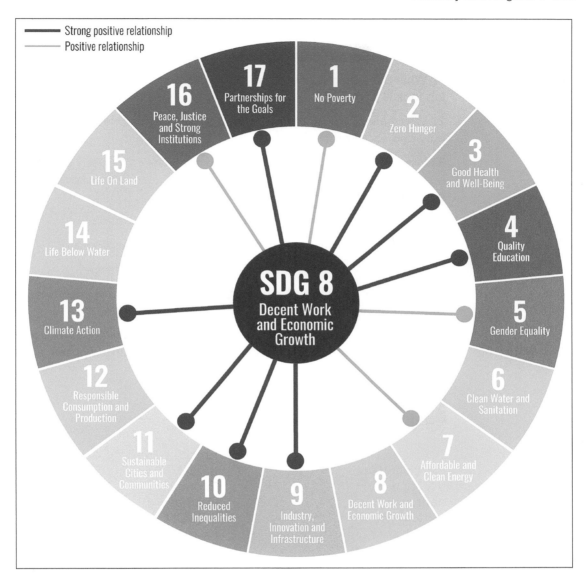

FIGURE 8.2 Interconnections in SDG 8 and the Other Goals
Decent work and economic growth connect to all of the SDGs in some way; this figure highlights several of the stronger connections.

(e.g., converting to climate-resilient agriculture practice) can also create jobs at the local level.

Finally, examining interconnections reminds us to be aware of counter-productive or cancelling outcomes. For example, increasing economic growth can enable governments to increase spending on health care, including toward providing universal health coverage (SDG 3 Good Health), but economic growth can be associated with adverse effects on the environment, including water, air, and soil pollution, and ecosystem change, which can increase the risk of communicable disease, illness, and death (SDG 14 Life Below Water and SDG 15 Life on Land).

QUESTIONS FOR DISCUSSION AND ACTIVITIES

1. What does decent work mean to you personally? What would a decent job look like to you?

2. Why have some countries, even the highest-income countries, not provided social protections?

3. Why has child labor been difficult to eliminate? What can you do as a consumer to ensure child labor has not been used in the products you buy?

4. What is the difference between free trade and fair trade?

5. Explain two positive and two negative impacts of ecotourism.

6. Which of the targets for SDG 8 are the biggest challenge?

7. Are green growth and sustainable employment compatible with a capitalist economic system?

8. What would "degrowth" look like in practice? How would a society transition from high to low levels of production and consumption?

TERMS

child labor
debt bondage
decent work
degrowth
ecotourism
forced labor
forced marriage
formal economy
green growth

green job
hazardous child labor
informal economy
International Labour Organization (ILO)
modern slavery
social protection
worst forms of child labor

RESOURCES USED AND SUGGESTED READINGS

Fiorino, D. *A Good Life on a Finite Earth: The Political Economy of Green Growth*. New York: Oxford University Press (2018).

Forstater, M., and Murray, M. *The Job Guarantee: Toward True Full Employment*. New York: Palgrave Macmillan US (2013).

Ghosh, P., and Ghosh, A. "Is Ecotourism a Panacea? Political Ecology Perspectives from the Sundarban Biosphere Reserve, India." *GeoJourna l* (84): 345–66 (2019). https://doi.org/10.1007/s10708-018-9862-7

Graber, D. *Bullshit Jobs: A Theory*. New York: Simon and Schuster (2018).

Hawkens, P., Lovins, A., and Lovins, L. H. *Natural Capitalism*. New York: Earthscan (2010).

Hunt, C. A., Durham, W. H., Driscoll, L., and Honey, M. "Can Ecotourism Deliver Real Economic, Social, and Environmental Benefits? A Study of the Osa Peninsula, Costa Rica." *Journal of Sustainable Tourism* 23 (3): 339–57 (2015). https://doi.org/10.1080/09669582.2014.965176

International Ecotourism Society. "What Is Ecotourism?" (2022). https://ecotourism.org/what-is-ecotourism/

International Labour Organization. "Global Estimates of Child Labour." Geneva, Switzerland: ILO (2017). https://www.ilo.org/wcmsp5/groups/public/@dgreports/@dcomm/documents/publication/wcms_575499.pdf

International Labour Organization. "World Employment and Social Outlook 2018; Greening With Jobs" (2018). https://www.ilo.org/global/research/global-reports/weso/greening-with-jobs/WCMS_628708/lang--en/index.htm

International Labour Organization. "2021 Declared International Year for the Elimination of Child Labour." Geneva, Switzerland: ILO (2019). https://www.ilo.org/global/about-the-ilo/newsroom/news/WCMS_713925/lang--en/index.htm

International Labour Organization. *What Works: Promoting Pathways to Decent Work*. Geneva, Switzerland: ILO (2019).

International Labour Organization and United Nations Children's Fund. *Child Labour: Global Estimates 2020, Trends and the Road Forward*. New York: ILO and UNICEF (2021). https://data.unicef.org/resources/child-labour-2020-global-estimates-trends-and-the-road-forward/?utm_source=newsletter&utm_medium=email&utm_campaign=childlabour_report

Jackson, T. *Prosperity without Growth: Foundations for the Economy of Tomorrow*. 2nd ed. New York: Routledge (2017).

Kallis, G., Paulson, S., D'Alisa, G., and Demaria, F. *The Case for Degrowth*. Cambridge: Polity Press (2020).

Meadows, D. *The Limits to Growth: A Report for the Club of Rome's Project on the Predicament of Mankind*. New York: Universe Publications (1972).

Miller, E. *Reimagining Livelihoods: Life beyond Economy, Society, and Environment*. Minneapolis: University of Minnesota Press (2019).

Muro, M., Tomer, A., Shivaram, R., and Kane, J. "Advancing Inclusion through Clean Energy Jobs: A Brookings Report" (2019). https://www.brookings.edu/research/advancing-inclusion-through-clean-energy-jobs/

OECD. "Green Growth and Sustainable Development" (2022). https://www.oecd.org/greengrowth/

Pawel, M. *The Crusades of Cesar Chavez: A Biography*. New York: Bloomsbury (2014).

Rai, S. B., Ruwanpura, D., and Ruwanpura, K. "SDG 8: Decent Work and Economic Growth—A Gendered Analysis." *World Development* 113: 368–80 (2019). https://doi.org/10.1016/j.worlddev.2018.09.006

Sahlins, M. *Stone Age Economics*. New York: Routledge (1972).
Stoknes, P. E. *Tomorrow's Economy: A Guide to Creating Healthy Green Growth*. Cambridge, MA: MIT Press (2021).
UNICEF. "Child Labour Rises to 160 Million—First Increase in Two Decades" (2021). https://www.unicef.org/press-releases/child-labour-rises-160-million-first-increase-two-decades

The International Labour Organization provides research and reports on targets in SDG 8. See https://www.ilo.org

- Yearly general economic outlook: World Employment and Social Outlook Trends 2020. https://www.ilo.org/global/research/global-reports/weso/2020/WCMS_734455/lang--en/index.htm
- Child Labor Global Estimates of Child Labour: Results and Trends 2012–2016. https://www.ilo.org/wcmsp5/groups/public/@dgreports/@dcomm/documents/publication/wcms_575499.pdf
- Modern Slavery: https://www.ilo.org/global/topics/forced-labour/lang--en/index.htm
- Trends for Women: World Employment and Social Outlook: Trends for Women 2018—Global snapshot. https://www.ilo.org/global/research/global-reports/weso/trends-for-women2018/WCMS_619577/lang--en/index.htm
- Green Jobs; World Employment and Social Outlook 2018: Greening with Jobs. https://www.ilo.org/global/research/global-reports/weso/greening-with-jobs/lang--en/index.htm

In addition, these organizations also play a role in advancing SDG 8 and publish periodic reports and updates:

- UN Development Program: https://www.undp.org/content/undp/en/home/
- Economic and Social Commission for Asia & the Pacific: https://www.unescap.org/
- Economic and Social Commission for Western Asia: https://www.unescwa.org/
- Economic and Social Commission for Africa: https://www.uneca.org/
- Economic and Social Commission for Europe: https://www.unece.org/info/ece-homepage.html
- Economic and Social Commission for Latin America and the Caribbean: https://www.cepal.org/en
- IMF World Economic Outlook database: https://www.imf.org/en/Publications/SPROLLS/world-economic-outlook-databases#sort=%40imfdate%20descending
- UN Capital Development Fund: https://www.uncdf.org/

Infrastructure, Industry, and Innovation

LEARNING OBJECTIVES

After reading this chapter, you should be able to:

- Identify the types of infrastructure that supports economic development
- Explain where infrastructure needs are greatest
- Explain the role of research and development in the context of sustainability
- Identify and explain the various targets and goals associated with SDG 9
- Explain how the three Es are present in SDG 9
- Explain how SDG 9 strategies connect to other SDGs

It is said that roads are "arteries of development." The tiny, landlocked nation of Nepal lies between two of the world's most populous countries, China and India. Despite being small in comparison, Nepal is an important trading partner for China and India and provides a vital transport connection across the continent for these much larger nations. But Nepal has distressingly poor and often dangerous road infrastructure. There is only one dependable road link between India and Nepal. Many roads are dirt or gravel, and have potholes, poor drainage, sinkholes, and deep ditches (Photo 9.1). An estimated one-third of the population live at least two hours, walk from a road. Being isolated has consequences for access to markets and education, and life-saving health care.

One major reason for the poor road infrastructure is the country's geography. The Himalayan range lies in the north of the country, including eight of the world's ten highest mountains, and these present huge physical obstacles. In the southern lowland plains of the country, rivers descending from the mountains change course frequently, presenting another challenge. Many of the bridges suffer intense wear, from the seasonal monsoons and frequent vehicle overloading.

A second reason for poor road infrastructure is political: inadequate funding and corruption. The lack of timely road safety audits, the use of substandard building materials, and noncompliance in the construction process contribute to the poor state of roads.

In the last decade Nepal has built about 4,300 miles (7,000 km) of roads and more is under construction. However, some warn the boom in road infrastructure could have unintended consequences. In areas where traditional livelihoods and trek tourism rely on unspoiled nature, the roads may negatively impact these local economies. There could also be an increase in damage to sacred sites, an expansion of timber cutting in previously inaccessible woodlands, and a growing trade in endangered and threatened species. Nepali architect and filmmaker, Sonam Lama, is a critic of

PHOTO 9.1 A Mountainous Road in Nepal
Many of Nepal's roads are dirt or gravel, which makes them more susceptible to being washed out or damaged by spring melting of the glaciers, or rains. This photo shows a waterfall cascading down a hill, which may lead to faster erosion of the roadbed.

the road construction. He says Nepalese should ask some basic questions such as "What do we mean by 'development'? For whom? Who is making the decisions? And on whose terms are we going to transition into the future?"* For countries such as Nepal, road construction can present a rural development conundrum.

WHAT IS INFRASTRUCTURE?

Governments have long recognized the vital role that modern telecommunications, transport, energy, and water services play in economic growth and poverty alleviation. But providing such infrastructure services is challenging and expensive.

Sustained investment in infrastructure is an important driver of economic growth and development. Yet, many countries have not invested enough in infrastructure. A 2018 OECD Business and Finance report concluded that between $3 trillion and $6 trillion must be invested annually to build the infrastructure needed for sustainable development.

"Infrastructure" is a broad term that originates from the French *infra*, meaning below, and the term "structure," which refers to an arrangement of or elements of something complex. Examples of infrastructure include transportation systems, communication networks, and sewage, water, and electric systems. Such systems tend to be capital-intensive and high-cost investments yet are vital to a country's economic development and prosperity. Large-scale infrastructure such as roads or bridges are usually produced by the public sector or publicly regulated monopolies, but at a smaller scale, infrastructure is often produced by private firms or through local collective action.

There are two types of infrastructure: soft and hard. **Soft or social infrastructure** refers to the social, political, and cultural institutions that support the social services in a community or country. These include healthcare systems, housing, education systems, parks and playgrounds, financial institutions, and emergency services.

---

*Woof, M. "Nepal Plans Road Infrastructure Expansion," *World Highways*, March 2014. https://www.worldhighways.com/wh10/wh8/feature/nepal-plans-road-infrastructure-expansion; Coburn, B. "Nepal's Road-Building Spree Pushes into the Heart of the Himalayas," *Yale Environment 360*, January 2020. https://e360.yale.edu/features/paving-the-himalayas-a-road-building-spree-rolls-over-nepal

Hard infrastructure are the large physical systems necessary to running a modern, industrialized economy. Hard infrastructure systems can be below ground (waste systems) or highly visible (bridges). Examples of hard infrastructure include transportation (roads, airports, water ports, railways, and subways), energy (the delivery of electricity, dams, utility systems), telecommunications, and solid waste (collection and disposal, hazardous waste). Additionally, as introduced in Chapter 6, water and sanitation infrastructure includes the complex network of water supply, gray and green wastewater systems, irrigation systems, dikes and levees, and canals.

Countries in the Global North are characterized by high levels of quality infrastructure. However, in the United States, decades of disinvestment in these public services has resulted in the deterioration of important infrastructure. Every four years, the American Society of Civil Engineers' *Report Card for America's Infrastructure* depicts the condition and performance of American infrastructure in the familiar form of a school report card—assigning letter grades based on the physical condition and needed investments for improvement. The most recent report in 2021 noted the overall "grade" for infrastructure was a C–. The report stated that the roads, bridges, public drinking and water systems, dams, airports, and mass transit systems in the United States require massive restoration. US infrastructure systems are failing to keep pace with the current and expanding needs, and investment in infrastructure is woefully inadequate (see Figure 9.1). *Understanding the Issue* 9.1 highlights some of the most problematic infrastructure issues in the United States.

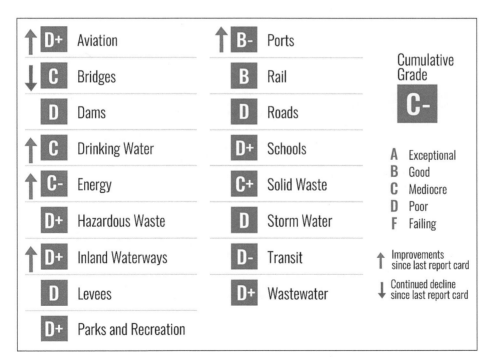

FIGURE 9.1 The 2021 Report Card for America's Infrastructure
US infrastructure systems are failing to keep pace with the current and expanding needs, and investment in infrastructure is woefully inadequate.

UNDERSTANDING THE ISSUE 9.1

Failing Infrastructure Investment in the United States

Unlike many other countries in the Global North, the United States has failed to invest in infrastructure maintenance and improvement. Decades of underinvestment in infrastructure has social and environmental impacts. For more than twenty years, the American Society of Civil Engineers graded infrastructure in the United States in the "D" range. Finally, in 2021, incremental progress toward restoring infrastructure resulted in an overall grade of "C–." According to the American Society of Civil Engineers' *2021 Infrastructure Report Card*, some of the most problematic infrastructure areas include the following.

Bridges: The United States has 617,000 bridges, four in ten of which are fifty years or older. A total of 46,154—7.5 percent—of the nation's bridges were structurally deficient in 2020, and on average, there were 188 million trips across structurally deficient bridges each day. The average age of America's bridges keeps going up and many of the nation's bridges are approaching the end of their design life. The most recent estimate puts the nation's backlog of bridge rehabilitation needs at $125 billion.

On August 1, 2007, during rush hour, the central span of the I-35W Mississippi River Bridge located in Minneapolis, Minnesota suddenly gave way, followed by the adjoining spans (see Photo 9.2). The structure and deck collapsed into the river and onto the riverbanks below. Thirteen people were killed and many more injured. As the investigation into the collapse unfolded, it was reported that the bridge had been rated as "structurally deficient" for several years prior to its collapse. Within days, bridge inspections were stepped up throughout the United States. The images of the collapsed bridge became a symbol of the consequences of the lack of investment in road infrastructure.

Airports: On average, US airports served more than two million passengers every day. The aviation industry is marked by technologically advanced and economically efficient aircraft; however, the associated aviation infrastructure of airports and air traffic control systems is not keeping up. Terminal, gate, and ramp availability is not meeting the needs of a continually increasing passenger base. Congestion at airports is growing;

PHOTO 9.2 The Collapse of the I-35W Mississippi River Bridge
Navy divers survey and assess the wreckage of the bridge collapse. The collapse started a larger conversation about the consequences of decades of underinvestment in infrastructure.

it is expected that twenty-four of the top thirty major airports may soon experience "Thanksgiving-peak traffic volume" at least one day every week. With a federally mandated cap on how much airports can charge passengers for facility expansion and renovation, airports struggle to keep up with investment needs, creating a $111 billion funding shortfall.

Schools: Every school day, nearly fifty million K–12 students and six million adults occupy close to 100,000 public school buildings on an estimated two million acres of land. While state and local governments make significant investment in public K–12 school infrastructure and schools play important civic, educational, and public safety roles in communities, the nation continues to underinvest in school facilities, leaving an estimated $38 billion annual funding gap. As a result, 24 percent of public school buildings were rated as being in fair or poor condition. While there have been a number of insightful reports in recent years, state and local governments are plagued by a lack of comprehensive data on public school infrastructure as they seek to fund, plan, construct, and maintain quality school facilities.

Dams: Dams provide vital service and protection to our communities and the economy. The average age of the 91,000 dams in America is fifty-seven years. Due to the lack of investment, the number of deficient high-hazard potential dams has doubled over the last twenty years. It is estimated that it will require an investment of nearly $45 billion to repair aging, yet critical, high-hazard potential dams.

Source: American Society of Civil Engineers. *2021 Report Card for America's Infrastructure* (2021, June 16). https://infrastructurereportcard.org/ and https://infrastructurereportcard.org/wp-content/uploads/2020/12/2021-IRC-Executive-Summary-1.pdf

In the Global South, from 1960 to the 1990s, most governments entrusted delivery of infrastructure to state-owned entities (often monopolies). But the results were disappointing. Public-sector monopolies were plagued by inefficiency. Many were strapped for resources because governments held prices below costs. Publicly owned utilities failed to expand services to meet rapidly growing demand and did not do a good job of providing service to poor and rural households. This led to a widespread move in the 1990s to engage the private sector in the provision and financing of infrastructure. However, this too has largely failed. Chronic inefficiency, poor pricing policies, and corruption meant that these companies were unable provide adequate services to existing consumers, let alone expand services. Some projects were canceled or renationalized. As in the Global North, many countries in the Global South have been scaling up infrastructure investment through a combination of public spending and public private partnerships in the last two decades.

TRENDS IN INFRASTRUCTURE INVESTMENT

Globally, investment in infrastructure has been increasing but not at the pace needed. A 2018 OECD report has prioritized the largest needs in infrastructure investment as:

1. Road transport,
2. Energy supply,
3. Rail transport,
4. Information and telecommunications (ICT), and
5. Water infrastructure.

Despite some improvements, the quality, quantity, and accessibility of infrastructure in low-income countries is far behind that in the Global North. Although there has been expanded service in information and communication technology, basic infrastructure like roads, sanitation, electrical power, and water remains scarce in many of the lowest-income countries. According to Sumeep Bath, in much of the Global South, especially in rural communities, many people still lack access to new technologies that could improve their livelihoods and standard of living. Around 2.6 billion people in the Global South still face difficulties in accessing electricity full time and many do not have access to water or basic sanitation. Economists say that poor access to electricity and transportation are major constraints to business activity and economic growth. The lack of existing infrastructure in many of the lowest-income countries represents one of the most significant limitations to achieving sustainable development.

Many countries and regions have a gap between the infrastructure they need to supply and the financing to build it. The infrastructure financing gap in Asia, for example, is around $228 billion. In Latin America, around $71 billion, or 3 percent of GDP, would need to be invested in infrastructure to satisfy needs. With costs this large, many governments have turned to private partnerships to construct and maintain infrastructure.

The Privatization of Infrastructure

Infrastructure is built and provided at different scales: local, national, and transnational. As a result, infrastructure often requires coordination among scales and different actors. For example, countries that share a border will need to coordinate the building of a regional railway, while at the local scale, there will be need for coordination of smaller-scale infrastructure such as hospitals or ICT.

For much of the twentieth century, the public sector (the government) played the leading role in constructing and financing infrastructure. Infrastructure investments and maintenance can be very expensive, costing hundreds of millions, or even billions of dollars.

However, in the last several decades, the public sector in many countries has lacked either sufficient resources or political will and has pursued policies to involve the private sector in the delivery and financing of infrastructure services. These are referred to as **public private partnerships** or PPPs. Generally, most roads, major airports and other ports, water distribution systems, and sewage networks are publicly owned, whereas most energy and ICT networks (cell phone or internet) are privately owned.

However, some political economists and political ecologists are highly critical of public services being privatized. The *Critical Perspectives* 9.1 box discusses the drawbacks and possible injustices that can arise when public goods are privatized.

Clearly, more investment is needed in infrastructure over the coming decades. Because Chapter 6 examines access to water and sanitation and Chapter 7 examines the lack of electricity, this chapter will focus on transportation and information and communication technology (ICT).

CRITICAL PERSPECTIVES 9.1

The Privatization of Infrastructure

While water is often abundant, millions of people struggle daily for access to potable water. In the 1990s, many governments in the Global South were strapped for cash and could not pay for expanding water infrastructure to meet increasing demand. At the same time, many public utilities were criticized for corruption and inefficiency. As a result, governments turned to privatization as a solution that could bring water to the poorest communities not connected to the public water supply.

The privatization of water production and delivery services has allowed companies to generate massive profits and economic growth. The private water business is valued at more than $45 billion. Trying to sell water for a profit, especially to those in poverty, however, has been a much more difficult and controversial project than first imagined.

Geographer Erik Swyngedouw has been highly critical of privatization of water resources. Swyngedouw argues privatization has made services *less* affordable and adversely affected access by the poor. He points out that wealthy neighborhoods located closer to the water mains enjoy household water connections that provide water sold at a highly subsidized price, while the poorest must buy water from private vendors that have created distribution monopolies. In Guayaquil, Ecuador, the lowest-income pay up to four hundred times more per liter than those connected to the public network.

Swyngedouw explains that the "water business" is driven by considerations of competitiveness, profitability, and the customer's ability to pay. He says that humanitarian motivations such as providing water to the poor, improving life expectancy or health, and contributing to overall development are no longer main objectives in private management contracts.

According to Swyngedouw, "Servicing urban residents with reliable potable water services is not an easy business. It requires significant long-term investment and complex organizational and management arrangements. And profitability is by no means assured, particularly in urban environments where many people have difficulty paying and problematic access conditions. As a result, private companies only really go for the nice bits—those that have some meat on the bone. That means areas with high-income residents with proven ability to pay are of course the valued customers of the privatized utilities."

Further, Swyngedouw points out that privatization de facto means taking away some control from the public sector and transferring this to the private sector. This may result in diminished transparency of decision making and an undermining of traditional channels of democratic accountability. Privatization, he claims, is a legally and institutionally condoned, if not encouraged, form of theft. Swyngedouw says that for many years, the World Bank and the IMF failed to critically analyze the results of privatization and instead encouraged it through their programs and financing. For example, in Guayaquil, the Inter-American Development Bank provided a $40 million loan under the condition that almost half of it be spent on preparing the privatization bid of the public water utility.

Scholarly research has shown that the privatization of water has exacerbated inequality. This research, along with growing community dissatisfaction about private contracts for public utilities, has led to a reversing of privatization: Since 2010, dozens of cities around the world have taken water service back into public control.

Sources: Swyngedouw, E. *Social Power and the Urbanization of Water: Flows of Power.* Oxford: Oxford University Press (2004).

Swyngedouw, E. "Dispossessing H2O: The Contested Terrain of Water Privatization." *Capitalism Nature Socialism* 16 (1): 8198 (2005). 10.1080/1045575052000335384

Transportation Infrastructure

Roads are important for development because they provide access to resources, jobs, and markets. Paved roads are preferred for transporting products and services and connecting the economic centers across a region. Improvements to road conditions can also improve access to education, health care, and a higher quality of life. More than 90 percent of all new road expansion is occurring in the Global South.

In many low-income countries, the main road network carries about 80 to 90 percent of passenger and freight transport and it is, therefore, of key importance to the national economy. The development gap between rural and urban areas in many of the regions in the Global South means that rural areas suffer from poorest road infrastructure. Rural roads may make up more than 80 percent of the road network length but are given lower priority in the allocation of funding because they carry much lower volumes of motorized traffic. As result, most rural roads are unpaved roads. Dirt roads can be washed away by rains or become impassable due to mud or long-term erosion. The need for paved roads is greatest in Southeast Asia, sub-Saharan Africa, and North Africa.

It is estimated that in Asia, more than half the roads are not paved. Bhutan, Vietnam, and Laos have not invested adequately in developing road infrastructure to make travelling easy around the country. Cambodia has sporadic road development in both rural and urban areas. The Philippines' roads are less developed compared to other East Asian countries, but Nepal, Mongolia, and Bangladesh's roads are also poorly developed.

Information and Communications Infrastructure

There is no doubt that the twenty-first century is digitally connected in important ways. Yet there are many parts of the world that experience a digital divide. **Information and communication technology (ICT)** form the backbone of today's digital economy. This includes artificial intelligence, the internet, and 5G. These new technologies hold great potential for human progress. They also raise complex questions about trust and privacy and pose a number of challenges, including the future of work, child online protection, digital sexual violence, and electronic waste.

Many countries and businesses are investing in key areas of ICT, including digital infrastructure, digital literacy, affordability, digital safety, and cybersecurity. The good news is that digital inclusion of the population has become a priority for most countries, as inadequate ICT infrastructure is a significant impediment to economic growth and poverty alleviation.

Access to mobile technology has spread rapidly around the world (see Maps 9.1 and 9.2). In 2021, 85 percent of the world's population had access to mobile broadband networks. As expected, however, this is uneven at other scales; access to mobile broadband is less in rural areas than in urban areas. For example, internet access at home in urban areas is twice as high as in rural areas. There is also an ICT gender gap: Currently 55 percent of the male population uses the internet, compared with 48 percent of the female population.

However, more than half the world's population—3.6 billion—still lack internet access, which is limited and unaffordable for many, particularly in the lowest-income countries. In sub-Saharan Africa, only 6 percent of rural households and 28 percent of urban households have internet access. With

more than half the world's population still offline, billions are unable to benefit from the positive impact that ICT could have on their lives. The positive
news is that almost 70 percent of the world's youth are using the internet.
Digital infrastructure and internet access must be a priority—especially for
those living in rural and remote areas. Bridging this digital divide is crucial

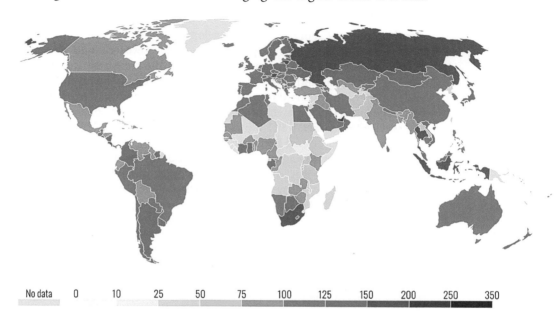

No data   0   10   25   50   75   100   125   150   200   250   350

MAP 9.1 Mobile Cellular Subscriptions per 100 People (2017)
Access to mobile technology has spread rapidly around the world, although there are still regions
and countries where access is limited.

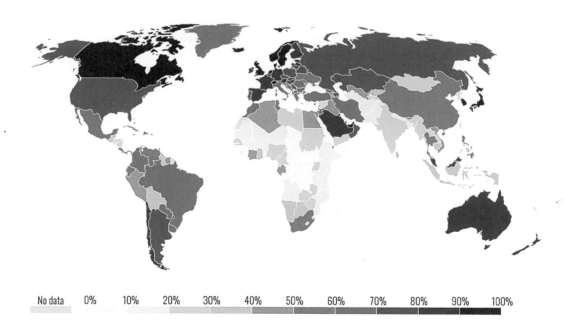

No data   0%   10%   20%   30%   40%   50%   60%   70%   80%   90%   100%

MAP 9.2 Share of the Population Using the Internet (2017)
More than half the world's population—3.6 billion—still lack internet access, which is limited and
unaffordable for many, particularly in the lowest-income countries in Africa and Central Asia.

to ensure equal access to information and knowledge and consequently foster innovation and entrepreneurship.

WHAT IS INNOVATIVE RESEARCH AND DEVELOPMENT?

Another component of sustainable development is a focus on increasing investment in research and development. The term **research and development (R&D)** is widely linked to innovation both in the corporate and government world or the public and private sectors. R&D includes activities that companies undertake to innovate and introduce new products and services. It is often the first stage in the development process. The goal is typically to take new products and services to market. R&D allows a company to stay ahead of its competition. The benefits of successful R&D can improve existing products and services and gain a competitive advantage. Innovative products can transform the fortunes of a business, giving it market leader status quickly. Additionally, even just one best-selling product idea may make it worthwhile to a company's profits.

There are three different types of R&D: basic, applied, and experimental. "Basic research" involves a department staffed by engineers who are tasked to develop new products. This typically involves research with no specific goal or application in mind (think of it as research for the sake of research). In contrast, there is applied research—research done with a specific goal, use, or product in mind. This usually involves a department composed of interdisciplinary teams of scientists who are tasked with specific research tasks such as the development of future products or the improvement of current products and/or operating procedures. Experimental development is systematic work, drawing on knowledge gained from research and practical experience and producing additional knowledge, which is directed toward producing new products or processes or to improving existing products or processes.

Companies that set up and employ entire R&D departments often commit substantial money to the effort. It also inevitably involves risk of capital—because there is no immediate payoff, and the return on investment is not guaranteed. For this reason, R&D is unusual among business activities in that it does not *always* result in a direct return on investment for the company.

Companies in different sectors and industries conduct R&D; pharmaceuticals, semiconductors, and technology companies generally spend the most. But other firms, including those that produce consumer products, invest time and resources into R&D as well. For example, a potato chip brand's many variations on the original product—"salt and vinegar," "nacho flavor," and "hot and spicy"—are the results of extensive R&D.

Companies spend billions of dollars on R&D to produce the newest, most sought-after products and, overall, investment in R&D has been growing. In 2010, $1.4 trillion was invested in R&D globally; by 2017, it increased to $2.2 trillion. According to the professional services firm PricewaterhouseCoopers, in 2018 (the most recent year of available data), Amazon spent $22.6 billion, Volkswagen $15.8 billion, Samsung $15.3 billion, and Intel $13.1 billion on R&D. Many other companies invested billions into R&D as well, including Microsoft, Merck, Apple, and Johnson & Johnson.

While many companies in the Global North invest substantial amounts of capital in R&D, countries and companies in the Global South, notably the lower- and lowest-income countries, need to scale up investment in scientific research and innovation.

WHAT IS SUSTAINABLE INDUSTRIALIZATION?

According to the UN Industrial Development Organization (UNIDO), there are many reasons why industrialization is fundamental for human development. First, it is an essential means for incorporating technological progress and innovation and promoting its use. Second, it offers a unique opportunity for learning, improvement, and transformation. Third, industrialization empowers people to access productive resources by expanding human capabilities through education, skills development, and sociocultural changes as well as to produce goods that are essential for nutrition, health care, and other human needs to improve the quality of life.

Although industrialization contributes to economic growth, its impact differs depending on the country's stage of development. In high-income economies, industrial growth is reflected in achieving higher productivity, embracing new technologies and intelligent production processes, and reducing the effects of industrial production on the environment and climate. For many of the emerging and lower-income economies, industrialization entails structural transformation of the economy from traditional **primary sectors** such as agriculture and fishery to modern manufacturing industries fueled by innovation and technology. Such an expansion of the manufacturing sector creates jobs, helps improve incomes and thus reduces poverty, introduces and promotes new technologies, and produces essential goods and services for the market. Of course, such economic transitions must take into account protecting the environment and human health to be sustainable.

World manufacturing is geographically uneven. While there is still manufacturing in the Global North, this sector has declined since the 1980s as manufacturing jobs and facilities shifted from the Global North to the Global South, and to China in particular. China, a leading industrial economy, has seen meteoric economic growth since the 1990s. Table 9.1 lists the top ten largest manufacturing countries, which together account for more than 70 percent of global manufacturing. Lower-income countries in the Global South account for less than 1 percent of total worldwide manufacturing currently, despite being home to more than 12 percent of the world's population. The lowest levels of manufacturing are mostly located in Africa and Central Asia.

There is no doubt that industrialization helped improve the income and living standards of those in the Global North. Yet the process of industrialization in the twentieth century is largely responsible for much of the environmental destruction and degradation such as pollution, climate change, habitat destruction, and overexploitation of natural resources. Can there be sustainable industrialization?

UNIDO defines **sustainable industrial development** as industrial development that is linked to formal job markets and health, safety, and environmental standards. Sustainable industrial development may have a positive impact on job creation, sustainable livelihoods, technology and skills development, food security, and equitable growth—some of the key requirements for eradicating poverty by 2030.

However, given the long history of unsustainable industrial development and its legacy of negative

TABLE 9.1 Top Ten Largest Manufacturing Economies since 2010

1	China
2	United States
3	Japan
4	Germany
5	India
6	Republic of Korea
7	Italy
8	France
9	Brazil
10	Indonesia

Source: Facevicova, K., and Kynclova, P. *How Industrial Development Matters to the Well-Being of the Population: Some Statistical Evidence*. Vienna, Austria: United Nations Industrial Development Organization (2020).

environmental impacts, it is crucial that programs, projects, and financing support genuine sustainable industrial development.

The Fourth Industrial Revolution?

The process of industrialization has significantly changed in recent years. Many people refer to this as the **Fourth Industrial Revolution**.

The First Industrial Revolution started in Britain around 1760. It was powered by a major invention: the steam engine. The steam engine enabled new manufacturing processes, leading to the creation of factories. The Second Industrial Revolution followed about 100 years later and used electric power to create mass production and used electronics and information technology to automate production in new industries like steel, oil, and electricity. Key inventions included the internal combustion engine, the light bulb, and the telephone. The Third Industrial Revolution is based on the semiconductor, personal computer, and the internet and is also called the "Digital Revolution."

A Fourth Industrial Revolution is currently emerging and is characterized by the fusion of the digital, biological, and physical worlds, as well as the growing utilization of new technologies such as artificial intelligence, cloud computing, robotics, 3D printing, the Internet of Things, and advanced wireless technologies. These technologies are rapidly changing the way people create, exchange, and distribute value.

The Fourth Industrial Revolution may have the potential to raise global income levels and improve the quality of life for populations around the world. However, economists Erik Brynjolfsson and Andrew McAfee counter that it could yield greater inequality because of its potential to disrupt labor markets. As automation substitutes for labor across the entire economy, the net displacement of workers by machines may exacerbate the gap between returns to capital and returns to labor. It is also possible that the displacement of workers by technology will result in a net increase in safe and rewarding jobs.

Those in the highest-income countries already enjoy some of the benefits of the Fourth Industrial Revolution. Workplaces and organizations are becoming "smarter" and more efficient as machines and humans start to work together, and connected devices enhance supply chains and warehouses. However, Africa has been left behind during the past industrial revolutions. Will this time be different?

In his book, *The Fourth Industrial Revolution*, Dr. Klaus Schwab, founder and executive chairman of the World Economic Forum, describes the enormous potential for the technologies of the Fourth Industrial Revolution as well as the possible risks. He writes, "The changes are so profound that, from the perspective of human history, there has never been a time of greater promise or potential peril. My concern, however, is that decision-makers are too often caught in traditional, linear (and non-disruptive) thinking or too absorbed by immediate concerns to think strategically about the forces of disruption and innovation shaping our future."

SOLUTIONS FOR INFRASTRUCTURE, INDUSTRY, AND INNOVATION

Table 9.2 lists the targets and indicators for SDG 9. This section will highlight three strategies: expanding and improving transportation infrastructure, building resilient infrastructure, and advancing more creative and inclusive financial services.

TABLE 9.2 SDG 9 Targets and Indicators to Achieve by 2030

Target	Indicators
9.1 Develop quality, reliable, sustainable, and resilient infrastructure, including regional and transborder infrastructure, to support economic development and human well-being, with a focus on affordable and equitable access for all	**9.1.1** Proportion of the rural population who live within 2 km of an all-season road **9.1.2** Passenger and freight volumes, by mode of transport
9.2 Promote inclusive and sustainable industrialization and, by 2030, significantly raise industry's share of employment and GDP, in line with national circumstances, and double its share in least developed countries	**9.2.1** Manufacturing value added as a proportion of GDP and per capita **9.2.2** Manufacturing employment as a proportion of total employment
9.3 Increase the access of small-scale industrial and other enterprises, in particular in developing countries, to financial services, including affordable credit, and their integration into value chains and markets	**9.3.1** Proportion of small-scale industries in total industry value added **9.3.2** Proportion of small-scale industries with a loan or line of credit
9.4 Upgrade infrastructure and retrofit industries to make them sustainable, with increased resource-use efficiency and greater adoption of clean and environmentally sound technologies and industrial processes, with all countries taking action in accordance with their respective capabilities	**9.4.1** CO_2 emission per unit of value added
9.5 Enhance scientific research, upgrade the technological capabilities of industrial sectors in all countries, in particular developing countries, including, by 2030, encouraging innovation and substantially increasing the number of research and development workers per one million people and public and private research and development spending	**9.5.1** Research and development expenditure as a proportion of GDP **9.5.2** Researchers (in full-time equivalent) per million inhabitants
9.A Facilitate sustainable and resilient infrastructure development in developing countries through enhanced financial, technological, and technical support to African countries, least developed countries, landlocked developing countries, and small island developing states	**9.A.1** Total official international support (official development assistance plus other official flows) to infrastructure
9.B Support domestic technology development, research, and innovation in developing countries, including by ensuring a conducive policy environment for, *inter alia*, industrial diversification and value addition to commodities	**9.B.1** Proportion of medium- and high-tech industry value added in total value added
9.C Significantly increase access to information and communications technology and strive to provide universal and affordable access to the internet in least developed countries by 2020	**9.C.1** Proportion of population covered by a mobile network, by technology

Source: United Nations Sustainable Development Goals. https://sustainabledevelopment.un.org/sdg9

Expand and Improve Transportation Infrastructure

As noted earlier, in many countries in the Global South, transportation infrastructure is absent or inadequate. In Africa, only a quarter of the roads are paved. The Programme for Infrastructure Development in Africa (or PIDA) is a regionwide plan that develops transport, ICT, and energy infrastructure. PIDA is funded mainly by African governments and international bodies such as banks, governments, and funding agencies. It was launched in 2010 and is due for completion in 2040. Transport makes up 30 percent of the current budget and roads are a big part of this investment. The plan is to expand the existing 6,200-mile-long network of major roads to between 40,000 and 60,000 miles—either by upgrading existing unpaved roads or building new ones. The result would be nine arteries, some following Africa's entire coastline, and others strategically crossing the continent. Some 150,000 miles of smaller roads will be built or upgraded to connect smaller cities and rural areas to the main arteries. Some of the priority routes include those in landlocked countries such as Rwanda, Uganda, and the Democratic Republic of the Congo.

There has been more investment in secondary rural roads, which is crucial given that many rural and remote areas are cut off from markets and public services such as health and education. Investing in secondary rural roads may help to reduce rural-urban inequities and reduce poverty. Such projects may also increase economic productivity, school enrollment, access to health services, economic growth, and women's empowerment.

However, recent studies are finding that constructing rural roads alone is not sufficient to promote local development in rural areas. Researchers Asher and Novosad found that, in India, a large rural roads project had no impact on the local economy except for increasing access to jobs outside the community. Another study found that improved rural roads in Ecuador had a positive impact on health and that it increased enrollment in secondary education, but there was no evidence that the program had any positive effect on overall household income, female empowerment, and food security.

There has also been an increased emphasis on **last mile** infrastructure requirements. The term "last mile," in transportation, refers to the final leg of moving people and goods from a transportation hub to a final destination. The term used in ICT refers to the telecommunications networks that deliver services to retail end users. In some areas, the telecommunications industry may find it too expensive to connect individual homes to the main network. In this context, the last mile refers to a gap in the service providers infrastructure.

One example is Project Last Mile that has supported work in eight countries in Africa. Project Last Mile is a public-private partnership between Coca-Cola Company, the Gates Foundation, and USAID to improve rural health systems by increasing access to medicines and services that go that last mile. In Tanzania, for example, Project Last Mile used Coca-Cola's supply chain and logistics experts to get medicines to isolated communities by optimizing routes for medical supply delivery.

Today road networks are expanding worldwide at an unprecedented rate. In Sumatra, Indonesia, the Trans-Sumatran Highway is planned for construction. When completed, it will traverse more than 1,500 miles and through numerous cities. India, a country that has laid 17,000 miles of highways at a cost of nearly $44 billion, is expected to add a further 52,000 miles of roads by 2025.

⚖ SDGs AND THE LAW 9.1

Environmental Impact Assessment

In 1969 the US Environmental Protection Agency created the Environmental Impact Assessment (EIA). Environmental Impact Assessment is a tool used to identify the environmental, social, and economic impacts of a project prior to decision making. In the context of infrastructure, EIA is the process of examining the anticipated and unintended environmental consequences of a proposed project—such as a highway, dam, bridge, or port facility. Typical environmental consequences of ill-planned projects include soil erosion, desertification, and the spread of water-borne diseases, as projects will often involve the clearing of land, diversion of water, or the creation of new water bodies. The objective of EIA is not to force decision makers to adopt the least environmentally damaging alternative, but rather to make explicit the environmental impact of the development, so that the environment is taken into account in decision making.

The environmental impact statement (EIS) is the document that is presented to the decision-making body, alongside the application for development consent. Most EIA systems make some provision for the involvement of the public. Public meetings, exhibitions, and displays provide a means for the public to be informed about projects that will affect the environment in which they live and to voice their concerns. The EIS and the consultation and public participation process factor in to whether the project should be permitted to proceed.

EIAs are used around the world; currently, more than 100 countries have legislation mandating the implementation of an EIA when a development/project is deemed to potentially have considerable impacts on environmental and social contexts. Organizations such as the World Bank and the UNDEP require the use of EIAs to minimize the negative externalities of funded construction projects.

But as countries try to "pave their way out of poverty," planners must realize that paved roads have some downsides, such as increasing access to remote areas, natural resource extraction, and hunting and illegal harvesting, all of which can have effects on wildlife. It is vital to ensure that road expansion does not lead to massive deforestation, habitat destruction, or other environmental impacts. In some countries there are legal mechanisms in place to ensure that negative environmental impacts are part of the infrastructure planning process, as discussed in *SDGs and the Law* 9.1. However, not all infrastructure projects follow a country's established environmental impact laws, particularly if the myth of jobs/infrastructure/growth versus the environment is introduced to avoid environmental protection measures or social equity concerns.

China's Belt and Road Initiative

The world's most ambitious infrastructure investment is China's **Belt and Road Initiative**, sometimes referred to as the New Silk Road. The vision is a vast network of railways, energy pipelines, highways, and streamlined border crossing that will forge new trade networks in Central Asia, South Asia, Europe, and Africa (see Map 9.3).

There are several driving internal forces behind China's initiative. First, China is now a major importer of primary products and raw materials for its factories and wants to expand and deepen the trade networks and relationships.

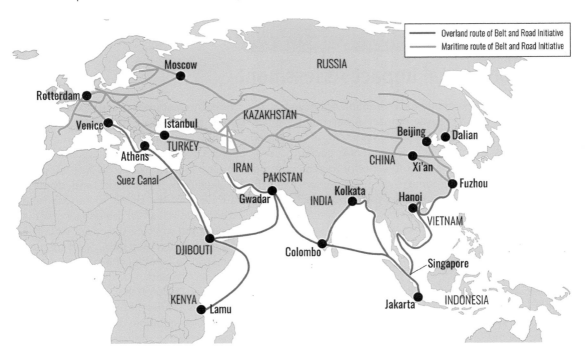

MAP 9.3 Countries Involved in the Belt and Road Initiative
This map highlights countries involved in China's Belt and Road initiative, which includes maritime infrastructure investments (ports and shipping facilities) and overland infrastructure (railways, pipelines, and roads).

Second, it has plans to move away from coal as its dominant source of energy and is seeking cleaner energy resources such as natural gas because it does not have these energy resources itself. Third, it needs to improve its own food security. Fourth, it wants to advance economic development in its western provinces, which are very poor relative to the eastern regions of China. There are also external reasons—China is interested in cultivating export markets and boosting global economic links. China also sees infrastructure investment as a way to improve geopolitical relationships.

China is now Africa's biggest trading partner and imports $200 billion worth of goods, including minerals, oil, and agricultural goods each year. It is the biggest importer of iron ore and coal from Australia. Central Asia and Europe are seen as a future source of natural gas, food, and luxury consumer goods.

In 2014, China pledged $40 billion to a Silk Road Infrastructure Fund to improve land-based connectivity and communication across Central Asia and Europe. The fund will be used to invest in roads, railways, and airports to link Asia and Europe.

In addition to the overland network, there is a Maritime Silk Road, launched in 2015, which involves investments in ports throughout Africa and South Asia. One report has predicted that China's overall investment for the Belt and Road Initiative could reach $1.2 trillion by 2027. Table 9.3 lists examples of Belt and Road Initiative projects and Photo 9.3 shows a bridge project in Laos. When completed, China's Belt and Road will link seventy countries comprising 30 percent of the world's GDP.

Country	Projects
TABLE 9.3 Selected Examples of Belt and Road Projects	
Azerbaijan	• Baku-Tbilisi-Kars Railway • Baku International Sea Trade Port
Djibouti	• Doraleh Multipurpose Port • Hassan Gouled Aptidon International Airport • Ahmed Dini Admed International Airport
Cambodia	• Lower Sesan Two Hydropower Dam
Greece	• Port of Piraeus Redevelopment
Indonesia	• Jakarta-Bandung High Speed Rail
Kenya	• Mombasa-Nairobi Standard Gauge Railway
Laos	• Boten-Vientiane Railway
Nigeria	• Abuja-Kaduna Railway Line
United Kingdom	• Yiwu-London Railway Line
Multiple countries	• Central Asia-China Gas Pipeline

Source: The Belt and Road Initiative. "BRI Projects" (2020, October 19). https://www.beltroad-initiative.com/projects/

Belt and Road Initiative projects are not "gifts" or direct aid from China to other countries; rather, several Chinese development banks such as the Silk Road Fund, China Development Bank, and Agricultural Development Bank of China provide low-interest loans to other countries to build this new infrastructure. They offer lower-interest rate loans than private banks, making

PHOTO 9.3 Bridge Construction in Laos
An example of a project of the Belt and Road Initiative. This bridge, under construction on the Mekong River in Luang Prabang Province, will be a part of the Boten-Vientiane railway. The banner in the forefront features Chinese language characters, to highlight the partnership project.

them attractive. Collectively, these three development banks have invested more than $150 billion to date.

Some economists are critical about the Belt and Road Initiative, concerned that lower-income countries may end up with high levels of debt. For example, when the Sri Lankan government failed in 2017 to keep up with payments on a Chinese-built and -financed port, it was pressured into turning over control of the port to China for the next ninety-nine years. Similarly, Montenegro took out a $1 billion loan from China to build a highway, but within a few years asked the European Union for help to repay the debt. Critics contend that by creating these loans, China is engaging in "debt-trap diplomacy"—a strategy of extracting political concessions out of a country that owes it money.

American officials have also criticized the Belt and Road Initiative as a worrisome extension of China's rising power and see China using its money to leverage political gains and increase its global power as it entices countries to "pivot away from the US to Asia." Finally, some critics worry about the environmental impacts that such colossal infrastructure projects may have. For example, some of the large highways may cause landslides or floods; the development of port facilities could impact sensitive species and marine habitats.

Build Resilient Infrastructure

In both the Global North and Global South, there is much work to be done to create community resilience, particularly in the aftermath of disasters such as hurricanes, tornados, earthquakes, or fires. SDG 9 has several targets that focus on developing resilient infrastructure.

According to the United Nations, **resilience** requires community capacity to plan for, respond to, and recover from stressors and shocks. Shocks are major disruptions such as storms, heat waves, derechos, or other extreme weather events—often intensified by climate change—that can disrupt a variety of critical systems. Resilience efforts are varied but include increasing local capacity to cope effectively with and learn from adversity. Planning for resilience is another growing trend.

However, it is crucial to understand that the concept of sustainability is different from resilience. Considerations of resilience—planning for and learning from adversity—should be a part of achieving sustainability. Resilience planning, however, may or may not holistically balance considerations of environment, economy, and equity, an essential value of sustainability. Today, many cities, regions, and countries are planning for resilience to climate change, conflict, or other potential hazards and disasters.

One example of resilient infrastructure is what the United Nations terms "Resilience Hubs" (see Figure 9.2). **Resilience Hubs** are enhanced community centers that can support residents and coordinate communication and resources before, during, and after disruption or disasters. In addition to hosting supplies, communications, and resources in the event of an emergency, Resilience Hubs serve community members year-round as a center for community-building and community revitalization. Some of the services of a Resilience Hub that are activated during an emergency may include:

- Supply of and access to freshwater and resources such as food, ice, refrigeration, charging stations, basic medical supplies, and other supplies needed in the event of an emergency;

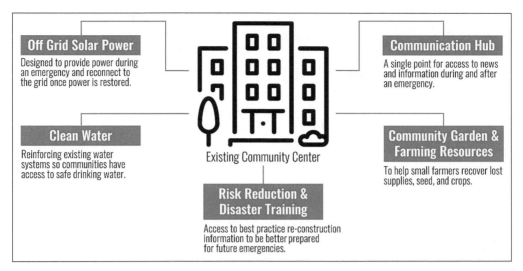

FIGURE 9.2 A Resilience Hub
Resilience Hubs are enhanced community centers that can support residents and coordinate communication and resources before, during, and after disruption or disasters.

- Energy systems that can provide extended power during power outages (typically a hybrid system that incorporates solar with back-up generation from natural gas);
- A place to grow fresh and local food, increased tree canopy for shade and cooling, and resources for residents who choose (or are forced) to shelter in place rather than evacuate; and
- Critical communication and information functions that help educate community members about hazards.

Resilience Hubs can provide support in the event of an emergency. However, they can also enhance community connectivity and offer translational services, access to health services like flu shots and diabetes screening, job opportunities or centers, locations for growing local food, spaces for before and after school programs, job training programs, and more.

In the United States, the city of Baltimore was one of the first cities to build Resilience Hubs, starting in 2014. The city has since built seven, including one renovated abandoned fire station turned into community space. Kristin Baja discusses how Resilience Hubs may assist in health emergencies such as fighting COVID-19 in *Expert Voice* 9.1.

Use Creative Financial Services: Microfinance

The World Bank says that inadequate, inaccessible financial services help keep many trapped in poverty. Without access to finance, the poor are unable to invest in tools to increase productivity, start a microenterprise, invest in education or health, or even take time to search for better opportunities. Many of the world's poor do not have formal bank accounts.

According to the World Bank's Global Findex, which gathers data on how adults save, borrow, and make payments, 1.7 billion adults globally are financially excluded, living without formal credit or savings. Many of the world's poor or low-income families may not qualify for traditional forms of credit—a

⬤ EXPERT VOICE 9.1

Kristin Baja on Resilience Hubs

Ms. Kristin Baja, programs director of Climate Resilience for the Urban Sustainability Directors Network, has written about the role of Resilience Hubs for communities. She makes an argument for how Resilience Hubs could provide important services during health emergencies, such as the COVID-19 pandemic:

"COVID-19 is quickly exposing countries' fragile social fabric. As individuals struggle to stay afloat, a lack of healthcare, childcare, living wages, and paid leave are worsening glaring inequalities around the world. And because governments prioritize response once a crisis hits, rather than strengthening communities beforehand, many cities lack needed resources and support. Anticipating disruptions more effectively—including outbreaks like COVID-19 or disasters such as floods, hurricanes or wildfires—requires a rethink in how we proactively prepare for crises. That's why many cities are now working to set up Resilience Hubs, to better build community resilience.

Here's the concept: Resilience Hubs are partnerships between local governments and community-based organizations (sometimes partially funded by foundations) that provide services such as job training and childcare, community programming, resource distribution, and communications coordination and generally enhance quality of life. Based in trusted community-serving facilities (e.g., recreation centers or faith-based institutions), they enhance capabilities in crisis, including solar and battery backup systems, access to potable water and healthy food, and supply distribution. Resilience Hubs can be activated to serve vulnerable communities without overloading local governments.

So that's the concept. In practice, how would they handle the coronavirus?

First, Resilience Hubs could provide community-based testing sites for medical personnel. One of the primary concerns for many cities, for example, is overloading emergency rooms.... Because Hubs are local by design, they could provide free testing in a trusted space, minimizing public transportation and worry of health insurance or citizenship proof. With widely available testing this would increase access to testing and reduce the number of carriers spreading COVID-19.

Second, Hubs could be neighborhood distribution centers that provide residents with access to healthy food, clean water, soap, and items that are routinely needed. ...

Third, as schools close around the world, Hubs could be locations to coordinate childcare and meals to ensure children have access to healthy food and clean water throughout the day, while supporting parents that still have to work. They could act as spaces to bring in people that are without work to fill the gaps for those who still have to go to work—a win-win.

Fourth, Hubs could organize virtual platforms to connect neighborhoods, coordinate volunteers, and share response funds among individuals who lost jobs or small businesses forced to close. ...

Fifth, Hubs could offer equity-centered proactive planning, providing opportunities to anticipate what's next, while prioritizing the needs of those with the most risk and least resources—all the while supporting different language, ethnic, and cultural differences. Currently, community members are stepping up to fill this gap.

Lastly, Hubs could provide redundancy. Many systems we rely on to support physical distancing—power, water, internet, etc.—are intact in developed countries (less so in developing countries). In crises such as earthquakes or hurricanes, these systems are often undermined, creating a need for Hub-like backup. In Puerto Rico, a community hub is developing community capacity and response to hurricanes and is outfitting the space with solar, water purification, communications, and other elements of disaster support.

If we had Resilience Hubs in underserved communities, we could have this more coordinated approach. We could reduce the strain on our medical and emergency management system. We could improve dissemination of accurate information, resources, and supplies. We could enhance community connectivity and social cohesion. Now, as governments invest in recovery, let's prioritize holistic solutions that build stronger communities rather than rebuilding broken systems. That's how you build resilience."

Source: Baja, K. "How Resilience Hubs in Cities Could Help Coronavirus Response" (2020, March 27). https://news.trust.org/item/20200327105242-opnw2

bank loan, because they may not be considered a good credit risk as they lack collateral (or assets) such as a house or car, that can be forfeited to the bank should the borrower default. Another reason is that most banks do not have small loan programs (under $500). The rigidity of commercial banks has resulted in financial exclusion of millions of people.

Financial inclusion means that individuals and enterprises can access and use a range of appropriate and responsibly provided financial services offered in a well-regulated environment. There is a growing evidence that increased levels of financial inclusion—through the extension of savings, credit, insurance, and payment services—contributes to sustainable economic growth. Financial inclusion is an important part of sustainable development, and many believe that reducing extreme poverty will not be possible if the poor cannot save or have access to credit. One solution has been **microfinance**.

Microfinance, also called microcredit, is a type of banking service provided to unemployed or low-income individuals or groups who otherwise would have no other access to financial services. Microloans can range from being as small as $100 to as large as $25,000. Very often, people need access to capital to grow their informal or formal businesses, and in many cases, a few hundred dollars is all they need. Like conventional lenders, microfinanciers may charge interest on loans and institute specific repayment plans.

Those seeking credit receive it on the basis of their membership in self-regulating "solidarity" groups. The idea is that these networks will not only regulate investments and loan paybacks but also increase the "social capital" of its members. In many instances, people seeking help from microfinance organizations are first required to take a basic money management class. Lessons cover topics like understanding interest rates, the concept of cash flow, how financing agreements and savings accounts work, how to budget, and how to manage debt. Once educated, customers may apply for loans. Just as one would find at a traditional bank, a loan officer helps borrowers with applications, oversees the lending process, and approves loans. The typical loan, sometimes as little as $100, may not seem like much, but for many impoverished people, this amount often is enough to start a business or engage in other profitable activities.

Microfinance often targets low-income women through group loans, or "village banking" and provides different loans for different groups. It offers larger loans for individual businesses, agricultural loans to assist farmers in purchasing seeds, fertilizer, livestock, and equipment, and energy loans help clients to purchase or lease clean energy systems (such as solar) or to build savings accounts.

The modern form of microfinancing became popular on a large scale in the 1980s. The first organization to receive attention was the Grameen Bank in Bangladesh, which was started in 1983 by Muhammad Yunus. Grameen Bank works on the premise that even the lowest-income individuals can manage their own financial affairs and development when given suitable conditions that include small long-term loans on low interest terms.

Since the establishment of the Grameen Bank, more than nine million people have borrowed from it, 97 percent of borrowers being women. What makes the Grameen Bank different is that it deliberately offers credit to formerly underserved people: the poor, women, illiterate, and unemployed people. The Grameen Bank and Muhammad Yunus were recognized when they were awarded the 2006 Nobel Peace Prize.

The success of the Grameen Bank has inspired similar projects in more than sixty-four countries around the world, including a World Bank initiative to finance Grameen-type schemes. Examples of other microfinance organizations include the US-based Kiva.org, a nonprofit organization that allows people to lend money using the internet to low-income entrepreneurs in seventy-seven countries. Since 2005, Kiva has crowdfunded more than 1.6 million loans, totaling more than $1.3 billion, with a repayment rate of between 96 and 97 percent. They offer a category of loans called "Green Kiva" to help fund solar panels, organic fertilizers, high-efficiency stoves, drip irrigation systems, and biofuels. The World Bank estimates that more than 500 million people have benefited from microfinance-related operations.

In addition to individuals and family units, there are many small businesses and local governments that are not able to attract traditional development finance. In lower-income countries, only 35 percent of small businesses have access to traditional credit and finance services, making it more difficult for them to compete, grow, and create more jobs in the formal sector. One response has been the UN Capital Development Fund (UNCDF), which provides seed capital to support the development of a pipeline of bankable "demonstration projects." These could include small but transformational infrastructure projects at the local level such as feeder roads, bridges, or micro-energy grids. The idea is that such projects will demonstrate to private- and public-sector investors that the project is feasible and worthy of additional and more conventional investment. As highlighted in *Solutions* 9.1, Kenya's M-PESA is an example of innovation in microfinancing.

A Critical Perspective on Microfinance

Microfinance is not without its critics. Some economists have noted microcredit banks depend on subsidies to operate, thus acting as another example of welfare, while other researchers have noted that it is not proven that microcredit has definitively reduced poverty. Philip Mader, in his book, the *Political Economy of Microfinance*, argues that rather than reducing poverty, microfinance encourages and even normalizes debt. He says that microfinance turns poverty into a problem of finance (access to money)—rather than seeing the

⊜ SOLUTIONS 9.1

M-Pressive M-PESA in Kenya

Target 9.3: Increase the access of small-scale industrial and other enterprises, in particular in developing countries, to financial services, including affordable credit, and their integration into value chains and markets.

Target 9.C: Significantly increase access to information and communications technology and strive to provide universal and affordable access to the internet in least developed countries by 2020.

The mobile phone has changed lives in Kenya. It has replaced networks of fixed line communication, which were often inadequate or unreliable. It has allowed millions to leapfrog the landline and become connected to services in a different way.

Mobile phones have opened up financial services for millions of people who previously lacked bank accounts or access to credit. One example is M-PESA (Pesa is a Swahili word for money), which allows users to transfer small amounts of money. In short, it's a digital wallet. Kenyans use M-PESA and its PIN-secured SMS text messages to transfer money from one account to another, pay utility bills, or deposit money into an account stored on their cell phones. M-PESA has been credited with reducing crime by substituting cash for pin-secured virtual accounts and eliminating the need for people to take bundles of cash with them in person. It has freed up people to spend time doing other things instead of standing in long lines at the bank or traveling long distances to get to the bank, both of which can be problematic in infrastructure-constrained countries such as Kenya.

M-PESA began as a joint partnership between the Kenyan telecom company Safaricom and a $1.5 million grant from the UK Department of International Development. Originally, it was conceived as a system to allow microfinance-loan repayments to be made by phone. But it has since become a larger financial platform. One research study found that 83 percent of Kenya's population fifteen years and older has access to mobile phone technology. As a result, more than twenty million Kenyans use M-PESA, including 70 percent of the country's poor. Some $2 billion are

PHOTO 9.4 A Kiosk Featuring M-Pesa, in Rural Kenya
Kiosks like this are found every few miles on main roads in rural Kenya, making this system more accessible than traditional banks.

exchanged each month, which represents about 25 percent of Kenya's GDP.

M-PESA is not a traditional bank—it does not pay interest on deposits or make loans and it does not have traditional bank offices. Instead, it relies on thousands of towers for network coverage. Kiosks shown in Photo 9.4 are found in in every village and town, allowing rural populations access to financial services that would have otherwise been impossible. These kiosks and the platform's development have created tens of thousands of jobs.

The World Bank says that inadequate, inaccessible financial services help keep the poor trapped in poverty. Without access to finance, the poor are unable to invest in tools to increase productivity, start a microenterprise, invest in education or health, or even take time to search for better opportunities. Many of the world's poor do not have formal bank accounts. While M-PESA does not replace all the functions of a bank, it has revolutionized access to financial services for the poor, empowering them in new ways and supporting entrepreneurial creativity.

Technology like M-PESA is increasing the transparency of financial transactions, spinning off new loans, and catalyzing new ways financial services are delivered to smallholder farms. It may even have the power to increase the net household savings. M-PESA and other mobile applications have shown the transformational power to reach the underserved, including women, who are important drivers for sustainable poverty eradication. These financial services allow households to save in secure instruments to enlarge their asset base and escape cycles of poverty.

M-PESA is considered one of the most successful mobile phone–based financial services in the region. It certainly caught the attention of *The Economist* magazine that noted that M-PESA "made paying for a taxi ride easier in Nairobi than in New York City." M-PESA is a good example of how a relatively small investment in innovation transformed millions of lives. Pretty M-pressive.

Source: T. S. "Why Does Kenya Lead the World in Mobile Money?" *The Economist*, March 2015.

underlying structure of capitalism as the root cause of poverty. Mader suggests microfinance is just a newer form of capitalist global finance in the neoliberal model, one that focuses on the poor. He rejects the idea that microfinance is an economics of liberation or transformation, but instead is a politics of repression, reinforcing economic dependence through debt. Similarly, Katharine Rankin argues that many microfinance programs engage the collective only in the most instrumental manner—reducing administrative costs and motivating repayment—at the expense of the more time-consuming processes of consciousness-raising and empowerment. Her work found little in the way of female empowerment as a result of microfinance.

SUMMARY AND PROGRESS

The most recent UN SDG Progress Report notes that global investment in R&D and financing for economic infrastructure in developing countries increased from $741 billion in 2000 to $2.2 trillion in 2017. While Europe and North America lead in R&D, East and Southeast Asia are catching up, increasing their share in R&D investment from 22.6 percent in 2000 to 40.4 percent in 2017. Still, regional disparities are stark—sub-Saharan Africa, North Africa, and Latin America lag in investments in R&D. The COVID-19 pandemic has underscored the need for increased investment in R&D in the pharmaceutical industry and in emerging technologies such as artificial intelligence that can assist in development drugs and vaccines.

In other targets, progress has been made in mobile connectivity. In 2020, almost all the world's population (97 percent) lived within reach of a mobile cellular signal.

The 2020 coronavirus pandemic has dealt a severe blow to manufacturing and transport industries, causing disruptions in global value chains and the supply of products as well as job losses and declining work hours in these sectors. In low-income countries, manufacturing jobs are an essential source of income and are key to poverty reduction. The growth of manufacturing has decelerated and industrialization in many countries in the Global South is too slow to meet GDP targets in SDG 8.

According to the most recent UN SDG progress report, the effects of COVID-19 have been so destabilizing that they threaten to halt or even reverse progress toward several SDG 9 targets. For example, air transport is a driver of economic development. The direct and indirect global economic impact of air transport was estimated at $2.7 trillion in 2016, the equivalent of 3.6 percent of the world's GDP. However, for the last half of 2020, travel demand and air transport had decreased by 90 percent, which could cost between $300 and $400 billion in lost revenues for airlines. Likely air transportation will return quickly as vaccination rates increase around the world, but the losses incurred may linger.

Many experts say that small-scale industrial enterprises are major sources of employment in low-income countries and will play a significant role in the recovery of the global economy post-COVID-19. However, the UN SDG report finds that they are more vulnerable due to their small size and do not have the capacity to deal with these unexpected shocks unless governments step in to provide increased access to credit.

The impact of COVID-19 has also once again highlighted the digital divide. With many forced to work, learn, and even seek health care from home, digital technologies and internet connectivity have never been more crucial. Although access to mobile cellular signals is almost universal, only 54 percent of the global population use the internet and in the lowest-income countries, only 19 percent use the internet. However, in many countries the COVID-19 pandemic has accelerated the roll-out of digital technologies and services. Universal access to digital infrastructure and broadband connection have become priorities: to increase access to services and as tools for a robust and resilient public health system response.

SDG 9 has interlinkages with many other SDG targets, including industry-related targets associated with job creation, sustainable livelihoods, and food security. In addition, innovation is required for the delivery, distribution, and consumption of energy, food, water, and housing, while access to ICT is critical for achieving SDG 4 Quality Education, SDG 8 Decent Work and Economic Growth, and SDG 10 Reduced Inequalities, among others. Other examples of linkages between SDG 9 and other SDGs are elaborated in the text that follows and highlighted in Figure 9.3.

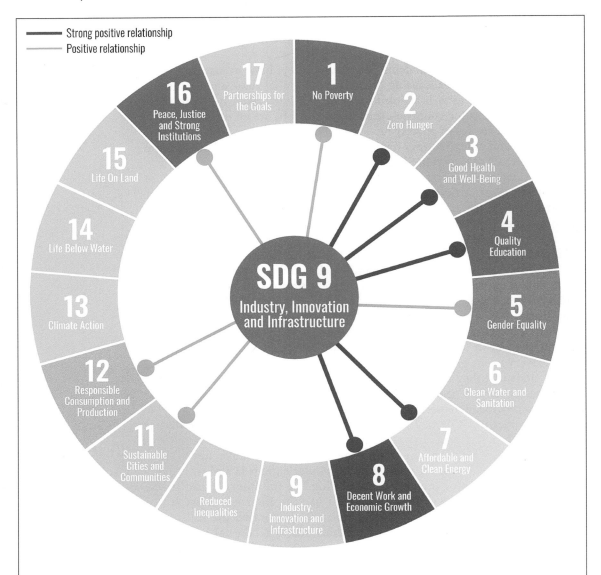

FIGURE 9.3 Interconnections in SDG 9 and the Other Goals

SDG 9 connects to all of the SDGs in some way; this figure highlights several of the stronger connections.

QUESTIONS FOR DISCUSSION AND ACTIVITIES

1. What is the state of soft infrastructure in your own community? Are there specific needs for investment, upgrading, or new infrastructure?

2. What is the state of hard infrastructure in your own community? Are there specific needs for investment, upgrading, or new infrastructure?

3. Explain the types of infrastructure development most needed in the Global South.

4. Explain obstacles to infrastructure expansion and development.

5. Is the Fourth Industrial Revolution as sustainable? Explain why or why not.

6. Where do you see the most need for R&D investment?

7. Why has it been easier to provide mobile cellular service compared to internet service?

8. Building transportation infrastructure is vital, yet it can be unsustainable. Describe three ways that reaching goals around road and transportation infrastructure could have negative impacts on other SDGs.

9. Research a specific project in the Belt and Road Initiative and provide a progress report. Discuss whether this project accounts for the three Es. Go here to select a project but be sure to use a diverse range of articles assessing the project (https://www.beltroad-initiative.com/projects/).

10. Which target in SDG 9 do you see as most challenging to achieve?

TERMS

Belt and Road Initiative
financial inclusion
Fourth Industrial Revolution
hard infrastructure
information and communication
    technology (ICT)
last mile
microfinance

primary sector
public private partnerships
research and development
    (R&D)
resilience
Resilience Hubs
soft infrastructure
sustainable industrial development

RESOURCES USED AND SUGGESTED READINGS

Ali, S. M. *China's Belt and Road Vision: Geoeconomics and Geopolitics (Global Power Shift)*. New York: Springer (2020).

Bath, S. "Ensuring Access to Water and Sanitation: What SDG 6 Wants to Achieve and Why It Matters" (2018). https://www.iisd.org/articles/insight/ensuring-access-water-and-sanitation-all-what-sdg-6-wants-achieve-and-why-it

Bossink, B. *Eco-innovation and Sustainability Management*. New York: Routledge (2012).

Brynjolfsson, E., and McAfee, A. *The Second Machine Age: Work, Progress, and Prosperity in a Time of Brilliant Technologies*. New York: W. W. Norton (2014).

Corral, L., and Zane, G. "Chimborazo Rural Investment Project: Rural Roads Component Impact Evaluation." *Inter-American Development Bank Technical Note IDB-TN-02116* (2021). https://publications.iadb.org/publications/english/document/Chimborazo-Rural-Investment-Project-Rural-Roads-Component-Impact-Evaluation.pdf

Grigg, N. *Infrastructure Finance: The Business of Infrastructure for a Sustainable Future*. New Jersey: John Wiley & Sons (2010).

"Home Page." United Nations Industrial Development Organization (2021). www.unido.org

"Home Page." United Nations Office for Disaster Risk Reduction (2022). https://www.undrr.org/

International Bank for Reconstruction and Development/World Bank. *Belt and Road Economics: Opportunities and Risk of Transport Corridors*. Washington, DC: World Bank (2019).

International Telecommunication Union. "Fast Forward Progress: Levering Tech to Achieve the Global Goals" (2017). https://www.itu.int/en/sustainable-world/Pages/report-hlpf-2017.aspx

Karim, L. *Microfinance and Its Discontents*. Minneapolis: University of Minnesota Press (2011).

Lieberman, I., Di Leo, P., Watkins, T., and Kanze, A. (eds.). *The Future of Microfinance*. Washington, DC: The Brookings Institution (2020).

Mader, P. *The Political Economy of Microfinance*. New York: Palgrave (2015).

OECD/The World Bank/UN. *Environment Financing Climate Futures: Rethinking Infrastructure*. Paris: OECD Publishing (2018). https://www.oecd.org/environment/cc/climate-futures/policy-highlights-financing-climate-futures.pdf

Ragnedda, M., and Gladkova, A. (eds.). *Digital Inequalities in the Global South: Global Transformations in Media and Communication Research*. Cham, Switzerland: Palgrave Macmillan (2020). https://doi.org/10.1007/978-3-030-32706-4_2

Rankin, K. "Social Capital, Microfinance, and the Politics of Development." *Feminist Economics* 8 (1): 1–14 (2011). https://doi.org/10.1080/13545700210125167

Rowley, A. *Foundations of the Future: The Global Battle for Infrastructure*. Singapore: World Scientific Publishing Co. (2020).

Schwab, K. *The Fourth Industrial Revolution.* New York: Crown/Penguin Publishing (2016).

The Programme for Infrastructure Development in Africa. "Closing the Infrastructure Gap Vital for Africa's Transformation" (2021). https://www.afdb.org/fileadmin/uploads/afdb/Documents/Generic-Documents/PIDA%20brief%20closing%20gap.pdf

United Nations. "The Sustainable Development Goals Report 2022" (2022, July). https://unstats.un.org/sdgs/report/2022/

United Nations Department of Economic and Social Affairs. "Micro-, Small and Medium-Sized Enterprises (MSMEs) and Their Role in Achieving the Sustainable Development Goals" (2021). https://sustainabledevelopment.un.org/content/documents/26073MSMEs_and_SDGs.pdf

Reduce Inequalities

LEARNING OBJECTIVES

After reading this chapter, you should be able to:

- Identify and explain trends and patterns of inequality
- Differentiate between vertical inequality and horizontal inequality
- Explain how inequality is measured
- Describe various types of inequality
- Explain the consequences of inequality
- Explain the strategies for SDG 10
- Explain how reducing inequality connects to other SDG targets

Almost one million Rohingya Muslims have fled Burma's Rakhine State to escape the military's large-scale campaign of ethnic cleansing over the last decade. The Burmese military has engaged in violence against Rohingya civilians, killing people and burning entire villages to the ground.

Burma (also called Myanmar) is a Buddhist-majority country. The majority of Burma's Rohingya are Muslim, living predominately in the western state of Rakhine. They differ from the dominant Buddhist groups ethnically, linguistically, and religiously.

The Rohingya have faced decades of discrimination and repression under successive Burmese governments, which have perpetuated a deep-seated culture of pervasive prejudice. Burma's Citizenship Act of 1982 denies citizenship to Rohingya on the basis of their ethnicity; as a result, they are one of the largest stateless populations in the world.

A growing Buddhist nationalism in Burma has led to a number of laws on religion and an increase in ethnic violence. The Burmese government has imposed segregation and discrimination against Muslims, particularly the Rohingya. Laws, policies, and practices have denied the Rohingya human rights, including restrictions on access to medical assistance, and other services. The government also has restricted their movements, requiring special permits to travel between towns and villages, and has imposed curfews.

Rohingya have fled mainly to Bangladesh, but also to Malaysia, Thailand, India, and Indonesia (Photo 10.1). Discriminatory practices have led to violence and ethnic cleansing, which in turn has created a massive refugee crisis.

PHOTO 10.1 Rohingya Muslim Refugees in Jakkur, India
They wait in line at a camp for ration. Many Rohingya have fled their homes with few possessions, creating a dire refugee crisis. The Rohingya have faced decades of discrimination and repression under the Buddhist-majority country.

UNDERSTANDING INEQUALITY

An underlying premise of this book is that sustainable development is a counter to development that has created inequalities in outcome and opportunity. With a rallying cry of "leave no one behind," a number of the SDGs include a cross-cutting focus on inequalities and the investment in and advancement of some communities that have historically experienced discrimination. The world is far from reaching the goal of equal opportunity for all: Circumstances beyond an individual's control, such as gender, race, ethnicity, migrant status, and, for children, the socioeconomic status of their parents, continue to affect one's chances of succeeding in life.

Chapter 1 on poverty focused discussion on how uneven development has led to disparities in income. Tackling poverty is an important part of reducing inequality, although it may not necessarily result in a completely equal distribution of income, nor in the reduction of other systemic inequalities. There are numerous ways in which inequality can manifest itself and SDG 10 looks beyond the narrower lens of income inequality to consider other types of inequalities.

The development economist Frances Stewart introduced and advanced the concepts of "vertical inequality" and "horizontal inequality". She defines **vertical inequality** as the inequality between individuals, typically measured by income. **Horizontal inequality**, however, occurs between culturally defined groups. Traditional development interventions tend to focus on vertical inequalities and the link between economic status, income, and assets. However, it is often horizontal inequalities, linked to social status and identity, that lock individuals into marginalization and poverty. Both vertical and horizontal inequalities are embedded in structural systems and cultural norms. This chapter focuses on the horizontal inequality: inequalities based on sex, age, disability, sexual orientation, race, class, ethnicity, and religion continue to persist across the world, within and among countries.

HORIZONTAL INEQUALITY

Social inequalities occur in nearly all countries and at different spatial scales, which include sex, age, disability, sexual orientation, race, class, ethnicity, and religion. The sociologist Manuel Castells has used the term **fourth world** to refer to stigmatized minority groups that are denied a political voice all over the globe (e.g., Indigenous minority populations, prisoners, and the unhoused). Chapter 5 detailed gender inequality in depth. This chapter will examine other types of horizontal inequalities.

Race and Ethnicity

Race and ethnicity continue to be the most significant reason for discrimination and treatment that can determine an individual's ability to be successful. Race is linked with physical characteristics such as skin color. Ethnicity is linked with an identity mostly on the basis of language and shared culture (religion, history, geography/territory). Ethnicity is something that can be chosen by an individual. Both race and ethnicity are complex and problematic concepts.

The idea of race originated among anthropologists and philosophers in the eighteenth century, who used geographic location and skin color to place people into different racial groupings. It was a flawed principle from the beginning because the basis for that premise (different and distinct genes) is not supported by genetic science.

In their book *Racial Formation in the United States*, sociologists Michael Omi and Howard Winant deconstruct the definition of race. Their racial formation theory highlights the ways that "race" is socially constructed. That is, how processes connected to social, economic, and political forces shape the way in which racial categories and hierarchies are formed. They also note that race is not an absolute, but rather fluid definition. Who is "Black" or "White" or "Asian" are values that we have chosen to ascribe to ourselves or each other, reinforced over time by institutional practices.

Cedric Robinson introduced the idea of racial capitalism. He argues that our modern system of capitalism could not have proceeded without colonial expansion, the slave trade, and plantation slavery. He also writes that capitalism was "racial" not because of some conspiracy to divide workers or justify slavery and dispossession, but because racialism had already permeated Western feudal society. The first European proletarians were racial subjects (Irish, Jews, Roma or Gypsies, Slavs, etc.), and they were victims of dispossession (enclosure), colonialism, and slavery within Europe. Robinson's work sheds light on how deeply implicated capitalism is in racial subjugation in many countries in the world.

Like race, ethnicity is also socially constructed. It too has been used to oppress different groups, as occurred during the Holocaust. An interesting example of how race and ethnicity are socially constructed can be found in the country of Rwanda. Under Belgian colonial rule from the 1880s to the 1950s, Belgians regarded the Tutsi minority as racially superior, and the Tutsis enjoyed preferential access to privilege and to positions of authority. In contrast, the Hutu majority were considered inferior. Yet, it was Belgium that first identified physical characteristics as defining features: Tutsis were characterized as taller, more intelligent, while Hutus were smaller and more ignorant. As Deborah Mayersen has written,

> The distinction between the Hutu majority and Tutsi minority subgroups has been varyingly described as one of race, tribe, caste, class, domination and subjugation, ethnicity and political identity. Each descriptor appears

to have more than a kernel of truth, but also elements of distortion and inaccuracy. Whereas today these identities are commonly referred to as ethnic identities, for much of Rwanda's history they were considered racial. First German then Belgian colonial authorities considered the Hutu, Tutsi and Twa as distinct races, which came to have profound consequences for the Rwandan people.

Mayersen notes that Rwanda's colonizers ranked each "race" hierarchically and, over time, this racialized hierarchy was institutionalized within Belgian colonial policies and internalized by the Rwandan population. Mayersen and other scholars have noted that, like many European colonizers, racial discrimination was formed by Belgium as a process of distinguishing and controlling other races. In the case of Rwanda, Belgium was able to "divide and conquer" within a country dominated by Black Africans who were not all that different from each other.

Although race and ethnicity may be socially constructed, and largely abstract, this does not negate their potent real-world influence. It is not simply that we have racial categories, but that we have constructed racial hierarchies that create layers of inequalities. Table 10.1 lists some of the ways in which racial and ethnic discrimination manifest themselves. Racism is not just a matter of individual injustice: In many places, deep-seated structural racism continues against various people, including Indigenous peoples and people of African descent. In addition, refugees and migrants have been systematically denied their rights and unjustly and falsely vilified as threats to the societies they sought to join.

Racism can also be seen at different spatial scales. In many cities, the lowest-income residents are minority populations. This is true even in the United States; indeed, race is often the single most defining factor of inequality in the United States. *Understanding the Issue* 10.1 describes race and health disparities in Washington, DC.

TABLE 10.1 Examples of Racial and Ethnic Discrimination

- Racial and ethnic profiling in law enforcement;

- Institutional racism and racial discrimination;

- Incidents of contemporary forms of racism and racial discrimination against Africans and people of African descent, Arabs, Asians and people of Asian descent, migrants, refugees, asylum-seekers, persons belonging to minorities and Indigenous peoples;

- The persistent denial of individuals belonging to different racial and ethnic groups of their recognized human rights, as a result of racial discrimination, constitutes gross and systematic violations of human rights;

- Continued anti-Semitism, Christianophobia, Islamophobia in various parts of the world, and racist and violent movements based on racism and discriminatory ideas directed at Arab, African, Christian, Jewish, Muslim, and other communities;

- The resurgence of neo-Nazi views;

- Laws and policies glorifying all historic injustices and fueling contemporary forms of racism, racial discrimination, xenophobia, and related intolerance and underpinning the persistent and chronic inequalities faced by racial groups in various societies;

- The sharp increase in the number of political parties and movements, organizations and groups that adopt xenophobic platforms and incite hatred, taking into account the incompatibility of democracy with racism;

- The impact of some counterterrorism measures on the rise of racism, racial discrimination, xenophobia, and related intolerance, including the practice of racial profiling and profiling on the basis of any grounds of discrimination prohibited by international human rights law.

UNDERSTANDING THE ISSUE 10.1

Race and Health Inequalities in Washington, DC

Washington, DC, has been home to some of the highest HIV/AIDS infection rates of any major US city. Currently, 1.9 percent of the population—13,000 people—living in the District are diagnosed with HIV. At its peak in 2007, the HIV rate in the District stood at approximately 3 percent, a rate higher than in many parts of Africa. Even in the present day, one report notes that if Washington, DC, were a country in sub-Saharan Africa, it would rank twenty-third out of fifty-four countries in the percentage of people with HIV.

But the real story about HIV/AIDS in DC is racial disparity. Communities of color experience the highest rates of HIV infection. According to the 2020 DC Department of Health's *Annual Epidemiology & Surveillance Report*, African Americans account for 71 percent of newly diagnosed HIV cases while only accounting for 47.4 percent of the DC population. Four percent of the District's Black male population is living with HIV, followed by 2.1 percent of Latino/Hispanic men and 1.7 percent of African American females. According to the World Health Organization, a generalized epidemic rate is 1 percent; communities of color in DC are all above that rate, signifying they are disproportionately impacted by HIV.

The rate of Black males living with an HIV diagnosis is 2.9 times that of White males, but even more alarming, the rate of Black females living with an HIV diagnosis is almost 29 times that of White females.

Disparities also play out spatially. Wards 5, 6, 7, and 8, where the majority of Blacks live, have the highest rates of HIV, and, not coincidently, the highest poverty rates in the city (Map 10.1). HIV/AIDS infection ranks are linked to poverty: The two poorest wards have an average household income of $44,000 while Ward 3—a

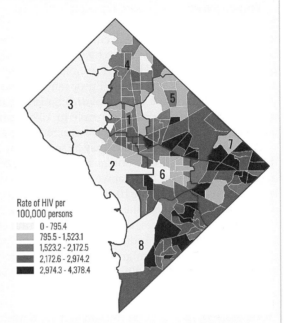

Rate of HIV per 100,000 persons

- 0 - 795.4
- 795.5 - 1,523.1
- 1,523.2 - 2,172.5
- 2,172.6 - 2,974.2
- 2,974.3 - 4,378.4

MAP 10.1 Rate of HIV Cases Living in the District of Columbia by Census Tract (2019) Note that Wards 2 and 3 are the highest-income wards, while Wards 7 and 8 are predominately low income and African American. Rates of HIV are highest in Wards 7 and 8.

predominately White population—has an average household income of $257,000.

In one of the wealthiest countries in the world, the long legacy of racial discrimination continues to impact the health of many African Americans. It also underscores that race is a leading factor in inequality.

Source: District of Columbia Department of Health, *Annual Epidemiology & Surveillance Report*. Washington, DC: Government of the District of Columbia (2020). https://dchealth.dc.gov/sites/default/files/dc/sites/doh/publication/attachments/2020-HAHSTA-Annual-Surveillance-Report.pdf

Discrimination against Indigenous Peoples

Discrimination based on ethnicity has been particularly pronounced among many of the world's Indigenous people. Consider the example of Indigenous Papuans living in Indonesia. In 2019, the UN Office of the High Commissioner on Human Rights (UNOHCHR) released news about disturbing trends in the discrimination of Indigenous Papuans by Indonesia. There had been numerous reported cases of alleged killings, unlawful arrests, and cruel, inhumane, and degrading treatment of Indigenous Papuans by the Indonesian police and military in West Papua and Papua provinces.

In one example, a video was circulated online of a handcuffed Indigenous Papuan boy being interrogated by Indonesian police holding a snake. The boy, who was arrested for allegedly having stolen a mobile phone, is heard screaming in fear while the laughing police officers push the snake's head toward his face.

A group of five human rights experts, including the UN Special Rapporteur on the rights of Indigenous people, noted that the case reflects a widespread pattern of violence, alleged arbitrary arrests, and detention as well as methods amounting to torture used by the Indonesian police and military in Papua. "These tactics are often used against indigenous Papuans ... and this latest incident is symptomatic of the deeply entrenched discrimination and racism that indigenous Papuans face, including by Indonesian military and police," said the working group, who stressed that "we are also deeply concerned about what appears to be a culture of impunity and general lack of investigations into allegations of human rights violations in Papua."

The incident in which the boy was mistreated comes amid an ongoing military operation in Papua, which became part of Indonesia in 1969. For more than forty years, the Papuan people have been one of the marginalized minorities of the Indonesian Republic. But the islands' remote and isolated location at the far eastern end of the Indonesian archipelago means that Indonesia's policies are played out with little oversight, and it has not attracted the attention it deserves in terms of human rights issues.

The International Convention Elimination of All Forms of Racism and Discrimination

In the global community, the UN Charter is founded on the principles of the dignity and equality inherent in all human beings; technically any member state that has adopted the Charter agrees to these principles. The International Convention Elimination of All Forms of Racism and Discrimination (ICERD), created by the United Nations in 1965, is tasked with monitoring racial and ethnic discrimination. The Convention defines racial and ethnic discrimination as:

> Any distinction, exclusion, restriction or preference based on race, color, descent, or national or ethnic origin which has the purpose or effect of nullifying or impairing the recognition, enjoyment or exercise, on an equal footing, of human rights and fundamental freedoms in the political, economic, social, cultural or any other field of public life.

Map 10.2 shows membership of the Convention. Despite more than fifty years of work in the global community to end discrimination, racism continues to plague societies around the world. This is a reminder that governments do not always live up to the aspirations and expectations of international agreements.

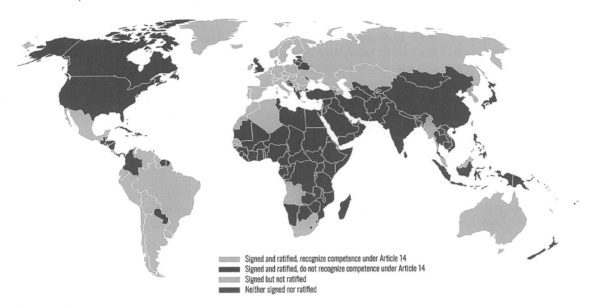

MAP 10.2 Membership of the Convention on the Elimination of All Forms of Racial Discrimination
All nations that signed and ratified are in blue, with different shades denoting application of Article 14. Article 14 establishes a mechanism for individuals to file complaints. Yellow denotes signatories that have not ratified. Red denotes countries that have neither signed nor ratified.

Age

People worldwide are living longer, and the world is becoming older. Today, for the first time in history, most people can expect to live into their sixties and beyond. By 2050, it is estimated that there will be some two billion people over the age of sixty and that 80 percent of them will live in low- and middle-income countries. Sixty-five percent of people over the age of sixty currently live in the Global South and by 2050 this number will have risen to 80 percent. While this shift in distribution of a country's population toward older ages—known as population aging—started in high-income countries (e.g., in Japan, 30 percent of the population is already more than sixty years old), it is now low- and middle-income countries that are experiencing the greatest increase in population age. By the middle of the century many countries such as Chile, China, the Islamic Republic of Iran, and the Russian Federation will have a similar proportion of older people to Japan.

Mary Robinson, the first woman president of Ireland and a former UN High Commissioner for Human Rights, wrote that "it is a sad irony at a time when the world has more older people than ever before—living longer with even greater wisdom and experience to offer—that they are often not respected as they have been in the past."

The practice of discriminating against a person because of their age, or ageism, is widespread in the world, and many have been marginalized because of their age. Older people are often regarded as victims of declining mental and physical capacity or threats to the opportunities of younger people, outdated but persistent stereotypes.

In most countries, it is still considered acceptable to deny people work, access to healthcare, education, or the right to participate in government because of their age. Emerging studies have shown that ageism holds back

💬 EXPERT VOICE 10.1

Dr. Isabella Aboderin on Aging in Sub-Saharan Africa

Dr. Isabella Aboderin is a Senior Research Scientist and Head of the Programme on Aging and Development at the African Population and Health Research Center (APHRC) in Nairobi, Kenya, and an Associate Professor of Gerontology at the Centre for Research on Ageing at University of Southampton. She notes that although we tend to focus on population growth and a relatively young population in sub-Saharan Africa, increasing life expectancy will mean an aging population over the long term. She writes:

> The growth of sub-Saharan Africa's older population this century will outstrip that of any other world region. By 2100, Africa will see a 15-fold growth in the number of older adults, from 46 million today to 694 million.

Partly in recognition of these trends, sub-Saharan Africa has made considerable strides in seeking to address older people's vulnerabilities and secure their basic rights. In recent years, a small but growing number of countries have adopted national policy frameworks on aging, and some are implementing or piloting social protection programs for older people. At a regional level, the African Union has endorsed an Africa Common Position on the Rights of Older People (2013) and is due to ratify a "Protocol on the Rights of Older Persons in Africa."

Despite these advances, sub-Saharan Africa's current older population continues to be viewed as, at best, marginal to the broader efforts to achieve economic and social development in the region. Despite the obvious importance of youth for building African economies, it may only be part of the story—and it is important to consider how older people fit into the equation. The lack of such consideration, thus far, reflects widely-held assumptions about old age as a period of "unproductivity" and economic dependence.

Sub-Saharan Africa's older people fulfill specific roles that are directly relevant to creating three conditions needed to realize a demographic dividend:

1. Their substantial economic activity, which is concentrated in small-holder agriculture. In most sub-Saharan African countries, more than 60 percent of older men and 50 percent of older women continue to work, with the share rising to more than 70 percent for men in twenty-four countries, and more than 60 percent for women in thirteen. An overwhelming majority of older workers are engaged in small-scale farming where they constitute a significant share of the overall labor force and land-holding population.

2. Their extensive intergenerational connections to children or adolescents within households and families and their consequent influence on the level or quality of financial or social investments that families make in the education and health of the young. In a range of the region's countries, around 20 percent to 30 percent of all children and adolescents live with an older person, with the share usually higher in poor population groups.

3. Their significant representation as "elders" among civic, political, and religious leaders at community and national levels, as well as among the business and professional elite. In these roles, older Africans actively and passively shape the conditions for—and the attitudes of younger generations toward—entrepreneurship, political and societal stability, and good governance.

The ways in which older people presently fulfill the previously mentioned functions may be conducive to or may hinder the forging of enabling conditions for a demographic dividend. A central challenge for policy and thought leaders then is to understand better—and to address the capacities, opportunities, and perspectives of older people in their varied roles—to create the most favorable foundation for a demographic dividend in sub-Saharan Africa.

"As policy makers set their sights on the promise of the region's youthfulness, they must not lose focus on the strategic relevance of the older population for realizing this potential."

Source: Aboderin, I. "Ageing in Sub-Saharan Africa." Age International, August 2018. https://www.ageinternational.org.uk/policy-research/expert-voices/ageing-in-sub-saharan-africa/

more older women and men in the world from living well and with dignity than any other single factor. Other recent studies have shown that households with older heads or members tend to be poorer than other households, and that ageism marginalizes and excludes older people in their communities.

The World Health Organization notes that ageism is the most socially "normalized" of any prejudice and is not widely countered—like racism or sexism. These attitudes lead to the marginalization of older people within our communities and have negative impacts on their health and well-being.

It is not uncommon for older people to struggle to secure loans or face barriers to getting the pensions they are already owed. In addition, there is a growing awareness that older people are often highly vulnerable during conflicts and disasters. Older people may find it more difficult to move, reach aid, and have their specific health and nutrition needs met; what's more, very few aid organizations consider this demographic in their aid packs. Finally, as many societies become "older" there is growing concern about adequate geriatric healthcare (such as providing cataract surgery, glasses, hearing aids, or walking canes) and how to support long-term care and independent living for older people. Dr. Isabella Aboderin provides perspective on aging in sub-Saharan Africa, a region we often neglect when considering aging populations because we tend to think of it as a region of young people (See *Expert Voice* 10.1).

Disability

Issues of disability have largely been absent from debates over equality, despite the fact that many countries have signed the International Convention on Disabilities (Map 10.3). So severe is the issue worldwide that in 2014 the

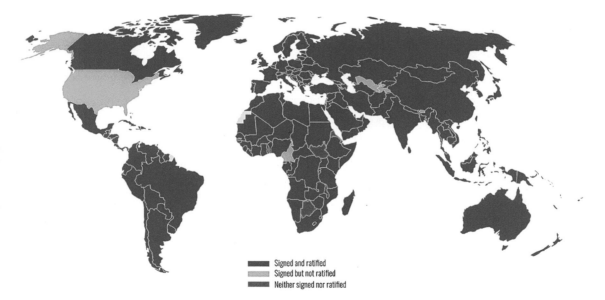

Signed and ratified
Signed but not ratified
Neither signed nor ratified

MAP 10.3 Membership of the Convention on the Rights of Persons with Disabilities
Of note is the United States, which has signed, but not ratified, the Convention. A state may have a delegate sign the Convention, which expresses the intention to comply with the agreement. However, this expression of intent is not binding. Once a country has signed an agreement, it will need to be formally ratified to make it binding. For example, in the United States, the Convention was signed by President Obama in 2009, but the Senate has failed to ratify the agreement several times since. Ironically, the Convention on the Rights of Persons with Disabilities was based on the US Americans with Disabilities Act of 1990.

United Nations created a Special Rapporteur position to examine the problem, which affects the one billion people—about 15 percent of the world's population—with some form of disability. There is a growing body of empirical data demonstrating the link between disability and poverty, including the World Bank study of fifteen lower-income countries that found that people with disabilities were significantly worse off, were more likely to experience multiple deprivations, and had lower educational attainment and employment rates than nondisabled people. Robert Saba, professor of constitutional law and human rights, notes that barriers can be both invisible, in the form of attitudes or assumptions held by others, and physical, as when steps or staircases literally prevent people with disabilities from accessing public spaces, offices, and transport.

Some countries have taken action. For example, the US Americans with Disabilities Act assures equal opportunities in education and employment for people with disabilities and prohibits discrimination on the basis of disability. However, there is a difference between legal equality and real equality; people with disabilities remain overrepresented among America's impoverished and undereducated. According to the US Department of Labor's Office of Disability Employment Policy, the labor force participation rate for people with disabilities (including physical, intellectual and developmental, sensory, and other disability categories) aged sixteen and over is 20.1 percent as compared to 68.6 percent for people without disabilities of the same age. For individuals who are blind and visually impaired, unemployment rates exceed 70 percent; for people with intellectual and developmental disabilities, the unemployment rate exceeds 80 percent. The US Census Bureau counted 38 million people in the United States with a disability.

Under international human rights law, governments have to respect and protect all citizens, including those with disabilities. That means undertaking all possible measures to dismantle barriers to equality within their borders.

Discrimination against LGBTQ+

Although many gay and lesbian people—and to a lesser extent trans people—have seen significant gains over the past twenty years, many of these gains are fragile and face backlash (Photo 10.2). Lesbian, gay, bisexual, and transgender, queer, intersex, and asexual (LGBTQ+) people suffer a crucible of egregious violations, including killings, rape, mutilation, torture, arbitrary detention, abduction, harassment, and physical and mental assaults. In some countries LGBTQ+ individuals have been subjected to lashings and forced surgical interventions, bullying from a young age, incitement to hatred, and pressures leading to suicide. More than seventy countries around the world today still criminalize same-sex relations, and in twelve countries the death penalty may be applied. In countries such as Afghanistan, Pakistan, and Qatar, these harsh measures tend not to be enforced even if they are legally permissible, but Iran still regularly executes LGBTQ+ individuals. Additionally, in geographic areas beyond the reach of governments, terrorist organizations such as the self-proclaimed Islamic State perpetrate anti-LGBTQ+ violence.

LGBTQ+ people also face blatant bias in housing, access to healthcare, employment, and opportunities in academic settings; in the ability to buy goods and services; and in the opportunities to participate meaningfully in a society's decision-making processes.

PHOTO 10.2 An LGBTQ+ Pride Parade in Istanbul, Turkey
The city denied permission to organize the Pride Walk, so activists celebrated the
event with a walk outside the city. Pride events are one way the LGBTQ+ community
has asserted their voice and demanded an end to intolerance and discrimination.

Although civil society organizations have lobbied the United Nations for
recognition of human rights on the basis of sexual orientation and gender
identity since the body's founding in 1945, it was not until 1994 that the
discrimination of individuals based on their sexual orientation became a recognized violation in international human rights law.

In the United States, a 2020 *Report of the State of the LGBTQ+ Community* found that one in three LGBTQ+ Americans faced discrimination
of some kind during 2020, including more than three in five transgender Americans. The report also notes that to avoid the experience of
discrimination, more than half of LGBTQ+ Americans report hiding a
personal relationship. Finally, many LGBTQ+ people report altering their
lives to avoid this discrimination and the trauma associated with unequal
treatment.

Urban–Rural Inequality

Another way to envision inequality within a country is to consider the spatial scale. In some countries, particularly the lowest-income countries, the
starkest disparity is often found between rural and urban life. More than 250
years ago, before the Industrial Revolution, more than 90 percent of people
lived in rural areas. Many were smallholder farmers, growing most of their
food for their own consumption, with some perhaps selling excess food in a
marketplace.

Many countries in the Global South are in the midst of the transition
from predominately rural to predominately urban, yet there are still substantial differences in quality between rural and urban life. Chapter 11

(SDG 11 Sustainable Cities) will discuss urbanization trends and patterns in more detail. For this discussion, it is sufficient to identify the inequalities between rural and urban life. First, income per person tends to be higher in urban areas than in rural areas. Second, there are notable geographic differences. Urban areas are often located along coastlines and waterways, where access to markets and transportation (water) is more easily available; rural areas tend to be located more inland. Third, as a result of settlement patterns, urban areas tend to have high population density, while rural areas are less dense. Fourth, infrastructure—for example, transportation, electrical—and public services such as hospitals and universities tend to be less available in rural areas. This is due to said lower population density, which makes it more difficult and less cost effective to invest in infrastructure and public services in rural areas. Fifth, because of all the aforementioned differences, there are social and health differences between urban and rural areas. Residents in rural areas tend to have poorer health but higher fertility rates (the average number of children per woman). For families living in rural areas, children are often seen as farm labor and there is incentive to have larger families.

Urbanization is associated with higher levels of development: better infrastructure, higher incomes, smaller family size, and better access to health and education services.

MEASURING INEQUALITY

The challenge of measuring who is being left behind is complicated by the fact that adequately measuring who is being left behind requires data from a variety of sources that are different in scope and purpose. National population censuses and some internationally standardized surveys are available for a large number of countries and are fairly comparable across countries.

However, none of them alone allows for a comprehensive international assessment of disadvantage or social exclusion. Assessing changes over time presents additional challenges because some data sources are available for only one point in time and comparability issues arise even between censuses or surveys of the same type.

Ideally, empirical analyses should determine which individual characteristics or combinations of characteristics increase the risk of disadvantage. However, most studies preselect some criteria that have been proven, empirically, to affect inequality—typically age, sex, ethnic background, income, nationality, and place of birth. Analyses based on these traditional criteria run the risk of overlooking new forms of inequality.

An additional challenge to measuring who is being left behind is that groups at high risk of poverty and exclusion are often statistically "invisible." Household surveys inevitably omit homeless persons, people in institutions, including prisons and refugee camps, and mobile and nomadic populations. In practice, they also tend to underrepresent populations in urban slums, those in insecure and isolated areas, and atypical households. While population censuses do not omit any of these groups by design, they often underenumerate them.

In addition, the definitions used to classify a population by nationality or by migrant, ethnic or disability status, vary across countries. While statistical groups are useful analytical categories, it is important to note that they are not necessarily entities with common agency or even common purposes. Some groups of people have shared beliefs and values and act in collective ways (such as religious and many ethnic groups). Other groups are defined on the basis of some shared characteristics (such as migrant status), but in reality, have little in common, aside from the discrimination they often face.

For these reasons, we must recognize the limitations of data used to measure inequality. At the same time, in a data-driven world, measures of inequality are often used to generate awareness and to prioritize interventions.

There are several different measures of inequality. For example, we can measure poverty using the World Bank's poverty categories; we can also use the Global Multidimensional Poverty Index (see Chapter 1). To measure gender inequality, most use the Gender Inequality Index or the Gender Empowerment Index set (see Chapter 5).

The most often used measure of general inequality is the Gini index or **Gini coefficient**, a statistical measure of distribution developed by the Italian statistician Corrado Gini in 1912. It is often used as a gauge of economic inequality, measuring income distribution or, less commonly, wealth distribution among a population. The Gini coefficient allows us to compare inequality among countries and within a country and shows different outcomes of development.

The coefficient ranges between 0, which reflects complete equality, and 1, which indicates complete inequality (one person has all the income or consumption, all others have none).

The Gini coefficient is widely used in fields as diverse as sociology, economics, health science, ecology, engineering, and agriculture. There are also Gini coefficients for education that estimate the inequality in education for a given population and for opportunity that measure inequality of opportunity. The Gini for income mobility, called Shorrocks index, estimates whether the income inequality Gini coefficient is permanent or temporary and to what extent a country or region enables economic mobility to its people so that they can move from one (e.g., bottom 20 percent) income quantile to another (e.g., middle 20 percent) over time.

Map 10.4 is a map of Gini coefficients for general inequality around the world. A Gini index value above 50 is considered a high level of inequality; countries including Brazil, Colombia, South Africa, Botswana, and Honduras can be found in this category. A Gini index value of 30 or above is considered a medium level of inequality; countries including Vietnam, Mexico, Poland, the United States, Argentina, Russia, and Uruguay can be found in this category. A Gini index value below 30 is considered a low level of inequality; countries including Austria, Germany, Denmark, Slovenia, Sweden, and Ukraine can be found in this category. The countries with the lowest inequality countries are located in western Europe, particularly in Scandinavia where the average Gini is about 0.25.

Notably, the United States does not compare well with western Europe, with a recently measured Gini of 0.40. This reflects that the United States has greater inequality than most other high-income nations (see *Critical Perspectives* 10.1). The United States may be home to 25 percent of the world's billionaires, but it is also home to approximately forty million poor. Much of the income growth of the past decades has accrued to the top 1 percent of the income bracket, while income growth for middle and lower quintiles has grown much more slowly. The average income of the top 1 percent has risen by 242 percent over the last forty years, about six times the growth for middle-earners. A 2018 study for the UN Human Rights Council called the United States the country with the

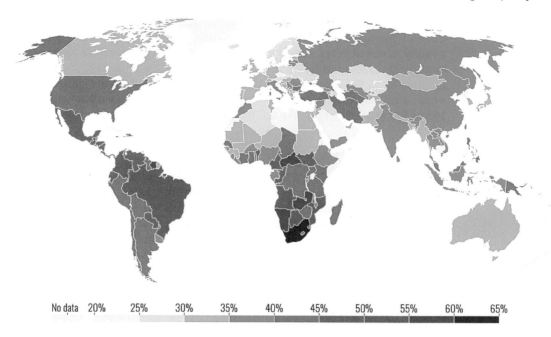

| No data | 20% | 25% | 30% | 35% | 40% | 45% | 50% | 55% | 60% | 65% |

MAP 10.4 Income Inequality Gini Coefficients (as %), 2018
Income inequality is highest in regions such as sub-Saharan Africa, Central America, and South America.
Countries such as South Africa, Namibia, and Zambia are among the highest, as is Suriname.

CRITICAL PERSPECTIVES 10.1

Inequality in the United States

The United States is a land of stark contrasts. It is one of the world's wealthiest societies, and a global leader in many areas. But its immense wealth and expertise stand in shocking contrast with the conditions in which vast numbers of its citizens live. About 40 million live in poverty, 18.5 million in extreme poverty, and 5.3 million in absolute poverty.

The United States has the highest youth poverty rate in the Organization for Economic Cooperation and Development (OECD) and the highest infant mortality rates among comparable OECD States. Its citizens live shorter and sicker lives compared to those living in all other rich democracies, eradicable tropical diseases are increasingly prevalent, and it has the world's highest incarceration rate, one of the lowest levels of voter registrations among OECD countries, and the highest obesity levels in the developed world.

The United States has the highest rate of income inequality among Western countries. The $1.5 trillion in tax cuts in December 2017 overwhelmingly benefited the wealthy and worsened inequality. The consequences of neglecting poverty and promoting inequality are clear. The United States has one of the highest poverty and inequality levels among the OECD countries, and the Stanford Center on Inequality and Poverty ranks it eighteenth out of twenty-one highest-income countries in terms of labor markets, poverty rates, safety nets, wealth inequality, and economic mobility. There is thus a dramatic contrast between the immense wealth of the few and the squalor and deprivation in which vast numbers of Americans exist. For almost five decades, the overall policy response has been neglectful at best.

The equality of opportunity, which is so prized in theory, is in practice a myth, especially for minorities and women, but also for many middle-class White workers.

_____

Sources: Human Rights Council. *Report of the Special Rapporteur on Extreme Poverty and Human Rights on His Mission to the United States of America.* United Nations (2018). https://undocs.org/A/HRC/38/33/ADD.1 Stanford Center on Inequality and Poverty. "Poverty and Inequality Trend Data" (2022). http://web.stanford.edu/group/scspi/cgi-bin/charts/poverty-inequality-trend-data.html

highest income inequality in the Global North. The report also noted that "the share of the top 1 percent of the population in the United States has grown steadily in recent years. In relation to both wealth and income the share of the bottom 90 percent has fallen in most of the past 25 years." A 2021 study by the Institute for Policy Studies noted there are 719 billionaires in the United States, worth a collective $4.56 trillion—more than four times the wealth held by the 165 million people who make up the lower 50 percent of Americans.

Trends in Reducing Inequality

Recent statistics have shown that reducing inequality is possible and efforts have been made in some countries to reduce income inequality. Between 2010 and 2016, in sixty out of ninety-four countries with data, the incomes of the poorest 40 percent of the population grew faster than those of the entire population.

There is an important difference between **"relative"** and **"absolute" inequality**. Economists Miguel Niño-Zarazúa and Finn Tarp explain the difference:

> Take the case of two people in Vietnam in 1986. One person had an income of U.S. $1 a day and the other person had an income of $10 a day. With the kind of economic growth that Vietnam has seen over the past 30 years, the first person would now in 2016 have $8 a day, while the second person would have $80 a day. So if we focus on "absolute" differences, inequality has gone up, while a focus on "relative" differences suggests that inequality between these two people has remained the same.

Niño-Zarazúa and his colleagues argue that if we focus on "absolute" differences, inequality has increased, but if we focus on "relative" differences, inequality between these two people has remained the same.

The positive news is that relative global inequality among countries, as measured by the Gini coefficient, has declined steadily over the past few decades. Despite this positive trend, absolute disparities among countries are still very large. The average income of people living in the European Union is eleven times higher than that of people in sub-Saharan Africa; the income of people in North America is sixteen times higher than that of sub-Saharan Africans.

There has been considerable variation within regions with respect to levels of, and changes in, domestic inequality over the period of analysis. For example, in Europe, the United Kingdom experienced an increase in relative inequality of 38 percent, while France saw a reduction of 16 percent; in Latin America, relative inequality in Argentina increased by 25 percent, while Brazil managed a reduction of 10 percent; in South Asia, Bangladesh experienced an increase of 60 percent, while Nepal saw a reduction of 38 percent. Inequality levels and trends differ among countries that are at similar levels of development and equally exposed to trade and technological innovation, underscoring that national policies and institutions do matter.

The decrease in the Gini was driven primarily by decreasing inequality between countries, notably by the rapidly growing economies of China, Brazil, and India. However, there has been a trend of *increasing* inequality within countries. For example, Omar Shahabudin McDoom and his colleagues studied inequality in the Philippines. The Philippines has made impressive progress in reducing disparities in education and access to basic public services, but McDoom found stark subnational income differences between regions and ethnoreligious groups in levels and trends. When examining the disparities more closely, progress is considered less positive because of significant differences within and between three salient ethnoreligious groupings—Muslims, Indigenous persons, and everyone else.

Complicating the picture is the fact that a sizeable part of income inequality can be attributed to inequality among social groups, although large differences are found across countries. In many countries the multidimensional poverty index is higher than average among ethnic minorities. For instance, inequality among racial groups accounted for an estimated 50 to 70 percent of total inequality in South Africa. Yet trends in these two key components of inequality—across groups and within groups—do not always go hand in hand. South Africa has seen the Gini coefficient of income inequality increase rapidly since the end of apartheid while racial inequality has declined.

The decline in interracial inequality has been driven largely by faster income growth in non-White lower-income households and by a growing percentage of non-White, middle- and upper-income households. Opportunities have gradually opened up in the civil service, business, and education to Black South Africans, who made up 6 percent of top management and company executives in 2001 and 14 percent in 2017. Efforts to reverse the effects of decades of segregation have been facilitated by antidiscriminatory legislation and by a constitution establishing citizens' rights to food, water, social security, and social assistance. Through a system of progressive taxation and social programs, the South African government has expanded provision of basic public services, achieved near-universal access to primary education, and implemented cash transfer programs for older persons, families with children, and persons with disabilities. Primary healthcare is free, while public hospital services are relatively low-cost or free for disadvantaged/vulnerable persons. These measures have contributed to reducing both absolute and multidimensional poverty. Despite this progress, inequality within racial groups has worsened.

SOLUTIONS TO REDUCING INEQUALITY

Table 10.2 outlines the numerous targets and indicators for reducing inequality in SDG 10. The targets for SDG 10 emphasize that equality of opportunity is a critical factor in reducing inequality. For example, the ability to access quality education, healthcare, energy, infrastructure, and so forth are important aspects of opportunity that will allow all people to work toward fulfilling their potential. There are numerous solutions for achieving SDG 10; this section looks briefly at increasing investment in rural development, eliminating discrimination, and wealth redistribution.

Increase Investment in Rural Development

Four out of every five people who face extreme poverty around the world live in rural areas. Strategies for rural development will likely be country-specific, grounded in the country's conditions regarding natural and human resources. Many countries, but particularly low-income countries, need additional public investment in rural basic infrastructure (including broadband connectivity).

The 2021 UN Department of Economic and Social Affairs report suggests there are several priority strategies to improve rural development. For example, rural areas may take advantage of new digital technologies to make

TABLE 10.2 SDG 10 Targets and Indicators to Achieve by 2030

Target	Indicators
10.1 Progressively achieve and sustain income growth of the bottom 40 percent of the population at a rate higher than the national average	**10.1.1** Growth rates of household expenditure or income per capita among the bottom 40 percent of the population and the total population
10.2 Empower and promote the social, economic, and political inclusion of all, irrespective of age, sex, disability, race, ethnicity, origin, religion, or economic or other status	**10.2.1** Proportion of people living below 50 percent of median income, by age, sex, and persons with disabilities
10.3 Ensure equal opportunity and reduce inequalities of outcome, including by eliminating discriminatory laws, policies, and practices and promoting appropriate legislation, policies, and action in this regard	**10.3.1** Proportion of the population reporting having personally felt discriminated against or harassed within the previous twelve months on the basis of a ground of discrimination prohibited under international human rights law
10.4 Adopt policies, especially fiscal, wage, and social protection policies, and progressively achieve greater equality	**10.4.1** Labor share of GDP, comprising wages and social protection transfers
10.5 Improve the regulation and monitoring of global financial markets and institutions and strengthen the implementation of such regulations	**10.5.1** Financial soundness indicators
10.6 Ensure enhanced representation and voice for developing countries in decision making in global international economic and financial institutions to deliver more effective, credible, accountable, and legitimate institutions	**10.6.1** Proportion of members and voting rights of developing countries in international organizations
10.7 Facilitate orderly, safe, regular, and responsible migration and mobility of people, including through the implementation of planned and well-managed migration policies	**10.7.1** Recruitment cost borne by employee as a proportion of yearly income earned in country of destination **10.7.2** Number of countries that have implemented well-managed migration policies
10.A Implement the principle of special and differential treatment for developing countries, in particular least developed countries, in accordance with World Trade Organization agreements	**10.A.1** Proportion of tariff lines applied to imports from least developed countries and developing countries with zero-tariff
10.B Encourage official development assistance and financial flows, including foreign direct investment, to states where the need is greatest, in particular least developed countries, African countries, small island developing states, and landlocked developing countries, in accordance with their national plans and programs	**10.B.1** Total resource flows for development, by recipient and donor countries and type of flow (e.g., official development assistance, foreign direct investment, and other flows)
10.C Reduce to less than 3 percent the transaction costs of migrant remittances and eliminate remittance corridors with costs higher than 5 percent	**10.C.1** Remittance costs as a proportion of the amount remitted

Source: United Nations Sustainable Development Goals. https://sustainabledevelopment.un.org/sdg10

agricultural production more efficient and profitable. Farmers can access services that help increase their yields and productivity, including detailed weather forecasts, mobile payment systems, crowdfunding platforms for access to finance, extension services for technical advice, and many others. These digital technologies may accelerate rural economic transformation and reduce the need to transition away from agricultural livelihoods. Governments can also encourage investment, entrepreneurship, and job growth in the nonfarm economy in rural economies.

Another strategy is the idea of *in situ urbanization*, under which the rural population reaches an urban standard of living without having to migrate to urban areas. This entails providing basic infrastructure, such as health services, electricity, water, and sanitation in rural areas, improving rural life and decreasing the pull to migrate to cities. This "place-based" model has proved effective in China and Sri Lanka in fostering long-term economic growth and spatial equity between rural and urban areas.

Other strategies for improving rural development include fairer distribution of land, secure access to land (such as land tenure), and ensuring rural women have equal access to land and resources.

Eliminate Discrimination

Discrimination remains a pervasive driver of inequality. As noted earlier in the chapter, many societies continue to make distinctions based on ethnicity, race, sex, and other characteristics. Historically, many laws and policies have explicitly limited or denied rights to specific groups; to move toward achieving sustainability will require governments to eliminate discriminatory laws, policies, and practices. Almost 200 ethnic or religious minorities worldwide face some form of overt political discrimination. For example, according to the World Bank, 104 countries have laws restricting the types of jobs that women can perform, and some groups have been denied citizenship on the basis of ethnicity.

The 2021 UN Department of Economic and Social Affairs report notes that beyond repealing discriminatory laws and introducing preventive measures, many countries have resorted to affirmative action to favor groups that were discriminated against in the past. This includes quotas or reservations to improve the representation of women or minority ethnic groups in decision-making roles, quotas and scholarships to improve access to education, and preferential treatment in hiring for certain jobs.

International migration can widen prospects for poverty reduction and social mobility as well. However, many migration policies have maintained or entrenched inequalities within and among countries, rather than contributing to their reduction. Governments need to do more to ensure migrant workers can use their skills productively and send remittances to their home country at a low cost. *SDGs and the Law* 10.1 considers the objectives of International Convention on the Protection of the Rights of All Migrant Workers and Members of Their Families. Countries of destination must also do more to promote the integration of migrants, uphold their rights, provide access to social services, and address discrimination against them. The *Solutions* 10.1 box looks at one strategy that US cities are implementing to deal with this issue.

In the United States the 2020 protests over the police killings of George Floyd and Breonna Taylor and others have renewed national conversations on

SDGs AND THE LAW 10.1

International Convention on the Protection of the Rights of All Migrant Workers and Members of Their Families

Target 10.7: Facilitate orderly, safe, regular, and responsible migration and mobility of people, including through the implementation of planned and well-managed migration policies.

Target 10.C: Reduce to less than 3 percent the transaction costs of migrant remittances and eliminate remittance corridors with costs higher than 5 percent.

The International Convention on the Protection of the Rights of All Migrant Workers and Members of Their Families. It aspires to guarantee dignity and equality in an era of globalization.

The Convention aims to protect migrant workers and members of their families; its existence sets a moral standard and serves as a guide and stimulus for the promotion of migrant rights in each country.

The Convention does not create new rights for migrants but aims at guaranteeing equality of treatment, and the same working conditions, including in cases of temporary work for migrants and nationals. The Convention relies on the fundamental notion that all migrants should have access to a minimum degree of protection.

Currently, only fifty-five countries have ratified the treaty (the United States has not, nor has Australia, both of which are major receivers of migrants). Countries that have ratified the Convention are primarily countries of origin of migrants (such as Mexico, Morocco, and the Philippines). For these countries, the Convention is an important vehicle to protect their citizens living abroad.

SOLUTIONS 10.1

The Welcoming City Initiative

Target 10.7: aims to facilitate orderly, safe, regular, and responsible migration and mobility of people, including through the implementation of planned and well-managed migration policies.

In the United States, some mayors have established a "Welcoming City" Initiative. The city of Anaheim, California, has taken steps to establish itself as a "Welcoming City" for immigrants and refugees. With foreign-born individuals making up nearly 40 percent of the city's population, the city identified strategies that would help create a supportive and accessible community for all. To do so, the mayor's "Welcoming America Task Force" was established. The task force has brought together residents and leaders from the community and the business sector to provide insight on proactive steps to build community engagement and intercultural exchange.

The program is modeled on the national Welcoming America initiative, which has been building a network of inclusive communities across the country since 2009. Welcoming City works across sectors to plan for a climate that embraces long-term integration, communicates messages of shared values, and commits institutions to policies and practices that promote inclusive and positive interactions between long-term and new residents.

The Welcoming America Task Force recommendations to the city include adding additional language translations to city publications, creating a "welcome kit" that informs residents of where resources and city services can be found, and providing opportunities for city staff to learn about cultural competency. Programs such as these confirm the importance of the city's efforts to welcome residents of all nationalities.

Source: National League of Cities. *Mayoral Views on Racism and Discrimination* (2018). https://www.nlc.org/sites/default/files/201809/CSAR_BostonU_REAL_Report_FINAL_small.pdf

structural racism and racial inequity. A recent survey of US mayors reported that the four groups most discriminated against in their cities and across the country are immigrants, transgender individuals, Blacks and Muslims. Access to affordable healthcare, primary and secondary education, and safe and affordable housing continue to favor White people. More recently, there has been a rise of anti-Semitism, anti-Asian hate crimes have been directed at women, and some Southern states are enacting new laws restricting voting and healthcare restrictions for transgender people. A 2021 Sustainable Development Solutions Network report noted that "every State in the Union would receive a failing grade of 'F' for their performance toward reducing inequalities for Black, Hispanic, Indigenous, Asian and Multiracial or 'Other' communities." Ensuring equal access to justice for all will involve, among other things, promoting campaigns to enhance legal awareness and literacy, scaling up services to provide advice and assistance, developing alternative dispute resolution mechanisms, and, ultimately, improving the institutional framework for resolving disputes, conflicts, and crimes. Discrimination also challenges the ability of those affected to have their voices heard. A key step to promote their inclusion is to remove obstacles to political participation, including the right to vote. Finally, addressing the root causes of discrimination calls for structural reforms of the justice system and other national institutions.

Wealth Redistribution: A Political Economy Perspective

A political economy or political ecology perspective would recognize that ending inequality will require more than legal reforms, education, and awareness, which are seen as just modest changes to a capitalist system not capable of righting inequalities. Incremental change will not be sufficient to address the distributional implications of existing inequalities. The more deeply structural the drivers of inequality are, the more radical the needed changes that rebalance the asymmetrical power relations.

Solutions to numerous inequalities—including race, gender, age, ethnicity, and so forth—will require massive economic redistribution. This could include new banking regulations, tuition-free higher education, and redistributive policies such as increasing taxes on the wealthiest. All these suggestions, however, have generated heated debate in the United States, where political rhetoric has framed these solutions as "socialist" and "un-American." Political economists advocate for public spending directed at the lower and middle parts of the income distribution through broad-based taxes such as the value-added tax (VAT), which are easy to collect. Others prefer to redress inequality at the very top using wealth taxes and more progressive income taxes. It is likely a combination of both will be needed.

Taxes on wealth and property can play an important role in increasing redistribution and have gained traction in recent political debates. Addressing inequality also calls for lessening the tax burden on people at the bottom of the income distribution. Raising minimum income tax thresholds and reducing the burden of indirect taxation can help make tax systems more progressive. Lower tax rates on basic goods, such as staple foods, may be helpful as well.

Other policies could intervene at the production stage of an economy, affecting the composition and organization of labor and production. Incentives in hiring a more diverse workforce, raising minimum wage, and creating incentives for R&D that focuses on the needs of those most marginalized have also been discussed. Political economists who research the global economy have called for more research into asymmetries in the role of transnational corporations in supply chains.

As discussed in Chapter 1 on poverty, another solution is to strengthen social protections by increasing cash transfers and public services. Making healthcare universal will also help reduce inequalities, as will expanding food stamp programs and negative earned income tax.

Political economists also say ending inequalities will require fundamental political transformations too. Racial and ethnic discrimination is structurally supported by laws and political systems that allow discriminatory treatment to persist. If the very wealthy exert too much political influence, one strategy may be to prevent wealth accumulation by raising taxes on the wealthiest. If the poor are disenfranchised and therefore have little voice in determining the economic policies that affect them, we should contemplate changes in political rules, such as making it easier to vote or restricting campaign finance. Additionally, the United States should amplify workers' bargaining power by increasing fines for illegal antiunion behavior, encouraging minority unions, and reversing state laws that undermine unions and prevent them from collecting dues for benefits they provide workers at unionized workplaces. Finally, in the wake of the George Floyd killing and the Black Lives Matter protests across cities in the United States, there is a growing recognition that significant transformations to the criminal justice system are needed to end mass incarcerations that disproportionately impact African Americans. Doing so may also help remove barriers to economic mobility.

SUMMARY AND PROGRESS

Some have criticized the SDG 10 for failing to inadequately address many of the most pressing inequalities. As a counterpoint, a recent report by the UN Development Strategy and Policy Analysis Unit argued that it is virtually impossible to imagine a society that managed to make all these improvements and yet remained more unequal. Improving access overall means a reduction in both horizontal and vertical inequality, as it reduces both disparities between groups and discriminatory practices, as well as improving equality between individuals across all groups and incomes.

There have been some positive signs of progress. A recent *UN Sustainable Development Goals Report* 2020 noted that in some countries there has been a decline in inequality with regard to income. In a majority of countries, the poorest 40 percent of a country's population has seen growth in household income. In about fifty of the ninety countries that had data, the bottom 40 percent experienced income growth higher than the overall national average, indicating lower levels of inequality. Progress in shared prosperity has been strongest in East Asia and Southeast Asia, with the bottom 40 percent of the population growing annually by about 5 percent on average. The Gini index has declined for more than thirty-eight countries over the last ten years, meaning those countries saw a reduction in inequality. Despite this, more than sixty countries still have a Gini index value of more than 40, which means inequality levels remain high. And, unfortunately, twenty-five countries have seen an increase in inequality, erasing progress toward the SDG 10.

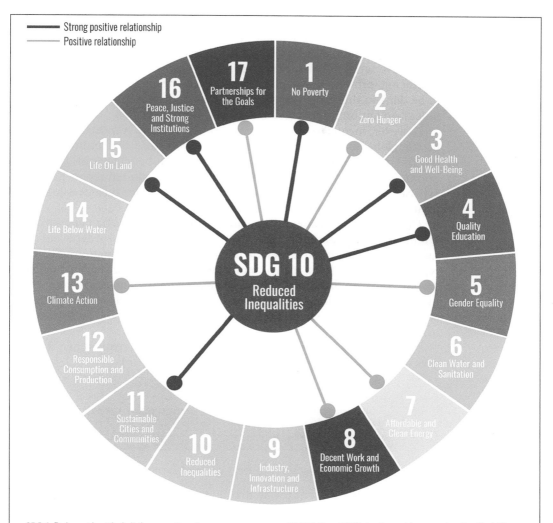

FIGURE 10.1 Interconnections in SDG 10 and the Other Goals
Advancing targets in SDG 10 can have positive impacts on other SDGs.

In addition, about half of all countries have a set of policies measures to facilitate orderly, safe, regular, and responsible migration. Central Asia, Southern Asia, and Latin America have the highest share of countries with a comprehensive set of policies. However, migrant rights and their socioeconomic well-being form the lowest proportions of policies measures and more needs to be done in this area. The rise of political populism in many countries has unleashed antiimmigrant and antirefugee sentiments.

However, inequality in its various forms persists. Almost 20 percent of people globally report having personally experienced discrimination in the last five years. Further, women are more likely to be victims of discrimination than men. Among those with disabilities, three in ten report personally experiencing discrimination. This points to the need for measures to combat multiple and intersecting forms of discrimination.

The COVID-19 pandemic may further exacerbate existing patterns of discrimination. It is impacting the most vulnerable people hardest, and those same groups are often experiencing increased discrimination. There are reports emerging from many countries of discrimination against different groups, such as those of Chinese origin, migrants, and racial and ethnic minorities. In addition, the impact of the pandemic is likely to reduce the flow of aid to low-income countries, leaving them with less capital for investments in the SDGs.

Ensuring inclusiveness and equality is also found in many other SDG targets; a few (but not all) of those are shown in Figure 10.1.

QUESTIONS FOR DISCUSSION AND ACTIVITIES

1. Is inequality an inescapable part of the human condition?

2. Is inequality different today than in the past?

3. Describe the patterns of inequality in major world regions.

4. Explain how inequality is measured.

5. Have you experienced or know anyone who has experienced discrimination based on age, ability, or identity?

6. Where do you see progress on inequality in the United States? Where do you see progress on inequality more globally?

7. In the United States, the number of wealthy is a small percentage of the population. Why do you think taxing the wealthy has been difficult to achieve? Which type of wealth redistribution might be possible in the current political climate in Congress?

8. Investigate some of the strategies for wealth distribution. Which do you think have the most potential in the Global South?

9. In what ways do you see SDG 10 targets connecting with targets in SDG 6, 9, 12, and 14?

10. A major ideology in the United States is the idea of the "American Dream." Debate the topic "The American Dream is alive and well." Each student will do his or her own research outside of class to support their position (either that the American Dream is alive and well or that it is not).

TERMS

absolute inequality
fourth world
Gini coefficient

horizontal inequality
relative inequality
vertical inequality

RESOURCES USED AND SUGGESTED READINGS

Age International. "Home Page." https://www.ageinternational.org.uk

Ambrosini, M., Cinalli, M., and Jacobson, D. "Research on Migration, Borders and Citizenship: The Way Ahead." In M. Ambrosini, M. Cinalli, and D. Jacobson (eds.). *Migration, Borders and Citizenship: Migration, Diasporas and Citizenship* (pp. 295–305). Cham, Switzerland: Palgrave Macmillan (2020). https://doi.org/10.1007/978-3-030-22157-7_13

Bourqia, R., and Sili, M. "A Kaleidoscope of Ideas for Rethinking Development in the Global South." In R. Bourqia and M. Sili (eds.), *New Paths of Development: Sustainable Development Goals Series* (pp. 1–21). Cham, Switzerland: Springer (2021). https://doi.org/10.1007/978-3-030-56096-6_1

Castells, M. "The Rise of the Fourth World." In D. Held and A. McGrew (eds.). *The Global Transformation Reader: An Introduction to the Globalization Debate* (pp. 348–54). Cambridge: Polity Press (2000).

Collins, C., Ocampo, O., and Paslaski, S. "Billionaire Bonanza 2020: Wealth, Tumbling Taxes and Pandemic Profiteers." Institute for Policy Studies (2020). https://ips-dc.org/billionaire-bonanza–2020/

Desmond, M. *Evicted: Poverty and Profit in the American City*. New York: BDWY (2017).

Disability Visibility Project. https://disabilityvisibilityproject.com/about/

Gayle-Geddes, A. *Disability and Inequality: Socioeconomic Imperatives and Public Policy in Jamaica*. Cham, Switzerland: Palgrave Macmillan (2015).

Human Rights Council. *Report of the Special Rapporteur on Extreme Poverty and Human Rights on His Mission to the United States of America*. United Nations (2018). https://undocs.org/A/HRC/38/33/ADD.1

Lynch, A., Bond, H., and Sachs, J. *In the Red: The US Failure to Deliver on a Promise of Racial Equality*. New York: SDSN (2021).

Mangharam, M. L. "Global Inequality." In E. Hadley, A. Jaffe, and S. Winter (eds.). *From Political Economy to Economics through Nineteenth-Century Literature*. Palgrave Studies in Literature, Culture and

Economics. Cham, Switzerland: Palgrave Macmillan (2019). https://doi.org/10.1007/978-3-030-24158-2_10

Mayerson, D. *"Race Relations in Rwanda: An Historical Perspective."* University of Wollongong Research Online (2010). https://ro.uow.edu.au/cgi/viewcontent.cgi?article=2306&context=artspapers

McDoom, O. S., Reyes, C., Mina, C., et al. "Inequality between Whom? Patterns, Trends, and Implications of Horizontal Inequality in the Philippines." *Soc Indic Res* 145 (2018). https://doi.org/10.1007/s11205-018-1867-6.

Milanović, B. *Worlds Apart: Measuring International and Global Inequality.* Princeton: Princeton University Press (2005).

Mitra, S. *Disability, Health and Human Development.* New York: Palgrave (2017).

National League of Cities. *Mayoral Views on Racism and Discrimination* (2018). https://www.nlc.org/sites/default/files/2018-09/CSAR_BostonU_REAL_Report_FINAL_small.pdf

Niño-Zarazúa, M., Roope, L., and Tarp, F. "Income Inequality in a Globalizing World" (2016, September 20). https://cepr.org/voxeu/columns/income-inequality-globalising-world

Omi, M., and Winant, H. *Racial Formation in the United States.* New York: Routledge (2014).

Robinson, C. *Black Marxism: The Making of the Black Radical Tradition.* Chapel Hill: The University of North Carolina Press (1983).

Robinson, M. "Do We Respect Our Elders?" *Age International* (2018). https://www.ageinternational.org.uk/policy-research/expert-voices/do-we-respect-our-elders/

Saba, R. "Around the Globe, People with Disabilities Face Unseen Discrimination. We Must Do Better." *The Conversation* (21 February 2017). http://theconversation.com/around-the-globe-people-with-disabilities-face-unseen-discrimination-we-must-do-better–70235

Stewart, F. *Horizontal Inequalities: A Neglected Dimension of Development.* Queen Elizabeth House Working Paper Series (2002).

Stewart, F. *Horizontal Inequalities and Conflict: Understanding Group Violence in Multiethnic Societies (Foreword by Kofi Annan).* Basingstoke, UK and New York: Palgrave Macmillan (2008).

Sustainable Development Solutions Network. "In the Red: The US Failure to Deliver on a Promise of Racial Equality" (2021, May). https://www.sdgindex.org/reports/in-the-red-the-us-failure-to-deliver-on-a-promise-of-racial-equality/

United Nations. "World Income Inequality Database and Interactive Map." United Nations University (2019a). https://www.wider.unu.edu/project/wiid-world-income-inequality-database

United Nations. "Exclusive Interview Kartik Sawhney." United Nations (2019b). https://www.un.org/sustainabledevelopment/blog/2019/01/exclusive-interview-kartik-sawhney/

United Nations. "Shattering Stereotypes—Jillian Mercado, Model and Disability Rights Advocate." United Nations (2019c). https://www.un.org/sustainabledevelopment/blog/2019/01/shattering-stereotypes-jillian-mercado-model-and-disability-rights-advocate/

United Nations Department of Economic and Social Affairs. *World Social Report 2020: Inequality in a Rapidly Changing World* (2020, p. 39).

https://www.un.org/development/desa/dspd/wp-content/uploads/sites/22/2020/01/World-Social-Report-2020-FullReport.pdf

United Nations Department of Economic and Social Affairs. *World Social Report 2021: Reconsidering Rural Development* (2021). https://www.un.org/development/desa/dspd/world-social-report/2021-2.html

United Nations Office of the High Commissioner on Human Rights. "Indonesia: UN Experts Condemn Racism and Police Violence against Papuans, and Use of Snake against Arrested Boy" (2019). https://www.ohchr.org/EN/NewsEvents/Pages/DisplayNews.aspx?NewsID=24187&LangID=E

United States Department of Labor. "Disability Employment Statistics Resources" (2022). https://www.dol.gov/agencies/odep/research-evaluation/statistics

Welch, F. *The Causes and Consequences of Increasing Inequality.* Chicago: University of Chicago Press (2001).

Wong, A. (ed.). *Disability Visibility: First Person Stories from the Twenty-First Century.* New York: Vintage Books (2020).

World Health Organization. (n.d.) "Ageing and Life-Course." https://www.who.int/ageing/en/

Sustainable Cities

LEARNING OBJECTIVES

After reading this chapter, you should be able to:

- Identify urbanization trends

- Describe the emergence of megacities and their challenges

- Explain slums and their challenges

- Explain the impact of deindustrialization on Global North cities

- Identify and explain the environmental impacts of urbanization

- Identify and explain the various targets and solutions associated with SDG 11

- Explain how the three Es are present in SDG 11

- Explain how SDG 11 strategies connect to other SDGs

In the Global South, millions of people live in substandard housing, called slums. It is estimated that hundreds of millions of slum dwellers do not have piped water supplies and thus have no alternative but to use contaminated water, or water whose quality is not guaranteed. In Libreville, Gabon, for example, only 58 percent of the population have access to clean water. This is not to say they have no access to water—they do. In some slums, there are public stand posts or public fountains from which residents can fill buckets and other containers. People spend a significant amount of time to obtain water, and often long distances must be traveled to collect it. In addition, there are water vendors (usually private firms) that sell water to the poor, but they frequently charge five to ten times more than the rates for water delivered by a public water system (Photo 11.1). Many of the urban poor cannot afford private water sources and hence have an inadequate supply.

Many slum residents must use whatever water they can find, which is usually contaminated. Water-borne diseases take a tremendous toll on human health. Diarrheal diseases affect an estimated 700 million people each year and account for most water-related infant and child deaths. A high proportion of slum residents have intestinal worms that cause severe pain and malnutrition. Among slum residents, neonatal deaths are two times higher, mortality from respiratory disease six times higher, and mortality from septicemia eight times higher than among the middle-class or wealthy in that same city.

Poverty exacerbates these issues and, as a result, infants and children do not always receive their vaccines for measles, whooping cough, and diphtheria. An infant is forty to fifty times more likely to die in a slum than in a city in the Global North.* There are an estimated 900 million people living in slums today, and many struggle to access water, electricity, and other basic services.

---

*Hackenbroch, K., and Hossain, S. "'The Organized Encroachment of the Powerful'—Everyday Practices of Public Space Water Supply in Dhaka, Bangladesh." *Planning Theory and Practice* (2012). https://doi.org/10.1080/14649357.2012.694265

PHOTO 11.1 Water Distribution in Nairobi, Kenya
Residents of the Kibera slum in Nairobi, Kenya, line up to fill jerrycans with water. In Kibera, as in many slums, residents lack access to fresh, clean water, safe electricity, and other basic services. Valuable time each day is spent securing water. If water is purchased from a private source, residents will pay considerably more for the water than in areas of the city where water services are provided by the local governments.

URBANIZATION: UNDERSTANDING THE ISSUE

A little more than a decade ago the world passed a major milestone of more than one-half of the world's residents living in urban areas. Today, that figure is almost 56 percent, or 4.4 billion people (Map 11.1). By 2050, more than two out of every three people on the planet will live in cities—some six billion people (Map 11.2). Table 11.1 shows the percentage of urban residents in 1950, 2010, and estimates for 2050. It is a fact that in many areas of the world, a majority of the population lives in urban areas, and this is only going to increase, with the fastest rates of urban growth occurring in the Global South. For many reasons, urbanization is a key challenge to sustainability. Karen Seto, an urban scholar, reflects on the promise and pitfalls of urban sustainability in the *Expert Voice* 11.1 box.

The reasons for **urbanization** have remained constant for a century and are mainly connected to urban-rural inequality. **Push factors** are those conditions that often make an individual or family migrate to an urban area out of desperation. These may include a lack of rural investment in infrastructure, lack of economic opportunities in rural areas, higher levels of inequalities such as poverty, and lack of access to health care and education. In many countries in the Global South, urban-rural inequities can be significant, reflecting long-term patterns of disinvestment in rural areas.

Pull factors are those that draw people to cities: greater interconnectivity, increased access to services and amenities, and increased opportunities to improve one's life. Cities have long been areas of concentrated economic activity, where job creation has increased the most and standards of living are comparatively better than in rural areas. Local governments can provide a wide variety of services to urban populations due to

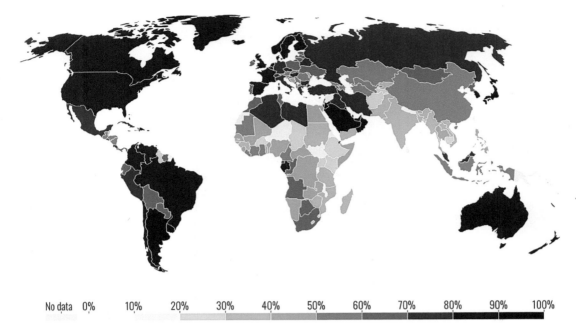

MAP 11.1 Share of People Living in Urban Areas (2017)
Countries in the Global North and South America have among the highest percentage of their population living in urban areas. South Asia, Central Asia, and sub-Saharan Africa have the lowest percentage of their population living in urban areas.

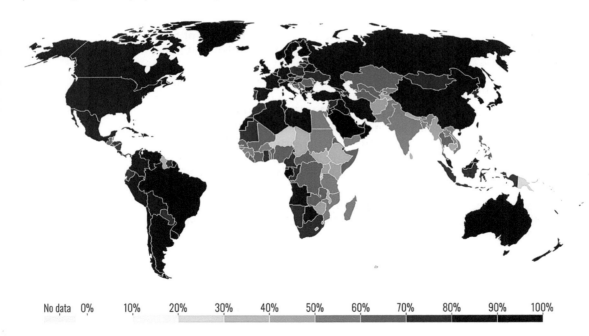

MAP 11.2 Share of People Living in Urban Areas (2050)
By 2050, most countries in the Global North, Middle East, and South America will have more than 80 percent of their population living in urban areas. However, the greatest increases in urbanization rates are expected to occur in sub-Saharan Africa, South Asia, and countries such as China and Indonesia.

TABLE 11.1 Percentage Urban, 1950–2050			
	1950	**2010**	**2050**
World	29.4	51.6	68.2
Global North	54.5	77.5	85.9
Global South	17.6	46.0	64.1

Source: Population Division of the Department of Economic and Social Affairs of the United Nations Secretariat. *World Population Prospects: The 2018 Revision* of *World Urbanization Prospects* (2018). https://www.un.org/development/desa/publications/2018-revision-of-world-urbanization-prospects.html

TABLE 11.2 Percentage Urban Population for Selected Countries, 1950–2050				
	1950	**2010**	**2020**	**2050**
Afghanistan	5.8	23.2	26.2	43
Brazil	36.2	84.3	86.4	90
China	11.8	49.2	60.1	80
Kenya	5.6	23.6	26.3	45

Source: Population Division of the Department of Economic and Social Affairs of the United Nations Secretariat, *World Population Prospects: The 2018 Revision* of *World Urbanization Prospects* (2018). https://www.un.org/development/desa/publications/2018-revision-of-world-urbanization-prospects.html

economies of scale, centralized resources, and shared networks. Both push and pull factors can operate simultaneously; the overall result has been increased urbanization over the twentieth century and into the twenty-first century.

Rates of urban growth have declined in many cities in the Global North because they have been at higher levels and are slowly approaching their urban maximums. For example, more than 80 percent of the US and Canadian population lives in urban areas. In Australia, more than 90 percent of the population already lives in six urban areas: Adelaide, Brisbane, Canberra, Melbourne, Perth, and Sydney.

Urbanization is occurring more rapidly in the Global South. In 1900, only 10 percent of Mexicans lived in cities; by 2020, that figure increased to 79 percent. In the Sudan, just 6.8 percent of the population lived in urban areas in 1950; by 2020 it grew to 35.5 percent. However, even among lower-income countries, there are substantial differences. Table 11.2 highlights four examples. Afghanistan and Kenya have fairly rapid urban growth rates, yet their urban population will not even reach 50 percent by 2050 if current trends continue. This may be due to lower rates of economic growth. Meanwhile, the rapidly growing economies of Brazil and China have experienced much higher levels of urbanization. In these countries, large-scale urbanization and rapid industrialization continue to go hand in hand.

There is another key factor that leads to high levels of urbanization in the Global South. In the postwar years (1950–80), large-scale development

EXPERT VOICE 11.1

Karen Seto on Cities and the SDGs

Karen Seto is Professor of Geography and Urbanization at Yale. She is an expert on urbanization in China and India, forecasting urban growth, and climate change mitigation. She reflects on why the Global Agenda needed a stand-alone goal for cities:

> We will build more urban areas during the twenty-first century than in all of human history. Simply stated, we cannot afford to create twenty-first century cities with outdated ideas and technology. Yet, that's what we're on the trajectory to do if we do not transform the way in which we build new and rebuild existing cities. The Urban SDG has the opportunity to be a catalyst for changing how we conceive, design, and manage cities. We need to urgently work toward establishing a plan of action for implementing and monitoring progress toward these goals.
>
> I see three potential pitfalls with the Urban SDG. The first is that it sits on a shelf like a family portrait: it's a snapshot that's static, collecting dust over the years, but represents happy times and great potential. There has been a lot of energy and effort toward establishing the SDGs, but that is only the first step. The next phase will be even more challenging—implementation. Here, the second pitfall is that implementation and monitoring falls well short of the target. If the Urban SDG becomes a way to repackage and rebrand existing efforts, then it will not achieve its goals. No single city is safe, resilient, and sustainable. Substantive effort, by way of science, policy, and financing, is necessary to make the Urban SDG a living process

> that can achieve its goals. The third potential risk is that cities emulate strategies from other cities that are not appropriate for their context, constituents, or needs. This would be a triple loss in effort, time, and opportunity. Cities will need to carefully identify sister cities with similar challenges and opportunities, from whom they can learn what works and what doesn't.
>
> How can we avoid these dangers? There is a long list of things that should be done, but there is at least one thing that must be done in order for the Urban SDG to be achieved and that is the coupling of strategies across scales. Many of the conditions, processes, and policies that affect urban areas occur outside of urban areas, be it the political economy or regional or national contexts. Cities cannot achieve the goals of the Urban SDG if they act alone. They must have the support of regional and national governments and institutions. However, support is also not enough. Cities must work together to ensure that efforts undertaken at the local scale are not subverted by strategies at other scales or by other actors. This will require a lot of coordination and sustained dialogue among diverse institutions, leaders, and communities.

Source: Seto, K. "Roundtable: Why We Need an Urban Sustainable Development Goal." *The Nature of Cities* (2015). https://www.thenatureofcities.com/2015/12/08/an-explicitly-urban-sustainable-development-goal-has-been-adopted-by-the-un-11-now-what-how-can-it-be-effective-in-the-ways-it-was-intended-and-in-the-ways-that-we-need-where-could-it-go-wrong/

projects, funded in part through the World Bank, were channeled to create urban infrastructure for an industrial economy. Development projects were centered in cities such as the building of water and sanitation systems, electrical systems, roads, factories and warehouses, port facilities, and the construction of government infrastructure such as courthouses and parliaments. Although there were many rural development projects such as dams and large highways, significant amounts of development loans and funds ended up concentrated in urban areas. The high levels of **urban primacy** in many developing countries also reinforced growth in selected urban centers. The net result was the encouragement of massive rural to urban migration. To those living in rural areas, cities were places of economic opportunity, and because many rural areas remained without water or electricity or sanitation, cities also seemed to be better places to live.

NEW URBAN FORMS

As the world has become more urban over the last century, the form and function of cities has also changed. In this section, we look at the emergence of new urban forms: megacities, slums, and postindustrial cities. These new forms have some unique challenges to achieving sustainability.

The Rise of Megacities

Throughout the world, cities have continued to grow larger. In 1800 there were only two cities—London and Beijing—that had more than one million inhabitants: by 1900, there were thirteen. In 2000, there were 370 cities that had one million or more inhabitants; today this number has increased to 550 and is projected to increase to 706 cities by 2030.

One of the more visible aspects of global urbanization has been the rise of **megacities**, large urban agglomerations with more than ten million inhabitants. They are a recent addition to the urban scene; in 1950, only New York City and Tokyo had populations of more than ten million. Today, there are thirty-three cities with at least ten million people, and it is estimated that there are another forty-eight cities with populations between five to ten million. Lagos is one example. In 1950, the Nigerian city had a population of only 320,000. By 1965, it surpassed one million and in 2002 it became the first sub-Saharan African megacity when its population topped ten million. Lagos is one of the fastest-growing cities in the world with an annual population growth rate of 9 percent. By 2030, it may have as many as twenty million residents.

Table 11.3 lists the top ten megacities, as of the most recent data. There are differences in city population totals in this table compared to UN sources because of differences in definition of metropolitan regions. Most of the megacities are located in the Global South and are where most of the growth in megacities occurs; China is already home to six megacities, while India has five (see Photo 11.2).

Slums

The term **slum** (also called shantytown, informal housing, and squatter housing) refers to unplanned, illegal, informal housing. Initially the term referred to run down parts of cities across the world, but in recent years it is commonly used to refer to informal settlements in the cities of the Global South. This type of housing is considered illegal because the occupiers hold no title to the land, do not pay taxes, and have constructed some form of shelter that does

TABLE 11.3 Ten Most Populous Megacities in 2018

Rank	City	Population in Millions
1	Tokyo	37.4
2	Delhi	28.5
3	Shanghai	25.5
4	Sao Paulo	21.6
5	Mexico City	19.1
6	Cairo	20.2
7	Mumbai	19.9
8	Beijing	19.6
9	Dhaka	19.5
10	Kyoto-Osaka	19.2

Source: United Nations. *The World's Cities in 2018* (2018). https://www.un.org/en/events/citiesday/assets/pdf/the_worlds_cities_in_2018_data_booklet.pdf

not meet building code. Because these are illegal structures, they usually lack government provisions such as sewage and sanitation infrastructure or services such as clean water. The United Nations defines a slum household as one in which the inhabitants suffer one or more of the following five deprivations:

1. Lack of access to improved water source (water is insufficient, not affordable, and difficult to obtain);
2. Lack of access to improved sanitation facilities (lack of a private toilet or public one shared with a reasonable number of people);
3. Lack of sufficient living area (more than three people sharing a room);
4. Lack of housing durability (no permanent structure that provides protection from extreme climate conditions); and
5. Lack of security of land tenure (no protection against eviction).

The most recent global survey of slums was published in 2018 by the United Nations. The survey estimated that around 880 million people lived in slums, up from 790 million in 2000. The good news is that the overall percentage of people in the world that live in slums has been declining (although there are regions where few people live in slums and regions where significant numbers of people live in slums). In many Asian and Latin American countries, between 10 to 30 percent of urban populations live in slum households. In sub-Saharan Africa, most cities have more than half of their urban populations living in slum households, and in some countries such as Sudan, South Sudan, and Central African Republic that figure is more than 90 percent.

Slums arise due to the inability of formal markets and public authorities to provide enough affordable and accessible housing. They grow in size because

PHOTO 11.2 Urbanization in China
This photo shows many high-rise buildings under construction in Shanghai,
China. Rapid urbanization has resulted in large-scale construction of buildings
and roads and has relied on excessive land conversion, inefficient urban
sprawl, and often wasteful real estate development.

of continuing rural to urban migration. In one sense, they represent endemic
poverty. For example, in Lagos, Nigeria, and Monrovia, Liberia, around 50 and
60 percent of the populations live below the poverty line, respectively. The World
Health Organization estimated that currently more than 860 million urban dwellers in developing cities have no access to clean water, sanitation, or drainage. In
areas with limited access to clean water and sanitation, the child mortality rate is
many times higher than in areas with adequate water and sanitation services. The
primary barriers to accessing water in slums are not solely monetary or technical
but also legal, institutional, and political. Those who live in slums face insecure
land tenure, exposure to hazards, and often lack political voice.

While people residing in slums face harsh living conditions, in many cases
this may represent an improvement compared to conditions in rural areas.
Millions move from the countryside to the city to improve their living conditions, and gain better access to employment, services, and the possibility of a
brighter future for their children.

There are different types of slums. One type contains buildings of permanent material and basic infrastructure and only differs from the formal
areas of the city by its illegal land occupation status. Sometimes distinctions are made between early self-help housing, which has communal
provision of services, and unauthorized housing illegally occupying land.
Another type of slum has buildings made of nonpermanent materials and
low levels of infrastructure. In the worst areas, slums feature makeshift
buildings and have no infrastructure. The worst slums typically occupy
the most hazardous sites, for example on steep slopes or areas vulnerable to flooding, landslides, and other environmental and social hazards.

Slum dwellers may be exposed to a myriad of environmental and social problems that include:

- Lack of infrastructure providing water, sewage, electricity, or trash collection;
- Disease-causing agents (pathogens) in air, food, water, or soil that impact human health and that are exacerbated by higher-density living;
- Pollutants in air, food, and water that impact human health in both the short and long term;
- Congestion on the roads and footpaths; and
- Physical hazards such as accidental fires, floods, mudslides, or landslides.

The urban population explosion of the past fifty years has forced expansion onto new spaces on more vulnerable sites such as steep hillsides, flood plains, or in areas with unstable soil conditions. Many slum structures are erected on such marginal lands because standard, legal housing long ago claimed the best, most secure land in the city. The poor are usually forced to settle on land subject to higher risks. Heavy rains often sweep away slum homes perched precariously on hillsides or flood slums located on floodplains. Although heavy rains contribute to the landslides and the flooding, the "real causes" of the destruction of slums are the inability of low-income groups to find safe living sites and the failure of government to create safer sites or to make existing sites safer.

Some slums face unique environmental issues, such as indoor air pollution. One of the major sources of indoor air pollution comes from indoor smoke, a result of cooking over open wood or dung fires. Most slums lack fans or exhaust systems, and because such structures typically have no electricity or gas, cooking takes place over open fires. The health impacts of consistent exposure to smoke and fumes have been underestimated and understudied, but burning coal, wood, or other biomass fuels can cause serious respiratory and eye problems. Research has shown that concentrations of total suspended particulates are ten to one hundred times higher in indoor dwellings. Chronic effects of exposure include inflammation of the respiratory tract, which in turn increases vulnerability to acute respiratory infections such as asthma, bronchitis, and pneumonia. Women are often heavily exposed because they spend several hours a day at the stove; infants and young children may also be heavily exposed because they remain close by their mothers. The exposure of infants and young children to indoor air pollution, combined with malnutrition, leads to a greater prevalence of long-term chronic bronchitis.

The Postindustrial City

Another important urbanization trend has affected primarily cities in the Global North: deindustrialization. Deindustrialization, which started around the mid-1970s and continues today, refers to the global shift in manufacturing leading to the rise of new industrial cities in countries such as China and the decline of industrial cities in Europe and North America. Deindustrialization resulted from the growing efficiency of industrial production that prompted massive job losses and a shift in corporate investment away from the older industrial cities of North America and Europe toward the cheaper labor in cities in the Global South.

In the last forty years, cities in the Global North have shifted from a manufacturing to a service economy. Even perennially successful cities such as New York witnessed massive industrial job loss. Between 2002 and 2010, for example, New York City lost close to 50 percent of its manufacturing jobs. However, cities such as New York with a wide and more varied job base were able to move from industrial to postindustrial relatively easier and quicker than cities with a heavy and single reliance on manufacturing. In cities such as Pittsburgh, Syracuse, Buffalo, Akron, Cleveland, and Detroit, when companies fired or relocated workers, closed factories, and moved out of the region or country, there were fewer economy and job alternatives. These cities were transformed from vibrant "industrial" manufacturing centers to ghost towns of despair, as many people moved elsewhere for jobs (see Photo 11.3). Even growth cities such as Los Angeles and San Francisco struggled to cope with the social and economic consequences of a decline in manufacturing-based employment. The loss of manufacturing employment marked a critical shift in the North American economy.

For many former industrial cities, high unemployment rates continue to impact local economies. In traditional industrial cities, political and economic leaders have tried strategies to replace lost jobs and investment by attracting other economic sectors such as services or tourism. The process of **deindustrialization** allows opportunities for urban redevelopment as factories are abandoned and new geographies of production and circulation leave old docks and railway lines economically redundant.

PHOTO 11.3 The Former Fisher Body Plant in Detroit
The plant is now shut down and covered in graffiti but was used in automotive manufacturing from 1919 until 1984. It is a good example of how deindustrialization in the Global North has left behind a polluted and unsightly mess in many cities.

THE ENVIRONMENTAL IMPACTS OF URBANIZATION

Urbanization has considerable impacts. Some of the more obvious effects on environmental change include pronounced land-use changes and increased environmental impact, especially along the frontiers of urban expansion. More specific changes include an increase in the number of urban heat islands, an increase in the number of impermeable surfaces, and more polluted runoff. Rapid and large-scale urbanization puts extra pressure on physical systems such as air, land, and water. In many cities in the Global South, rapid urban growth was and is still often associated with the creation of a more toxic urban environment.

Urban Footprints

Megacities and **metacities** (cities of more than twenty million) consume large amounts of resources. Manila and Mexico City, for example, consume vast quantities of water and are already dangerously depleting their groundwater supplies. There are also indirect consequences of megacities. As megacities grow, their peripheries enlarge, consuming agricultural land, forests, and wetlands. Dhaka, Bangladesh, currently has a population of 22 million, and is forecast to increase to 31 million people by 2035. Where will new residents go? Dhaka is bounded on the west and south by the flood plain of the Burhi Ganga River and on the east by the flood plain of the Balu River. Both areas are flooded up to four months of the year. Land above the flood plain is high-value agricultural land but is rapidly being converted to urban uses as Dhaka expands.

The city can be modeled as an ecosystem with inputs of energy and water and outputs of noise, sewerage, garbage, and air pollutants. One way to think about these relationships is to consider the ecological footprint of a city, which is defined as the amount of land required to meet the resource needs of a city and absorb its waste. US cities, which outpace most of the world in energy usage and waste production, have a larger footprint than cities of the Global South (*Key Terms and Concepts* 11.1 introduces three different urban footprints). At the same time, many dense urban areas in the United States have lower ecological footprints per capita than surrounding suburban areas because of factors such as the higher use of public transportation, resulting in fewer vehicle miles traveled per person.

Modifying the Environment

Cities also modify the environment. The most obvious example of this is the **urban heat island effect**. Cities tend to be warmer than surrounding areas because of the amount of extra heat produced in the city, the lower amounts of evapotranspiration, and the heat absorption of man-made materials such as tarmac, asphalt, and concrete. Heat is absorbed by these surfaces during the day and released at night. Human activity in the city also produces pollutants. Industrial processes and auto engines emit substances that include carbon oxides, sulfur oxides, hydrocarbons, dust, soot, and lead.

There is a large set of case studies that document some of the changes associated with rapid urbanization around the world that include land-use cover changes; ecosystem change; ecosystem fragmentation; increased resource use;

KEY TERMS AND CONCEPTS 11.1

Urban Footprints

Ecologists have developed the notion of **ecological footprint** to refer to the total area of productive land required to support an ecosystem. The ecological footprint measures how much land and water area a human population requires to produce the resources it consumes and to absorb its wastes. It is measured in global hectares (gha) per capita. The ecological footprint of an urban region includes all the land necessary to support the resource demands and waste products of a city. The global average is around 2.6 gha.

London's ecological footprint was measured at 4.54, slightly lower than the UK average of 4.64. More people in London use public transport than almost any other city in the United Kingdom, reducing the relative size of the footprint. On the opposite end is Calgary, Canada, with a calculated ecological footprint of 9.8 gha. If everyone on earth had the same ecological footprint as the average Calgary resident, we would need five Earths to maintain that level of resource consumption. Cold winters and a large city sprawl mean that many people use cars to get around Calgary. A city that is more reliant on private autos than public transport has greater energy needs and thus has a larger ecological footprint. In cities, individual and national ecological footprints combine and interact to produce distinctly urban regional effects. Calgary's ecological footprint analysis demonstrates that the city's greatest challenge is mobility and highlights the need for a multimodal transportation system and improved jobs-to-housing balance to reduce its ecological footprint.

Comparing city and national footprints and biocapacity can shed more light on potential leverage points for improving sustainability. They can be useful in:

- Helping governments track a city or region's demand on natural capital and compare this demand with natural capital available;
- Informing a broad set of policies, ranging from transportation to building codes to residential development;
- Highlighting significance of long-term infrastructure decisions, amplifying future opportunities or risks;

- Adding value to existing data sets on production, trade, and environmental performance by providing a comprehensive framework to interpret them;
- Helping understand the link between local consumption and global environmental impact; and
- Raising sustainability awareness and engagement among citizens.

Another measure is a city's **carbon footprint**, which is the total amount of greenhouse gases it produces. The basic unit is kg or metric ton of CO_2. The global average is 1.19 metric tons per person. One study measures the carbon footprint of major cities. Seoul, Guangzhou, New York, Hong Kong, and Los Angeles had the largest footprints; London, Beijing, and Jakarta had smaller footprints. The study reports that 20 percent of global emissions come from just one hundred cities. Hence, curbing the absolute carbon levels may be more achievable because the power lies in the hands of a relatively small number of local mayors and governments.

Finally, there is also a **water footprint**. This is a measure of both direct and indirect water used in an area. The water footprint can tell us how much water is being consumed from a specific river basin or from an aquifer, and for what purpose the water is being used. The footprint can calculate all the freshwater used to produce all the goods and services consumed in a city. The water footprint leads to a set of broader questions for cities such as:

- How well are regulations protecting the water resource?
- How secure is this water resource?
- How can we reduce our water footprint?

All three footprints are imperfect measures, but they constitute a start, and they have produced some interesting findings.

To learn more, visit: https://www.footprintnetwork.org/our-work/cities/.

Source: Short, J. R. "How Green Is Your City: Towards an Index of Urban Sustainability." *The Conversation* (2015, March). https://theconversation.com/how-green-is-your-city-towards-an-index-of-urban-sustainability-38402

water quality changes; increased global climate change; air quality; and public health implications.

Many analysts view urban growth as detrimental to environmental quality and there are numerous studies that confirm this finding. However, there is often an implied assumption that the shift from rural to urban involves a downward slide in environmental quality. But this argument ignores the oft-disastrous effects of agriculture, especially traditional agricultural techniques such as slash and burn. The direct connection between urbanization and environmental deterioration is not a given; a country's early stages of rapid urban growth can have an enormous effect, but this effect can lessen as the experience of rapid urbanization awakens environmental sensitivity among a population. The growing debates on urban sustainability arose not just from a realization of rapid urbanization's deleterious impacts on the environment but also from a reconceptualization of urbanization to be more in line with ecological realities and long-term sustainability. Cities create parks and ecological reserves as well as parking lots and factories. Rapid and unplanned urban growth can overwhelm and destroy ecosystems, but sensitive and sustainable urbanization can protect and nurture.

There are both direct and indirect environmental consequences attributed to megacities and their magnitude of size. For example, because megacities are so large, the volume of pollutants is very high, which puts millions of residents at risk. Many megacities are located on the coast, such as Tokyo, Shanghai, Jakarta, Manila, Mumbai, Karachi, Istanbul, New York, Buenos Aires, Rio de Janeiro, and Lagos. Coastal megacities face special concerns such as use conflicts in coastal areas, coastal erosion, salt water intrusion, freshwater shortage, and the depletion of fishery resources. Rapid demographic and economic growth in Jakarta has created water pollution, coastal erosion, mangrove destruction, and the intrusion of salt water into freshwater supply. In Mumbai, high-polluting industries like chemicals, fertilizers, iron and steel, and petrochemicals have released semitreated or untreated waste material into the coastal waters, degrading the beaches and impacting the tourism industry. In Buenos Aires, untreated sewerage has been dumped into the River Plate and ultimately into the seas, creating serious coastal environmental problems.

Air Pollution

Air pollution can be particularly severe in megacities. Motor vehicle traffic is a significant source of air pollution in all megacities, and in nearly half the world's megacities, it is the single most important source. Emissions for power generation are also a problem. In China, Beijing and Shanghai are dealing with high levels of sulfur pollution that stems from the use of coal as a major energy source. China is now home to sixteen of the world's twenty most air-polluted cities and demand for fossil fuels (particularly coal) has only increased. However, on a positive note, some megacities have made vast improvement in air pollution in both the developed and developing world. Despite rapid growth, Mexico City has improved its air quality. In 1992, air pollution was severe, responsible for 1,000 deaths per year and thirty-five times as many hospitalizations. Since then, Mexico City has promoted public transport such as a free bicycle loan program, relocated polluting industries, and improved automobile exhaust systems. The results are dramatic. Lead in the air has been reduced by 90 percent, ozone levels have dropped by 75 percent, and suspended particulate matter has been reduced by almost 70 percent.

Environmental Hazards

Cities are places where the threat of hazards and the prospect of disaster are always a possibility. We are reminded of this whenever major disasters are televised and searing images are broadcast around the world: the devastation of Puerto Rico after Hurricane Maria in 2017 and again after Hurricane Fiona in 2022, the destruction of Fort Meyers due to Hurricane Ian in 2022, the 2018 wildfires that scorched the hills of Los Angeles, or the tsunami that battered Japan's Pacific coast in 2011.

Environmental hazards include floods, windstorms, landslides, heat waves and biting cold, earthquakes, and volcanic eruptions. There are also social hazards such as fire, nuclear plant meltdown, and infrastructure failure. The distinction between the two can be confusing because an environmental hazard can create a social hazard. In the case of the 2011 tsunami in Japan, the environmental hazard of an earthquake-induced tsunami triggered the spread of nuclear radiation fallout, a social hazard.

Urban environmental disasters show, in the starkest terms, the vulnerability of cities. For example, flooding, caused by increased rainfall due to climate change, disproportionately impacting slums, is one of the many connections between environmental issues, such as global warming and social justice. It is estimated that more than two billion people were affected by disasters between 2010 and 2020.

The Social Construction of Hazards: A Political Ecology Perspective

A political economy or political ecology framework is valuable for showing how vulnerability to hazards is also socially constructed. Poorly planned urban growth, deforestation, and poor medical provision are just some of the factors that increase the chances that hazards become disasters. Many cities, and particularly those in poverty in these cities, are vulnerable to environmental hazards that can turn into disasters. Socioeconomic status, wealth, and power all play an important role in how different people experience hazards.

Consider Hurricane Katrina, which hit New Orleans in 2005. It was not the ferocious winds that damaged the city but the storm surge that breached levees in the city. The city was flooded when parts of levees at 17th Street and Industrial Canal collapsed. Almost 80 percent of the city was flooded, in some cases by water more than twenty feet in depth. An estimated one thousand people were killed, most of them drowned by the rapidly rising floodwaters. Much of the city was destroyed in the flooding that followed the hurricane. Investigations showed that the city and state had failed to invest in upgrades to the levee system; poor design by the Army Corps of Engineers also played a factor in the flooding.

Sociologists Chester Hartman and Greg Squires showed that the impacts of Hurricane Katrina on the city of New Orleans were not a "natural disaster," but the consequences of decisions about infrastructure, urban growth patterns, and a long history of racial discrimination and injustice. A hurricane is a force of nature. But it was a force of nature whose impacts and effects were mediated through the prism of socioeconomic power structures and arrangements. The flooding of the city was caused by the poorly designed levees that could not withstand a predictable storm surge. It was not Katrina that caused the flooding but shoddy engineering, poor design, and inadequate funding of vital public works. Storm surges are neither unknown nor unpredictable in New Orleans. Yet the levees were poorly constructed with pilings set in unstable soils.

TABLE 11.4 SDG 11 Targets and Indicators to Achieve by 2030

Target	Indicators
11.1 Ensure access for all to adequate, safe, and affordable housing and basic services and upgrade slums	**11.1.1** Proportion of urban population living in slums, informal settlements, or inadequate housing
11.2 Provide access to safe, affordable, accessible, and sustainable transport systems for all, improving road safety, notably by expanding public transport, with special attention to the needs of those in vulnerable situations, women, children, persons with disabilities, and older persons	**11.2.1** Proportion of population that has convenient access to public transport, by sex, age, and persons with disabilities
11.3 Enhance inclusive and sustainable urbanization and capacity for participatory, integrated, and sustainable human settlement planning and management in all countries	**11.3.1** Ratio of land consumption rate to population growth rate **11.3.2** Proportion of cities with a direct participation structure of civil society in urban planning and management that operate regularly and democratically
11.4 Strengthen efforts to protect and safeguard the world's cultural and natural heritage	**11.4.1** Total expenditure (public and private) per capita spent on the preservation, protection and conservation of all cultural and natural heritage, by type of heritage (cultural, natural, mixed, and World Heritage Centre designation), level of government (national, regional, and local/municipal), type of expenditure (operating expenditure/investment), and type of private funding (donations in kind, private nonprofit sector, and sponsorship)
11.5 Significantly reduce the number of deaths and the number of people affected and substantially decrease the direct economic losses relative to global GDP caused by disasters, including water-related disasters, with a focus on protecting the poor and people in vulnerable situations	**11.5.1** Number of deaths, missing persons, and persons affected by disaster per 100,000 people **11.5.2** Direct disaster economic loss in relation to global GDP, including disaster damage to critical infrastructure and disruption of basic services
11.6 Reduce the adverse per capita environmental impact of cities, including by paying special attention to air quality and municipal and other waste management	**11.6.1** Proportion of urban solid waste regularly collected and with adequate final discharge out of total urban solid waste generated, by cities **11.6.2** Annual mean levels of fine particulate matter (e.g., PM2.5 and PM10) in cities (population weighted)
11.7 Provide universal access to safe, inclusive and accessible, green and public spaces, in particular for women and children, older persons, and persons with disabilities	**11.7.1** Average share of the built-up area of cities that is open space for public use for all, by sex, age, and persons with disabilities **11.7.2** Proportion of persons victim of physical or sexual harassment, by sex, age, disability status, and place of occurrence, in the previous twelve months
11.A Support positive economic, social, and environmental links between urban, preurban, and rural areas by strengthening national and regional development planning	**11.A.1** Proportion of population living in cities that implement urban and regional development plans integrating population projections and resource needs, by size of city

Target	Indicators
11.B Substantially increase the number of cities and human settlements adopting and implementing integrated policies and plans toward inclusion, resource efficiency, mitigation and adaptation to climate change, resilience to disasters, and develop and implement, in line with the Sendai Framework for Disaster Risk Reduction 2015–2030, holistic disaster risk management at all levels	**11.B.1** Proportion of local governments that adopt and implement local disaster risk-reduction strategies in line with the Sendai Framework for Disaster Risk Reduction 2015–2030. **11.B.2** Number of countries with national and local disaster risk-reduction strategies
11.C Support least developed countries, including through financial and technical assistance, in building sustainable and resilient buildings utilizing local materials	**11.C.1** Proportion of financial support to the least developed countries that is allocated to the construction and retrofitting of sustainable, resilient, and resource-efficient buildings utilizing local materials

Source: United Nations Sustainable Development Goals. https://sustainabledevelopment.un.org/sdg11

In addition to the poorly designed levees, the city's glaring racial and economic inequity was made visible. Those with cars were able to evacuate but there was little provision for the most vulnerable; those with no access to private transport were abandoned. While the more affluent could leave, the very poorest, the most disabled, the elderly, and infirm were trapped. They were the ones that perished in the aftermath of the hurricane.

The effects of Hurricane Katrina on the city were socially and racially determined. Flooding disproportionately affected the poorest neighborhoods of the city. The more affluent, predominantly White sections of the city, such as the French Quarter and the Garden District, were at a higher elevation and escaped flood damage. The flooded areas were 80 percent non-White. The hardest hit neighborhoods were non-White and most of the high-poverty tracts were flooded. The racial and income disparities in the city were cruelly reflected in the pattern of flood damage. A political economy perspective can highlight how an environmental disaster appears in closer detail as a social disaster, and New Orleans is but one example of many.

SOLUTIONS FOR MORE SUSTAINABLE CITIES

The adoption of the stand-alone urban goal—SDG 11—to make cities safe, inclusive, resilient, and sustainable firmly places urbanization at the forefront of international development policy. This recognition goes beyond viewing urbanization simply as a demographic phenomenon, but as a transformative process capable of galvanizing momentum for many aspects of global development. Many consider the agreement on a stand-alone goal in SDGs on cities was monumental, reflecting the increased attention on "urban" as a development theme. Table 11.4 summarizes the goals and targets for SDG 11.

There has been a slowly growing focus on cities and the recognition of their integral role in sustainability. However, moving toward sustainability is not simply "greening" the city. A genuinely sustainable city is a place where the environment is protected, the economy can sufficiently provide its residents with basic needs such as food and shelter, and the residents who live there have opportunities to live good lives. A sustainable city is also a just city.

There are several key solutions cities can implement to advance sustainability, including creating and implementing sustainability plans, upgrading slums, redeveloping brownfields, and planning for resilience.

Create Urban Sustainability Plans

The 1992 UN Rio Earth Summit, which highlighted the importance of local planning and efforts to increase the sustainability of development, was a key turning point. Cities and other local governments were each called upon to create a "Local Agenda 21" strategy for sustainability through a community participatory process that would prioritize action. These plans included vision statements that identified main sustainability issues, an action plan that included goals and actions, timelines to achieve those goals, and a discussion of how success would be measured. This initiative was the progenitor of modern sustainability plans.

The city scale can be a beneficial starting point for local activism and community involvement around sustainability. Local policy is important for many reasons, not least that cities have tremendous control over significant tools for change such as land use, public education, transportation, and economic development. Cities also already undertake a myriad of planning exercises. One of the biggest developments in the last ten to fifteen years is the profusion of cities around the world finding meaningful ways to address environmental, economic, and social equity within their cities and developing sustainability plans.

Municipal sustainability plans are comprehensive visions, goals, and priorities for sustainability set forth by a government or other civic organization. Such plans cover a diverse range of issues including but not limited to climate, energy, transportation, green jobs, housing, human health, recreation, and parks. These plans often inventory current problems and standings, identify solutions and priorities, and set indicators for measuring progress. Sustainability plans are more holistic than most other planning documents and consider multiple goals relating to improving environmental, economic, and social equity conditions simultaneously. Many plans are comprehensive in that they address a range and diversity of issues that include the "three Es." Among the most common issues in sustainability plans are:

- Food
- Air quality
- Climate change
- Water quality
- Water supply
- Parks and recreation
- Social justice and equity
- Green economy/green jobs
- Transportation
- Energy use
- Housing
- Health
- Waste, garbage, and recycling
- Risk and resilience to hazards

CRITICAL PERSPECTIVES 11.1

China's Failed Eco-Cities

Getting to urban sustainability is not going to be easy. Since 1980, China's urban population has quadrupled to about 800 million, and its cities have tripled in number to 657. Over the next twenty years, some predict 250 million people will move to one of China's cities. China has built hundreds of new cities and urban districts; hundreds more are set to be built by 2030.

Yet China's astonishing pace of urbanization has had catastrophic environmental consequences. Many of its cities have air pollution far above World Health Organization standards, newsworthy traffic jams, exposure to toxins and water pollution, and poorly constructed housing.

In response, China has pledged to construct 285 "eco-cities." The concept of an "eco-city" was coined by American urbanist Richard Register to describe an ecologically healthy city. Register argues that eco-cities should:

- Be designed from scratch to be compact;
- Be designed for living beings;
- Fit the bioregion and heal the biosphere;
- Reduce energy consumption;
- Promote social equity, community, and health;
- Prioritize pedestrians and bicycles; and
- Contribute to the economy.

In China, one much anticipated eco-city plan was Dongtan, designed in 2005. The plan called for the eco-city to house 10,000 residents by 2010 and reach 500,000 by 2050. Dongtan was to be located at the east end of Chongming Island in the mouth of the Yangtze River north of Shanghai. It was a joint project of Arup (a British transnational engineering and design firm), the Shanghai Industrial Investment Company, a Shanghai municipal government public-private pharmaceutical and real estate company, as well as Chinese and British state agencies, universities, and planning institutions.

The intent was to create an ecologically, socially, and economically self-sufficient city. Dongtan would have a 60 percent smaller ecological footprint, require 66 percent less energy, use 40 percent energy use from bio-energy, reduce landfill waste by 80 percent, and emit almost no carbon emissions. Its main source of energy would be electricity generated by burning rice husks and from solar panels and wind turbines. It would be car-free and recycle its water. Dongtan caught the attention of the media who hailed this as a "great green leap forward" for China.

Sadly, Dongtan was postponed indefinitely due to a corruption scandal and a feud over who would ultimately fund the project. It is now considered a failed project and joins several other failed eco-cities such as Huangbaiyu and Mentougou.

These failed or stalled projects have prompted a critique of China's planned eco-cities. Some scholars have derided eco-cities as a fantasy that exists only as drawings, a "greenwashing." They say that China never intended to build the project, noting that China tends to do everything by the rules handed down from the top, and that the plan for Dongtan challenged those rules. Still others note that the concept of eco-cities or sustainable communities is so loosely defined that no one really knows that the term means. There are no formal standards as to what an eco-city is or is not and there is no certification criteria. This may be why China claims 42 percent of all its cities are eco-cities, which Austin Williams calls "patent nonsense." Williams argues that eco-cities could only happen in China, a country with large areas of clear space (or the political will to create large areas of open space), a compliant population, an authoritarian regime, vast amounts of money, national pride, and a competitive spirit. He concludes that China's eco-cities are merely much-needed urban improvements and infrastructural development with an "eco-" prefix.

The bold visions are still on the drawing board: The dream of a new role model eco-city has yet to be realized.

Sources: Register, R. *Eco-Cities: Building Cities in Balance with Nature.* Gabriola, BC, Canada: New Society Publishers (2006).

Sze, J. *Fantasy Islands: Chinese Dreams and Ecological Fears in an Age of Climate Crisis.* Oakland: University of California Press (2015).

Williams, A. *China's Urban Revolution: Understanding Chinese Eco-Cities.* London: Bloomsbury (2017).

A sustainability plan is meant to provide both short and long-term guidance for current and future decision makers, city employees, city leaders, city residents, and other community groups and entities. They are a starting point for change. Visions, even general ones, can be powerful. Consider the vision statement in Washington, DC's sustainability plan:

> In just one generation—20 years—the District of Columbia will be the healthiest, greenest, and most livable city in the United States. An international destination for people and investment, the District will be a model of innovative policies and practices that improve quality of life and economic opportunity. We will demonstrate how enhancing our natural and built environments, investing in a diverse clean economy, and reducing disparities among residents can create an educated, equitable, and prosperous society.

Sustainability will require a substantial increase in the number of cities and human settlements adopting and implementing integrated policies and plans toward inclusion, resource efficiency, mitigation and adaptation to climate change, and resilience to disasters. Urban sustainability planning is now a growing and important trend. Importantly, sustainability planning should be inclusive, involving input from various municipal agencies, community organizations and nonprofit groups, residents, and other stakeholders. But planning for sustainability is only the first step: implementation is critical, and some cities have failed to realize their plans. The *Critical Perspectives* 11.1 box discusses China's failed eco-city plans.

European cities are leaders in sustainability planning, having embraced the UN Local Agenda 21 in the 1990s (which called for local-government-led sustainability efforts to affect global change) with EU support of these efforts. Paris, Freiburg, Helsinki, Oslo, and London are pioneers of sustainability planning and have spearheaded many urban innovations such as bikeable cities, car sharing, and climate action plans. Malmö and Copenhagen often rank among the top of many "green cities" lists. In the United States, nearly every major city now has a sustainability plan. Chicago, for example, stands out for leading the United States in green roofs, an outcome of their sustainability plan (see *Solutions* 11.1).

Much of the discussion of cities in the Global South tends to emphasize "brown" issues such as poor sanitation, water quality, air pollution, and housing problems. However, all is not doom and gloom. Sustainability planning has not been limited to the Global North. Many cities in the Global South including Singapore, Seoul, Bangkok, Rio, and Mexico City have all recently developed sustainability plans. Photo 11.4 shows a bikeshare station in Buenos Aires.

One of the most often-cited green cities in the Global South is Curitiba, Brazil, located near the coastal mountain range in the southern part of the country. Like many of Brazil's cities, it developed rapidly in the second half of the twentieth century, growing from 500,000 inhabitants in 1965 to 2.2 million in 2020. Such growth brought the typical urban problems of unemployment, slums, automotive gridlock, pollution, and environmental deterioration. Yet despite this, Curitiba is often cited as a model "Green City."

In the late 1960s and 1970s, Curitiba's political elites, led by its three-term mayor, Jaime Lerner, encouraged urban planners to think imaginatively, integrating social and ecological concerns. The planning process created

⊜ SOLUTIONS 11.1

Green Roofs in Chicago

Chicago mayor Richard M. Daley vowed to make Chicago "the greenest city in America." One of his earliest efforts was in 2001 when the city planted a lush garden on the rooftop of City Hall to combat the urban heat island effect and to improve urban air quality. At the time, it was the first municipal rooftop garden in the country. By 2016, Chicago led the United States with more than 350 green roofs covering 5.5 million square feet. Today, there are over 600 vegetated rooftop gardens.

Green roofs are rooftops that are partially or completely covered with vegetation and soil, or a growing medium, planted over a waterproof membrane. Green roofs are more than just a patch of green on the skyline. They can reduce carbon dioxide, reduce summer air conditioning needs, reduce winter heat demands, reduce stormwater runoff, provide songbird habitats, and reduce the heat island effect.

Green roofs can particularly help combat the heat island effect. Traditional building materials soak up the sun's radiation and reflect it back as heat, making cities at least 7 degrees Fahrenheit hotter than surrounding areas. By contrast, buildings with green roofs can be many degrees cooler than traditionally roofed buildings. This is critically important in Chicago, where summer heat waves exacerbated by the heat island effect can have disas-

trous impacts. In 1995, a five-day heat wave in Chicago caused 739 heat-related deaths. Most of the victims of the heat wave were elderly low-income residents of the city, who could not afford air conditioning and did not open windows or sleep outside for fear of crime.

The most notable green roofs in Chicago include:

- *City Hall*—This was the first green roof in Chicago and the most publicized. It has 150 plant species and covers more than 20,000 square feet. Additionally, City Hall has beehives that produce approximately 200 pounds of honey per year. Since the installation, Chicago gas saved about $25,000 in energy costs.
- *Millennium Park*—This covers a rail station, two parking garages, and the Harris Theater. It is arguably the largest green roof in the world covering 24.5 acres.
- *McCormick Place West*—At 20,000 acres, the McCormick Place West was created by the Chicago Botanic Garden. The largest "farm to fork" rooftop garden contains vegetables and herbs that yield more than 4,000 pounds of fresh produce rising to 12,000 pounds.

Source: Bertoni, V. "The Rise of 'Green Roofs' in Chicago." *US Green Technology* (2016). https://usgreentechnology.com/chicago-green-roofs/

innovative solutions in recycling, garbage collection, and green space expansion, many of them models of sustainable urban planning. But it is Curitiba's public transportation system that is the most well-known result.

Curitiba was faced with an inefficient public transportation system and an increase in the number of private automobiles. Initially, planners leaned toward the development of a subway system, at the cost of some $60 to $70 million per kilometer. Instead, they turned to modifying the bus system, at $200,000 per kilometer, or 1 percent of the cost of the subway. To meet the growing needs for transportation and to curtail the use of private automobiles, planners focused on encouraging Curitiba's physical expansion along linear axes, each with a central road that has a dedicated lane for express buses. This is what many cities now call **Bus Rapid Transit (BRT)** and it has become the model for the rest of the world.

PHOTO 11.4 A Bikeshare in Buenos Aires, Argentina
Many cities in have installed bikeshares, scooters, and other forms of low-cost, "green" transportation for both residents and tourists. This is a bikeshare in Buenos Aires.

The aim of BRT is to reduce congestion while returning the central area to the pedestrian. The use of express buses was far cheaper than subways or light railways and highlights a more practical and affordable solution to public transportation in the developing world. The city was able to keep public transportation affordable; on average, residents spend only 10 percent of their income on transport. In addition, modifications to the buses and boarding tubes have made the system very efficient. Although the city has more than 500,000 private cars (more cars per capita than most Brazilian cities), most residents use the bus system. In fact, the public transportation system is used by around 70 percent of the commuters—more than 1.3 million passengers each day.

Curitiba's long-standing innovations in transportation may be one model for cities looking to achieve Target 11.2—to provide access to safe, affordable, accessible, and sustainable transport systems for all. As a result of this and other sustainability projects, Curitiba consistently ranks among the world's ten "greenest cities."

Upgrade and Integrate Slums

Sustainability plans are just one strategy cities can use to make progress toward sustainability. Cities are also building on policies and plans that have been in place such as **slum upgrading** programs. This is critical because predictions are that the number of people living in slums will continue to increase (Map 11.3). Historically, governments tended to respond to slums by ignoring them, eradicating them, or relocating their residents. However, eradication, eviction, and involuntary relocation are insensitive to the loss of community. Often, slum dwellers have made both financial and social investment in their homes and communities, something that would be lost in an attempt to clear slum neighborhoods. This is one reason why slum upgrading gained prominence as a valid and cost-effective way to improve the living conditions of cities. This is reflected in SDG Target 11.1 that seeks to upgrade slums *in situ*, so they are safe and have basic services. Slum upgrading *in situ* is different from the traditional approach to upgrading slums that consists of razing and reconstructing without residents' consent of participation. Slum upgrading *in situ* refers to improvements to housing and/or basic infrastructure in slum areas. It can cover a wide range of possible interventions that can include:

- Installation of basic infrastructure such as water, sanitation, waste, roads, storm drainage, and electricity;
- Regularization or recognition of land tenure;
- Housing improvement;
- Construction or rehabilitation of community facilities such as nurseries, health facilities, and open spaces;

- Improved access to health care, education, and other social support services;
- Removal or mitigation of environmental hazards; and
- Relocation of and compensation for any residents dislocated by improvements.

Physical upgrades of slums have proven to make positive social and economic changes in many cities. Socially, upgraded slums improve physical living conditions, improve the general well-being of communities, strengthen local social and cultural capital networks, generate livelihood opportunities, improve quality of life, and increase access to services and opportunities. In many instances, upgrading slums started the processes of improving tenure conditions and security, a vital factor that must also be included. Economically, upgraded slums trigger local economic development, improve urban mobility and connectivity, and integrate an economically productive sphere into the physical and socioeconomic fabric of the wider city.

There are some success stories in slum upgrade programs, including Rio's Favela Bairro Program, housing improvements in Thailand's Baan Mankong, and Mumbai's community toilets program. In Indonesia's city of Surabaya, the Kampung slum area has been considered a best practice. Surabaya was able to raise outside funding from development organizations, including the World Bank, to provide basic infrastructure such as gutters, paved footpaths, stormwater drainage, public toilets, waste management, and primary schools. On top of physical interventions, the city empowered individuals to participate in planning processes and upgrading their homes. Communities typically contributed between one-third and one-half of the upgrade costs. Residents also helped with the ongoing operation and management of projects like piped water and roads, which created a greater sense of

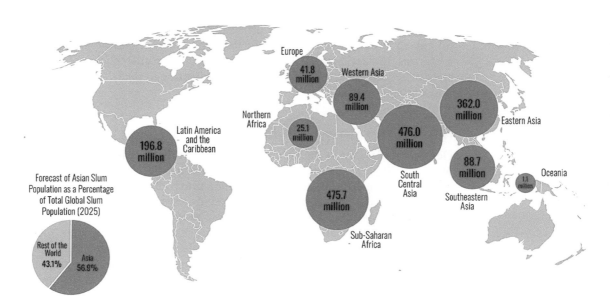

MAP 11.3 Projected Global Slum Population (2025)
The projected number of people living in slums will continue to grow significantly. Nearly half a billion people in sub-Saharan Africa and nearly half a billion combined in South Asia and Central Asia are expected to live in slums.

neighborhood ownership and agency. This upgrade project has not solved all the issues, but it has improved access to basic services and created a healthier local environment, while respecting personal and cultural preferences of the community.

However, slum upgrade programs face considerable challenges. First, mayors and local authorities will need to acknowledge the presence of slums and have the political will to improve them. Too often political promises for upgrading are made just before an election and never materialize. Second, cities must mobilize financial resources to design and implement upgrades; this will usually entail coordinating with national governments for resources. Third, cities must have capacity to plan and implement said plans at institutional and technical levels. Fourth, some studies have suggested that slum upgrades depend on participation of slum residents, and this is most effective when initiated at the neighborhood level through individual or community projects that are relatively limited in scale. The participation of slum dwellers and community organizations is critical to slum upgrade programs—this must be a bottom-up process.

More recently, urban planners have begun to use the term **slum integration**. Whereas upgrading is about improving access to basic urban services, slum integration is a social goal that embodies the idea of participation, inclusion, equality, agency, and sustainability. Slum upgrade programs by themselves do not necessarily address wider issues of housing affordability and land tenure. The lack of affordable housing is a main driver of high rates of illegal construction, and the creation of slums in the first place. Cities will need to shift housing policy, urban planning, and building practices. According to UN-Habitat, housing policies need to be closely harmonized with other development aspects such as economic, social, and environmental interests. For instance, beyond the mere provision of shelter, housing projects have to be understood as playing an active role in boosting employment and the economy, reducing poverty and improving human development. Likewise, housing policies have to include urban planning considerations, advocating for mixed urban uses and medium to high density, ensuring small urban footprints and rationalized mobility patterns. Unfortunately, few cities have approached the issue of land tenure (e.g., giving title to the land on which the house/structure sits).

Redevelop Brownfields

Deindustrialization in the Global North has left many cities with the legacy of a toxic environment. Industrial manufacturing processes in urban areas commonly generate hazardous waste. For much of the twentieth century, hazardous wastes were largely unregulated. A new term emerged to describe these abandoned but contaminated areas: **brownfields**. Although brownfields can be found anywhere, many are concentrated in cities that hosted manufacturing activities. A widely accepted definition of brownfields, provided by the US Environmental Protection Agency (EPA), is "real property, the expansion, redevelopment, or reuse of which may be complicated by the presence or potential presence of a hazardous substance, pollutant, or contaminant."

In the United States, there are an estimated 450,000 brownfields; in Germany, there may be as many as 362,000 sites covering about 317,000 acres in France there are about 200,000; in the United Kingdom some 105,000 sites; and, in Belgium some 50,000 sites covering at least 22,000 acres. Environmental contamination not only has an adverse effect on the environment but is also considered a significant barrier to the economic and social redevelopment of the city. In addition, there is growing evidence to suggest that many urban brownfields are situated in areas with higher concentrations of minority populations and households below the poverty level, raising important questions about environmental justice.

Where there is a functioning land market, brownfields are redeveloped in a variety of ways: for industrial reuse, for commercial or residential uses, and as green spaces such as parks, playgrounds, trails, and greenways. The projects can be modest—the reuse of a single isolated property—or more ambitious such as the revitalization of an entire distressed neighborhood. On the positive side, brownfield redevelopment is a form of land recycling that can restore and regenerate formerly derelict and toxic urban spaces. Brownfield initiatives have been integrated with community economic redevelopment and job creation. The US EPA estimates that each dollar spent on the program leverages another $18 and in total in the United States generates more than 76,000 new jobs. The program also improves health and safety issues, a vital part neighborhood restoration. On the negative side, the process can be regressive, in that low-income communities, affected worst by the pollution, do not benefit the most.

The processes of deindustrialization left many cities with abandoned warehouses and buildings and unused port facilities on their waterfronts. Cities in the United States, Canada, Europe, Australia, New Zealand, and Japan have transformed their once abandoned or polluted waterfronts into vibrant, public spaces that attract locals and tourists. The waterfronts of Boston, Pittsburgh, Toronto, and Vancouver have become the new festival spaces filled with sports stadiums, restaurants, and hotels. Even smaller cities such as Austin, Buffalo, Charleston, Cleveland, Ottawa, Savannah, Syracuse, and Victoria have transformed their harbors, lakes, or riverfronts. Similar trends are evident throughout the world. In London, the Docklands were transformed into a vast complex of multiuse spaces that include office buildings, shops, museums, and residences.

Such large-scale development is not without its costs and controversies. The reconstruction of Baltimore's Inner Harbor cost $2.9 billion, but there were also social costs. The diversion of funds to Baltimore's Inner Harbor contrasts with the city's underfunded public school system and the perceived decline in many public services. In some cases, waterfront development is part of a valorization of selected parts of the urban landscape that allow for the further enrichment of real estate interests sometimes at the expense of social welfare programs. While Baltimore's Inner Harbor flourishes, many inner-city neighborhoods continue to experience high crime rates, population loss, and housing abandonment. Balancing economic redevelopment and social equity is a complex and difficult challenge for many former industrial cities.

The city of Detroit has an example where brownfield redevelopment has been centered on issues of social equity. In 2021, the Detroit Black Community Food Security Network teamed with several other organizations to propose redeveloping 1.42 acres of land into a mixed-use commercial and retail development. The new development will house the Detroit Food Commons, a community-owned store, selling health and locally sourced foods. This redevelopment project aspires to help build wealth in a low-income neighborhood, improve local food security, and reinvigorate investment in an underserved Detroit community.

Plan for Resilience

Reducing vulnerability to the impact of disasters is connected to building and enhancing resilience. The term "resilient city" was coined by urban planners Lawrence Vale and Thomas Campanella, who write about the ability of cities to survive disasters and employ a model of recovery consisting of four stages:

1. *Emergency responses*: may last from days to weeks and involves rescue; normal activities cease.

2. *Restoration*: lasts two to twenty weeks; involves the reestablishment of major urban services and return of evacuees/refugees.

3. *Replacement and reconstruction*: lasts 10 to 200 weeks; the city is returned to predisaster levels.

4. *Development reconstruction*: lasts 100 to 500 weeks; commemoration and betterment.

In addition to developing effective response and recovery strategies, government must be involved in building resiliency, not just locally, but also nationally. One cost-effective prevention strategy is providing disaster-response kits in the existing infrastructure, such as schools and public buildings. In addition, cities and national governments must look critically at public policies that tend to highlight rescue and repair instead of prevention. Public policies, unless carefully monitored, can also encourage risky behaviors. In the richer countries, government-backed insurance policies for those in flood or disaster zones can perversely mean public underwriting of individual risky behaviors. In the United States specifically, such polices have resulted in taxpayer bailouts of affluent households building beach houses in hurricane zones. Resilience in this case leads to public underwriting of risky private behaviors.

Urban resiliency is often improved when biophysical systems are employed as risk reducers. Urban planners and landscape architects now realize the value of wetlands to buffer hurricanes and the importance of flood plains. In the Netherlands, rivers are now allowed to meander across their flood plain rather than being severely channelized. Regular small-scale flooding across the river flood plain is considered a more acceptable risk than the severe flood if the river breaks the banks. The policy can be roughly translated from the Dutch as: Better damp feet than a wet head. There is now a growing recognition that the physical world can be used to protect cities.

SUMMARY AND PROGRESS

Cities around the world are rapidly changing. Climate change, growing populations, demographic shifts, and other factors have resulted in challenges to the status quo and provide opportunities for innovative problem solving. Change at the city level will be essential for achieving sustainability. However,

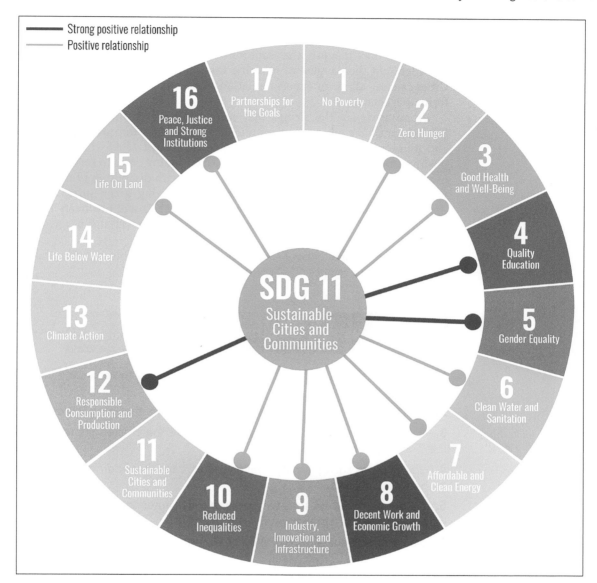

FIGURE 11.1 Interconnections in SDG 11 and the Other Goals
Moving toward sustainable cities may also advance targets in many other SDGs. This figure highlights several of the stronger connections.

there are many challenges. While European cities are making progress, a 2020 *US Cities Sustainable Development Report* found that no US city is on track to achieve the SDGs and that they lag behind most European cities. In many cities in the Global South, there has been less effective planning for sustainable development at the local level. The next few years are crucial for cities if they are to achieve the targets of SDG 11. In countries where a significant portion of the population lives in metro areas, progress in cities will be essential for the country to achieve the SDGs.

In many ways, achieving the overall goals of sustainable cities and communities will require implementation of the other sixteen SDGs. This is because the urban level is just one scale at which sustainability must be implemented. Cities are microcosms for the challenges for implementations in general; they magnify the importance of taking an integrated approach and highlight the dangers of ignoring the interlinkages among SDG issues. Cities can also allow for experimentation and act as testing grounds for transformative change.

Urban areas are the strings that connect all SDGs; more than half the SDG targets have an urban component. For example, improving air quality and municipal and other waste management (Target 11.6) will have direct positive impacts on human health (SDG 3); designing cities to serve as places of refuge for migrants and empowering people with disabilities (Target 11.1) will reduce inequalities (SDG 10); and incentivizing green infrastructure projects over unsustainable alternatives (Target 11.7) will positively impact SDG 15. Figure 11.1 highlights those goals with the strongest synergies with SDG 11.

No doubt, the challenge of sustainable cities is formidable, but there can be no sustainable development without sustainable cities.

QUESTIONS FOR DISCUSSION AND ACTIVITIES

1. Describe trends in urbanization in the Global North versus the Global South.

2. Describe the types of sustainability challenges that megacities face.

3. Explain the reasons for slums and describe the challenges to each of the Three E's of sustainability.

4. Explain the sustainability challenges as a result of deindustrialization in the Global North.

5. What are some reasons why people may prefer to live in cities rather than rural areas or small towns?

6. Are suburbs sustainable? Why or why not?

7. How might investment in rural development reduce the growth of slums? Describe the types of actions for rural development that would reduce rural-to-urban migration.

8. Research the sustainability plan for the city or community where you live. Describe some of the priority targets in that plan. Look to see if there has been a progress report and note the progress made (or not).

9. LEED (Leadership in Energy and Environmental Design) (https://www. usgbc.org/leed/rating-systems/leed-for-cities) has resources that include best practices and solutions, often connected to other SDGs such as energy, climate change, and health. Select a city to learn more about their LEED project and describe what other SDGs this project may positively impact.

10. Investigate a best practice or project facilitated by the C40 Cities on Climate Change here: https://www.c40.org/cities/

TERMS

brownfield
bus rapid transit (BRT)
carbon footprint
deindustrialization
ecological footprint
environmental hazard
megacity
metacity
pull factor

push factor
slum
slum integration
slum upgrading
urban heat island
urban primacy
urbanization
water footprint

RESOURCES USED AND SUGGESTED READINGS

Benton-Short, L., and Short, J. R. *Cities and Nature*. 2nd ed. New York: Routledge (2013).

Boone, C., and Moddares, A. *City and Environment*. Philadelphia: Temple University Press (2006).

Caat, N., Graamans, L., Tenpierik, M., and Dobbelsteen, A. "Towards Fossil Free Cities—A Supermarket, Greenhouse and Dwelling Integrated Energy System as an Alternative to District Heating: Amsterdam Case Study." *Energies* (Basel) 14 (2):347 (2021) .

Gandy, M. *Concrete and Clay: Reworking Nature in New York City*. Cambridge, MA: MIT Press (2002).

Hartman, C., and Squires, G. (eds.). *There Is No Such Thing as a Natural Disaster*. New York: Routledge (2006).

Lynch, A., LoPresti, A., and Fox, C. *The 2019 US Cities Sustainable Development Report*. New York: Sustainable Development Solutions Network (2019).

Rosenzweig, C., Solecki, W. D., Romero-Lankao, P., Mehrotra, S., and Dhakal, S. (eds.). *Climate Change and Cities, Second Assessment Report of the Urban Climate Change Research Network*. Cambridge: Cambridge University Press (2018).

Rothstein, R. *The Color of Law: A Forgotten History of How Our Government Segregated America*. New York: W. W. Norton & Company, 2017.

Short, J. *The Unequal City: Urban Resurgence, Displacement and the Making of Inequality in Global Cities*. New York: Routledge (2018).

Turley, R., Saith, R., Bhan, N., Rehfuess, E., and Carter, B. "Slum Upgrading Strategies Involving Physical Environment and Infrastructure Interventions and Their Effects on Health and Socio-economic Outcomes." *Cochrane Database of Systematic Reviews* 1: Art. No.: CD010067 (2013). https://doinorg https://doi.org/10.1002/14651858.CD010067.pub2

United Nations Department of Economic and Social Affairs, Population Division. *The World's Cities in 2018—Data Booklet* (2018). https://www.un.org/en/events/citiesday/assets/pdf/the_worlds_cities_in_2018_data_booklet.pdf

United Nations Human Settlements Programme. "A Practical Guide to Designing, Planning and Implementing Citywide Slum Upgrading Programs." Nairobi, Kenya: UN Human Settlements Programme (2014). https://unhabitat.org/a-practical-guide-to-designing-planning-and-executing-citywide-slum-upgrading-programmes

Vale, B., and Vale, R. *Living within a Fair Share Ecological Footprint.* Abingdon, UK, and New York: Routledge (2013).

Vale, L. J., and Campanella, T. J. *The Resilient City: How Modern Cities Recover from Disasters.* New York: Oxford University Press (2005).

Vojnovic, I. *Urban Sustainability: A Global Perspective.* East Lansing: Michigan State University Press (2012).

Wheeler, S. *Planning for Sustainability: Creating Liveable, Equitable, and Ecological Communities.* 2nd ed. New York: Routledge Press (2013).

Production and Consumption

LEARNING OBJECTIVES

After reading this chapter, you should be able to:

- Describe global trends and patterns in consumption and production and discuss how these may impact the environment
- Describe the types of waste
- Explain the role society plays in defining waste
- Identify and explain the various targets and goals associated with SDG 12
- Identify and explain some of the strategies for achieving targets associated with SDG 12
- Explain how the three Es are present in SDG 12
- Explain how SDG 12 strategies connect to other SDGs

"Fast fashion" describes brands who mass produce clothing, footwear, and accessories to sell current trends in cheap retail stores. Large retailers such as H&M, Zara, Forever 21, and Uniqlo are examples of fast fashion retailers. While these brands allow consumers to afford current styles, fast fashion has a significant impact on the environment.

Mass production of clothing accounts for between 5 to 10 percent of worldwide greenhouse gas emissions, placing fast fashion in the top five of the most polluting industries in the world (Photo 12.1). It is also a top producer of water pollution. Even more troubling is the waste in fast fashion production.

Textiles that use polyester also have a negative environmental impact. Polyester in fast fashion is like high-fructose corn syrup in fast food: Both are cheap materials used to reduce costs. Consumers throw away polyester clothing due to its poor quality and short life span. Polyester, which is composed of acrylic and nylon mixed with plastic (derived from fossil fuels), takes hundreds of years to decompose in a landfill. An estimated 60 percent of all fabric fibers are synthetic. Fast fashion brands pump out new products much quicker than traditional fashion houses. In some cases, fast fashion may introduce 500 to 1,000 new items each month, most of which end up being thrown away.

In addition to environmental costs are social costs, as the fast fashion industry is infamous for a culture of exploitative labor, wage theft, and general mistreatment of workers abroad.*

---

*Ganz, G. "How Fast Fashion Causes Environmental Poverty." *Borgen Magazine*, October 25, 2020. https://www.borgenmagazine.com/fast-fashion-causes-environmental-poverty/

PHOTO 12.1 A Protest against Fast Fashion
The production of clothing contributes 1.2 billion tons of CO_2 per year, directly and indirectly. Synthetic fibers such as polyester and nylon are derived from fossil fuels. Each year an estimated 342 million barrels of crude oil are used in the production of synthetic fibers, a direct cause of CO_2 emissions. Indirectly, fast fashion produces CO_2 through supply chain transportation.

CONSUMPTION AND PRODUCTION: UNDERSTANDING THE ISSUE

Each person on this planet is a consumer. We consume food and water daily. We buy clothes, shoes, books, cell phones, cars, and thousands of other products that are based on natural resources. We are not producing or consuming goods in a sustainable way. Overuse of natural resources, pollution, and disruption of the planet's natural process is a direct reflection of what gets produced and consumed, and in what amounts.

Sustainable consumption is about decoupling economic growth from environmental degradation, increasing resource efficiency, and promoting sustainable lifestyles. SDG 12 aims to "ensure sustainable consumption and production patterns."

The efficient management of shared natural resources and the way we dispose of toxic waste and pollutants are important targets to achieving this goal. So too is encouraging industries, businesses, and consumers to recycle and reduce waste. Finally, shifting cultural values around how much we consume in the first place will be needed to achieve the targets in SDG 12.

Defining Waste

Waste is a fundamental outcome of both production and consumption. There are two forms of waste. The first type of waste is **solid waste** (such as garbage) that consists of items that are thrown away or taken to landfills. The second type of waste is a less visible form of resources that have been used extravagantly or expended carelessly and reflects the inefficient production or consumption of a good.

Waste is defined as "material that has no apparent, obvious, or significant economic or beneficial value to humans." This suggests waste is a socially constructed concept and can change over time as different societies assign value or lack of value to either certain goods or certain practices that create "waste." Such a definition makes waste a relative concept: It is possible one person can find value in what another considers worthless.

In the United States, economic prosperity starting in the 1960s created an enduring "throwaway" culture that has generated increased volumes of waste. New materials such as plastics, other synthetic products, and toxic chemicals made their way to landfills and industries redesigned products for short-term use. The packaging industry alone has created innumerable goods with very short lives. More recently, the "planned obsolescence" of electronics such as cell phones and computers has created an entirely new set of e-waste management challenges. Rather than seeking long-term reuse of limited natural resources, consumable or nondurable goods are designed for being thrown away, following the American preference for convenience over conservation.

Not surprisingly, Americans discard more solid waste per capita than those of other prosperous nations, and far more than those in the Global South. In some regards, the production of garbage reveals much about a society's levels of affluence, household formation, commercial activity, and values. Our "garbage crisis" is as much social and political as physical.

Garbage, particularly household refuse, is one pollutant that is often overlooked by most consumers. We throw away our unwanted food scraps and household items into bags, then into bins, which are then wheeled to the sidewalk or curbside. A large collection truck comes by, usually when we are away from the home or apartment, and it disappears. Garbage is effectively removed from our lives. Few people ever visit landfills and see the vast accumulation of trash (Photo 12.2). And unlike air and water pollution, which is often visible on a daily basis, garbage is something we rarely see or think of as impacting the environment.

PHOTO 12.2 A Landfill in Monterey, California
Most Americans often overlook solid waste and garbage as a form of pollution. Landfills are also symbolic of unsustainable consumption patterns; we consume products that have very short life spans.

Landfills can be considered silent "monuments" to our consuming lifestyle. Before it was closed, Fresh Kills Landfill in Staten Island, New York, received some 17,000 tons of trash per day. It was a putrid mountain of waste, was the largest human-made structure in the history of the world, and could be seen from orbiting satellites. The Great Pacific Garbage Patch can also be seen from space and is said to be composed of 3.5 million tons of garbage, mostly floating plastics.

Many of the highest-income economies, especially in the United States, favor convenience over conservation and short-term needs over long-range resourcefulness. The fast food of McDonald's is both indicative of cultural values that embraces convenience and accepts the short-term duration of packaging. Food arrives within minutes. The food wrappers and drink cups have a commercial life span of less than one hour, yet they survive in landfills for years. The fast-food society is a disposable society. This extends beyond the fast-food restaurants into many aspects of our culture, such as fast fashion. It is often less expensive to purchase a new television or toaster oven rather than repair a broken one. We now purchase purposely "disposable" products such as razors, toothbrushes, paper plates and cups, and writing pens. Yet all this is unsustainable consumption and most consumers are rarely confronted with the impacts their consumption patterns can have on the environment.

A Political Ecology Perspective on Waste

Political ecologists have paid attention to the concept of waste and waste management. Objects such as trash or pollutants have agency. They influence the things around them in relation to each other. Objects also have politics and are entangled in struggles of power and meaning.

The field of "discard studies" uses a critical framework to question premises of what seems normal or given, and analyzes the wider role of society and culture, including social norms, economic systems, forms of labor, ideology, infrastructure, and power in definitions of, attitudes toward, and behaviors around waste. Political ecologists ask: Who carries the responsibility for waste (businesses, citizens, waste creators, or waste managers)? Without paying close attention to the ways materials are located within wider material and social systems, we will struggle to find solutions to our waste problem.

As its starting point, discard studies holds that waste is not produced by individuals and is not automatically disgusting, harmful, or morally offensive, but that both the materials of discards and their meanings are part of wider sociocultural-economic systems. Research in discard studies interrogates these systems for how waste comes to be, and offers critical alternatives to popular and normative notions of waste. It considers the complex politics of waste management, focusing on the environmental, social, and cultural dynamics of working with and engaging with waste products at a number of different scales. Managing waste (collecting it, sorting it, and recycling it) is an infrastructural service that is critical to daily life and municipal governance, and a site from which larger questions about politics, relationships, and environmental dynamics can be discussed. In her book *Garbage Citizenship*, Rosalind Fredericks investigates the political ecology of waste in Senegal. She argues that the responsibility for dealing with waste in Dakar has largely fallen on the urban poor. Her main argument is that the city's garbage infrastructure has become the stage for struggles over government, the value of labor, and the dignity of the working poor in Senegal. How trash is dealt with is another manifestation of inequality.

Recycling—a form of waste management—can also be contextualized within a wider system of political economy. Most cities/localities recycle as long as there is a profitable market for recyclables. Recycling rates vary by material, with high rates of recycling for paper, metals, and some plastics, and relatively low rates for glass and lower-quality plastics. As most recycling is determined by profitability at the global and local scale, recycling rates are largely dependent on the value of specific materials and related transportation costs in the context of global and local price fluctuations related to currency, the price of oil, and the price of virgin materials. As a result, recycling will vary from place to place in relation to these systems.

It is tempting to see waste as a neutral everyday process. A political ecology perspective is valuable for highlighting the more complex institutional processes that produce waste, encourage consumption, and then manage waste by-products. Waste is less an individual failing, than a broader consequence of institutions and systems that have avoided responsibility for the production and disposal of waste, and instead shifted responsibility to the consumer.

WASTE IN PRODUCTION

The process of production necessitates the use of raw material or natural resources. Whether this is lumber for furniture, corn for food consumption, or coltan for cell phones, all products start with natural resources. Additionally, almost everything produced requires the use of energy applied to materials.

Many industrial processes have the potential to produce significant and hazardous waste. The construction, demolition, and renovation industry generates numerous wastes including debris, asphalt wastes, and an array of chemicals found in flooring, paints, and adhesives. Dry-cleaning businesses generate spent solvents, many of which are designated as hazardous. Furniture manufacturing generates acetone, alcohols, and volatile organic compounds in the process of surface preparation, staining, and painting. Leather manufacturing generates high volumes of wastewater and suspended solids and the process of tanning generates chromium and acids.

Waste in production includes both by-products (such as pollution) and inefficiencies in resource use. Inefficiencies would include overproduction, defects, waiting, and transportation.

Typically, consumers rarely consider the impacts of production on issues of the environment or equity. However, the production process can have significant impacts, and it is often less visible to consumers, who may not understand the environmental impacts of the products they purchase. Consider the example of the production of metals.

Pollution: The Production of Metals

Metals are used in many essential items such as construction of buildings (columns and beams), telecommunication (towers and dishes that feed signals), transportation (trains, planes, and automobiles), and more everyday items like cutlery, appliances, coins, jewelry, batteries, and coins.

Metals such as copper, nickel, lead, zinc, and aluminum are examples of metal ores. The production of copper, for example, impacts the environment at numerous stages in production, generating waste in the form of pollution. Metal ores like copper are extracted by mining, which involves removal of rock from the ground. Copper mining operations cause physical disturbances to the landscape, contaminate soil and water and air, and can have health and safety impacts on laborers and surrounding communities. Mines produce large amounts of waste because the metals extracted are only a small fraction of the total volume of the mined material.

At the mine site, open pits and waste rock disposal areas are common. Waste piles from processing, such as tailings impoundments, leach piles, and slag piles vary in size, but can be very large. Erosion of mineralized waste rock into surface drainages may lead to concentrations of metals in stream sediments.

Copper is taken from the mine to undergo pyrometallurgical processing such as smelting and converting. Slag is a by-product of the smelting process and is considered "waste." Slag may contain remnant minerals that can be a potential source of metal release to the environment. At some sites, gas and particulate emissions that were released to the atmosphere from historical smelting operations have been a source of human health concerns and environmental impacts. Some mining processes have an impact on air quality.

Mining also can have indirect impacts such as the consumption of raw materials and the use of electricity to process metal ores. Even infrastructure built to support mining activities, such as roads, ports, railway tracks, and power lines, can affect migratory routes of animals and increase habitat fragmentation.

Other negative impacts include those on human health and living standards. Mining is also known to affect traditional practices of Indigenous peoples living in nearby communities and conflicts in land use are also often present. The impact of mining includes environmental damage to traditional lands in addition to loss of culture, traditional knowledge and livelihoods, often resulting in conflict and forced displacement, further marginalization, increased poverty, and a decline in the health of Indigenous peoples.

Production of another metal, steel, is one of the most energy-consuming and CO_2-emitting industrial activities in the world. Steel production has increased since 1950. In 1950, 189 million tons of steel were produced; in 2018 it was 1.8 billion tons. Today more than two billion tons of iron ore is mined each year—about 95 percent is used by the steel industry. Steel production has a number of impacts on the environment, including air emissions (CO, SOx, NOx, PM2), wastewater contaminants, hazardous wastes, and solid wastes. On average, 1.83 tons of CO_2 are emitted for every ton of steel produced, making steel production a major contributor to climate change, adding more than 3.3 billion tons annually to global emissions.

Of the materials used by society, metals have the greatest potential for unlimited recycling, as they are not biodegradable. Aluminum is one of the better recycled metals, but many metals are not recycled as much as they should be. It is also important to note that while minerals and metals are recyclable, they are still nonrenewable resources because their supply is finite.

Inefficiency in Production

Efficiency really refers to a lack of waste. An inefficient washing machine operates at higher cost, while an efficient washing machine operates at lower cost because it is not wasting water or energy. An inefficient organization operates with long delays and high costs, while an efficient organization is focused, meets deadlines, and performs within budgets.

There are inefficiency issues within production processes, including the extraction and refinement of resources, manufacturing of products, and transportation of products to market. Examples of inefficiency can include overproduction of the product; excess inventory waiting on a machine or part, adding costs and time to production; or reworking or scrapping parts of a product, yielding solid waste and wasting time redoing the work. The "8 wastes" commonly observed in manufacturing facilities include waste from overproduction, defects, inventory, unnecessary motion and transportation, waiting times, overprocessing, and unused time and creativity of employees. Overproduction, for example, is the most obvious form of waste in production that leads to depleted raw materials, but also to wasted storage and excess capital tied up in unused products.

There are inefficiencies and loss at each step in the global food system. For example, the growing of crops leaves behind roots and residues when the entire plant is not used. There may be unharvested crops or crops lost during harvest. Inefficiencies also occur during processing, as do losses to spillage, contamination, and degradation during storage and distribution. Consumers often waste food by not finishing their meals or allowing food to spoil before it can be consumed. Some of these losses or inefficiency may be unavoidable (such as leftover roots or leaves) but others may be avoidable (such as consumer buying habits).

Waste is often the result of inefficiency. Few countries have implemented strong water conservation regulations, yet we are consuming fresh water faster than nature can recycle it. It is estimated that one-third of all food produced—equivalent to 1.4 billion tons, worth—ends up rotting in the bins of consumer and retailers or spoiling due to poor transportation and harvesting practices. And while there have been good gains in energy efficiency, nonrenewable energy consumption continues to increase in some places. We literally waste natural resources; such inefficient use of natural resources is both unsustainable and unjust.

WASTE IN CONSUMPTION

While we tend to think of "population" as being an issue of too many people for the carrying capacity of the earth, the real issue is how much a given population consumes. Consumption is a global problem yet there are dramatic differences between consumption patterns in the Global North and Global South.

The impacts of consumption can be calculated as an ecological footprint. The **ecological footprint** calculates human consumption of resources in relation to the Earth's renewable capacity. The ecological footprint approximates how much productive land and water are needed to provide for a population. The ecological footprint can be measured at any scale: individuals, cities, and countries (Chapter 11 introduced the ecological footprint at the urban scale). The land used is then compared to the land occupied.

On average, there are 4.5 acres of productive land per person in the world. The average productive land needed for most high-income countries is 15.8 acres, for middle-income countries 5 acres, and for low-income countries 2 acres—but Americans require 24 acres. Calculating the number of acres multiplied by the population gives the number of Earths necessary (as if we had more than one) to live in that fashion. For example, if everyone on Earth were to consume at the American rate, there would need to be 5.3 Earths.

The ecological footprint does not propose that all people live exactly the same, but it does allow people to see where they are in relation to the rest of the world population. For example, if everyone lived at the same level, dividing up all resources equally, the standard of living on Earth would be something like the following:

- Strict vegetarian diets that only allow consumption of eggs a few times a week,
- No processed food,
- Two people living in 500 square feet,
- No running water,
- No electricity,
- Public transportation or bicycle, and
- No airline travel.

The ecological footprint is a useful concept and widely used, but political ecologists caution that it puts the focus on individuals rather than large-scale actors (such as companies) and institutions.

While the ecological footprint equation has been criticized, it is an approximate measure of the demands on nature caused by a large population consuming its resources. The human consumption of natural resources by the current world population is approximately 140 percent more than the Earth can sustain. It should be noted that the ecological footprint does not calculate all environmental impacts, but it is a representative measure. Generally, the ecological footprint of countries in the Global North is far higher than in the Global South.

A large share of the population in the Global South is consuming far too little to meet even their basic needs, while many of the richest countries consume global resources at unsustainable rates that deplete supply. Globally, 20 percent of the world's people in the highest-income countries account for 76 percent of total private consumption expenditures, the poorest 20 percent a minuscule 1.5 percent.

Although the United States constitutes about 5 percent of world population, Americans consume about one-third of all processed minerals and a quarter of nonrenewable energy and create about one-third of global pollution. The United States remains the world's largest producer of garbage and industrial waste. Each American uses between 100–175 gallons of water daily, while more than half the world's population lives on twenty-five gallons each.

Consumption in the Global North degrades resources worldwide because of dependence on imported goods, often exporting its raw materials overseas to be processed in countries with low wages and fewer environmental regulations. In addition, most consumer goods are produced, transported, and sold in a complex global supply chain that involves multiple countries.

A major challenge in dealing with waste on the consumer end is that reuse, recycling, and other forms of waste disposal are highly local and can vary tremendously. In the United States, what can be recycled will vary from one city to the next, leaving consumers confused or uninterested in recycling.

Another major challenge for most societies is that little regulation exists to stop the production of waste in the first place; most initiatives focus on cleaning up waste once it has been created. However, the most effective way to

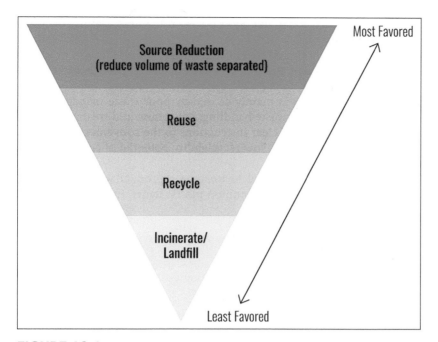

FIGURE 12.1 The Waste Hierarchy
The most effective way to address the problem of waste is to stop creating it. This requires prioritizing the reduction of consumption and reuse over recycling.

address the problem of waste is to stop creating it. This requires prioritizing the reduction of consumption and reuse over recycling. Source reduction is the preferred method because it reduces the use of raw materials and avoids the energy and pollution associated with transportation and even recycling. Figure 12.1 highlights the **waste hierarchy**. Unfortunately, underlying social and cultural values challenges efforts to reduce and reuse and these continue to lag behind efforts to recycle. In reality, the waste hierarchy is practiced upside down: Most waste goes to landfills, while few countries have robust reuse programs, and still fewer have any regulations that incentivize lower consumption rates.

THE PROBLEM OF PLASTICS

Like metals, plastic is a valuable resource that is used for many applications in our life. A 2020 UN Environment Programme Report estimated that about 348 million tons of plastics were produced globally, of which about one-third is used in single-use plastic products.

Single-use plastic is an umbrella term for different types of products that are typically used once before being thrown away or recycled and includes food packaging, bottles, straws, containers, cups, cutlery, and shopping bags. However, they are rarely recycled and are prone to becoming litter. Single-use plastic is emblematic of a "throwaway" cultural mentality.

Many single-use plastics are found in the food business and are common in "takeaway orders." Takeaway food is food that is sold for immediate consumption after purchase and is consumed away from the food outlet such as the home, work, or on the street. There are a diverse range of food takeaway packaging that are used today like food boxes, containers, clamshells, trays, crates, and food savers.

Estimates are that about 100 to 150 million tons of plastics are produced for single-use purposes and about eight million tons of plastics are dumped into the oceans every year.

The dumping of plastics is mainly caused by poor waste management in certain countries (including waste handling, collection, and treatment), poor packaging design, and lack of clear instructions to the consumers about suitable ways of disposal, leading to low recyclability potential of certain types of packaging.

Plastics are pervasive. It is hard to imagine a product without a plastic. Table 12.1 lists some of the most common plastic items. In 1945, the United

TABLE 12.1 Common Plastic Used in Everyday Items	
Items made with Polyethylene Terephthalate (PET or PETE)	• Soft drink bottles • Juice bottles • Water bottles • Shampoo/conditioner bottles • Liquid hand soap bottles • Carry-home food containers
Items made with High-Density Polyethylene (HDPE)	• Toys • Food storage containers • Waste and recycling receptacles • Outdoor signage • Reusable water bottles
Items made with Polyvinyl Chloride (PVC)	• Rigid pipes • Wire insulation • Residential flooring • Automobile instrument panels • Building siding
Items made with Low-Density Polyethylene (LDPE)	• Disposable shopping bags • Juice boxes • Wire insulation • Plastic film • Plastic baggies
Items made with Polypropylene (PP)	• Carpet fibers • Plastic containers • Reusable water bottles • Medical components • Outdoor furniture • Toys • Luggage • Car parts
Items made from Polystyrene (PS)	• Household appliances • Car parts • Instrument panels • Foam in child car seats • IT equipment • TVs and computers • Medical test tubes, culture trays, and petri dishes

Source: American Chemistry Council. "How Plastics Are Made" (2021). https://plastics. americanchemistry.com/How-Plastics-Are-Made/

States produced some 400,000 tons of plastic products. In 2018, the United States generated almost thirty-six million tons, only three million tons of which was recycled. Figure 12.2 shows the growth in the production of plastics over the last several decades.

Most plastics are derived from fossil fuels such as natural gas, coal, and crude oil through a process called polymerization. The production of plastics begins with the distillation of crude oil in an oil refinery. This separates the heavy crude oil into groups of lighter components, called fractions. Each fraction is a mixture of hydrocarbon chains (chemical compounds made up of carbon and hydrogen) that are refined into ethane and propane. Ethane and propane are then treated with high heat in a process called "cracking" that turns them into ethylene and propylene. These materials are then combined to create different polymers—such as polystyrene, nylons, and polyesters. Different elements can be attached to the carbon-to-carbon backbone. Polyvinyl chloride (PVC) contains attached chlorine atoms. Teflon contains attached fluorine atoms and so on.

Ridding ourselves of plastic is a major challenge because it is not simply about dealing with the end waste products. Plastic must be confronted at the front end of production. This is not a problem that consumers alone can solve. On the one hand, many plastics companies are investing in R&D to find better ways to recycle or deal with plastic waste. On the other hand, producing plastic is a multi-billion-dollar business that implicates many companies including ExxonMobil, Chevron, Shell, Dow Chemical, and DuPont, some of which receive government subsidies that keep the costs of plastic production artificially low and highly profitable.

Many plastic products are not properly disposed of and only a small percentage of plastics are recycled. As a result, many plastics find their way into sewage systems, which empty into streams and rivers and into the oceans. Plastics in the ocean break up into small particles. Plastic debris that is less

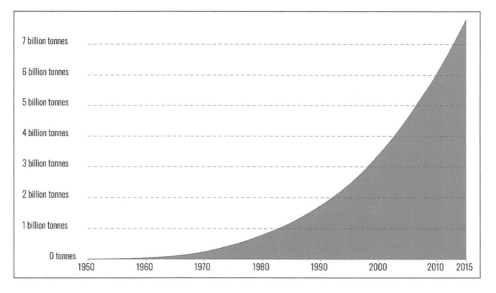

FIGURE 12.2 Cumulative Global Plastics Production, 1950 to 2015
Although plastics has increased over the last seventy years, there has been a significant acceleration of the production of plastics since 1990. This may be because plastics are present in so many products—from toothpaste tubes to residential flooring, toys and nylon backpacks.

than five millimeters in length (the size of a sesame seed) is called a **microplastic**. While the full extent of micropolastics on water ecosystems is not understood, there are noticeable negative impacts: Aquatic life and birds can mistake microplastics for food and choke on them, for example.

DISPOSING OF WASTE

The US Environmental Protection Agency (EPA) classifies different types of solid waste sources: residential, industrial, commercial, institutional, and construction and demolition. The most popular form of disposal is land disposal and the destination of much of these wastes is a **landfill**. Contemporary landfills are huge engineering feats that involve layers of nonpermeable clay or synthetic lining and a network of pipes to collect leachate and methane as an energy source. Landfills are sited according to drainage patterns, wind, and distance from the city or community, rainfall, soil types, and the depth of the water table. Most contemporary landfills operate under established standards for use and disposal; however, they are not without environmental consequences. Landfills contribute to climate change by emitting large amounts of methane gas (a by-product of anaerobic bacterial processes) and they can also pollute groundwater. Furthermore, landfills waste precious resources and require a great deal of energy and money to manage not only during operation but also long after they have closed.

While landfills remain the management solution of choice for much of the world, other options exist. Waste-to-energy incineration generates electricity through the controlled combustion of organic material at very high temperatures. This process allows for some energy recovery from waste, but also contributes to air pollution by releasing fine particulates, heavy metals, and carcinogenic substances like dioxin into the atmosphere. After incineration, toxic residues must be disposed of as hazardous waste. Other concerns include the cost of constructing incineration plants, the long-term contracts that cities must sign to get them built, and the potential disincentive placed on recycling goods in ways that recoup more than just imbedded energy.

Another landfill alternative is large-scale anaerobic digestion, a process that purposefully creates conditions for bacterial communities to break down biodegradable waste and sewage sludge to produce biogas for energy. The advantages are such systems can reduce waste while generating a renewable energy, and any remaining sludge can be used as fertilizer. However, a disadvantage is they are highly technical systems that require experts to design, construct, and operate them.

Typically, the **waste stream** or waste composition in Global North countries is more than one-third paper, 28 percent organic materials such as food wastes and yard trimmings, and 9.9 percent plastics. In contrast, the waste composition in the lowest income is anywhere between 40 to 85 percent organic compostable matter and only 5 percent paper. India and China diverge from this trend because they use coal as a household fuel source, thereby generating large quantities of dense ash. The percentage of consumer packaging (plastic, paper, glass, and metal) as part of waste composition tends to be lower in Global South countries; packaging wastes correlate with the nation's degree of wealth and urbanization. However, as lower-income countries become richer and more urbanized, they may see a significant increase in paper and packaging as there are more newspapers and magazines, fast-service restaurants, and single-serving beverages.

Global North countries have extensive networks of solid waste collection that include curbside pick-up of residential garbage, recycling, and other wastes to be transported to landfills and recycling centers. In most countries, waste and refuse are handled as a local issue and waste must actively be collected and disposed of by local governments. Coordinating regular collection over an entire city is a logistical feat. New York City generates fourteen million tons of waste and recyclables every year, which are hauled by 2,000 city government and 4,000 private trucks.

In the Global South, however, many cities and communities lack landfills and institutionalized waste diversion programs for dealing with solid waste. There is a marked contrast between solid waste service in the Global North and Global South. For example, 95 percent of Australians are served by municipal waste collection, compared to 73 percent in North Africa, 66 percent in South Asia, and 44 percent in sub-Saharan Africa. Corrupt bureaucracies, poor planning, and lack of funding render the few official facilities inadequate to handling the volume of waste generated in Global South countries. In recent years, many cities and communities in the Global South have significantly improved waste collection, particularly in commercial districts and tourist areas; however, there remain significant gaps in service.

Open Dumps

In many Global South cities, particularly megacities, waste is transferred to open dumps. In Brazil, only 10 percent of solid waste is dumped in sanitary landfills; 76 percent is dumped in illegal landfills and another 13 percent in open dumps. Beijing is said to have more than 461 illegal open dumps. Open dumps are unsanitary and lack liners, methane collections systems, and other mechanisms, all of which are now mandated in higher-income countries. This presents a pressing problem as most open dumps contaminate nearby surfaces and ground water, create explosive methane gas, and allow for the breeding and thriving of disease-bearing insects and rats.

Open dumps, like Payatas in Manila, Dhapa in Calcutta, Bantar Gebang in Jakarta, and Matuail just outside of Dhaka, are more than eyesores. They are hazards. In Malaysia's capital Kuala Lumpur, the waste dump Taman Beringin is a smelly, twelve-hectare dump for the city's garbage. Nearby low-income residents put up with flies, rats, disease, and the reek of rotting garbage. When the dump caught fire in 2004, it burned for more than two weeks. Illegal open dumps create risk for the water supply and the health of the urban population. Few cities have the resources to monitor the environmental effects of waste disposal in open dumps.

Cities in Asia generate about 760,000 tons of municipal solid waste per day; by 2025, this figure could increase to 1.8 million tons and cities will likely need to double their current $25 billion per year spending on solid waste management. Urban populations in the Global South will likely triple their current rate of municipal solid waste generation over the next twenty years, while cities in Nepal, Bangladesh, Vietnam, Laos, and India might see a four to six times increase. Such dramatic rises in municipal solid waste generation will place enormous stress on already limited financial resources. How will these cities and countries deal with increasing waste and the need for more landfills or disposal systems? Will rising affluence equate to more waste generation, as it has in the Global North?

EXPERT VOICE 12.1

Dr. Aman Luthra on the Role of Waste Pickers in the Informal Economy

Dr. Aman Luthra is a geographer who studies informal recycling in India. The following are his observations on the crucial role informal recyclers play in the waste system (see Photo 12.3).

The informal economy of waste includes a range of actors and units engaged in the entire value chain of waste management and recycling activities—pickers, traders, dealers, suppliers, and re-processors. Waste pickers—those who collect, segregate, and supply recyclable materials—are the foundation of India's vibrant recycling industry. By the International Labor Organization's estimates, waste pickers alone represent 0.1 percent of the Indian urban workforce—about 1 million workers—a figure that many consider a severe underestimate. How do cities benefit from their work? As they make a living, workers

PHOTO 12.3 Waste Pickers Search for Recyclable Waste at the Bhalswa Landfill Site in New Delhi, India
Recycling is part of the informal market; however, many workers are poor and marginalized. They are often exposed to hazards that affect their health and well-being. Many waste pickers have been organizing to advocate for wages, protection, and other services.

in the informal economy of waste provide crucial municipal and environmental services.

Waste collection as municipal service: Although regulations mandate that municipalities provide doorstep waste collection in their jurisdictions, this ambitious goal remains a distant dream for many Indian cities. A 2014 Government of India study showed that municipalities are only able to provide waste collection services to about 50 percent of residents in urban areas and less than 5 percent in rural areas. Even in areas where government provides collection services, these are often limited to collecting waste from neighborhood or community bins and not directly from households. Fortunately, the informal sector has historically stepped in to take on this role and provide doorstep waste collection services to city residents. In doing so, waste pickers reduce municipal solid waste collection and disposal costs for cities across the world. Another study on informal waste collectors in six cities around the world showed that they saved those cities $44.5 million annually.

Recycling as environmental service: Although estimates of their contribution to the recycling economy vary by city and according to the particular research objectives of the study, a survey of six cities showed that the informal sector recycles as much as 66 percent of solid waste in cities. Benefits of recycling aside, their work diverts recyclable materials from the waste stream and therefore reduces the waste burden and costs for managing this burden. In addition to the economic benefits, solid waste collection and recycling is a crucial environmental service reducing greenhouse gas emissions and the urban environmental footprint. Another study that estimated greenhouse gas emissions reductions through recycling of materials alone, leaving out other contributions such as the use of non-mechanized transport, found that the informal sector in Delhi reduces emissions by an estimated 962,133 tons of carbon dioxide equivalent each year.

Despite their crucial contribution to the functioning of Indian cities, informal workers remain socially, economically, and politically marginalized and face many challenges. First, privatization of waste management services constantly threatens to displace and has already displaced many waste pickers across the country. Second, they are often harassed by municipal officials, the police, and the general public. Third, their social and political marginalization hinders their access to basic government services. Fourth, daily exposure to hazards due to the nature of their work takes a toll on their health and well-being. In response to some of the challenges, waste pickers across the world, including in India, have been organizing to advocate for themselves at the local, national, and international levels to secure safe and stable livelihoods and access to government social security programs and services. Although the work done so far has been commendable and has achieved positive policy results, much work remains to be done if waste pickers are to secure their place in 21st century India.

Sources: Luthra, A. "Efficiency in Waste Collection Markets: Changing Relationships between Firms, Informal Workers and the State in Urban India." *Environment and Planning A: Economy and Space* 52 (7): 1375–94 (2020, March). https://doi.org/10.1177/0308518X20913011
Luthra, A. "Municipalization for Privatization's Sake: Municipal Solid Waste Collection Services in India." *Society and Business Review* 14 (2): 135–54 (2019, May). https://doi.org/10.1108/SBR-11-2017-0102

Informal Recycling

Many low-income countries do not have a long history of dealing with significant amounts of waste, so they often lack established infrastructure. In many cases, waste programs are characterized by inconsistent collection, a lack of equipment, and open air dumps. But perhaps one of the most distinguishing features of waste in low-income countries is the fact that most of the recycling is in the informal sector.

In many Global South cities and communities, open-air dumps are located near slums or low-income residential areas that are home to a growing number of collectors and scavengers who fill an important need for waste collection, waste separation, and recycling. In these countries, many residential areas near open-air dumps have narrow or unpaved streets, making door-to-door collection by garbage trucks difficult, necessitating an informal sector of recyclers. Dr. Aman Luthra highlights the role of the informal sector in recycling in India in *Expert Voice* 12.1.

SOLUTIONS TO CREATE SUSTAINABLE CONSUMPTION AND PRODUCTION

Changing production and consumption processes and patterns is a major challenge that requires social and psychological changes as well as economic and technical innovations. We should be cautious about the "easy technical fix" that allows us to continue to think and act in the way we always have without having to radically transform values and practices. The **Kuznets Curve** was once considered a way to predict consumption patterns but has been criticized (*Key Terms and Concepts* 12.1).

KEY TERMS AND CONCEPTS 12.1

The Environmental Kuznets Curve

In the 1950s, economist Simon Kuznets advanced what became known as the Kuznets Curve. The Kuznets Curve graphs the hypothesis that as an economy develops, market forces first increase, then decrease, economic inequality. It was subsequently adapted for environmental applications. The Environmental Kuznets Curve suggests that the early stages economic development will initially lead to a deterioration in the environment, but at a certain level of economic development, society begins to improve its relationship with the environment (Figure 12.3). In other words, as societies get wealthier, they take better care of the environment (or so we tell ourselves). The Kuznets Curve's hypothesis allows us to imagine we can achieve sustainability without a significant deviation from business as usual.

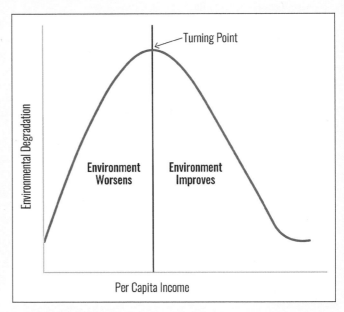

FIGURE 12.3 The Kuznets Curve

Most countries in the Global North have experienced the curve and have seen improved environmental conditions. For the Global North, the turning point was the process of deindustrialization—a transition away from industrial and manufacturing economies that began in the 1970s. This coincided with the rise of an environmental movement and the creation of the US EPA in the United States, for example.

However, deindustrialization did not end industrialization and manufacturing. It just shifted the location of these processes to countries in the Global South. Which means that the Global North has benefited from being able to "export" pollution to other countries and "import" the goods consumers want.

The optimism of the Environmental Kuznets Curve is too simplistic and not supported by evidence, leading to the following criticisms:

1. It can misleadingly suggest that economic growth is good for the environment ("the solution to pollution is economic growth").
2. An improved environment is unlikely to occur without government regulations and changing attitudes. Pollution is not simply a function of income, but of many factors.
3. It does not account for the fact that many lower-income countries are addressing pollution already.
4. It does not account for a global scale of pollution. Pollution may reduce in Germany and Canada but may increase in countries such as Brazil and Indonesia. Higher-income countries may implement policies to stop deforestation, but at the same time import meat and furniture from countries who are converting forests to farmland. Globally, net pollution reduction may not be occurring.
5. Countries with the highest GDP have the highest levels of CO_2 emissions. This would suggest that economic growth can lead to greater resource use in some circumstances.

Some critics have noted that high-income countries with superior scientific and technological capacity have far higher per-capita environmental impacts than lower-income countries. This seems to indicate that technological prowess does not ensure sustainable consumption and production patterns but can have the opposite effect.

Source: Stern, D. "The Rise and Fall of the Environmental Kuznets Curve." *World Development* 32 (8): 1419–39 (2004).

According to the United Nations, **sustainable consumption and production** is defined as "the use of services and related products, which respond to basic needs and bring a better quality of life, while minimizing the use of natural resources and toxic materials as well as the emissions of waste and pollutants over the life cycle of the service or product so as not to jeopardize the needs of future generations." Given the trends and patterns of consumption and waste currently, this is an ambitious goal.

Table 12.2 lists the specific targets for SDG 12. Some targets focus on promoting more efficient production methods and products, mainly through technological improvement and informed consumer choice. Other targets stress the need to consider overall volumes of consumption, distributional issues, and related social and institutional changes. Several targets emphasize increasing efficiency in resource use, including water and energy, and reducing food waste. Other solutions encourage rethinking "waste" as both a physical and social issue with the adoption of zero waste programs, eliminating toxic chemicals, adopting Life Cycle Assessment, a Circular Economy, and more comprehensive sustainable production and consumption plans.

Move to Zero Waste

Garbage is a challenge to sustainability. There are problems of collection and disposal and environmental hazards associated with refuse. Unlike air and water pollution, solid waste—garbage—must be collected to be disposed of. Most of our garbage goes to landfills. But in the Global North nearly all of what we throw away could be reused, recycled, or composted.

Zero waste is a set of strategies that involves changing consumer responsibility (buying less) and producer responsibility (eliminating waste, toxicity, and packaging in production), encouraging repair, reuse, and donation sites, and increasing rates of recycling and composting.

The good news is that material discarded in US landfills peaked around 2005 and has declined as cities around the United States have implemented more effective recycling and composting programs. No US city has done this more effectively than San Francisco.

In 2002, San Francisco pledged zero waste by 2020, meaning that by 2020, the city would send zero discards to the landfill. To achieve this ambitious goal, the city passed a series of ordinances to deal with solid waste in a variety of ways. It passed a Mandatory Recycling and Composting Ordinance that requires everyone in San Francisco to separate recyclables, compostable items, and landfill-bound trash. Another ordinance requires every event held in San Francisco to offer recycling and composting at the event.

One of the less visible but important categories of waste encompasses all the debris produced by the construction industry. In 2006, the city passed an ordinance requiring the building trade to recycle at least two-thirds of its debris such as concrete, steel, and timber at a registered facility. Companies failing to comply run the risk of their registration being suspended for six months. At the same time, the city undertook to only use recycled materials for public works such as asphalting or gutters.

San Francisco's three-bin system, policies, financial incentives, and extensive outreach to residents and businesses helped the city achieve the highest diversion rate—80 percent—of any major city in North America (where the average is about 35 percent). San Francisco has shown that political determination and changing behaviors in how residents and businesses deal with garbage is effective.

TABLE 12.2   SDG 12 Targets and Indicators to Achieve by 2030

Target	Indicators
12.1 Implement the ten-year framework of programs on sustainable consumption and production, all countries taking action, with developed countries taking the lead, taking into account the development and capabilities of developing countries	**12.1.1** Number of countries with sustainable consumption and production (SCP) national action plans or SCP mainstreamed as a priority or a target into national policies
12.2 Achieve the sustainable management and efficient use of natural resources	**12.2.1** Material footprint, material footprint per capita, and material footprint per GDP **12.2.2** Domestic material consumption, domestic material consumption per capita, and domestic material consumption per GDP
12.3 Halve per capita global food waste at the retail and consumer levels and reduce food losses along production and supply chains, including postharvest losses	**12.3.1** Global food loss index
12.4 Achieve the environmentally sound management of chemicals and all wastes throughout their life cycle, in accordance with agreed international frameworks, and significantly reduce their release to air, water, and soil to minimize their adverse impacts on human health and the environment	**12.4.1** Number of parties to international multilateral environmental agreements on hazardous waste, and other chemicals that meet their commitments and obligations in transmitting information as required by each relevant agreement **12.4.2** Hazardous waste generated per capita and proportion of hazardous waste treated, by type of treatment
12.5 Substantially reduce waste generation through prevention, reduction, recycling, and reuse	**12.5.1** National recycling rate, tons of material recycled
12.6 Encourage companies, especially large and transnational companies, to adopt sustainable practices and to integrate sustainability information into their reporting cycle	**12.6.1** Number of companies publishing sustainability reports
12.7 Promote public procurement practices that are sustainable, in accordance with national policies and priorities	**12.7.1** Number of countries implementing sustainable public procurement policies and action plans
12.8 Ensure that people everywhere have the relevant information and awareness for sustainable development and lifestyles in harmony with nature	**12.8.1** Extent to which (i) global citizenship education and (ii) education for sustainable development (including climate change education) are mainstreamed in (a) national education policies; (b) curricula; (c) teacher education; and (d) student assessment
12.A Support developing countries to strengthen their scientific and technological capacity to move toward more sustainable patterns of consumption and production	**12.A.1** Amount of support to developing countries on research and development for sustainable consumption and production and environmentally sound technologies

Target	Indicators
12.B Develop and implement tools to monitor sustainable development impacts for sustainable tourism that creates jobs and promotes local culture and products	**12.B.1** Number of sustainable tourism strategies or policies and implemented action plans with agreed monitoring and evaluation tools
12.C Rationalize inefficient fossil-fuel subsidies that encourage wasteful consumption by removing market distortions, in accordance with national circumstances, including by restructuring taxation and phasing out those harmful subsidies, where they exist, to reflect their environmental impacts, taking fully into account the specific needs and conditions of developing countries and minimizing the possible adverse impacts on their development in a manner that protects the poor and the affected communities	**12.C.1** Amount of fossil-fuel subsidies per unit of GDP (production and consumption) and as a proportion of total national expenditure on fossil fuels

Source: United Nations Sustainable Development Goals. https://sustainabledevelopment.un.org/sdg12

In late 2019, it was clear that the city would not achieve its zero-waste goals by 2020. The problem? Plastics and electronics. You cannot recycle what is not recyclable. In 2017, China stopped accepting contaminated plastics and electronics for recycling, upending recycling programs around the United States. It reminded many in the waste industry that a city can only be as ambitious as its global supply chains allow. In response, San Francisco set new goals to achieve by 2030 and now aims to *approach* zero waste, even if it has learned that it may be difficult to achieve it.

Eliminate Persistent Organic Pollutants

Many persistent organic pollutants (POPs) were widely used during industrial production after World War II, when thousands of synthetic chemicals were introduced into commercial use. Many of these chemicals proved beneficial in pest and disease control, crop production, and industry. These same chemicals, however, have had unforeseen effects on human health and the environment.

Many people are familiar with some of the most well-known POPs, such as PCBs, DDT, and dioxins. PCBs, for example, have been useful in a variety of industrial applications (in electrical transformers and large capacitors, as hydraulic and heat exchange fluids, and as additives to paints and lubricants) and DDT is still used to control mosquitoes that carry malaria in some parts of the world.

Exposure to POPS can lead to serious health effects including certain cancers, birth defects, dysfunctional immune and reproductive systems, greater susceptibility to disease and damages to the central and peripheral nervous systems. The so-called dirty dozen is a group of twelve highly persistent and toxic chemicals that include aldrin, chlordane, DDT, dieldrin, endrin, heptachlor, oxaphenebenzene, mirex, polychlorinated biphenyls (PCBs), polychlorinated dibenzo-p-dioxins, polychlorinated dibenzofurans, and oxaphene. Some countries have banned or reduced the use of these dirty dozen, but many have not. DDT, which is banned in the United States, is still in use in other countries.

The Stockholm Convention on Persistent Organic Pollutants was signed in 2001 and became effective in 2004. It is a global treaty to protect human health and the environment from chemicals that remain intact in the environment for long periods, become widely distributed geographically, accumulate in the fatty tissue of humans and wildlife, and have harmful impacts on human health or on the environment. A major impetus for the Stockholm Convention was the finding of POPs contamination in relatively pristine Arctic regions—thousands of miles from any known source. Given their long-range transport, no one government acting alone can protect its citizens or its environment from POPs; the need for an international treaty became apparent.

Key elements of the convention include the requirement that rich, industrial countries provide new and additional financial resources and measures to eliminate production and use of intentionally produced POPs, eliminate unintentionally produced POPs where feasible, and manage and dispose of POPs wastes in an environmentally sound manner.

Use Life Cycle Assessment to Understand a Product

Life Cycle Assessment (LCA) is a methodology used for understanding the environmental impact of a product throughout its entire life cycle: from raw material extraction through production, use, and waste treatment, to final disposal. LCA can quantify potential impacts at different scales—global or regional—and for different environmental impacts such as climate, acidification, eutrophication, and resource use. LCA allows the environmental comparison of different products, services, or technological systems. In addition, LCA has gained acceptance as a holistic decision-making tool in industry, procurement, and policy making. An LCA can help product designers, service providers, and individuals to make more informed choices.

LCA implies that everyone in the whole chain of a product's life cycle, from cradle to grave, has a responsibility to prevent waste, and to consider all the relevant impacts on the economy, the environment, and the society. Most consumers focus on the use of the product and are less aware of the other four phases of the life of a product. Thus a LCA approach may help to shift some responsibility back on producers, who deal with a product at different phases. At each stage in a product's life cycle, there is the potential to reduce resource consumption and improve a product's performance. Businesses that incorporate LCA may experience environmental, occupational health and safety, and quality-management benefits, which could motivate them to develop and apply cleaner process and product options. Incorporating life cycle and sustainability management will improve image and brand value for both world market players as well as smaller suppliers and producers. This allows individuals to make more informed decisions about what they purchase. LCA consists of several phases as highlighted in Figure 12.4.

Single-use plastic bags are one of the most consumed items globally and yet also criticized for their environmental impact. However, the alternatives for shopping such as cotton and paper bags also come with an environmental footprint and do not necessarily outperform plastic bags in all environmental categories.

A 2020 UNEP report by Tomas Ekvall and colleagues found there are many complex factors to consider in an LCA that compares plastic, cotton, and paper bags. These include the production process of each type of bag, the ingredients

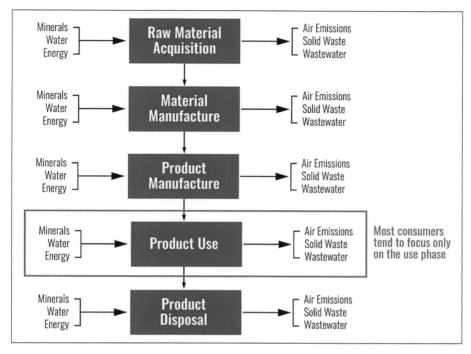

FIGURE 12.4 Life Cycle Assessment
The first phase sets the goals and scope of the LCA. For example: what the product is, what environmental impacts will be analyzed (climate change, soil erosion, etc.). The second phase is inventory analysis. This starts with a construction of the life cycle flow chart and the collection of data for relevant inputs (energy and materials) and outputs (emissions and waste). This is followed by classifying and characterizing the impact regarding the inputs and outputs. Finally, the results are analyzed and interpreted.

comprising each bag by weight and composition, the number of times a bag is used, and how the bags are disposed. For example, paper bags may seem like a more natural and better solution to single-use plastic bags, but the production of the bag requires trees to be felled and logged, and processing at a paper and pulp mill uses energy and may discharge pollutants in the water.

The LCA results show different outcomes depending on the scenario. For instance, paper bags that end up in landfills cause emissions of methane that impact climate change, while plastic bags are relatively inert in landfills. However, the incineration of used plastic bags adds carbon dioxide emissions, while combusting paper bags emits carbon dioxide that was already in the carbon system that does not add to total emissions. There are other considerations: A cotton reusable bag needs to be used 50 to 150 times to have less impact on the climate compared to one single-use plastic bag. Paper bags contribute less to the impacts of littering, but can have a larger impact on eutrophication and acidification of waterways than single-use plastic bags.

Reducing environmental impacts of bags is not just about choosing, banning, recommending, or prescribing specific materials or bags but is also about changing producer processes and consumer behavior to increase the reuse rate and to avoid littering. The LCA comparing plastic, cotton, and paper bags reminds us that often we embrace a new product or a supposed "green solution" without realizing that it, too, may have equally negative environmental impacts. The report concluded that the shopping bag that has the least impact on the environment is any bag the consumer already has at home and can reuse.

Create a Circular Economy

There is an extensive history of rethinking the relationship between economic growth and the environment and the relationship between production, consumption, and waste. In 1970 the social scientist Kenneth Boulding made a distinction between a *cowboy economy* and a *spaceship economy*. A cowboy economy assumes infinite resources and unlimited production and consumption. Growth is not only possible and attainable but also desirable. In contrast, a spaceship economy assumes there are finite resources and operates accordingly with limits, recycling, and careful management of said resources. In this simple but provocative metaphor, Boulding captured the sense of limits, interdependence, and planetary vulnerability.

There has been a great deal more discussion about limits on consumption and production since Boulding's article. In 2002, architect William McDonough and chemist Michael Braungart introduced the term **cradle-to-cradle (C2C)** that presented an integration of design and science. C2C is a term for a product that is taken back by the manufacturer at the end of its useful life, at which point its components or materials are used to make new products of equal or higher value. This is different from cradle to grave, where manufacturers and producers have no responsibility for dealing with a product at the end of its life. C2C is a concept of total recycling through design. C2C production mimics natural cycles, minimizes environmental impact, and aims for more sustainable production and increased social responsibility. Along with a myriad of other benefits. C2C moves us beyond our disposal-oriented society to a more sustainable closed-loop recirculation of materials.

In their 1999 book *Natural Capitalism: Creating the Next Industrial Revolution*, Paul Hawken, Amory Lovins, and L. Hunter Lovins describe a global economy in which business and environmental interests overlap, recognizing the interdependencies that exist between the production and use of human-made capital and flows of natural capital. **Natural capital** refers to the world's stocks of natural assets including soil, air, water, and all living things.

More recently, the term **circular economy** has emerged as a concept that integrates reusing the waste of a company as a resource for another one (Figure 12.5). The Circular Economy is based on the principles of designing out waste and pollution, keeping products and materials in use, and regenerating natural systems, where the emphasis is on reuse. This is in contrast to a linear economy where we take resources from the ground to make products, which we use, and, when we no longer want them, throw them away.

In a Circular Economy, there are three key objectives:

1. Design out waste and pollution: waste and pollution are not accidents, but the products of decisions made at the design stage.

2. Keep products and materials in use: design products and materials to be used and reused and take materials back.

3. Regenerate natural systems: figure out how to return valuable nutrients or materials to nature.

Of note is the EU adoption of a Circular Economy plan, which has begun to see results (discussed in *Making Progress* 12.1).

A Critical Perspective of the Circular Economy

There are and have been many attempts to create "alternative economies" that are more sustainable. However, a key point is to reduce consumption and production *in the first place*, something that most modern societies have failed to do in regulation or practice. Some experts caution about a circular economy as the "cure" for unsustainable production and consumption.

Kevin Moss, of the World Resources Institute, reflects on some of possible unintended consequences that may occur. He asks: Will the jobs generated by circular innovations be better jobs? Who will get them? Circularity could reduce material extraction and waste through reuse—but at the cost of what other resources? Will circularity reduce consumption, or just maximize use of existing products?

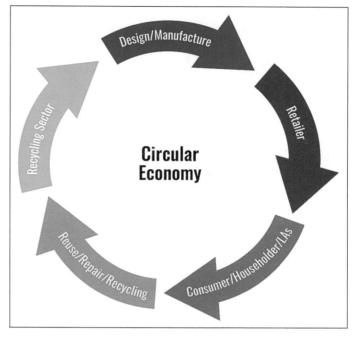

FIGURE 12.5 Visualizing the Circular Economy
The Circular Economy is based on the principles of designing out waste and pollution, keeping products and materials in use, and regenerating natural systems, where the emphasis is on reuse.

Moss says that circularity could fundamentally change the employment patterns, and it is imperative to measure impact for those who lose from the transition. For example, the jobs created by the circular economy may be different from existing jobs in scope and location. Perhaps repair will create more creatively fulfilling roles than mass production, but in an economy that values refurbishment over mass production, work may move closer to the consumer, potentially causing dislocation in regions that can least afford the job losses.

Moss also says that zero waste and 100 percent circularity have become popular goals—but achieving them can be resource intensive and could lead to actions that compromise overall sustainability.

Finally, Moss says that cultural norms will have to shift. Lower consumption will only be realized if we value longevity and reuse over the purchase of virgin material-based goods. The phenomenon calls to mind the **Jevons paradox**: Increasing efficiency doesn't lead to less consumption—it can lead to more. In this case, the consumer feels good about resource savings in one area (contributing to the circular economy by donating old clothes), allowing them to rationalize using more resources in another (restocking the closet with new purchases).

The circular economy is not a silver bullet for employment, sustainability, and prosperity. Companies and governments must carefully measure the anticipated and actual impact of these actions and ensure they take us in the right direction—not into a circular but even less sustainable future.

MAKING PROGRESS 12.1

The European Union's Circular Economy Action Plan

The European Union's Circular Economy Action Plan, adopted in 2015, aims to transition the European economy from a linear to a circular economy model. The action plan mapped out fifty-four actions, as well as four legislative proposals on waste. Example actions include:

- Set an overall 70 percent recycling rate target to be met by 2030 for all packaging materials.
- Implement a recycling target per material type; the targets are 85 percent recycling rate for paper and cardboard packaging, 55 percent for plastics, and 30 percent for wood.

These legislative proposals were put forward by the European Commission along with the action plan and included targets for landfill, reuse, and recycling, to be met by 2030 and 2035, along with new goals for separate collection of textile and biowaste. The action plan covered several policy areas, material flows, and sectors alongside cross-cutting measures to support this systemic change through innovation and investments. More than EUR 10 billion (USD $13 billion) of public funding was allocated to the transition between 2016 and 2020. The action plan has also encouraged at least fourteen member states, eight regions, and eleven cities to put forward circular economy strategies. Some of the results include:

- The European Union's overall circularity rate, the percentage of recovered and recycled materials used in production, increased from 3.4 percent to 11.7 percent between 2004 and 2016.
- According to Eurostat, jobs related to circular economy activities increased by 6 percent between 2012 and 2016 within the European Union.
- France adopted an antiwaste law in early 2020 that bans the destruction of unsold goods, encourages donations, and fosters secondary markets.

The European Union is now recognized as a leader in circular economy policy making globally.

Source: "The EU's Circular Economy Action Plan: Setting the World's Largest Single Market on a Path towards a Circular Economy." Ellen Macarthur Foundation and the European Commission (2020, June 26). https://www.ellenmacarthurfoundation.org/case-studies/the-eus-circular-economy-action-plan

Establish Sustainable Consumption and Production Plans

In 2012, the United Nations adopted the 10-Year Framework of Programmes on Sustainable Consumption and Production Patterns (SCP). The framework is built around three main objectives.

First, SCP aims to decouple environmental degradation from economic growth. This involves doing more and better with less and increasing net welfare gains from economic activities by reducing resource use, degradation, and pollution along the whole life cycle while increasing quality of life. "More" is delivered in terms of goods and services, with "less" impact in terms of resource use, environmental degradation, waste, and pollution.

Second, SCP applies LCA to increase the sustainable management of resources and achieve resource efficiency along both production and consumption phases of the life cycle. This includes resource extraction, the production of intermediate inputs, distribution, marketing, use, waste disposal, and reuse of products and services.

Third, SCP seeks opportunities such as investments and new technologies for lower-income countries to "leapfrog" to more resource efficient, environmentally sound, and competitive technologies. This allows lower income countries to bypass the inefficient, polluting, and costly phases of development that the highest-income countries experienced.

Many of these programs may already be a part of sustainability plans for communities, cities, or countries. For example, **sustainable procurement** integrates environmental and social considerations into the procurement process, with the goal of reducing adverse impacts upon health, social conditions, and the environment. Sustainable procurement means making sure that the products and services we buy are as sustainable as possible, have the lowest possible environmental impact, and create the most positive social results. For example, a company can look for ways their product may be suitable for composting, or a product can be designed specifically to be taken apart so that materials can be more easily recycled. Other examples include minimizing the use of virgin and nonrenewable materials, or to minimize packaging.

There are several reasons why a company may see advantages in adopting sustainable procurement. First, they can reduce the risk that their products will use child labor or contribute to pollution, which can have a negative impact on brand value. Second, in some cases, they can reduce production costs. Third, having a product certified as sustainable can bring in higher revenue. For example, according to the World Economic Forum, engaging in sustainable procurement practices provides businesses with a 15 to 30 percent measurable increase in brand equity, which fosters revenue growth. Fourth, they may create new markets. Fifth, they provide the wider community benefits.

SUMMARY AND PROGRESS

Many of the SDG 12 targets encourage businesses to find new solutions that enable sustainable consumption and production patterns and governments to support such efforts. Understanding a product's life cycle and the environmental and social impacts of products is a responsibility of the producer, not the consumer. Identifying opportunities within the value chain is a first step that businesses can take.

Many businesses now embrace **Corporate social responsibility (CSR),** a self-regulated business model that helps any company be socially accountable to itself, its stakeholders, and the public. CSR is not mandated by national laws and is voluntary. A company that engages in CSR is aiming to operate in ways that enhance society and the environment. The idea is that a company can do well and can do good.

There are four general categories of CSR initiatives: philanthropy, volunteer efforts, ethical labor practices, and environmental efforts (such as reducing the carbon footprint or sourcing sustainable materials). Some examples of CSR include Ben & Jerry's Ice Cream, which supports LGBTQ+ equality and climate justice; TOMS shoe company that donates one-third of its net profits to various charities that support physical and mental health; and Starbucks that has implemented a socially responsible hiring process to diversify their workforce, focusing on hiring more veterans, young people, and refugees.

LEGO is considered a world CSR leader. As part of its commitment to providing all children access to healthy play materials, LEGO donated 580,000 used bricks in 2018, then matched a MacArthur Foundation $100 million grant to Sesame Workshop to bring the power of play to children affected by the Rohingya and Syrian conflicts. On the environmental side,

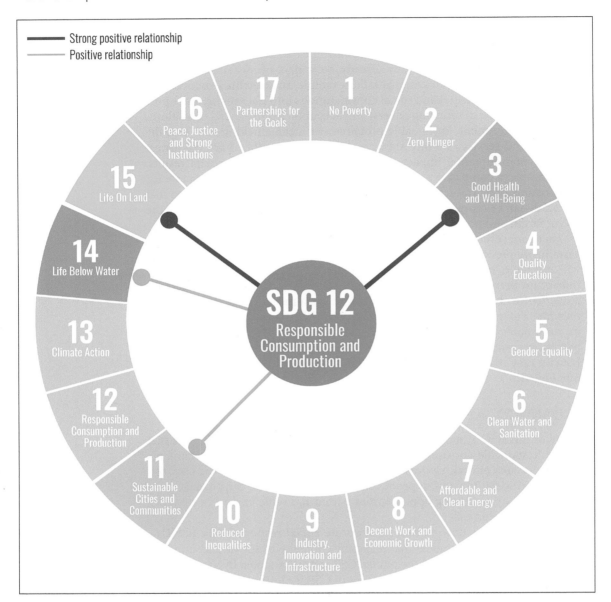

- Strong positive relationship
- Positive relationship

FIGURE 12.6 Interconnections in SDG 12 and the Other Goals
Moving toward sustainable production and consumption connects to many of the SDGs; this figure highlights several of the stronger connections.

the company is using renewable energy in its factories and introducing a plant-based alternative to the polyurethane used in its toys, showing that even a company that relies on plastic can begin to transition to more sustainable production.

While there has been some good progress, challenges remain. Some CSR programs are initiated and run in an uncoordinated way by a variety of internal managers. Studies suggest that companies need to develop coherent strategies and that this should be an essential part of the job of every CEO and board.

While many of the targets of SDG 12 focus on businesses and governments, individual consumers still play a critical role in practicing responsible consumption and production. SDG 12 challenges an individual consumer to reduce consumption, reduce waste, and be more thoughtful about what he or she buys. For example, reducing consumption of plastic will reduce one of the main pollutants of the ocean. Making informed purchases and searching for sustainable options can also help make a difference.

Still another challenge is that many experts argue there is a need for more profound changes in patterns of consumption as well in production and argue that many of the targets for SDG 12 are not sufficiently bold to achieve such a transformation.

It is not clear that the COVID-19 pandemic has set back the targets in SDG 12 to the same degree it has with other SDGs. Some experts suggest the COVID-19 pandemic offers an opportunity to promote sustainable consumption and production in recovery plans, but there are few examples that have been implemented.

An analysis of the interlinkages between SDG 12 and the rest of the SDGs indicates that sustainable consumption and production can influence progress on targets in other SDGs. For example, sustainable management of natural resources (Target 12.2), including resource use efficiency of water–energy–material flows, may improve water quality (Target 6.3), promote the use of renewable energy (Target 7.3), and encourage waste minimization through 3Rs (Target 12.5) and sustainable use of ecosystems (Target 15.1).

The introduction of a circular economy (Target 12.5) would accelerate the shift away from fossil fuel to renewables, and create new jobs (Target 8.5) and business opportunities, thus reducing poverty and inequalities (Target 10.3). Finally, waste reduction and prevention of plastic and hazardous chemicals waste (Target 12.5) will reduce contamination of marine and terrestrial ecosystem and animal habitats (SDG 14 and 15), with impacts on human health as well (SDG 3). Figure 12.6 highlights the stronger reinforcing interconnections SDG 12 has with other SDGs.

QUESTIONS FOR DISCUSSION AND ACTIVITIES

1. Describe global trends and patterns in consumption and production and how these may impact the environment.

2. Explain what is meant by the idea that waste is socially constructed.

3. Of what does your personal waste stream consist? How much of this waste stream do you recycle? How much of this waste stream do you reuse?

4. Research the LCA of an item you use frequently—such as your cell phone or laptop— to determine its environmental and social impact.

5. If sustainability requires us to reduce consumption, would this inevitably lead to an economic crisis?

6. Much of the efforts around sustainable consumption and production have occurred voluntarily, rather than through government laws and regulations. Explain whether you think government should or should not become more involved in creating incentives or regulations to curb consumption.

7. As a consumer, do you buy brands from companies that have established sustainability sourcing? Why or why not?

8. Discuss how the three Es are present in SDG 12.

TERMS

circular economy
corporate social responsibility (CSR)
cradle-to-cradle (C2C)
ecological footprint
Environmental Curve
Jevons paradox
landfill
Life Cycle Assessment (LCA)
microplastic

natural capital
single-use plastic
solid waste
sustainable consumption and production
sustainable procurement
waste
waste hierarchy
waste stream
zero waste

RESOURCES USED AND SUGGESTED READINGS

Bartley, T. *Looking Behind the Label: Global Industries and the Conscientious Consumer*. Bloomington: Indiana University Press (2015).

Bennett, J. *Vibrant Matter: A Political Ecology of Things*. Durham, NC: Duke University Press (2009).

Boulding, K. "The Economics of the Coming Spaceship Earth," in G. Debell (ed.). *The Environment Handbook* (pp. 96–101). New York: Ballantine (1970).

Dauvergne, P. *The Shadows of Consumption: Consequences for the Global Environment*. Cambridge, MA: MIT Press (2008).

Ekvall, T., Liptow, C., and Miliutenko, S. "Single-Use Plastic Bags and Their Alternatives: Recommendations from Life Cycle Assessments." United Nations Environment Program (2020). https://wedocs.unep.org/20.500.11822/31932

Engler, M. *Designing America's Waste Landscape*. Baltimore and London: The Johns Hopkins University Press (2004).

Eriksen, M., Thiel, M., Prindiville, M., and Kiessling, T. "Microplastic: What Are the Solutions?" in M. Wagner and S. Lambert (eds.).

Freshwater Microplastics: The Handbook of Environmental Chemistry. Vol 58 (pp. 273–98). Cham, Switzerland: Springer (2018). https://doi.org/10.1007/978-3-319-61615-5_13

Fredericks, R. *Garbage Citizenship: Vital Infrastructures of Labor in Dakar, Senegal.* Durham, NC: Duke University Press (2018).

Gandy, M. *Recycling and the Politics of Urban Waste.* New York: St. Martin's Press (1994).

Geels, F. W., McMeekin, A., Mylan, J., and Southerton, D. "A Critical Appraisal of Sustainable Consumption and Production Research: The Reformist, Revolutionary and Reconfiguration Positions." *Global Environmental Change* 34: 1–12 (2015).

Hawken, P., Lovins, A., and Lovins, H. *Natural Capitalism: Creating the Next Industrial Revolution.* Boston: Little, Brown and Company (1999).

Hawkins, G. *The Ethics of Waste: How We Relate to Rubbish.* Lanham, MD: Rowman and Littlefield (2005).

Lacy, P., and Rutquvit, J. *Waste to Wealth: The Circular Economy Advantage.* New York: Springer (2015).

Lebreton, L., Slat, B., Ferrari, F. et al. "Evidence That the Great Pacific Garbage Patch Is Rapidly Accumulating Plastic." *Sci Rep* 8: 4666 (2018, March 22). https://doi.org/10.1038/s41598-018-22939-w

Long, J., Lacy, P., and Spindler, W. *The Circular Economy Handbook: Realizing the Circular Advantage.* London: Palgrave Macmillan (2020).

Mancini, C. *Garbage and Recycling.* New York: Greenhaven Press (2010).

McDonough, W., and Braungart, M. *Cradle to Cradle: Remaking the Way We Make Things.* New York: Farrar, Straus and Giroux (2002).

Melosi, M. *The Sanitary City: Urban Infrastructure in America from Colonial Times to the Present.* Baltimore: The Johns Hopkins University Press (2000).

Moss, K. "Here's What Could Go Wrong with the Circular Economy—and How to Keep It on Track," World Resources Institute (2019, August 28). https://www.wri.org/insights/heres-what-could-go-wrong-circular-economy-and-how-keep-it-track

Owen, D. *The Conundrum: How Scientific Innovation, Increased Efficiency, and Good Intentions Can Make Our Energy and Climate Problems Worse.* New York: Riverhead Books (2011).

Redclift, M. *Wasted: Counting the Costs of Global Consumption.* London: Earthscan (1996).

Reisch, L., and Thøgersen, J. (eds.). *Handbook of Research on Sustainable Consumption.* Northampton, UK: Edward Elgar (2015).

Stahel, W. *The Circular Economy: A User's Guide.* New York: Routledge (2019).

Strasser, S. *Waste and Want: A Social History of Trash.* New York: Owl Books (2000).

Suzman, J. *Affluence without Abundance: The Disappearing World of the Bushmen.* New York: Bloomsbury (2017).

Vogel, D. *The Market for Virtue: The Potential and Limits of Corporate Social Responsibility.* Washington, DC: The Brookings Institution (2006).

Weetman, C. *A Circular Economy Handbook: How to Build and More Resilient, Competitive and Sustainable Business.* London: Krogran Page (2020).

World Economic Forum. "Social Innovation: A Guide to Achieving Corporate and Societal Value" (2016). https://www3.weforum.org/docs/WEF_Social_Innovation_Guide.pdf

World Economic Forum. "Platform for Accelerating the Circular Economy" (2018). https://www3.weforum.org/docs/WEF_PACE_Platform_for_Accelerating_the_Circular_Economy.pdf

Climate Change

After reading this chapter, you should be able to:

- Explain the causes of climate change

- Describe the numerous impacts of climate change

- Describe efforts at climate mitigation and climate adaptation

- Explain international efforts to address climate change

- Identify and explain the various targets and goals associated with SDG 13

- Explain how the three Es are present in SDG 13

- Explain how SDG 13 strategies connect to other SDGs

It is increasingly likely that hundreds of thousands of people will be forced from their homes due to climate change in the next several decades.

The small, low-lying island nations of the Pacific are most vulnerable to sea-level rise as well as the increased cyclonic activity associated with climate change. The highest point in the Marshall Islands is only thirty-three feet above sea level, Kiribati has an average elevation of only seven feet, and the Maldives averages four feet in elevation (Photo 13.1). On many islands, the majority of people live close to the shore and most of the infrastructures on the island chains are within a quarter of a mile of the coastlines. Recent climate models predict waves will become higher and more of the land will be inundated with salt water, even during normal tides and especially during storms, killing off vegetation, plants, and crops, and polluting fresh water sources. And there are no hills to escape to higher ground.

For Kiribati, Samoa, Vanuatu, and many of the other islands in the South Pacific, these changes could force the evacuation of the entire island population. The former president of the Maldives, Mohamed Nasheed, announced plans to purchase new land in India, Sri Lanka, or Australia, using funds generated by tourism. He explained his intentions: "We do not want to leave the Maldives, but we also do not want to be climate refugees living in tents for decades." Similarly, the government of Kiribati has purchased twelve square miles of land in Fiji as a possible future settlement in the event of forced evacuation, but this is a less than ideal solution.

A major challenge for the small island countries of the South Pacific is that they are small in population, resources, and wider political influence. They are so small and so distant from centers of political power that they have little voice in climate change discussion and negotiations. With the lowest carbon footprints on the planet, they may pay the heaviest price: the real threat of territorial extinction.

PHOTO 13.1 Malé, Maldives
This is a low-lying atoll subject to potentially devastating impacts with sea-level rise. Will the residents of the Maldives become climate refugees?

CLIMATE CHANGE: UNDERSTANDING THE ISSUE

In its 2021 report on climate change, the **Intergovernmental Panel on Climate Change (IPCC)** states: "It is unequivocal that human influence has warmed the atmosphere, ocean and land. Widespread and rapid changes in the atmosphere, ocean, cryosphere and biosphere have occurred." **Climate change** has gone from an inconvenient truth to an undeniable reality.

The IPCC is the UN body that assesses the science related to climate change. It was established by the UN Environment Program (UNEP) and the World Meteorological Organization (WMO) in 1988 to provide policy makers with regular scientific assessments concerning climate change, its implications, and potential future risks, and to suggest adaptation and mitigation strategies. The IPCC issues comprehensive assessments every five to six years, most recently in 2021.

The IPCC does not carry out new research. Instead, experts nominated by governments from around the world synthesize hundreds of methodology reports, technical papers, and published and peer-reviewed scientific technical literature. Each assessment is reviewed by hundreds (sometimes thousands) of expert reviewers, including scientists, industry representatives, and nongovernmental organization experts who offer a wide range of perspectives.

For more than twenty years, the IPCC has provided balanced information for policy makers by rigorously assessing published scientific studies on climate change. Their 2021 report includes stronger evidence of the many ways the planet is already experiencing climate change such as rising average land and ocean temperatures, sea-level rise, shrinking glaciers, decreasing snow and ice cover, hydrologic system changes such as droughts and floods, and increased frequency of extreme weather events.

Climate Science Basics

Climate change is linked to **greenhouse gas (GHG)** emissions. In report after report, the IPCC has expressed increasing certainty over the connections between climate change and anthropogenic (human-caused) GHG emissions. The 2021 IPCC report states "observed increases in well-mixed greenhouse gas (GHG) concentrations since around 1750 are *unequivocally* caused by human activities."

GHGs absorb some of the energy that is radiated from the surface of the earth and delays it in the atmosphere like a blanket, making the earth's surface warmer than it would be otherwise. This is called the **greenhouse effect** (Figure 13.1). Most GHGs are naturally occurring and are necessary for life; without them, the planet's surface would be about 15°C (60°F) colder, enough to freeze the oceans. According to NOAA and NASA data, the Earth's average surface temperature has increased by about 1°C (about 1.9°F) (Figure 13.2). While this increase may seem small, climate change is not just about warming temperatures but also about the instability that these changes cause in the systems that humans depend on, such as our hydrologic system.

It is important to distinguish between the terms "global warming" and "climate change." **Global warming** refers to the long-term rise in global temperatures due mainly to the increasing concentrations of GHGs in the atmosphere. Climate change encompasses global warming but refers to a broader range of changes occurring, such as precipitation, wind patterns, temperature, rising sea levels, shrinking mountain glaciers, accelerating ice melt, and shift in flower/plant blooming times.

GHGs include water vapor, methane, nitrous oxide, ozone, chlorofluorocarbons, and carbon dioxide (CO_2). Chlorofluorocarbons such as CFCs (used as refrigerants, solvents, and fire retardants) are entirely anthropogenic (man-made). Of the GHGs, CO_2, methane, nitrous oxide, and **fluorinated gases** are the most concerning; normally, these gases are continuously emitted into and removed from the atmosphere by natural processes on

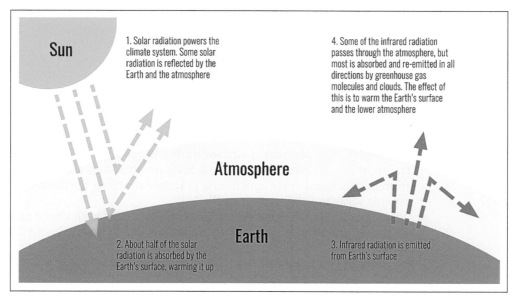

FIGURE 13.1 The Greenhouse Effect

The greenhouse effect occurs when GHGs absorb some of the energy that is radiated from the surface of the earth and delays it in the atmosphere like a blanket, making the earth's surface warmer than it would be otherwise. This process has been established scientific fact for many decades.

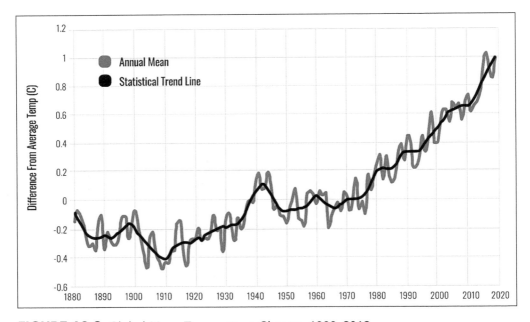

FIGURE 13.2 Global Mean Temperature Change, 1880–2019
This shows clear upward trends. According to NOAA and NASA data, the Earth's average surface temperature has increased by about 1°C (about 1.9°F). While this increase may seem small, it can generate significant instability in the climate systems that humans depend on, such as our hydrologic system.

Earth. Anthropogenic activities, however, can increase the amount of GHGs emitted and changes the global average atmospheric concentrations of each GHG. Most anthropogenic emissions come from the combustion of carbon-based fuels, principally coal, oil, and natural gas. The primary sources of each GHG are:

- Carbon dioxide (CO_2): CO_2 enters the atmosphere through burning fossil fuels (coal, natural gas, and oil), solid waste, trees and wood products, and certain chemical reactions (e.g., manufacture of cement). CO_2 is removed, or sequestered, from the atmosphere when it is absorbed by plants and oceans as part of the biological carbon cycle. Currently CO_2 accounts for about 82 percent of all US GHG emissions from human activities.
- Methane (CH_4): Methane is emitted during the production and transport of coal, natural gas, and oil. Methane emissions also result from livestock and other agricultural practices and from the decay of organic waste in solid waste landfills.
- Nitrous oxide (N_2O): **Nitrous oxide** is emitted during agricultural and industrial activities, as well as during combustion of fossil fuels and solid waste.
- Fluorinated gases: Hydrofluorocarbons, perfluorocarbons, and sulfur hexafluoride are three synthetic (man-made), powerful GHGs that are emitted by a variety of industrial processes.

GHGs have been increasing since the beginning of the Industrial Revolution in 1800 and have intensified over the last two hundred years. Figure 13.3 shows CO_2 emissions for the past 800,000 years. Scientists measure the history of Earth's atmosphere and climate by examining ancient air bubbles trapped

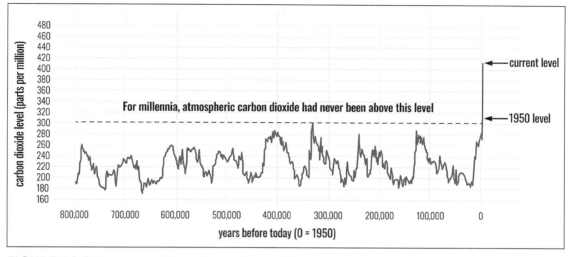

FIGURE 13.3 Increase in Atmospheric Carbon Dioxide
This graph, which compares atmospheric samples from ice cores to more recent, direct measurements, is evidence that atmospheric CO_2 has increased since the Industrial Revolution. According to the 2021 IPCC report, concentrations of CO_2 in the atmosphere are higher now than they have been at any time in the past two million years.

 ## KEY TERMS AND CONCEPTS 13.1

Past Climates

A key part of understanding climate change today is to understand climate changes in the past. However, a systematic record of weather (temperature, precipitation, etc.) has only been kept for about a hundred years and the direct and continuous measurements of CO_2 in the atmosphere only extend back to the 1950s.

To investigate climate change further back in time, experts use "proxies"—records that can provide data on secondary factors such as temperature, ice cover, and precipitation. These can include tree rings, pollen deposits, changes in landscapes, marine sediments, corals, and ice cores.

Ice cores are cylinders of ice drilled out of an ice sheet or glacier. These can be analyzed for the thickness, gas content (specifically CO_2 and methane), and isotopes of oxygen and hydrogen, all of which provide a direct archive of past atmospheric gases. This is because air is trapped in large ice sheets, and when the compacted snow turns to ice, the air is trapped in bubbles.

High rates of snow accumulation in Greenland, Antarctica, and the Arctic have preserved actual samples of the world's ancient atmosphere and have ice sheets that are several kilometers (miles) thick. Seasonal differences in the snow properties create layers—just like rings in trees.

By looking at past concentrations of GHGs in layers in ice cores, paleo climatologists can calculate how modern amounts of CO_2 and methane compare to those of the past and compare past concentrations of GHGs to temperature.

Ice core records show that the Earth's climate has oscillated between ice ages and warm periods. They provide direct information about how GHG concentrations have changed in the past and direct evidence that the climate can change abruptly under some circumstances. In the 800,000 years of ice core data, there is no period that contains the concentrations of CO_2 or methane comparable to those of the current period. The magnitude and rate of the recent increases of these gases are unprecedented.

in ice (see *Key Terms and Concepts* 13.1). According to the 2021 IPCC report, concentrations of CO_2 in the atmosphere are higher now than they have been at any time in the past two million years.

During the Ice Ages, CO_2 levels were around 200 parts per million (ppm), and during the warmer interglacial periods, they hovered around 280 ppm (shown in the fluctuations in the graph). In 2013, CO_2 levels surpassed 400 ppm; today, CO_2 levels are approximately at 420 ppm, the highest level in four million years. It is estimated that human activity adds forty billion metric tons of CO_2 pollution to the atmosphere each year. This relentless rise in CO_2 reveals a constant, direct relationship between fossil-fuel burning and CO_2 levels, based on the understanding that about 60 percent of fossil-fuel emissions stay in the air.

So, is there a genuine "climate debate"? Yes and no. The 2021 IPCC report states unequivocally that climate-warming trends over the past century are due to increases in human activities, and nearly all the leading scientific organizations worldwide have issued public statements endorsing this position. Climate scientists have used computer models to link anthropogenic GHG emissions to climate changes with very high levels of confidence. Scientists are increasingly certain that human activities are changing the composition of the atmosphere and thus the planet's climate; but they are less certain by how much, at what rate, or what (and where) the exact effects will be. Table 13.1 summarizes some of the facts researchers confidently claim about climate change.

There are several unanswered questions about climate change that require more research. These include:

- How much will climate change?
- How quickly will climate change?
- What will the impacts of climate change be?
- Where will the impacts of climate change be felt?

TABLE 13.1	Climate Change: The Facts
Fact #1	The greenhouse effect is established scientific fact.
Fact #2	The burning of fossil fuels releases CO_2, sulfur oxides, and nitrogen oxides.
Fact #3	Agriculture and industrial activity releases methane into the atmosphere.
Fact #4	Atmospheric CO_2 levels are now 46 percent higher than before the Industrial Revolution. This increase is due to fossil fuel usage and deforestation.
Fact #5	Methane shows a huge and unprecedented increase in concentration over the last two centuries.
Fact #6	The magnitude and rate of the recent increase are unprecedented over the last 800,000 years that scientists have analyzed using ice core samples.
Fact #7	Human-caused contributions of GHG emissions have dramatically increased in the last two hundred years and are the cause of increased GHG emissions.

Source: By author, compiled from several resources.

The Geography of Greenhouse Gas Emissions

Scientists are certain that the burning of fossil fuels is the major cause of the increase in GHG emissions and climate change. However, there is an uneven geography to the production of GHG emissions. Many high-income countries such as the United States, Canada, the United Kingdom, and Australia rely heavily on fossil fuels, which explains why these countries contribute significantly to climate change. For example, currently fossil fuels produce about 60 percent of US electricity and are responsible for 32 percent of CO_2 emissions (about 787 metric tons released in 2020), two-thirds of SO_2 emissions, and almost a quarter of NO_2 emissions. Another major source is transportation; planes, trains, and cars all use petroleum.

Rapidly growing economies in countries such as China, India, and Brazil similarly have increased emissions of GHGs. Since 2006, China has been emitting more CO_2 than any other country—however, it is also by far the most populous country (Map 13.1). Furthermore, if we examine total CO_2 emission per capita, the United States is the largest CO_2 emitter (Map 13.2). Figure 13.4 compares total emissions versus per capita emissions. The data only considers CO_2 emissions from the burning of fossil fuels and cement manufacture, but not emissions from land use such as deforestation.

The top ten largest emitter countries account for more than 67 percent of the global total. The global disparity in carbon footprints is profound (Figure 13.5). However, it is also important to understand that countries such as China and the United States emit on behalf of other countries, when factoring in the dependence on Chinese and American exports to the global economy. Recent research estimates that about 25 percent of emissions from China are due to the transportation of goods by rail or cargo ships going to other markets.

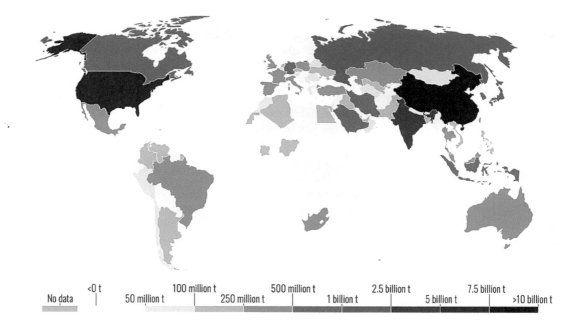

No data	<0 t		100 million t		500 million t		2.5 billion t		7.5 billion t	
		50 million t		250 million t		1 billion t		5 billion t		>10 billion t

MAP 13.1 Annual CO_2 Emissions (2019)
China and the United States are the two largest emitters of CO_2. Since 2006, China has been emitting more CO_2 than any other country—however, it is also by far the most populous country, which affects how we view the data. Also of note are those countries that emit very little—many countries in Africa, for example.

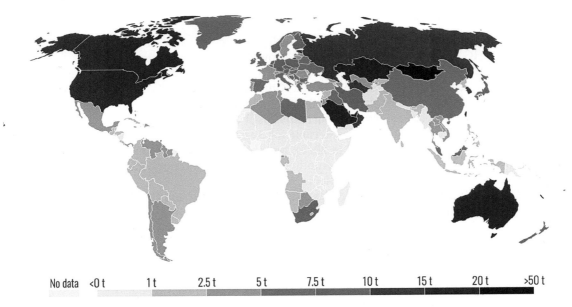

No data   <0 t   1 t   2.5 t   5 t   7.5 t   10 t   15 t   20 t   >50 t

MAP 13.2 Per Capita CO$_2$ Emissions (2019)
A map of emissions per capita highlights countries such as the United States, Canada, Australia, and Russia who have among the highest CO$_2$ emissions per capita. Other high emissions per capita include oil-producing countries such as Saudi Arabia, Oman, and the UAE, where small populations have access to very cheap oil resources. Mongolia and Kazakhstan are also high for different reasons. In Mongolia, large reserves of dirty coal supply most of the energy. In Kazakhstan, aging coal plants result in inefficient combustion of fossil fuels for energy.

While it is important to understand the geography of emissions, the interdependent nature of the global economy makes this a more nuanced issue. Countries with low emissions may depend on purchases of goods from the big economies of the United States and China. In addition, many countries whose emissions are low nevertheless are dependent on fossil fuels. At the same time, those countries with significantly large GHG emissions raise issues of equity, as climate change will impact all areas of the world, including countries that contribute very little GHGs.

THE IMPACTS OF CLIMATE CHANGE

Climate change is now affecting every country on every continent. Weather patterns are changing, sea levels are rising, weather events are becoming more extreme, and GHG emissions are now at their highest levels in history. These changes are already impacting natural and human systems and many land and ocean ecosystems. Climate change is a global challenge that does not respect national borders and requires solutions supported by international coordination.

The impacts of climate change are numerous and geographically diverse and may include:

- Increased extreme temperatures in many regions;
- Increases in frequency, intensity, and amount of heavy precipitation;
- Increases in intensity or frequency of droughts;
- Sea-level rise;
- Ocean warming;
- Species loss and extinction;

- Increases in poverty and disadvantage among some populations;
- Projected increases in risks from some vector-borne diseases, such as malaria and dengue fever; and
- Smaller net production in yields of maize, rice, wheat, and potentially other cereal crops, particularly in sub-Saharan Africa, Southeast Asia, and Central and South America.

Consider the evidence of climate change impacts in the United States. In 2021, a "heat dome" settled over the Pacific Northwest, trapping hot air and shattering temperature records in Oregon, Washington, and British Columbia. Seattle

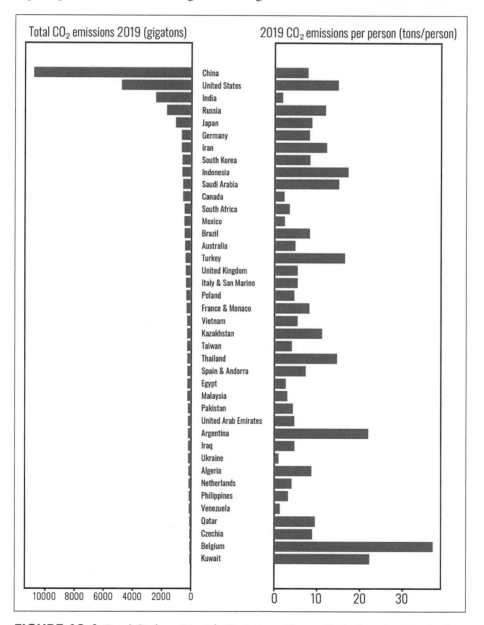

FIGURE 13.4 Total Carbon Dioxide Emissions Versus Emissions Per Capita for Selected Countries

The top ten largest emitter countries account for more than 67 percent of the global total.

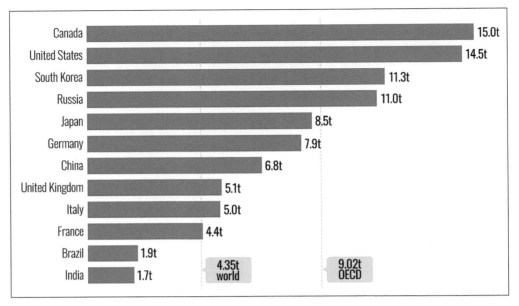

FIGURE 13.5 Largest Carbon Footprints for Selected Countries
The largest carbon footprints based on per capita CO_2 emissions. Access to abundant and cheap energy in countries such as Canada and the United States helps to explain the higher per capita emissions compared to other high-income countries such as Japan and Germany.

recorded three straight days of temperatures near or above 38°C (100°F), and in Lytton, British Columbia, temperatures soared above 49°C (121°F). It is estimated that several hundred people may have died in the heat wave. Over the last decade, parts of the United States have repeatedly experienced record-breaking floods, tornadoes, droughts, heat waves, and wildfires—sometimes in the same month.

Climate change is evident in many other countries and regions. For example, Australia has seen a poleward shift in the distribution of marine species, increases in the coral bleaching on the Great Barrier Reef, and increased firestorms. Like the firestorms that have recently destroyed parts of California, Australia has also been impacted by an increase in extreme fire risk days and bushfires (*Understanding the Issue* 13.1). In 2019, numerous bushfires burnt more than three million hectares and destroyed 700 houses. Some of these bushfires were so intense and expansive they were called "mega blazes."

The effects of climate change are not always as obvious in some places as it is in Australia. Bangladesh, for example, is one of the lowest-lying countries; nearly 25 percent of the country is less than seven feet above sea level. Recent measurements indicate that sea surface temperatures in the shallow Bay of Bengal have increased by 0.3°C in the last forty years. Floods and droughts have increased over the years and are a growing threat to plant and animal life. Higher ocean temperatures cause ocean acidification that affects numerous fish species that are a major part of the coastal food chain. Although the flooding has become pronounced, it is difficult to notice the difference between normal and abnormal changes because the low-lying Bangladesh typically experiences seasonal flooding associated with monsoons.

Some of the worst impacts are expected to be felt among populations who depend on agricultural and coastal economies such as fishermen, Indigenous

Q UNDERSTANDING THE ISSUE 13.1

Australia's Bushfires

Over the last several years, Australia has experienced record-breaking heat, drought, and devastating bushfires. In December 2019 and January 2020, fires scorched approximately sixteen million acres, an area the size of Belgium. Hot, dry weather combined with prolonged drought and strong winds created conditions for fire to spread rapidly. At one point, there were more than two hundred fires burning across the country (Photo 13.2). Traffic was gridlocked as people fled their homes; navy ships were called to rescue hundreds of people stranded on beaches.

The bushfires were so intense and generated so much heat that they created their own weather systems, spawning fire-generated thunderstorms, dry lightning storms, and fire tornadoes. The weather conditions are the results of the formation of pyrocumulonimbus clouds, created when intense heat from the fire causes air to rise rapidly, drawing in cooler air. As the cloud climbs and then cools in the low temperatures of the upper atmosphere, the collisions of ice particles in the higher parts of the cloud build up an electrical charge, which can be released as lightning. In turn, lightning strikes may ignite new fires. The rising air also spurs intense updrafts that suck in so much air that strong winds develop, causing a fire to burn hotter and spread further. In some places, the fires were traveling more than sixty miles per hour.

The bushfires killed animals and destroyed local habitats. An estimated 500 million animals died (not including frogs and insects). Some wildlife experts believe as many as 50 percent of all koalas may have died in the bushfires. Koalas live in the eucalyptus forests, where most of the fires occurred. These animals move slowly and were unable to escape death; 30 percent of koalas habitats may have been destroyed.

The fires also caused yet another disaster: massive air pollution. Air-quality index readings above 200 are considered hazardous; in early January 2020, readings at a Canberra monitoring site peaked at 7,700. The city all but shut down, experiencing the worst air quality in the world that

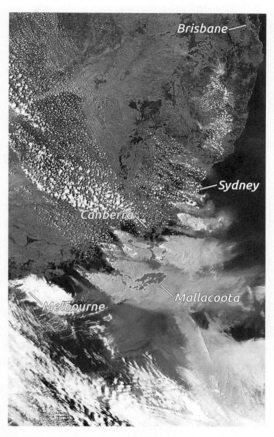

PHOTO 13.2 NASA Satellite Imagery of Bushfires in Australia
This shows satellite imagery on December 7, 2019, overlaid with markers showing bushfires across the east coast of Australia. Highlighted in red are fire detections, with notable landmarks labeled. Bush fires have become more frequent and more intense in Australia and the firestorms of 2019 were not an outlier.

week, far worse than Delhi, Lahore, and Shenyang, China (cities that typically rank high for the worst air quality). Health authorities advised people to stay inside, but smoke infiltrated everywhere, including schools and hospitals, where smoke particulate matter interfered with MRI machines.

Although bushfires are not uncommon, these fires were unprecedented. Some commentators describe this two-month period as a climate disaster; others spoke of climate grief, highlighting how difficult it can be to wrap your mind around enduring a months-long disaster.

Australian Prime Minister Scott Morrison was criticized for not handling the bushfire crisis effectively. His tenure as prime minister has been characterized by his refusal to acknowledge the link between the use of fossil fuels and climate change, and he has been generally supportive of the coal industry and fossil fuels. The fires have prompted a moment of political reckoning as Australians confront images of dead koalas and apocalyptic red-orange skies ablaze. As a result, anti-climate-policy politicians may not have enough support for reelection.

Source: Wikipedia. https://upload.wikimedia.org/wikipedia/commons/d/d2/2020-01-04_East_Australian_and_Mallacoota_Fires_Aqua_MODIS-VIIRS-LABELS.png Data captured from https://worldview.earthdata.nasa.gov

people, children and the elderly, poor laborers, poor urban dwellers in African cities, and people and ecosystems in the Arctic and Small Island Developing States.

A collaboration of climate scientists, economists, computational experts, and analysts at multiple universities are using data to calculate both costs and benefits of climate change and adaptation. The aim is to allow researchers to monetize those costs to society, referred to as the **social cost of carbon (SCC)**. SCC represents the monetary cost of the damages caused by the release of each additional ton of CO_2 into the atmosphere. For example, data has been used to estimate the long-term effects of climate change on mortality, labor productivity, coastal vulnerability, energy, and agriculture. In terms of mortality, for instance, air pollution from fossil fuels indirectly cause premature deaths each year; projections estimate an additional 85 deaths per 100,000 by 2100. The rising incidence of extreme heat may also cause substantial global losses in labor supply because many outdoor workers lack an effective means of adapting to such conditions. Calculations show that for every ton of carbon emitted, there are about $51 in "social" damages, such as increased mortality rates. This suggests that many governments, and the United States particularly, may be undervaluing the positive impacts of adopting more aggressive reductions in CO_2 emissions.

Humans are not the only ones impacted by climate change—ecological systems are also changing. Plant and animal communities are already experiencing shifts in ranges and abundance, and their ability to adapt to these changes is hampered by other human impacts such as land use change, invasive species, and overharvesting. We must be prepared to respond to changes in ecological communities.

Regional Variability

The 2021 IPCC report noted that the signs of climate change are increasingly apparent on smaller spatial scales. The high northern latitudes show the largest temperature increase with clear effects on sea ice and glaciers. The warming in the tropical regions is also apparent because the natural year-to-year variations in temperature there are small. Long-term changes in other variables such as rainfall and some weather and climate extremes have also now become apparent in many regions.

Rising temperatures amplify the exposure of small islands, low-lying coastal areas, and deltas to the risks associated with sea-level rise, including increased saltwater intrusion, flooding, and damage to infrastructure. Arctic ecosystems, dryland regions, and small island developing states are at a disproportionately higher risk of climate change.

A recent study showed that people of the Caribbean rely heavily on imported food and other goods and services, leaving them critically exposed to climate-related disruptions in transportation systems. Crop species crucial to the region's economies and food security—such as coffee, plantains, and mangoes—have evolved in narrower climatic niches relative to temperate crops and may be detrimentally affected by relatively small shifts in temperature, humidity, and rainfall. Disruptions from extreme climate-related events, such as droughts and hurricanes, can devastate large portions of local economies and cause widespread damage to crops, water supplies, infrastructure, and other critical resources and services. In small countries such as Dominica or the Bahamas, entire islands may be impacted by such changes.

There may also be regional variability of climate impacts within a country. Because a country like the United States is so geographically large, different regions have different vulnerabilities to climate change. For example, western states such as Arizona and Nevada may see an increased frequency of drought. States in the Midwest and Northeast may experience more frequent large storm events, wetter winters (more rain and snow), and heavier snowfall. Some cities located on rivers such as Kansas City, Cincinnati, Memphis, and Sacramento may witness rivers flood in the spring as more snowpack melts sooner but experience reduced river flows in the summer. Coastal cities such as Miami, San Diego, and New York City may experience sea-level rise resulting in increased flooding and storm surges. Map 13.3 highlights some of the regional differences with regard to climate change impacts and emphasizes how complex creating a "national" response can be when there are so many possible impacts and responses.

Despite the complexity of crafting a national response, every four years, the National Oceanic and Atmospheric Administration (NOAA) leads in preparing the US National Climate Assessment. Much like the IPCC reports, the National Climate Assessment is a high-level synthesis that analyzes the risks and vulnerabilities to the United States posed by climate change. The most recent National Climate Assessment was completed in 2018; the next report is currently underway. A major finding is that "human health and safety" and American "quality of life" is "increasingly vulnerable to the impacts of climate change, and without substantial and sustained global mitigation and regional adaptation efforts, climate change is expected to cause growing losses to American infrastructure and property and impede the rate of economic growth over this century."

Cities and Climate Change

As discussed in Chapter 11, cities around the world are home to about 4.4 billion people, and are already confronting climate change impacts, especially on their infrastructure systems. These include energy, water and wastewater, transportation, public health, banking and finance, telecommunications, food and agriculture, and information technology, among others. Already, climate change is causing damage to roads, buildings, and industrial facilities, and are posing an increasing risk to ports.

Globally, some 600 million people reside in coastal cities. These include cities such as Dhaka, Shanghai, São Paulo, and London. Coasts are affected by climate change in a variety of interconnected ways. Sea-level rise is caused

Northwest

Reduced snowpack and lower summer stream-flows impact water supplies

Sea level rise may increase erosion of coastlines, damaging infrastructure and ecosystems

Higher temperatures may increase pests and diseases, impacting forests, agriculture, and fish populations

Midwest

Water and wetter winters, springs with heavier precipitation, hotter and drier summers

Warmer temperatures may decrease agricultural yields

Reduced air quality and increased allergens

Northeast

More frequent heat waves will impact human health

Sea level rise and more frequent heavy rains may increase flooding and storm surges

Southwest

Increases in temperatures and more frequent and severe droughts

Greater water demand in growing cities could stress drinking water supplies

Higher temperatures will increase poor air quality, particularly ground-level ozone, and residents' vulnerability to heat-related illnesses

Drought, wildfire, invasive species, pests, and changes in species' geographic ranges will increase

Great Plains

Increases in temperature and drought frequency will further stress the High Plains Aquifer

An increase in frost-free days will lengthen the pollen season for common allergens, impacting public health

Southeast

Sea level rise may increase salinity of estuaries, coastal wetlands, tidal rivers, and swamps

Higher temperatures will strain water resources

Increase in extreme weather including hurricane activity

MAP 13.3 Impact of Climate Change in the United States by Region
The map highlights some of the regional differences with regard to climate change impacts on such a vast country.

by the thermal expansion of sea water due to warmer temperatures and a net increase of water due to melting ice. Sea-level rise will increase saltwater intrusion in proximate aquifers and groundwater and estuaries, affect coastal wetlands vegetation, and increase coastal erosion. The IPCC projects a global average rise in sea level of about two feet by 2100. The combination of sea-level rise and changing precipitation patterns, which include larger, more intense storms, is forcing many cities to prepare for more frequent and severe flood events.

Almost 40 percent of the US population, about 130 million people, lives in coastal counties that may be impacted by sea-level rise. A 2022 NOAA report projected that sea levels along the US shoreline will rise 25–30 centimeters (10–12 inches) by 2050. Sea levels will tend to be higher along the Atlantic and Gulf shores because of greater land subsidence there than in the Pacific. Sea-level rise is likely to result in more frequent episodes of coastal inundation associated with storm surges and high tides. Flooding along rivers, lakes, and in cities following heavy downpours, prolonged rains, and rapid melting of snowpack is exceeding the limits of flood protection infrastructure with outdated designs. Extreme heat is damaging transportation infrastructure such as roads, rail lines, and airport runways. The US National Climate Assessment Report notes that power outages and road and bridge damage are among the infrastructure failures that have occurred during extreme climate events. The study also notes that infrastructure is highly interconnected: climate-related disruptions of services in one infrastructure system will almost always result in disruptions of one or more other infrastructure systems. Miami Beach has launched a $400 million Sea-Level Rise Plan that requires construction of a series of storm water pumps, improved drainage systems, elevated roads, and higher seawalls. Currently, the city is raising one hundred miles of roads by two feet and the new wastewater treatment plant was constructed five feet higher than initial plans to account for climate change predictions.

Climate change is expected to increase the frequency and intensity of storms across the entire United States. However, the Northeast has recently experienced a greater increase in extreme precipitation than any other region in the United States with a more than 70 percent increase in the amount of precipitation falling in "very heavy events" (defined as the heaviest 1 percent of all daily events) between 1958 and 2010. The frequency of these heavy downpours is projected to continue increasing over the remainder of the century. More intense storms will overwhelm the nation's already aging water infrastructure, disrupt and damage city infrastructure, and degrade water quality in nearby waterbodies.

For example, flooding is the top hazard in Louisville due to its location on the banks of the Ohio River and proximity near eleven major stream systems (for a total of 790 stream miles). As a result, about 15 percent of the metro area lies within a floodplain. Between 2000 and 2017, the Federal Emergency Management Administration (FEMA) paid $35 million in flood-insurance payouts for Louisville infrastructure projects.

Concern about the impacts of climate change on the natural environment has motivated awareness of this issue in many places, particularly places that rely on environmental tourism. Juneau, Alaska has experienced a decrease in annual snowfall from approximately 109 inches to 93 inches and the rapid retreat of the Mendenhall Glacier. Beyond the impacts on tourism, Juneau is also concerned about impacts on its hydropower resources and on salmon and other marine fish and wildlife that are important to the economy.

The impacts of climate change are diverse and will vary from place to place, from region to region.

Interconnections: Climate Change and Water

Although many people associate climate change with rising temperatures, many places are most likely to feel climate change's first effects through water-based ecological changes. Climate change is already having measurable effects on the amount, distribution, precipitation, and availability of water. There are strong connections between SDG 13 and SDG 6 (Water and Sanitation).

In terms of water supply and availability, the water cycle is expected to undergo significant change because of climate change. For example, a warmer climate causes more water to evaporate from both land and oceans; in turn, a warmer atmosphere can hold more water. Changes like this are expected to lead to specific, and in many cases negative, consequences. Some areas can expect increased precipitation and runoff, especially in winter and spring, leading to increased flooding such as those that devastated Pakistan in 2022. Other areas can expect less precipitation, especially in the warm months, and longer, more severe droughts as storm tracks shift northward, leaving arid areas even drier. In some regions, droughts are exacerbating water scarcity and thereby negatively impacting people's health and productivity. Fresh water sources can also be impacted by floods that destroy water points and sanitation facilities and contaminate water sources.

The types of precipitation are also subject to change in response to warming; climate projections for areas of North America suggest less snow, overall, and more rain. In areas dependent on the gradual melting of snowpack to supply surface water through the warm months, this means lower flows and greater water stress in summer—a trend already underway in parts of the western United States. While the effects of climate change on groundwater are not fully understood, rising water competition and stress at the surface are likely to drive greater use—and overuse—of this resource.

In addition to water supply, climate may impact water quality. Water temperature, for example, usually rises in streams, lakes, and reservoirs as air temperature rises. This tends to lower levels of dissolved oxygen in water, hence more stress on the fish, insects, crustaceans, and other aquatic animals that rely on oxygen. Heavier rainstorms can increase surface runoff, or the water that flows over the ground after a storm. This moving water may strip nutrients from the soil and pick up pollutants such as sediments, nitrogen from agriculture, disease pathogens, pesticides, and herbicides. The pollution load in streams and rivers is often carried to larger bodies of water downstream—lakes, estuaries, and the coastal ocean—which can lead to one of the more dramatic consequences of heavy runoff: blooms of harmful algae and bacteria.

A Critical Perspective: Climate Justice

Climate change undermines efforts to achieve justice and reduce inequality across the world. There is evidence that climate change hurts those in poverty the most, both within and between countries, exacerbating inequality and hampering poverty reduction. Francesco Fuso Nerini and colleagues examined the ways in which climate-induced resource stresses—including on water, agricultural crops, or other biotic resources—could exacerbate competition and conflict, threaten the peace and inclusivity of societies, and undermine social justice. Climate change–related impacts and disasters are also key drivers of human displacement and mass migrations. Climate change is predicted to worsen gender inequalities, for example in cases in which girls are the first to be withdrawn from schooling in response to drought or other climate-related shocks. Climate-related disasters could lead to increased vulnerability of women and girls to violence, for example if they cause a shift in family power relations or lead to women and girls being vulnerably housed. Women's unequal access to economic resources can also compound their vulnerability to climate impacts.

In his book *Extreme Cities: The Peril and Promise of Urban Life in the Age of Climate Change*, Ashley Dawson examines the impacts of climate change on the world's megacities including Jakarta, Delhi, São Paulo, and New York. Using a political ecology perspective, he argues that neoliberalism and racial capitalism have already made these cities places of stark economic inequality (what he terms "extreme"); as a result vulnerable populations such as people of color and those experiencing poverty will be even more vulnerable to floods, sea-level rise, heat events, and so on. For example, in Indonesia the major brunt of climate change will be faced by the 26.5 million who live below the poverty line and have limited resources and capacity for resilience. Millions of these live in Jakarta. The climate shocks and stresses will also force the near-poor population hovering marginally above the national poverty line to fall into poverty. A political ecology perspective draws attention to the poverty climate nexus and underscores that climate actions need to be carefully designed so that they explicitly benefit the poor and near poor and do not inadvertently increase vulnerability and inequality.

In addition to the injustices that may fall disproportionately on the most vulnerable, the issue of climate change raises issues of intergenerational justice. The benefits for present generations seem to stand in conflict with the rights of future generations. Those who will feel the impacts of climate change the most have not yet been born.

The idea of **intergenerational justice** is that present generations have certain duties toward future generations. It can be difficult to think about the future beyond our lifetime and immediate descendants. Yet the idea of intergenerational justice is central to the most well-known definition of sustainable development. A key challenge for policy makers is to ensure that each generation can achieve a decent quality of life, not just the current generation.

An obstacle to intergenerational justice is that climate change can seem both distant and abstract, particularly by residents of more affluent countries, who perceive it as both distant in space (a threat to poorer nations) and distant in time (a threat to future generations). The impression that climate change is a relatively "long threat" compared with other kinds of intergenerational concerns, such as housing affordability, health care, and labor rights, is both a reflection of why it is perceived as a distant problem and of the time lag inherent in burning fossil fuels and the damage these emissions inflict over decades and centuries.

Research has shown that while many people feel strongly about protecting the future for their own children and grandchildren, extending intergenerational responsibility to encompass wider human groups and future generations is too abstract. Environmental issues such as climate change can seem "too big" and overwhelming.

In the last several years, the rights of children, youth, and future generations have become pivotal to the environmental movement's moral case for action on climate change. Young leaders such as Thunberg, Haven Coleman and others have gained international recognition for the School Climate Strike movement, under the name Fridays for Future. Recent school walkouts have seen young people calling for urgent action on climate change, while taking to task today's politicians for short-changing the future, claiming: "We are going to have to pay for the older generation's mistakes."

It is critical to understand that in a globally connected world, things that happen far away (in time or space) affect us. A strong appreciation for interdependency may help to develop a framework of intergenerational responsibility. It also challenges us to rethink how we talk about climate change, why we act, and for whom we act.

Inconsistent Global Leadership: The US Record on Climate Change

In the spring of 2017, the Trump administration announced it was withdrawing the United States from the **Paris Agreement**, turning its back on the fight against climate change. Currently, the United States is the only country to have rejected or withdrawn from the Paris Agreement. The US federal government has been lax in its commitment to accepting and arresting global climate change. As leader in GHG emissions, the failure of the United States to consistently address this issue in a significant way has been met with global disappointment and condemnation. The Trump administration's withdrawal from the Paris Agreement was only the latest step back in global leadership; previous administrations failed to act assertively as well (e.g., President George W. Bush announced the United States would not implement the Kyoto Protocol in 2001).

Following Trump's announcement, more than fifty US mayors immediately pledged to uphold the Paris Agreement. Governors from states like California and New York said they will keep pursuing their own programs to reduce emissions and the private sector is already shifting toward cleaner energy. This coalition of cities, states, and private-sector agents has been called "We Are Still In." As New York City stated: "Let others endlessly debate the causes (or even the existence) of climate change. New York City has chosen, once again, to act—by continuing to reduce its contribution to climate change and, at the same time, taking decisive and comprehensive steps to prepare and adapt." There are numerous organizations designed to help cities, states, and the private sector learn and adopt best practices. Other organizations besides the "We Are Still In" coalition include Beyond Carbon and the US Climate Alliance.

On his first day of office in 2021, President Biden rejoined the Paris Agreement. In April 2021, Biden announced a new target to achieve a 50–52 percent reduction in GHGs from 2005 levels. Biden's commitment to climate change seems serious, but how this will result in vigorous mitigation and adaptation over the next several years remains to be seen. No doubt, the inconsistency in US leadership over the last twenty years reminds us that climate change remains highly politicized. As a result, other countries are skeptical of the US commitment to taking action on climate change.

TAKING ACTION ON CLIMATE CHANGE

Table 13.2 presents the targets for addressing climate change.

A major challenge in preparing for the impacts of climate change is that it is a "wicked" complex problem. Many climate change impacts are cascading and interdependent: For instance, extreme heat events contribute to air-quality problems, while rising sea levels exacerbate the flooding problems caused by more intense storm events. Additionally, multiple climate impacts effect the same systems. For instance, in some places, climate resilient buildings must be built to withstand floods, use materials that ameliorate the urban heat island effect, and conserve both water and energy.

Governments at all scales (local, state, and national) are responding to climate change in two main ways: mitigation and adaptation. While mitigation addresses the causes of climate change, adaptation concerns preparing society, our infrastructure, and ecological systems for the effects. One of the most pressing questions today is: "Can places manage unavoidable changes and avoid unmanageable changes?"

TABLE 13.2   SDG 13 Targets and Indicators to Achieve by 2030

Target	Indicators
13.1 Strengthen resilience and adaptive capacity to climate-related hazards and natural disasters in all countries	**13.1.1** Number of deaths, missing persons, and persons affected by disaster per 100,000 people **13.1.2** Number of countries with national and local disaster risk-reduction strategies **13.1.3** Proportion of local governments that adopt and implement local disaster risk-reduction strategies in line with national disaster risk-reduction strategies
13.2 Integrate climate change measures into national policies, strategies, and planning	**3.2.1** Number of countries that have communicated the establishment or operationalization of an integrated policy/strategy/plan that increases their ability to adapt to the adverse impacts of climate change, and foster climate resilience and low GHG emissions development in a manner that does not threaten food production (including a national adaptation plan, nationally determined contribution, national communication, biennial update report, or other)
13.3 Improve education, awareness-raising, and human and institutional capacity on climate change mitigation, adaptation, impact reduction, and early warning	**13.3.1** Number of countries that have integrated mitigation, adaptation, impact reduction, and early warning into primary, secondary, and tertiary curricula **13.3.2** Number of countries that have communicated the strengthening of institutional, systemic and individual capacity building to implement adaptation, mitigation, and technology transfer, and development actions
13.A Implement the commitment undertaken by developed-country parties to the UN Framework Convention on Climate Change to a goal of mobilizing jointly $100 billion annually by 2020 from all sources to address the needs of developing countries in the context of meaningful mitigation actions and transparency on implementation and fully operationalize the Green Climate Fund through its capitalization as soon as possible	**13.A.1** Mobilized amount of United States dollars per year starting in 2020 accountable toward the $100 billion commitment
13.B Promote mechanisms for raising capacity for effective climate change–related planning and management in least developed countries and small island developing states, including focusing on women, youth, and local and marginalized communities * Acknowledging that the UN Framework Convention on Climate Change is the primary international, intergovernmental forum for negotiating the global response to climate change	**13.B.1** Number of least developed countries and small island developing states that are receiving specialized support, and amount of support, including finance, technology, and capacity building, for mechanisms for raising capacities for effective climate change–related planning and management, including focusing on women, youth, and local and marginalized communities

Source: United Nations Sustainable Development Goals. https://sustainabledevelopment.un.org/sdg13

Briefly, **mitigation** focuses on reducing the concentrations of GHGs in the atmosphere either by reducing their sources or increasing their sinks. Strategies include reducing fossil fuel use (by using renewables), promoting energy efficiency and conservation (encouraging weatherproofing and green buildings), and creating carbon sinks (carbon sinks—such as trees—sequester carbon). Many of the international agreements on climate change focus on mitigation efforts.

Adaptation, however, is the process of adjusting to our changing climate. In human systems, adaptation seeks to moderate or avoid harm from climate change. Examples include relocating critical infrastructure (e.g., electrical substations) to higher ground, enacting water conservation plans, and preparing disaster response and evacuation plans. Currently, climate adaptation is still an evolving field.

Take Action to Mitigate Climate Change

Around the world, countries, cities, and communities are acting on climate mitigation. Table 13.3 highlights some examples of mitigation. Mitigation focuses on reducing emissions. One strategy for this is for a country to introduce "carbon pricing" (charging of carbon), which promotes reduction by putting a price on CO_2 emissions and is considered to be an effective countermeasure to emissions. Sweden, Switzerland, Finland, and Norway have robust carbon taxes; the United States does not, but some US states such as California do. Carbon taxes can encourage the spread of products that emit less CO_2 and accelerate the development of energy-saving technologies.

Mitigation can also include anything that increases energy efficiency—such as retrofitting buildings to make them more energy efficient, moving toward electric vehicles and bus rapid transit. As explored in Chapter 7 on energy, GHG emissions can be reduced by making power on-site with renewables that

---

TABLE 13.3   Examples of Climate Mitigation

At the core of mitigation is stabilizing, then reducing, GHG emissions on a time scale that will prevent the substantial risks association with human-induced climate change. The following are actions toward mitigation.

- Set a cap on GHG emissions (this can be done at a variety of scales from a worldwide cap to countries, regions, and cities)
  - Example: the Paris Agreement on Climate Change
- Invest in and deploy an increasing percentage of energy in the form of renewable-energy technologies: wind, solar, geothermal, among others
- Remove fossil-fuel subsidies like depletion allowances, tax relief to consumers, support for oil and gas exploration
- Prevent tropical deforestation and encourage the planting of trees and other vegetation that can act as carbon sinks
- Increase energy efficiency by making energy conservation rules more stringent (e.g., building codes to require more insulation, use of energy-efficient lighting)
- Reduce the amount of fuels used in transportation by raising mileage standards
- Increase opportunities for mass transit
- Impose and increase carbon taxes on fuels
- Sequester CO_2 emitted by capturing it at the site of emission

Source: By author, compiled from several resources.

include rooftop solar panels, small-scale wind generation, and geothermal energy. The US EPA plans to lower GHGs by increasing car and truck fuel standards (how far a car or truck can travel on a gallon of fuel). These standards have been rising since 1975. In 1978, the vehicle fuel requirements of an average passenger car was about nineteen miles/gallon. Today, it is about forty miles/gallon for a car, thirty-one miles/gallon for a light truck.

Mitigation primarily focuses on reducing emissions, but it may also be achieved by increasing the capacity of carbon sinks (such as trees, bogs, coastal ecosystems, grasslands, and the ocean) that aim to remove CO_2 from the atmosphere. Carbon sinks can also be man-made, such as using underground formations to store CO_2. A Dublin-based company, Silicon Kingdom Holdings, announced its plans to install a pilot project of "mechanical trees" that will suck CO_2 from the air. The set of metal columns, or trees, are fitted with filters; it is hoped that the trees will capture about 36,500 metric tons of CO_2 a year, the equivalent of nearly 8,000 vehicles driven for a year. This technology is known as direct air capture.

Another mitigation strategy is to decarbonize agriculture and forestry. A 2020 McKinsey Global Institute study found that decarbonizing agriculture in Asia and preventing deforestation is a significant mitigation opportunity; agriculture and deforestation combined account for 10 percent of CO_2 emissions in Asia and more than 40 percent of methane emissions. Key strategies to reduce emissions in agriculture include promoting a shift from a diet rich in animal protein to plant-based protein, improving farming practices (such as dry direct seeding, improved rice paddy water management, and improved fertilization of rice), and promoting sustainable forestry (ending deforestation and scaling up reforestation). Another mitigation strategy to restore ecosystems is discussed in the *Solutions* 13.1 box.

 # SOLUTIONS 13.1

Ecosystem Restoration

Target 13.1: Strengthen resilience and adaptive capacity to climate-related hazards and natural disasters in all countries.

In early 2019, the Un General Assembly declared 2021–2030 the UN Decade on Ecosystem Restoration. The plan is to remove up to twenty-six gigatons of GHGs from the atmosphere by bringing at least 350 million hectares of degraded landscapes under active restoration by 2030.

Scientists say that restoring the world's forests by planting a trillion trees is a promising and cost-effective means of tackling climate change. But it has to be done with the right trees and at the right place and time.

Beyond sequestering carbon, these trees can guard against extreme weather events, protect endangered species, and bring shelter, food, money, and cultural preservation to communities around the world.

Ecosystem restoration can also have cobenefits. For example, it can create green jobs, restore biodiversity, help farmers make better incomes, stabilize water supply for big cities, and stabilize food supplies.

However, the way in which local or national governments perceive the financial costs (vs. benefits) may be one reason why ecosystem restoration projects have not been as widespread as they could be.

Source: UN Environment Program. "When We Protect Nature, Nature Protects Us" (2019, October). https://www.unenvironment.org/news-and-stories/story/when-we-protect-nature-nature-protects-us

In the United States federal-level measures to reduce GHG emissions include policies to promote advanced, low-carbon energy technologies and to increase energy efficiency. Federal measures also include regulations to phase down the use of hydrofluorocarbons and standards for reducing methane emissions from fossil fuel extraction and processing. In addition to mitigation efforts at the local or national scale, climate change mitigation has been the focus of several international agreements since the 1990s as discussed in *SDGs and the Law* 13.1.

SDGs AND THE LAW 13.1

The Paris Agreement on Climate Change

According to the IPCC, limiting global climate change to 1.5°C rather than 2°C above preindustrial levels would make it easier to achieve many aspects of sustainable development, especially eradicating poverty and reducing inequalities.

Climate change has been the subject of numerous international efforts, beginning in the 1990s with the creation of the IPCC. Afterward, the most significant effort was the 1997 Kyoto Protocol. The **Kyoto Protocol** set 2012 reduction targets for CO_2, nitrous oxide, methane, and sulphur emissions. Initially, the Kyoto Protocol focused on thirty-eight countries, many of which were the largest emitters. These countries agreed to reduce emissions of six GHGs to 5.2 percent below 1990 levels, by 2012.

Each country that signed the accord must also ratify the accord. The Kyoto Protocol required at least fifty-five countries to ratify it. The Kyoto Protocol was adopted in December 1997, but was not entered into force until February 2005. More than 182 parties signed the Kyoto Protocol. Although the United States signed the Kyoto Protocol, it was never ratified by the Senate. President George W. Bush opposed the Kyoto Protocol for two reasons: first, it exempted developing countries and second, it would cause serious harm to the US economy. Bush then retreated from his campaign promise to protect the environment. This move was seen by the international community as an abdication of leadership by the United States.

The follow-up to the Kyoto Protocol was the 2012 **Doha Agreement**. Many countries felt this agreement was weaker, and so declined to take on new targets. In addition, some countries felt any new agreement needed to include all countries, particularly the rapidly growing China, India, and Brazil.

The most recent international effort to address climate was the 2015 Paris Agreement. The **Paris Agreement** will work toward making sure the Earth's temperature does not rise more than 2°C (or 3.6°F) above preindustrial levels.

This temperature goal marks the upper limit that scientists and policy makers view as reasonable and acceptable risk; higher temperatures increase risks of irreversible change and even more dramatic effects of climate change. The Paris Agreement requires that ratifying nations "peak" their GHG emissions as soon as possible and pursue the highest possible ambition that each country can achieve. It established the so-called 20/20/20 targets:

• Reduce CO_2 emissions by 20 percent;
• Increase renewable energy's market share to 20 percent; and
• Increase energy efficiency by 20 percent.

Unlike the Kyoto Protocol, under the Paris Agreement, each country is required to create its own plan and regularly report on its efforts to mitigate climate change. A total of 195 nations—including the United States—signed the Paris Agreement, the first time that the world has collectively agreed on a path forward.

One of the arguments in favor of implementing climate mitigation is that it can have multiple benefits. For example, reducing GHG emissions means reducing air pollutants, which in turn improves air quality. Both mitigation and adaptation should be seen as opportunities for new economic strategies— sustainable green technologies and green jobs could counterbalance the costs of GHG reduction.

Take Action to Adapt to Climate Change

Climate change impacts vary by region, so adaptation efforts must be tailored to address localized impacts of climate change. This is especially challenging because current scientific modeling has trouble predicting the localized effects of climate change. The uncertainty and unpredictability of climate change impacts have led some to question if there is sufficient certainty to make major investments in adaptation efforts. Table 13.4 lists examples of climate adaptation.

Many countries and cities have developed climate adaptation plans. The United Kingdom was among the first countries in the world to take a long-term legislative approach to reducing GHGs through its Climate Change Act of 2008. Nearly every European city has a climate action plan that outlines

TABLE 13.4 Examples of Climate Adaptation	
Climate adaptation seeks to increase resiliency to climate change. The following are examples of climate adaptation:	
Policy for adaptation	• Develop climate adaptation plans and strategies at various scales (country, region, city) • Create emergency evacuation plans • Reconsider land-use zoning • Ensure new buildings and residences can withstand floods, use materials that ameliorate the urban heat island effect, and conserve both water and energy • Create disaster preparedness and response plans to better understand where people and structures are most vulnerable
Social-Economic investments	• Create community cooling centers, neighborhood watch programs • Build capacity to disseminate information to the community quickly • Help countries gain access to finance for building resilience and national capacity
Engineering approaches	• Relocate or elevate critical electrical, transportation, and water infrastructure to areas less vulnerable to flooding • Build floodwalls, dikes, and levees • Retrofit hospitals and nursing homes
Ecosystems-based adaptation	• Restore wetlands and sand dunes to better absorb storm intensity • Establish water conservation measures
Source: By author, compiled from several resources.	

both mitigation and adaptation measures and currently about one hundred US cities have developed plans specifically for climate adaptation. However, given funding and staff shortages and a multitude of other pressing demands, few of these plans are truly comprehensive and action-oriented. Instead, these plans tend to be overarching, big picture documents that do not yet propose a specific work plan.

Many cities approach climate change adaptation by focusing on disaster preparedness, either because they realize the importance of doing so or because there are state or national mandates to create disaster response plans. During this process, cities conduct climate change risk and vulnerability assessments that generally use regional climate modeling data to provide a more detailed review of the possible impacts to natural and human systems. Understanding areas of vulnerability can sometimes uncover the inequities of climate change: Some populations (like the elderly or those with asthma) might be more sensitive to exposures than others, and some populations (like low-income families) may have less resilience or adaptive capacity to recover after a disaster.

Cambodia, for example, is making investments in climate change adaptation projects. This is particularly important as climate change is likely to increase storms, cause deforestation, and raise sea levels, threatening communities living on the Cambodian coast and causing floods that destroy crops by swamping paddy fields and farmland with saltwater. At the same time, rising temperatures and increasing periods of drought are threatening people's livelihoods.

One project is attempting to protect low-lying rice fields and groundwater from saline intrusion and storms by building and strengthening dykes across the province. In one community, these efforts have stopped saltwater from entering an important rice-growing area by bolstering a seven-kilometer-long dyke that was breaking down due to sinking, thereby increasing flooding. The dyke has helped prevent seawater from getting into the fields and damaging the rice. As a result, rice yields have improved due to dykes rehabilitated by the project and from widespread training of the community on climate-resilient farming practices, salt-tolerant rice varieties, how to raise livestock and fish, and climate change threats. Another project aims to protect coastal land by supporting community-led planting of trees to prevent erosion, including mangroves—the first line of defense against the rising tide of climate change. Many countries in Asia are taking steps to implement climate adaptation, as discussed in the *Solutions* 13.2 box.

Not surprisingly, the most common climate change adaptation practices involve preparation for more frequent large storm events and flood prevention. Flooding poses many risks to a city such as public health and safety hazards, interruptions in key services, and damage to buildings and infrastructure. In addition, floods can disrupt transportation and hamper emergency services and evacuation efforts. Because gasoline and diesel pumps and sump pumps require electricity to operate, a power failure during a flood could limit access to fuel for operating cars and running generators and contribute to water damage in buildings.

The 2018 US National Climate Assessment report noted many states were adapting to climate change. Water managers in the Colorado River Basin have mobilized users to conserve water in response to ongoing drought intensified by higher temperatures. To address higher risks of

⚌ SOLUTIONS 13.2

Responding to Climate Change in Asia

Vast areas of East Asia and South Asia—and especially coastal areas—are exposed to physical climate risk. It is estimated that by 2050, between 600 million and one billion people in Asia will be living in areas with a very high annual probability of lethal heat waves. Additionally, damage from flooding may cost $1.2 trillion in a given year by 2050. Risks to infrastructure and supply chains will increase due to more frequent extreme precipitation events and typhoons. This is particularly important in China given its role in regional and global supply chains.

A 2020 McKinsey Global Institute report on climate risk and response in Asia highlighted some of the following climate adaptation projects that are already underway.

Protect people and assets
Measures to protect people and assets include hardening assets, such as reinforcing or elevating physical assets and infrastructure; building green defenses, such as restoring natural defenses and ecosystems; and building gray defenses that reduce the severity or duration of climate events, such as disaster-relief community shelters. For example, in a typical year, Kuala Lumpur experiences flash flooding. The Malaysian government has introduced flood controls by increasing river channel capacity, building a highway tunnel, and channeling water to hold-

ing ponds. The entire project provides storage for three million cubic meters of water, sufficient to offset most of the flooding in a typical year.

Build resilience
Yunnan and Guangxi provinces in southwest China are predominantly rural communities. Over the past ten years, pressure on water systems and frequent droughts have led to significant crop losses. One project to foster resilience helped farmers develop new maize varieties better adapted to drought and pests. In the Ladakh region of India, which relies on melting snow and ice from the Himalayas to irrigate its fields, as glaciers have shrunk, water supplies have declined. A solution was devised to store meltwater in huge standing structures, providing irrigation throughout the year.

Reduce exposure
One example of large-scale exposure reduction is the Indonesian government's 2019 decision to relocate the country's capital from Jakarta, parts of which may be submerged by 2050.

Source: McKinsey Global Institute. *Climate Response in Asia* (2022, November). https://www.mckinsey.com/business-functions/sustainability/our-insights/climate-risk-and-response-in-asia

---

flooding from heavy rainfall, local governments in southern Louisiana are pooling hazard reduction funds, and in Alaska, a tribal health organization is developing adaptation strategies to address physical and mental health challenges driven by climate change and other environmental changes. Forest managers in the Northwest are developing adaptation strategies in response to wildfire increases that affect human health, water resources, timber production, fish and wildlife, and recreation. After extensive hurricane damage fueled in part by a warmer atmosphere and warmer, higher seas, communities in Texas are considering ways to rebuild more resilient infrastructure.

In addition to national- or state-level plans, many cities are researching future scenarios of climate change. For example, San Francisco has assessed the impacts of sea-level rise on the city based on two different sea level rise scenarios over the next fifty years. Researchers have concluded that the city would experience an increased likelihood and intensity of storm surges, high tide flooding, and shoreline erosion. The report also estimates the economic impacts on people, transportation, and property. At the low end, a one-meter (three-foot) sea level rise could cause $50 billion in damages to property and displace about 220,000 residents; at the high end, a 1.4-meter (4.5-foot) sea-level rise could cause $62 billion in damages and displace almost 270,000 residents. The Embarcadero Waterfront, the city's commercial and tourist center, would be most affected out of all neighborhoods and the study predicts $4 billion worth of waterfront property to be at risk by 2100. Putting dollar amounts on these scenarios has helped to motivate endorsement for mitigation and adaptation investment.

Building Climate Resilience in Impoverished Communities

As we seek to mitigate and adapt to climate change, we must be mindful of the injustices that may fall disproportionately on the most vulnerable, those experiencing poverty. Climate change impacts on the livelihoods of both rural and urban poor require a combination of measures to strengthen resilience. A 2022 study by the Asian Development Bank on climate risk and poverty listed a combination of actions as critical for promoting resilient livelihoods for the urban poor. First, strengthen copying mechanisms, such as stockpiling food for flood seasons. Second, implement incremental adaptation to accommodate changes from climate change such as building higher dikes to protect slums from increased floods. Third, and more challenging, is to undertake transformational adaptation that introduces fundamental system change by addressing the root causes of vulnerability to climate change such as land-use changes that introduce nature-based solutions to manage flooding, and the involvement of local women in protecting natural resources.

The report suggested governments build resilience among the poor by increasing social protection measures, adapting public health systems, and creating policies that ensure safe housing and robust community infrastructure. For example, propoor policies that enhance social and economic safety nets and improve education and job skills can help strengthen resilience. In terms of public health systems, building resilience will require health policies and plans that recognize, predict, and provide services due to the likely health impacts of climate change (such as increased heat stress). Health-care systems could introduce new heat-stress-related programs that support those who work outdoors. In addition, safe housing will be critical if disasters and weather impacts damage the housing of low-income households that have used substandard materials or are more exposed to hazards. Programs that help the poor upgrade their homes or land purchase initiatives, for example, can build resilience. The report recommends governments adopt "no regret" or "low regret" solutions: In other words, see the value of poverty reduction strategies that generate social and economic benefits irrespective of how the future climate pans out.

SUMMARY AND PROGRESS

To date, progress on climate change is elusive. The last decade has been the warmest ever recorded, and CO_2 and other GHGs continue to rise to new

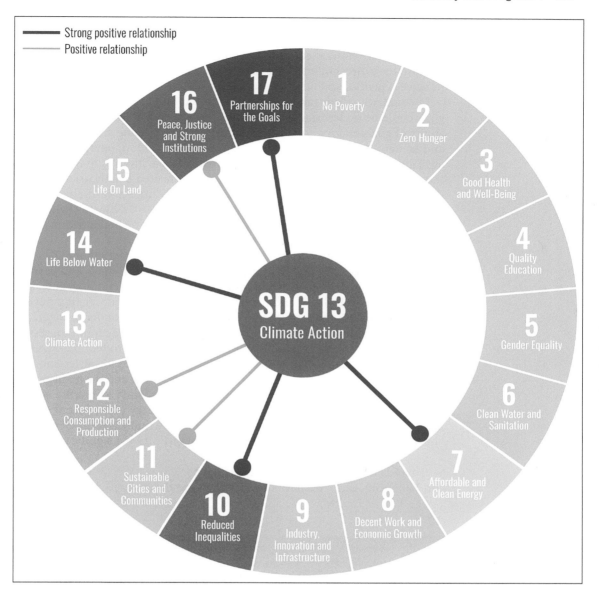

FIGURE 13.6 Interconnections in SDG 13 and the Other Goals
Addressing climate change may also advance targets in many other SDGs. This figure highlights several of the stronger connections. The connection with energy is among the strongest: Continued use of fossil fuels will accelerate climate change.

record levels. The Paris Agreement to hold global temperatures 2°C above preindustrial levels is a start, but by no means will it resolve the issue.

The COVID-19 pandemic lowered GHG emissions briefly due to travel bans and economic slowdowns. However, global emissions have generally returned to prepandemic levels. Unfortunately, few governments took the opportunity to shape recovery plans that include long-term systemic shifts to address climate change and lower GHGs. In fact, it appears that many encouraged a return to normal in terms of economic growth.

Much of the action taken to address climate change will also help advance SDG targets on renewable energy, human health, ecosystem health, and human rights (see Figure 13.6). Francesco Fuso Nerini and colleagues have examined the interconnections among climate change and the other SDGs. Their research shows that actions taken to mitigate or adapt to climate change can also have direct interactions with development goals that can involve both positive synergies and negative trade-offs.

If countries fail to take climate action, sixty-seven targets across sixteen goals could be undermined. Specifically, climate change will affect the achievability of goals relating to material and physical well-being such as prosperity and welfare, poverty eradication and employment, food, energy, and water availability, and health. For example, climate change impacts may exacerbate the distribution of disease vectors and disaster-related health risks (SDG 3). Climate change-driven water shortages can directly impact health by reducing access to clean drinking water and sanitation (SDG 6). Climate change may also impact the productivity of agricultural lands, causing malnutrition as well as loss of livelihoods and prosperity (SDGs 1 and 2).

However, acting on climate change can positively impact other SDGs. Climate action can enable and reinforce building prosperous, equal, and peaceful societies. It provides a foundation for building strong, functioning, and capable institutions (SDG 17), and has synergies with targets concerning poverty reduction, welfare, and jobs (SDG 1). Many of the targets on food, water, and energy systems are reinforcing or indivisible with climate action. Progress on several targets concerning sustainable consumption and production (SDG 12) will advance climate action by reducing emissions related to wastes and production. There are also synergies between climate action and the management and conservation of other environmental resources, such as marine ecosystems (SDG 14). Finally, the relationship between energy (SDG 7) and climate change is very strong. Energy resources are directly implicated in GHG emissions; moving toward clean, renewable energy will have a significant impact on reducing GHG emissions.

There is no doubt climate change is happening. As World Bank President Jim Yong Kim said, "We have to wake up to the fierce urgency of now."

QUESTIONS FOR DISCUSSION AND ACTIVITIES

1. Explain the causes of climate change since the 1850s.

2. Describe three impacts of climate change.

3. Why do you think the United States has lacked a strong and consistent national response to climate change?

4. What evidence of climate change have you observed in your local community?

5. What efforts have you observed around mitigation and/or adaptation in your community? If local efforts to address climate change are effective, to what degree do we need strong national leadership in the United States?

6. In what ways will climate change disproportionately impact the most vulnerable?

7. What steps have you taken to lower your carbon footprint?

8. Does the Paris Agreement do enough to combat climate change? If not, explain what an effective agreement would include.

9. How does a political economy/political ecology framework help explain climate injustice? Is climate change an inevitable result of capitalism?

10. Discuss the role of food systems in contributing to climate change. Discuss how climate change impacts food production and yield.

TERMS

adaptation
carbon dioxide (CO_2)
climate change
Doha Agreement
fluorinated gases
global warming
greenhouse gases (GHGs)
greenhouse effect
intergenerational justice

Intergovernmental Panel on Climate Change
Kyoto Protocol
methane
mitigation
nitrous oxide
Paris Agreement
social costs of carbon (SCC)

RESOURCES USED AND SUGGESTED READINGS

Asian Development Bank. *Building Resilience of the Urban Poor in Indonesia.* Manila, Philippines: Asian Development Bank (2022, January). https://www.adb.org/sites/default/files/publication/763146/building-resilience-urban-poor-indonesia.pdf

Dawson, A. *Extreme Cities: The Peril and Promise of Urban Life in the Age of Climate Change.* New York: Verso (2017).

Dessler, A., and Parson, E. *Science and Politics of Global Change.* 3rd ed. New York: Cambridge University Press (2019).

Emanuel, K. *What We Know About Climate Change.* Cambridge, MA: MIT Press (2018).

Fuso Nerini, F., Sovacool, B., Hughes, N., Cozzi, L., Cosgrave, E., Howells, M., Tavoni, M., Tomei, J., Zerriffi, H., and Milligan, B. "Connecting Climate Action with Other Sustainable Development Goals." *Nature Sustainability* 2: 674–80 (2019, August). https://doi.org/10.1038/s41893-019-0334-y

Gorman, S. "U.S. Sea Level to Rise by 2050 as Much as in Past Century, NOAA Says." *Environment* (2022, February 16). https://apple.news/AyVZPDldUThun46vfVwlH3w

Henson, R. *The Thinking Person's Guide to Climate Change.* 2nd ed. Boston: American Meteorological Society (2019).

"Infrastructure." National Climate Assessment (2014). https://nca2014.globalchange.gov/highlights/report-findings/infrastructure

Intergovernmental Panel on Climate Change. *Climate Change 2021: The Physical Science Basis. Contribution of Working Group I to the Sixth Assessment Report of the Intergovernmental Panel on Climate Change.* Masson-Delmotte, V., Zhai, P., Pirani, A., Connors, S. L., Péan, C., Berger, S., Caud, N., Chen, Y., Goldfarb, L., Gomis, M. I., Huang, M., Leitzell, K., Lonnoy, E., Matthews, J. B. R., Maycock, T. K., Waterfield, T., Yelekçi, O., Yu, T., and Zhou, B. (eds.). Cambridge: Cambridge University Press (2021).

Intergovernmental Panel on Climate Change. "Working Group I: The Physical Science Basis." (2021, August) https://www.ipcc.ch/working-group/wg1/

Intergovernmental Panel on Climate Change. Six summary reports and other special reports can be found at https://www.ipcc.ch/

Johnson, A. E., and Wilkenson, K. (eds.). *All We Can Save: Truth, Courage, and Solutions for the Climate Crisis.* New York: One World/Penguin (2020).

Rajalakshmi, P. R., and Achyuthan, H. "Climate Change as Observed in the Bay of Bengal." *Journal of Climate Change* 7 (3): 69–82 (2021).

US Congress House Committee on Oversight and Reform. Subcommittee on Environment. Hearing on "Economics of Climate Change." 116th Cong., 1st sess. (2019, December 19).

US Environmental Protection Agency. "Climate Change in Coastal Communities" (2020, November). https://www.epa.gov/cre/climate-change-coastal-communities

US Global Change Research Program. *National Climate Assessment.* Washington, DC (2018). https://nca2018.globalchange.gov/chapter/front-matter-about/

US Global Change Research Program. "US Caribbean." *National Climate Assessment.* Washington, DC (2018). https://nca2018.globalchange.gov/chapter/20/

United Nations Environment Programme. "Climate Change" (2020). https://www.unenvironment.org/explore-topics/climate-change.

United Nations Environment Programme. "From Rice to Riches: Adapting to Climate Change on Cambodia's Coasts" (2019, January 28). https://www.unenvironment.org/news-and-stories/story/rice-riches-adapting-climate-change-cambodias-coasts

The Ocean

LEARNING OBJECTIVES

After reading this chapter, you should be able to:

- Describe the challenges facing the ocean

- Explain the major causes of marine pollution

- Describe the major impacts of marine pollution

- Identify and explain the various targets and goals associated with SDG 14

- Explain the strategies for implementing the targets of SDG 14

- Explain how the three Es are present in SDG 14

- Explain how SDG 14 strategies connect to other SDGs

The Great Pacific Garbage Patch has an estimated 1.8 trillion pieces of plastic weighing a total of 80,000 tons.

The Great Pacific Garbage Patch is the name for a collection of marine debris that accumulated over time due to converging ocean currents called **gyres** (see Map 14.1). Gyres are unique forms of ocean currents because they are large systems that move in a swirling motion, leaving marine debris caught in their path to move and be trapped in the calmer, more stable middle of the gyre. Two segments of the Great Pacific Garbage Patch exist in the Pacific Ocean: the Western Garbage Patch, which is closer to Japan, and the Eastern Garbage Patch, which is closer to southern California and Mexico.

The Great Pacific Garbage Patch formed due to the buoyant and durable nature of plastic, and particularly microplastics, which can float for years in the ocean without breaking down. Some of the plastic in the patch is more than fifty years old and includes items (and fragments of items) such as water bottles, toothbrushes, pens, plastic containers, cell phones, and plastic bags. Larger and heavier pieces of debris sink, and there is more garbage below the surface of the two patches.

Marine debris can cause serious and harmful effects. Every year, more than a million seabirds and 100,000 mammals are killed by plastic debris. Lost fishing nets are especially dangerous. These are often called "ghost" nets because they continue to fish even though they are no longer under the control of a fisher. Ghost nets can trap or wrap around migratory animals like whales, turtles, or sharks, entangling them. Plastic debris with loops (such as six-pack rings and handles of plastic bags) can also get hooked on wildlife.

Sea turtles, fish, and sea birds often mistake plastic for food. The plastics take up room in their stomachs, making the animals feel full and stopping them from eating real food; some even die. Microplastics and other debris can also harm plankton and algae by blocking the sunlight they need to be able to undergo photosynthesis and produce energy for themselves, which causes harm throughout the marine food web.

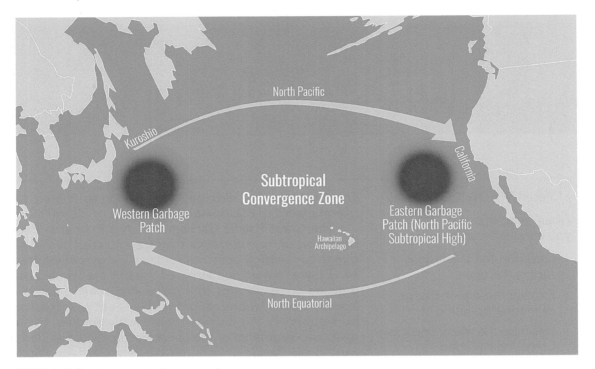

MAP 14.1 The Two Garbage Patches
Garbage and marine debris accumulate over time due to converging ocean currents called gyres.
Most of the garbage patch is not visible from the surface, as most of the heavier debris floats below.

THE OCEAN: UNDERSTANDING THE ISSUE

The ocean is the defining physical feature on our planet Earth—covering approximately 70 percent of the planet's surface (Photo 14.1). Waters flow and intermingle all around the globe. Technically, there is one global ocean with many ocean basins or regions, such as the North Pacific, South Pacific, North Atlantic, South Atlantic, Indian, Southern, and Arctic (see Map 14.2). The ocean is connected to major lakes, watersheds, and waterways because all major watersheds on Earth drain to the ocean. Rivers and streams transport nutrients, salts, sediments, and pollutants from watersheds to coastal estuaries and to the ocean.

The ocean transports heat from the equator to the poles and regulates climate and weather patterns in a global circulation system called the thermohaline. The **thermohaline circulation** system moderates global weather and climate by absorbing most of the solar radiation reaching Earth. Heat exchange between the ocean and atmosphere drives the water cycle and oceanic and atmospheric circulation and can result in dramatic global and regional weather phenomena, impacting patterns of rain and drought. Significant examples include the El Niño Southern Oscillation and La Niña, which cause important changes in global weather patterns because they alter the sea surface temperature patterns in the Pacific. It is possible climate change could dramatically alter the thermohaline circulation systems, discussed in *Understanding the Issue* 14.1.

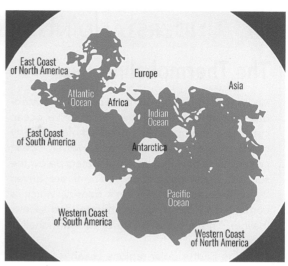

PHOTO 14.1 The Pacific Ocean
This infrared image from the GOES-11 satellite shows the vastness of the Pacific Ocean.

MAP 14.2 One Ocean
This map highlights the perspective of one ocean, comprised of numerous ocean regions that include the North Pacific, South Pacific, North Atlantic, South Atlantic, Indian, Southern, and Arctic.

The ocean also dominates Earth's carbon cycle. Half the primary productivity on Earth takes place in the sunlit layers of the ocean. It absorbs roughly half of all carbon dioxide and methane that are added to the atmosphere.

The ocean produces more than half the world's oxygen and stores fifty more times the world's carbon dioxide than the atmosphere. It is also the world's largest ecosystem, home to nearly a million known species and containing vast untapped potential for scientific discovery.

Billions of people rely on the ocean. Almost half the world's population lives within 150 kilometers (ninety miles) of the coast and, for many living in coastal communities, the ocean is not only a source of food and livelihoods but also it is an intrinsic part of culture and heritage. For humans, the ocean economy produces $3 trillion each year in goods and services, provides food and jobs for hundreds of millions, and transports 90 percent of international trade.

For too long, the ocean has been underrepresented in global environmental policy and protection. The ocean has absorbed the impacts of our pollution, marine debris, habitat degradation, and overfishing, threatening the health of the ocean in unprecedented ways. Map 14.3 highlights those marine ecosystems that have been impacted the most by human activities.

MARINE POLLUTION

Maintaining a healthy ocean is vital, yet pollution is threatening the health of the ocean. Virtually every pollutant present on land is also present in the ocean. In fact, more than 80 percent of all marine pollution originates on land. The ocean has long been used as a repository for sewage, nutrient runoffs, heavy metals, nuclear waste, persistent toxins, pharmaceuticals, garbage, and other land-based pollutants.

UNDERSTANDING THE ISSUE 14.1

The Thermohaline

An important function of the ocean is to distribute heat around the globe. Winds drive ocean currents in the upper one hundred meters (forty miles) of the ocean's surface. However, ocean currents also flow thousands of meters below the surface. These deep-ocean currents are driven by differences in the water's density, which is controlled by temperature (thermo) and salinity (haline). This process is known as thermohaline circulation.

In the Earth's polar regions ocean water gets very cold, forming sea ice. As a consequence, the surrounding seawater gets saltier because when sea ice forms the salt is left behind. As the seawater gets saltier, its density increases, and it starts to sink deep into the ocean where it moves along in a current until it reaches the equator. At the equator, heat from the sun then warms the cold water at the surface, and evaporation leaves the water saltier. The warm salty water is then carried northward; it joins the Gulf Stream, a large powerful ocean current that is also driven by winds. The warm salty water travels up the US East Coast, then crosses into the North Atlantic region where it releases heat and warms Western Europe. Once the water releases its heat and reaches the North Atlantic, it becomes very cold and dense again, and sinks to the deep ocean. The thermohaline circulation plays a key role in determining the climate of different regions of the earth.

It is unclear how the thermohaline circulation system may be impacted by climate change. Researchers are running models on how climate change could impact circulation patterns. In one scenario, increased temperatures and increased rainfall in the North Atlantic will increase the melting of glaciers and sea ice, and the influx of this warm freshwater onto the sea surface could block the formation of sea ice, disrupting the sinking of cold, salty water. This sequence of events could slow or even stop the conveyor belt, which could result in potentially drastic temperature changes in Europe. The thermohaline circulation system is also critical for moving nutrients from one part of the ocean to another; another scenario predicts a collapse in many marine species should the thermohaline circulation system slow or collapse.

Source: NOAA. "Thermohaline Circulation." https://www.climate.gov/climate-and-energy-topics/thermohaline-circulation

Pollutants tend to accumulate on land (in soils) and in water, and those places where they accumulate are referred to as **pollution sinks**. The Earth's largest sink for pollutants is the ocean, lakes, rivers, and reservoirs. Pollutants that enter a sink will take different paths, depending on their chemical and physical forms and the condition of the sink. Pollutants such as herbicides, lead, and mercury may be taken up by plants. Heavy metals such as lead and arsenic settle in the sediment. Recent excavations of bottom sediment from Boston Harbor found high concentrations of heavy metals even though these chemicals have been banned for many years. This makes cleanup of these types of pollutants more difficult and expensive, as the US Environmental Protection Agency (EPA) requires that these heavy metals be treated according to hazardous waste regulation.

Most pollutants enter the oceans from nonpoint sources via rivers, streams, and the atmosphere. Pollutants such as solid waste, oil spills, or industrial pollutants are often dumped directly into the oceans. As a result, some of the most polluted ocean waters are bays, estuaries, and the shallow waters of the continental shelves, where most marine life is concentrated.

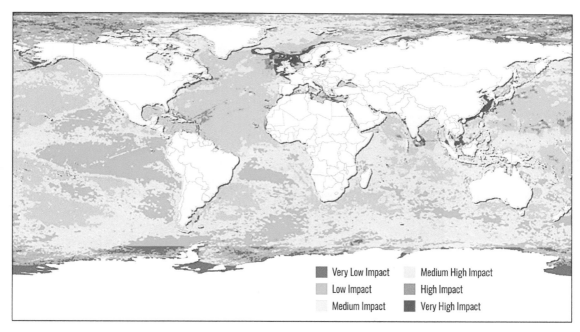

MAP 14.3 Human Impact on Marine Environments
This map shows the marine ecosystems that have been impacted the most by human activities on the ocean. There has been very high impact observed in the Atlantic in Northern Europe, parts of the Mediterranean, the Bay of Bengal, the South China Sea, and the East China Sea.

Land-Based Nonpoint Sources

A significant amount of nonpoint sources start on land and are transported via stormwater runoff. **Stormwater runoff** includes sediment and chemicals from farm fields, city streets, suburban lawns, and other dispersed sources. Stormwater makes its way to the ocean directly or indirectly, through rivers and lakes that eventually flow into the ocean. As discussed in Chapter 6, in urban areas, stormwater runoff can include inorganic and organic debris, fertilizers, heavy metals, and salts that wash off roads. In many cities partially treated or raw sewage is discharged into the sea during rains, adding the danger of disease-causing organisms. For example, more than 80 percent of sewage enters the Caribbean Sea untreated, making it the primary source of land-based marine pollution. More than half households in the region lack sewer connections.

Most of the sediment discharged into the oceans is deposited in shallow waters close to the shore and in and around large river deltas. **Sediment** includes soil, but such deposits can still have profound impacts. Sediment can come from farming, forest clearing, and construction. In urbanized coastal areas, sediment may also originate from the dredging of harbors and river mouths. Sediment loads from rivers such as the Ganges, Mekong, and Huang He (Yellow) are extensive and growing (Photo 14.2), but there are thousands of smaller watersheds where sediment is altering aquatic habitats. Coral reefs, for example, are especially susceptible to sediment damage because many coral species cannot survive under sediment in muddy water.

Many solid items, such as plastic bags, cigarettes, food wrappers, and glass bottles enter the ocean from stormwater runoff or from the dumping of municipal wastes (some cities or communities that do not have widespread waste-collection systems may dump their garbage into the ocean).

PHOTO 14.2 The Huang He (Yellow River) in China
The most sediment-filled river on Earth. Flowing northeast to the Bo Hai Sea from the Bayan Har Mountains, the Yellow River crosses a plateau blanketed with up to 300 meters (980 feet) of fine, wind-blown soil. The soil is easily eroded, and millions of tons of it are carried away by the river every year. Some of it reaches the river's mouth, where it builds and rebuilds the delta.

An additional type of nonpoint pollution happens during shipping and marine transportation. Oil and other pollutants are discharged into the oceans as tankers and cargo ships empty ballast and clean tanks. Accidental spills from oil tankers, barges, and offshore oil rigs also contribute to ocean pollution. However, some studies report that more oil is discharged into the oceans from stormwater than from accidental tanker spills as a result of millions of motor vehicle engines dropping small amounts of oil each day onto roads and parking lots, which is then taken up in stormwater.

Atmospheric Pollution

Some water pollution starts as air pollution, which settles into waterways and oceans as rainwater deposits these chemicals. The ocean is an integral part of the water cycle and is connected to all Earth's water reservoirs through evaporation and precipitation processes. **Atmospheric pollution** is part of the hydrologic cycle. Airborne pollutants combined with water vapor and cloud droplets are returned to the Earth's surface, on land, and into oceans, rivers, and other waterways. The atmosphere is a fast-moving part of the hydrologic cycle, dispersing pollutants over great distances. For example, air pollutants generated in Mexico have been found in the Great Lakes.

The transfer of chemicals such as nitrogen from the atmosphere to the ocean affects nutrient sources or can change the pH balance; these natural exchanges have been occurring for hundreds of thousands of years. However, human activity has added to the volume of chemicals such as nitrogen, iron, and phosphorus, and introduced new chemicals into the hydrologic cycle such as DDT, PCBs, and mercury. Iron is an essential micronutrient from marine photosynthetic organisms and the primary source for iron in the ocean is atmospheric deposition. But too much iron can cause algae blooms. The atmosphere is the main source of mercury in the ocean. Mercury is highly toxic and there have been a number of instances of its toxicity in coastal regions.

Large quantities of the toxic heavy metal lead have been emitted to the atmosphere as a result of human activities (e.g., leaded gas and smelters). Lead can be transported thousands of miles before depositing in the ocean. Additionally, as atmospheric carbon dioxide (CO_2) rises due to human activities, the amount of dissolved carbon dioxide in the oceans also increases. Since industrialization, about half the anthropogenic carbon dioxide emitted to the atmosphere has dissolved in the oceans.

The atmosphere is also the most important vector for distributing nitrogen. As a by-product of combustion, nitric oxide (NO) from cars, trucks, biomass burning, and energy production is emitted to the atmosphere, where it then falls in rainwater into the ocean. All organisms on earth require nitrogen. However, in aquatic systems, excess nitrogen can stimulate an explosive growth of plants and algae, which deplete oxygen levels when they die and decompose.

The Impacts of Pollution: Bioaccumulation and Biomagnification

The impacts of water pollution are numerous. Pollutants may settle into the physiological systems of organisms (and people) and the food chains or food webs of ecosystems. At the physiological level, chemicals enter the body by ingestion of food and water and by penetration through the skin. Chemicals are then absorbed into the blood system and distributed to tissues and organs by the bloodstream. Some chemicals may be altered and transformed into other chemical forms, most toxins are stored in organs and fatty tissue (fat soluble), and those that are water soluble such as acids, salts, or bacteria are eliminated with bodily wastes.

Pollutants that are stored in organs and tissue can accumulate in individual organisms over the span of its life, resulting in a higher concentration over time. This is known as **bioaccumulation**. For example, when an animal consumes food having DDT residue, the DDT accumulates in the tissue of the animal; if the animal continues to consume food having DDT residue, this will increase the concentration of DDT in the tissues over time.

The transfer of contaminants in aquatic food webs can result in **biomagnification**. This occurs when stored contaminants are not only retained in the food web but also accumulate in higher concentrations at higher trophic levels. When a fish is consumed by another fish, the containments in the fish's body are passed onto the organism that consumed it, affecting the food web. For example, when tiny phytoplankton take up contaminants such as mercury or pesticides, they pass this along to zooplankton, which in turn feed small fish. Larger fish, which feed on smaller fish, will see a dramatic increase in contaminant levels. One study found that PCBs are 50,000 times more concentrated in gull eggs than in phytoplankton.

Bioaccumulation and biomagnification are not necessarily negative processes. For example, omega-3 fatty acids can be accumulated; these are thought to have beneficial effects on cardiovascular disease.

The Impacts of Pollution: Eutrophication

An important impact of pollution in waterways is eutrophication. **Eutrophication** is the process in which lakes, rivers, coastal waters, and estuaries receive nutrients (phosphorus and nitrogen) and sediment from the surrounding watershed. Eutrophication sets off a chain reaction in the ecosystem, starting with an overabundance of algae and plants. The excess of nutrients can cause harmful algal blooms, dead zones, and fish kills, which are the results of a process called eutrophication—this begins with the increased load of nutrients to estuaries and coastal waters. Excessive nutrients lead to algal blooms and low-oxygen waters that can kill fish and seagrass and reduce essential fish habitats.

Nutrients feed algae, like they do other plants. Algae grows and blocks sunlight. Plants die without sunlight. Eventually, the algae die too. Bacteria digest the dead plants, using up remaining oxygen and giving off carbon dioxide. If they cannot swim away, fish and other wildlife become unhealthy or die without oxygen.

In ocean and freshwater environments, the term **hypoxia** refers to low or depleted oxygen in a water body. Hypoxia can lead to **dead zones**, areas where life cannot be sustained. Dead zones cause die-offs of fish, shellfish, corals, and aquatic plants. **Harmful algal blooms**, sometimes called a "red tide," occur when certain kinds of algae grow very quickly, forming patches, or "blooms," in the water (Photo 14.3). These blooms can emit powerful toxins that endanger human and animal health. The most conspicuous effects of harmful algal blooms are mass mortality of marine fauna such as fish and sea turtles, and reduction in the quality of recreational and shellfish harvesting areas, all of which have been documented. Blooms can lead to odors that require more costly treatment for public water supplies. Harmful algal blooms in the United States have caused an estimated $1 billion in losses over the last several decades to coastal economies that rely on recreation, tourism, and seafood harvesting. In 2018, a red tide outbreak in Florida led the authorities to declare a state of emergency in some counties as thousands of tons of dead fish washed up on shore.

PHOTO 14.3 Red Tide at Sechelt in British Columbia
A view of a red tide causing discoloration of coastal waters due to large algal blooms in the ocean waters of Sechelt in British Columbia on Canada's west coast. Red tide can cause die-offs of fish, shellfish, corals, and aquatic plants and are indicative of excessive nutrient runoff into waterways.

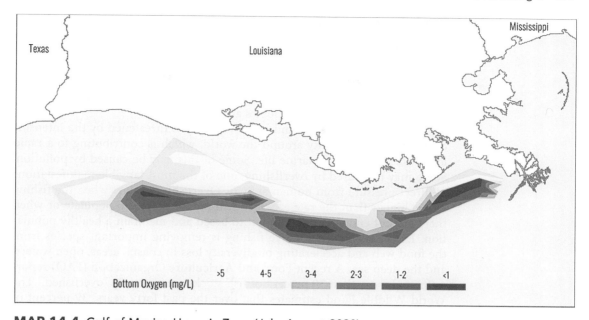

MAP 14.4 Gulf of Mexico Hypoxia Zone (July–August 2020)
The Gulf of Mexico dead zone in the summer of 2020 was the largest ever measured. Red areas are those with the lowest levels of oxygen in the water and suffer from hypoxia.

Since 1985, NOAA-sponsored research has monitored the largest dead zone in the United States, which forms every spring in the northern Gulf of Mexico. In 2020, the dead zone covered more than 22,800 square kilometers (8,776 square miles) of sea floor, the largest recorded (Map 14.4). Over the last five years, the average Gulf dead zone is about 14,000 square kilometers (5,380 square miles—about the size of the state of Connecticut). In the United States, hypoxia is also being studied in Lake Erie and the Chesapeake Bay. In Europe, the biggest dead zone is found in the Baltic, but there are others in the coastal areas of the Adriatic and Mediterranean seas. There are also significant dead zones in the Bay of Bengal, along the coastal waters of Japan and the South China Sea. Many of the dead zones in Asia are caused by increased intensive farming as well as population growth.

In addition to affecting the oxygen level of water, the excess algae and plant matter eventually decompose, producing large amounts of carbon dioxide. This lowers the pH of seawater, a process known as **ocean acidification**. Acidification slows the growth of fish and shellfish and can prevent shell formation in oysters, clams, sea urchins, shallow water corals, and deep sea corals. This leads to a reduced catch for commercial and recreational fisheries, meaning smaller harvests and more expensive seafood.

OVERFISHING

Scientists have confirmed there are more than 230,000 species of marine life, but there may be as many as 2.2 million marine species in the ocean, making it far more biodiverse than on land. For instance, of the thirty-four major animal phyla, only twelve are found on land, while thirty-three have been recorded in the ocean.

The ocean is a major source of food for more than half the planet—4 billion. Tuna, shrimp, oysters, mussels, and clams supply protein in the diet of millions.

Fish and marine animals contain several nutritional benefits. Rich in vitamins A and D, selenium, zinc, iodine, and iron, fish also contain essential omega-3 fatty acids, docosahexenoic acid, and eicosapentaenoic acid, which support proper brain functioning. In Asian and Nordic countries, where seafood is a dominant part of the cuisine, health studies have shown the life expectancy of both men and women is years longer and rates of obesity, cancer, cardiovascular disease, and diabetes are much lower.

However, the ocean's abundant biodiversity is threatened by the intensification of human activity around the world, which is contributing to a rapid loss of biodiversity in marine life. Some of this may be caused by pollution. Some may be caused by overfishing, one of the most tangible manifestations of direct pressure from human activity. Overfishing occurs when the fishing rate is higher than the rate at which fish reproduce and repopulate or when there are not enough adult fish left to breed and maintain a healthy population. Excessive and destructive fishing is removing important species from the food web and accelerating biodiversity loss in coastal areas, open waters, and the deep sea. A recent Food and Agriculture Organization (FAO) report estimated more than 30 percent of fish stocks are currently "overfished." The World Wildlife Fund estimates that over the past forty years, 39 percent of marine species have decreased.

Sharks, tuna, and billfish are especially depleted. A 2020 global shark survey found no sharks in almost 20 percent of the 371 surveyed reefs across fifty-eight nations, with levels of shark depletion being closely correlated to poor governance, the density of human population, and distance to the nearest market.

Overfishing can be exacerbated by government subsidies that increase the capacity of the fishing fleet and by **Illegal, Unreported, and Unregulated (IUU) fishing.** IUU fishing takes place when vessels or fishermen operate outside the laws of a fishery or nation. It is often conducted without concern for marine life or the environment. IUU fishing threatens the sustainability of fish populations, ecosystems, and the livelihoods of those who fish legally. A recent FAO report suggests that IUU fishing may account for twenty-six million tons of fish, valued at between $10–23 billion.

IUU fishing is also linked with labor and human rights abuses and crime. Human trafficking and forced labor onboard fishing vessels are often associated with IUU fishing. Drug trafficking, corruption, and other crimes are also linked to IUU fishing. Finally, IUU threatens to undermine national and regional efforts to manage fisheries, sustainability.

The Tragedy of the Commons

A 2020 report by the Ocean Panel noted that the "tragedy of the commons" applies in many parts of the ocean. In 1968 the ecologist Garrett Hardin developed the concept of the **tragedy of the commons** to explain why commonly shared resources, such as fisheries or livestock pastures, may be degraded over time. As a rational being, each fisherman seeks to maximize his or her gain. Therein is the tragedy, with each fisher locked into a system that compels him or her to increase the catch without limit—in a world that is limited. Hardin's solution was to regulate resources through government intervention or to privatize the resource.

In the case of overfishing, the tragedy of the commons is that too many boats pursue too few fish in ways that are destructive. Overfishing has become the number one driver of extinction risk for marine vertebrates (excluding birds). Many experts believe that if current fishing practices continue, the overall yield in 2050 could be 16 percent lower than it is today. In contrast, if fishing stocks were fished more sustainably, production could be 20 percent higher by 2050.

In contrast to Hardin, political scientist Elinor Ostrom has offered a counterargument. Her research of common resources in Maine, Indonesia, and Nepal revealed that individuals and communities could, in fact, manage their own collective resources very effectively. However, Ostrom notes that common resources are well managed when those who benefit from them the most are in close proximity to that resource. For her, the tragedy occurs when external groups exert their power (politically, economically, or socially) to gain a personal advantage. Her work also highlights that government intervention is often ineffective unless it is supported by individuals and communities. Ostrom's work has led to the development of a set of design principles that have supported effective mobilization for local management of common pool resources in a variety of areas.

DESTRUCTION OF MARINE HABITAT

There are many threats to marine habitat. Climate change is increasing sea surface temperatures, which can impact ocean temperatures. Climate change also generates stronger winds and more storms, and this could affect ocean circulation and ocean currents, which in turn can impact the larval dispersal among fish, or changes in the coastal upwelling that provides cold, nutrient-rich waters upon which many species feed. In addition, greenhouse gas emissions have accelerated ocean acidification. For example, between 1970 and 2000, seagrass meadows declined by roughly 30 percent, mangroves by 35 percent, and salt marshes by 60 percent, while between 11 percent and 46 percent of marine invertebrates are threatened. A 2016 report found key coastal habitats such as mangroves are being lost at a significant rate; global mangrove cover has declined by around 25–35 percent. Mangroves, seagrasses, and salt marshes are referred to as **"blue carbon" ecosystems** because they actively sequester and store organic carbon from the environment, meaning their loss increases emission.

Between 2014 and 2017, there was a massive global **coral bleaching** event, the longest, most extensive, and probably most damaging (in terms of coral mortality) ever recorded. When corals are stressed by changes in conditions such as ocean temperature, light, or nutrients (or pollution), they expel the symbiotic algae living in their tissues, causing them to turn completely white (Photos 14.4(right) and 14.4(left)). When a coral bleaches, it is not dead. Corals can survive a bleaching event, but they are under more stress and more likely to die.

Finally, invasive species are on the increase. Untreated ballast water from ships discharged into foreign ports is one way potentially invasive alien species are being introduced.

Many marine habitats are under threat from compounding and multiple reasons. For instance, coral reefs around the globe are impacted by warmer ocean temperatures and also overfishing and pollution. The decline of average hard coral cover on Caribbean reefs from 50 percent in the 1970s to 10 percent in the early 2000s, for example, was caused by the introduction of an invasive pathogen killing an important herbivore (sea urchin), on top of decades of overfishing of herbivores and grazers (parrotfish and multiple

PHOTO 14.4 Healthy Coral (left) Versus Bleached Coral (right) at Heron Island, the Great Barrier Reef in Australia

other species of fishes) as well as predators essential to the integrity of the system, sediment from deforestation on land, warmer water from climate change, and physical destruction.

SOLUTIONS FOR OCEAN SUSTAINABILITY

For far too long, the ocean has been neglected by policy makers. Nations usually manage their waters sector by sector, issue by issue, or by individual watersheds. The resulting hodgepodge of policies fails to consider collective impacts. Most countries today agree on what needs to happen—use marine resources responsibly and equitably and manage them sustainably, avoiding overfishing, pollution, and habitat destruction. But political action to deliver a healthy ocean has been lacking. Table 14.1 lists the targets for SDG 14.

In 2018, fourteen nations, led by Norway and Palau, commissioned a major science-based review of the challenges facing oceans. They worked with more than 250 scientists and ocean policy experts in much the same way as the Intergovernmental Panel on Climate Change works (reviewing and summarizing decades of research and science on the issue). The High Level Panel for a Sustainable Ocean Economy published its conclusions in 2020, and outlined key strategic priorities. These are to restore the ocean; protect the ocean; enhance ocean seafood operations, address climate change and the potential of the ocean as an energy source; regulate the ocean economy; and manage the ocean holistically. Table 14.2 highlights some suggested actions to achieve these priorities.

Restore the Ocean: Eliminate Pollution

A main solution is to eliminate pollution. The problem of ocean pollution starts on land. Plastic, along with numerous other pollutants, including pharmaceuticals and excess nutrients, enters the ocean because systems for their proper disposal on land are inadequate. The most effective way of stopping pollutants from entering the ocean is to tackle the root causes of pollution on land.

As discussed in the chapter on SDG 12 Responsible Consumption and Production, shifting to a "circular economy"—a system in which resources

TABLE 14.1 SDG 14 Targets and Indicators to Achieve by 2030	
Target	**Indicators**
14.1 Prevent and significantly reduce marine pollution of all kinds, in particular from land-based activities, including marine debris and nutrient pollution	**14.1.1** Index of coastal eutrophication and floating plastic debris density
14.2 Sustainably manage and protect marine and coastal ecosystems to avoid significant adverse impacts, including by strengthening their resilience, and take action for their restoration to achieve healthy and productive oceans	**14.2.1** Proportion of national exclusive economic zones managed using ecosystem-based approaches
14.3 Minimize and address the impacts of ocean acidification, including through enhanced scientific cooperation at all levels	**14.3.1** Average marine acidity (pH) measured at agreed suite of representative sampling stations
14.4 Effectively regulate harvesting and end overfishing, illegal, unreported, and unregulated fishing and destructive fishing practices and implement science-based management plans to restore fish stocks in the shortest time feasible, at least to levels that can produce maximum sustainable yield as determined by their biological characteristics	**14.4.1** Proportion of fish stocks within biologically sustainable levels
14.5 Conserve at least 10 percent of coastal and marine areas, consistent with national and international law and based on the best available scientific information	**14.5.1** Coverage of protected areas in relation to marine areas
14.6 Prohibit certain forms of fisheries subsidies that contribute to overcapacity and overfishing, eliminate subsidies that contribute to illegal, unreported, and unregulated fishing, and refrain from introducing new such subsidies, recognizing that appropriate and effective special and differential treatment for developing and least developed countries should be an integral part of the World Trade Organization fisheries subsidies negotiation	**14.6.1** Progress by countries in the degree of implementation of international instruments aiming to combat illegal, unreported, and unregulated fishing
14.7 Increase the economic benefits to Small Island Developing States and least developed countries from the sustainable use of marine resources, including through sustainable management of fisheries, aquaculture, and tourism	**14.7.1** Sustainable fisheries as a percentage of GDP in Small Island Developing States, least developed countries, and all countries
14.A Increase scientific knowledge, develop research capacity and transfer marine technology, taking into account the Intergovernmental Oceanographic Commission Criteria and Guidelines on the Transfer of Marine Technology, to improve ocean health and to enhance the contribution of marine biodiversity to the development of developing countries, in particular Small Island Developing States and least developed countries	**14.A.1** Proportion of total research budget allocated to research in the field of marine technology

TABLE 14.1	SDG 14 Targets and Indicators to Achieve by 2030 (Continued)	
Target		**Indicators**
14.B Provide access for small-scale artisanal fishers to marine resources and markets		**14.B.1** Progress by countries in the degree of application of a legal/regulatory/policy/ institutional framework that recognizes and protects access rights for small-scale fisheries
14.C Enhance the conservation and sustainable use of oceans and their resources by implementing international law as reflected in United Nations Convention on the Law of the Sea (UNCLOS), which provides the legal framework for the conservation and sustainable use of oceans and their resources, as recalled in paragraph 158 of The Future We Want		**14.C.1** Number of countries making progress in ratifying, accepting, and implementing through legal, policy, and institutional frameworks, ocean-related instruments that implement international law, as reflected in the UNCLOS, for the conservation and sustainable use of the oceans and their resources

Source: United Nations Sustainable Development Goals. https://sustainabledevelopment.un.org/sdg14

TABLE 14.2	Priority Actions to Achieve SDG 14
Reduce Ocean Pollution	• Incentivize the development, production, and use of viable and sustainable alternatives to plastics to enable the phaseout of problematic and unnecessary plastics, where warranted and where such alternatives exist. • Use financial incentives, trade opportunities, and extended producer responsibility to encourage sustainable product design and promote standards to maximize reduction, reuse, and recycling in pursuit of a circular economy, as well as research on new biodegradable materials that substitute plastics. • Enforce rules on waste shipments and illegal exports of plastic waste. • Promote a comprehensive life-cycle approach that includes improved waste management and innovative solutions toward reducing the discharge of marine plastic litter to zero. • Eliminate discharges of plastic litter and microplastics from sea-based sources including ships, offshore installations, and land-based sources including ports and bridges, through stronger regulations, technology development, training programs, and capacity building. • Eliminate ghost fishing gear through such means as reuse and retrieval, promoting gear marking and loss reporting, and supporting development of new environmentally friendly cost-effective gear. • Promote public and private awareness of and investment in sewage and waste management infrastructure in developing countries, including as a means to stop diseases. • Invest massively in waste collection and recycling technology and infrastructure. • Bring transparency and accountability to the flow of plastic polymers through the value chain.
Sustainable Ocean Food	• Eliminate illegal, unreported, and unregulated fishing by incentivizing the use of the latest innovations and technologies—such as digital traceability—to increase transparency; strengthening monitoring, control, and surveillance • Prohibit harmful fisheries subsidies that contribute to overcapacity, overfishing, and illegal, unreported, and unregulated fishing. • Promote more (sustainably) farmed finfish, seaweeds, and bivalves in diets. • Require transparency of seafood supply chains ensuring full ocean-to-plate traceability.

Ocean Energy and Climate Change	• Invest in research, technology development, and demonstration projects to help make all forms of ocean-based renewable energy—including wind, wave, tidal, current, thermal, and solar—cost-competitive, accessible to all and environmentally sustainable.
	• Work collaboratively with industry and other stakeholders to develop clear frameworks addressing environmental impacts of ocean-based renewable energy, enabling capacity, coexistence, and integration with other uses of the ocean.
	• Establish early national targets and strategies to support decarbonization of vessels.
	• Incentivize sustainable, low-carbon ports that support the transition to decarbonized marine transport and shipping fleets through renewable energy and zero-carbon fuel supply chains. Test and deploy low-carbon fuels.
	• Eliminate port air pollution through environmental regulations.
Ocean Economy	• Create national strategies for sustainable tourism growth.
	• Implement tourism taxes as payment for ecosystem services.
	• Invest in sustainable tourism that regenerates the ecosystems on which it depends, builds the resilience of coastal communities and Indigenous peoples, reduces inequality through promoting equal opportunity and equitable distribution of benefits, and addresses climate change and pollution.
	• Implement sustainable tourism management strategies that advance environmental, social, and economic priorities and enable monitoring and transparent reporting with the full participation of coastal communities and Indigenous peoples.
	• Implement mechanisms to increase the reinvestment of tourism revenue into local and Indigenous communities to build capacity and skills for increasing local employment in tourism, diversify economic opportunities, and increase resources for coastal and marine restoration and protection.
	• Accelerate financial incentives for including nature-based solutions in sustainable tourism infrastructure.
	• Invest in sewerage and wastewater infrastructure for coastal and marine tourism to improve the health of coastal communities and reduce the impacts on coastal and marine ecosystems.
Manage the Ocean Holistically	• At the country level, establish comprehensive integrated marine spatial plans for 100 percent of the areas under national jurisdiction.
	• Ensure continued funding and capacity for the ongoing implementation of ocean management plans.
	• Develop sustainable ocean economic zones as spatially defined "laboratories" for fully managed areas comprising various sectors, multisectoral projects, and fully protected areas.

Source: Stuchtey, M., Vincent, A., Merkl, A., Bucher, M., et al. "Ocean Solutions That Benefit People, Nature and the Economy." Washington, DC: World Resources Institute, p. 85 (2020). www.oceanpanel.org/ocean-solutions

are designed to be used continually and recovered or regenerated at the end of their service—would yield enormous benefits for the ocean economy. In addition, agricultural regulations aimed at reducing ocean dead zones could result in farmers adopting precision agriculture practices to reduce runoff, which would also improve the health of the soil and the quality of water in rivers and streams. Recent trends show that positive change is possible: 104 of 220 coastal regions improved their coastal water quality between 2012 and 2018.

In the United States, the EPA has addressed pollution reduction by implementing the **total maximum daily load or TMDL** (or "pollution diet").

A TMDL is a calculation of the maximum amount of a pollutant that a waterbody can receive and still safely meet water-quality standards. Once the TMDL has been established, discharge permits are issued for point sources. For nonpoint sources, actions can include a variety of regulatory (required) or incentive-based programs (e.g., a cost-share). This has been a key strategy for dealing with water pollution in the Chesapeake Bay, the largest estuary in the United States and the third largest in the world. The Chesapeake Bay watershed encompasses the entire District of Columbia, as well as parts of six states: Delaware, Maryland, New York, Pennsylvania, Virginia, and West Virginia. The bay region is home to 13.6 million people.

Increased agriculture and urbanization in the watershed have overwhelmed the Chesapeake Bay with excess amounts of nutrients. Nitrogen and phosphorus come from a wide range of point and nonpoint sources, including sewage treatment plants, industrial facilities, agricultural fields, and suburban lawns. As forests and wetlands have been replaced by farms, cities, and suburbs to accommodate a growing population, nitrogen and phosphorus pollution to the Chesapeake Bay has greatly increased. Almost half of the nitrogen and phosphorus pollution delivered to the Chesapeake Bay derive from agricultural sources, from both livestock production and row crop land.

Scientists have researched the causes and impacts of bay pollution for many years, and have a clear understanding of what actions to take to reduce pollution. However, the challenge is agreeing on how to implement changes and how to coordinate the various scales of governance (states and cities). Reducing bay pollution is less science and more politics. For example, a major challenge is that the majority of land in the Chesapeake Bay watershed is nonfederal land that private landowners, states, and local governments manage. Many are hoping that the TMDL strategy will effectively with the vast amount of nonpoint source pollutants and also with fragmented governance and management. The use of TMDL can be seen as a starting point or planning tool for restoring water quality.

Protect the Ocean

Another solution to protect the ocean through **marine protected areas (MPAs)**. An MPA is an internationally recognized area of ocean (or of land and ocean combined) where human activities such as tourism, development, and fishing are managed to ensure sustainability. An MPA is similar to a national park on land. In 2000, MPAs represented only two million square kilometers (1.2 million square miles) or 0.7 percent of the ocean. Today, MPAs cover about twenty-seven million square kilometers (sixteen million square miles) or 7.5 percent of the ocean.

Within an MPA, all human activities—from large-scale shipping to small-scale fishing—are regulated. MPAs come in many different forms with varying levels of protection. No-take MPAs (also known as marine reserves) have complete protection from human activities and help to restore and protect ecosystems. Some MPAs allow regulated human activity such as fishing.

A 2020 report by the Ocean Panel Expert Group calls MPAs one of the most effective tools for achieving the conservation of genetic diversity on an ecosystem scale. Fully or highly protected large-scale MPAs and networks of MPAs can encompass multiple sites of importance for the life cycle of marine species. Well-managed MPAs with adequate protection levels function as storehouses of genetic diversity that simultaneously serve as important reference points for understanding changes to the ocean. In areas where marine habitat is under dire threat, designating the area an MPA can halt the net loss, increase the extent, and improve the condition of coastal and marine ecosystems, in particular critical ecosystems such as mangroves, seagrasses, salt marshes, kelp beds, sand dunes, reefs, and deep ocean ecosystems.

Today there are more than 13,600 MPAs around the world that range in geography, size, and diversity. The Medes Islands Marine Reserve, northeast of Barcelona, Spain, is one of the best natural reserves in the western Mediterranean. Scuba divers come from all over Europe to see the abundant fish—including large Mediterranean dusky groupers and other predatory fishes, relict red coral populations, octopus, and hundreds of other marine species around these islands. The Medes MPA was created thirty-five years ago as a fifty-one-hectare no-take marine reserve that banned fishing but allowed diving, navigation, and moorings only on buoys. This protection proved successful on all fronts, even in this relatively small area. Fish populations have fully recovered, and six main species have almost reached the maximum carrying capacity of the ecosystem. In addition, the restored biodiversity and biomass have transformed the Medes Islands into a paradise for divers and snorkelers, supporting thriving ecotourism in the area. Two hundred full-time jobs are supported and $15 million in revenue is generated, compared with $0.5 million before the creation of the reserve.

There are nearly 1,000 MPAs in the United States. They include Monterrey Bay National Marine Sanctuary in California, the Yukon Delta Refuge in Alaska, and Thunder Bay National Marine Sanctuary in the Great Lakes. The largest MPA in the United States is the Papahānaumokuākea Marine National Area in Hawaii. This protected area encompasses 363,000 square kilometers and was established in 2014.

The Southern Ocean surrounding Antarctica is one of the most pristine places on Earth. Home to nearly 10,000 unique and diverse species, it remains largely unaffected by human activity. In 2016 Australia led an international effort to establish the first Antarctic and largest marine protected area in the world encompassing 1.55 million square kilometers (2,600,000 square miles) in the Ross Sea. While the majority of the Ross Sea MPA is fully protected, it also includes a Special Research Zone (SRZ) and a Krill Research Zone (KRZ) that allow limited fishing for krill and toothfish for scientific research (see Map 14.5).

There are proposals to establish other MPAs in Antarctica; however, some countries such as China and Russia have blocked the proposals because they do not endorse restrictions on commercial fishing in the nutrient- and fish-rich area.

Improve Ocean Seafood Management

More protein and essential nutrients will be needed to feed the world and reduce food insecurity, yet the ocean's ability to sustainably produce food is underrealized. Ocean food plays a critical role in feeding global populations by supplying an essential and accessible source of animal protein and micronutrients. Fish act as an important source of essential nutrients, including omega-3 fatty acids, iodine, vitamin D, and calcium, which are important during pregnancy and

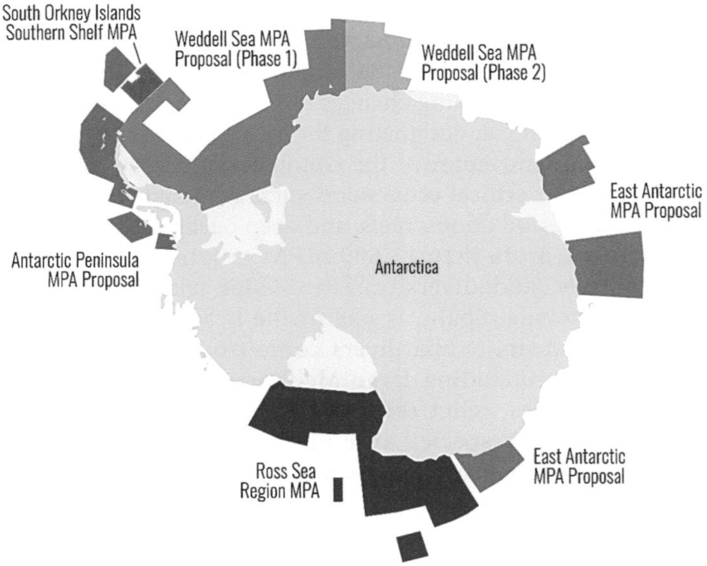

MAP 14.5 Marine Protected Areas of Antarctica

the first two years of a child's life. This is particularly important in low-income, food-insecure countries and Small Island Developing States.

Currently, fish, crustaceans, and mollusks provide only 17 percent of edible meat. This number is higher in lower-income countries such as Indonesia, Sri Lanka, and many small island states, which derive 50 percent or more of their animal protein from aquatic foods. There is potential for seafood to increase food supply. Seafood is nutritionally diverse and avoids or lessens many of the environmental burdens of terrestrial food production, so it could contribute to both food provision and future global food and nutrition security. If done correctly, sustainable fisheries and mariculture together could produce up to six times more food than it does today, feeding hundreds of millions more people.

Aquaculture and mariculture consist of cultivating aquatic products under controlled conditions. Both, if done sustainably, have potential to expand food production while avoiding harmful depleting of wild caught fish and crustaceans. To feed the growing human population, even more fish from aquaculture and mariculture will be required.

There are two types of aquaculture/mariculture: unfed (e.g., seaweed and filter-feeders) and fed (e.g., finfish and crustaceans). **Aquaculture**, which refers to fish farming in freshwater and landlocked ponds, indoor hatcheries, or culture tanks, has tremendous potential for expansion, notably with mollusks, including oysters, clams, and mussels, which obtain their food by filter feeding. Aquaculture accounts for about two-thirds of production.

Mariculture, which is aquaculture that occurs in saltwater environments, consists of cultivating marine organisms in the open ocean, or in an enclosed section of the ocean. Unfed mariculture holds great promise. Mariculture has been growing at a steady pace of about 6 percent annually, but finfish mariculture has been associated with unsustainable practices (e.g., overuse of antibiotics) and consumers who consider fish caught in the wild to be a higher quality than farm fish.

Aquaculture and mariculture must be done with an understanding of the ecological limits to how much fish and feed could be caught without depleting stocks. Opponents have questioned the sustainability of shrimp farming, for destroying habitats by converting sensitive mangrove areas into ponds. Opponents also note that waste from shrimp farms can lead to eutrophication. Bill McGraw argues that despite these issues, aquaculture, however, could be more sustainable if it uses lower amounts of fishmeal, incorporates all waste from the system into other forms of agriculture, has a small footprint, does not use antibiotics, and has no contamination.

A more sustainable form of aquaculture is called integrated multitrophic aquaculture. Multitropic aquaculture cultivates, in proximity, species from different trophic levels, and with complementary ecosystem functions in a way that allows one species' uneaten feed and wastes and by-products to be recaptured and converted into fertilizer, feed, and energy for the other crops. For example, a rice-fish system is stocking rice paddy fields with fish or prawns. The animals eat pests and fertilize the rice crop, increasing yields and providing an extra source of protein (or income) for small-scale farmers. Another example would be having blue mussels and cucumbers (both of which filter water) grow in proximity to fish ponds. Research is also being done to assess the benefits growing kelp in proximity to farmed fish. Kelp removes inorganic nutrients from water and also provides diverse commercial products. Bivalves (e.g., oysters and mussels) and seaweed can substantially increase the production of nutritious food and feed, with little negative impact on the marine environment. In some cases, this kind of mariculture could enhance wild fisheries by creating artificial habitats and nursery grounds for fish. In addition to these benefits, oysters can also help filter polluted water, as discussed in the *Solutions* 14.1 box. *Solutions* 14.2 explores another small-scale innovative approach to sustainable fisheries.

 ## SOLUTIONS 14.1

The Benefits of Bivalves

Target 14.1: Prevent and significantly reduce marine pollution of all kinds, in particular from land-based activities, including marine debris and nutrient pollution.

In recent years bivalve mollusks, such as oysters, clams, and scallops, have been used to help slow and, in some cases, reverse the process of eutrophication because they efficiently remove nutrients from the water as they feed on phytoplankton and detritus. Mollusks naturally reduce nutrients through their filter-feeding activities.

A project in Long Island Sound showed that the oyster aquaculture industry in Connecticut provides $8.5–$23 million annually in nutrient reduction benefits. The project also showed that reasonable expansion of oyster aquaculture could provide as much nutrient reduction as the comparable investment of $470 million in traditional nutrient-reduction measures, such as wastewater treatment improvements and agricultural best-management practices.

The benefits of bivalves extend beyond their ability to filter pollution. They may also contribute substantially to food security by providing relatively low-cost and accessible food because they have a high production potential at low costs compared to finfish production. Some experts say bivalves could contribute about a third of future aquatic food.

Source: NOAA. "Eutrophication" (2021). https://oceanservice.noaa.gov/facts/eutrophication.html

SOLUTIONS 14.2

Innovations in Seafood Production

A background in agricultural biochemistry led Michael Selden, co-founder and CEO of Finless Foods, to focus his environmental activism on the way we eat. He founded Finless Foods when he was twenty-six. Finless Foods is the world's first cell-based seafood company, growing real seafood directly from high-quality animal cell-stock without the need to raise and slaughter entire animals. Using cellular agriculture, Finless makes seafood without harvesting fish. In the following text, he reflects on the connection between fish, food, and sustainability.

How is Finless Foods addressing the UN Sustainable Development Goals (SDG), specifically SDG14: Life Below Water? And how do you measure progress in this respect?

"UN SDG 14 states "Conserve and sustainably use the oceans, seas and marine resources." Specifically, Target 14.4 pertains to fisheries, aiming that by 2020, globally we should "effectively regulate harvesting and end overfishing, illegal, unreported and unregulated fishing and destructive fishing practices and implement science-based management plans, to restore fish stocks in the shortest time feasible."

As demand for seafood continues to increase, especially for highly sought-after species such as tuna, and as the ocean continues to face pressure from overfishing and IUU fishing, Finless Foods aims to provide a new, innovative solution to supplying real, trusted, sustainable, and quality seafood to the world without relying on the continual harvest of fish from the ocean.

In providing a complementary solution to sustainable and responsible wild-caught fisheries and aquaculture, Finless Foods aims to reduce stressors on our global fisheries so that we can ensure a healthy, biodiverse, and thriving ocean in the future.

Though it is too early to use metrics to measure direct impact on our ocean, as Finless still is in the R&D phase, Finless views progress in terms of ensuring a healthy number of fish remain in the ocean, to support stable global fish stocks, biodiversity, and an overall healthy, balanced, thriving ocean."

How do you measure success at Finless Foods and what are your goals for the future?

"Finless Foods' mission is to create a future for seafood where the ocean thrives. We measure success by ocean health and so our goals all relate to changing the seafood supply chain for the better. Our near-term goals all center around putting out delicious seafood that doesn't hurt the ocean, and to try and stabilize the precarious situation our Earth is in."

There is good potential in Finless Foods but given that it is still in the R&D phase, and its scale of operation is likely to be small to begin, it is not likely to contribute significant amounts of food to the food system.

Source: Frisch, L. "A Complementary Solution to Sustainable Wild-Caught Fisheries." *The Source, Springer Nature* (2020, June 4). https://www.springernature.com/gp/researchers/the-source/blog/blogposts-communicating-research/finless-foods-complementary-sustainable-solution/18045512

It should be noted that integrated multitrophic aquaculture is not a new concept: It is an old practice in many of the world's regions, particularly among Indigenous groups, that has been rediscovered within a sustainability discourse.

Use the Ocean to Combat Climate Change and Provide Energy

The ocean holds tremendous potential to provide clean energy for the world. Scaling up ocean-based renewable energy may generate jobs and boost economic development while providing a pathway to decarbonization.

Ocean-based renewables offer varied options for power generation. Energy from wind, wave, tidal, current, thermal, and solar can be used in different places, depending on the geography and situation. Investing in ocean-based renewable energy could deliver climate benefits, reduce local and global pollution, and build energy security. Some projections suggest that this could be a $1 trillion industry that has the potential to deliver up to one million full-time jobs by 2050.

Investing in restoring and expanding the blue carbon ecosystems of mangroves, seagrass beds, and salt marshes will increase carbon storage. A recent example is the successful restoration of 3,000 hectares of seagrass beds in Virginia lagoons along the US eastern seaboard, which has resulted in sequestration of about 3,000 tons of carbon per year.

Many experts also encourage the decarbonizing shipping. More than 90 percent of global goods move across the seas, but ships use heavy fuel oils that release soot and sulfur as well as CO_2—which totals 18 percent of air pollutants and 3 percent of greenhouse gas emissions. Every $1 invested in decarbonizing international shipping and reducing emissions to net zero by 2050 is estimated to generate a return of $2–$5. Technology to decarbonize and minimize the negative environmental impacts of marine transport exists but must be brought to scale. Some strategies to decarbonize vessels and develop and adopt technologies for producing and storing new zero-emission fuels are currently in development but they remain at the small scale. Countries could also tighten and enforce energy efficiency of ships and provide incentives for decarbonization. They could eliminate port air pollution by tightening or enforcing environmental regulations.

Invest in the Ocean Economy

There are many diverse maritime economies, including, fisheries, aquaculture, shipping, submarine cables, port activities, offshore wind farms, offshore oil and gas, maritime manufacturing and construction, and tourism, all of which generate $2.5 trillion in the global economy each year. The ocean economy and "blue jobs" are predicted to double by 2030. More than three billion people depend on marine and coastal biodiversity for their livelihoods. Securing a sustainable ocean economy will be crucial.

There are no figures for global ocean tourism, but about 80 percent of all tourism takes place in coastal areas, with half all global tourists mentioning "the beach" or "coastal areas" as their reason for travel. Between 60 and 350 million people annually travel to the world's coral reef coast, generating an estimated $12 billion annually. Cruise tourism has grown strongly over the last decade. In 2019, more than thirty million people traveled across the ocean on a cruise ship. Tourism is projected to become the single-largest ocean-based industry by 2030. At the same time, coastal and marine tourism remains vital to the economic prosperity of island and coastal communities. The continued viability of this sector remains at risk from climate change, disasters, pollution, urbanization, and ecosystem degradation. Sustainable ocean-based tourism may restore and protect the ocean while delivering jobs and prosperity.

"Blue jobs" and tourism effort holds great potential for jump-starting economies. Marine fisheries provide fifty-seven million jobs globally, but that could increase with investment. Restoring coastal and marine ecosystems can create jobs and enhance tourism. Building and extending sewage and wastewater infrastructure will create jobs and improve health, tourism, and water quality. Investment in sustainable, community-led, nonfed mariculture such as shellfish, especially in low-income economies, will simultaneously encourage new jobs and improve water quality.

Also important is the need to more accurately value ocean **ecosystem services**. Consider the economic value of storm protection. More than 500 million people worldwide live in a coastal zone that is protected by coral reefs. Without their protection, flood damages from one-hundred-year storms would increase by 91 percent to $272 billion. Richard Costanza and colleagues calculated that the US coastal wetlands provide $23.2 billion a year in storm protection services. Another study found that mangroves reduce annual flooding globally by more than 39 percent per year for eighteen million people, and reduce annual property damage by more than 16 percent, or $82 billion. Similarly, the value of coastal ecosystems in terms of nursery and habitat for fishes and other marine species, regulation of water flow and filtration, carbon sequestration, and contaminant storage and detoxification has also been calculated at $100 to $10,000 an acre.

Manage the Ocean Holistically

The ocean has been plagued by complexity of governance (there are no fewer than 576 bilateral and multilateral agreements). Dozens of international frameworks, including multilateral environment agreements, programs, and initiatives, have been developed to focus on improved ocean governance. A few of these include the UNCLOS, International Maritime Organization, MARPOL International Convention for the Prevention of Pollution from Ships, and Global Programme of Action for the Protection of the Marine Environment from Land-Based Activities.

At the global level, the reach and authority of even well-established ocean organizations and treaties is often challenged. For example, some nations dispute the International Whaling Commission, or claim exceptions to specific articles of the convention.

UNCLOS is the legal basis for all ocean activities, and existing international ocean commitments must be implemented as a foundation for achieving a sustainable ocean economy. UNCLOS is referred to as the "constitution for the oceans." It sets out the legal framework within which all activities in the oceans and seas must be carried out, including with regard to the conservation and sustainable use of the oceans and their resources. UNCLOS sets out to promote the peaceful uses of the seas and oceans; the equitable and efficient utilization of their resources; the conservation of their living resources; and the study, protection, and preservation of the marine environment. Despite its global acceptance, UNCLOS has received continued criticism since its adoption.

According to Alexi Nathan one of the major problems with UNCLOS is the lack of legal rules regarding the high seas. The "high seas" make up the 64 percent of the ocean that is beyond the jurisdiction of any one country as defined by the UNCLOS. When UNCLOS was first negotiated in 1982, the high seas were largely ignored due to inaccessibility. However, after forty years of technological innovation, there is virtually nowhere that industrial fishing vessels cannot reach. This has led to increased mismanagement, despite the fact that the high seas play a major role in the overall health of our global oceans.

Nathan also notes that further problems arise from the fact that UNCLOS only applies to countries, leaving the activities of nonstate actors largely unregulated by international law. This affords private actors, such as shipping companies, a large amount of flexibility in their operations. These companies are often motivated by the desire to reap never-ending profits, which may lead to their general disregard for the health and maintenance of the high seas.

Despite the number of agreements—or maybe because of them—ocean development and governance has largely occurred ad hoc. For example, plans for a new port or tidal energy project might not consider the destruction of blue-carbon ecosystems or the impacts of shipping on fish. Similarly, multiple organizations that manage different oceans differ widely in funding, scientific capacity, relative authority with member states, and, ultimately, fishery outcomes. When communication among the food, energy, and shipping sectors does occur, it is more often about conflict resolution than collaboration. The absence of an overarching mandate toward a healthy ocean continues to challenge the development of ocean sustainability. At the same time, new technologies hold the promise of helping to improve our understanding and management of the ocean, as discussed in *Solutions* 14.3.

 SOLUTIONS 14.3

Using Technology to Better Manage the Ocean

Strategies for Target 14A: Increase scientific knowledge, develop research capacity and transfer marine technology, taking into account the Intergovernmental Oceanographic Commission Criteria and Guidelines on the Transfer of Marine Technology, to improve ocean health and to enhance the contribution of marine biodiversity to the development of developing countries, in particular Small Island Developing States and least developed countries.

We know far too little about the ocean. The data revolution could change the way informed decisions in the ocean realm are made, reshaping our understanding and management of the ocean. New sensing, data management, visualization, simulation, and modeling technologies could change that.

For example, every ship's journey and the nature of its business at sea could be public information. If so, lawbreakers such as illegal fishers, polluters, smugglers, or other violators would literally be on the public radar and subject to arrest.

New levels of transparency may enable us to better understand the state of the world's fish-

eries. Product tracking throughout the chain of custody could help brands embrace sustainable practices and would help small producers connect to global supply chains.

For ocean resource managers, replacing trial-and-error methods with reliable simulations lowers feedback and response times from years to hours, and allows quick insight on how the ocean reacts to specific inputs, rules, and incentives. A number of these "flight simulator" efforts are now in development for applications ranging from fishery management to ship routing and ecosystem conservation. The POSEIDON model, for example, simulates the feedback loop between fishery policies, fishing fleets, and ocean ecosystems, allowing for real-time testing of policy alternatives. These applications will allow managers to adjust to changing conditions, such as dynamic management of fishing areas and quotas, ship traffic adjustments, or avoidance of endangered species by catch.

Source: Stuchtey, M., Vincent, A., Merkl, A., Bucher, M., et al. "Ocean Solutions That Benefit People, Nature and the Economy" (p. 85). Washington, DC: World Resources Institute (2020). www.oceanpanel.org/ocean-solutions

Finally, to manage the ocean holistically, it must also be managed in a way that is equitable. According to the High-Level Panel for a Sustainable Ocean Economy, ocean equity must facilitate the equitable distribution of ocean wealth and ensures equality of opportunity for all. It must ensure that there are no labor rights abuses, child labor, forced labor, trafficking in persons and contraband, or tax evasion. Ocean equity also recognizes the specific climate vulnerabilities and financing and capacity constraints of lower-income countries, Indigenous communities, and Small Island Developing States. The following are priorities to realize ocean equity:

1. Require transparent, responsible business practices that engage and benefit coastal communities, including small-scale fishers, and protect the rights of all workers in ocean industries.

2. Create the conditions to facilitate the full engagement of women in ocean activities to help unlock their economic and social potential and empower them to safeguard natural resources while enhancing opportunities to access decent work.

3. Recognize and respect the interests of coastal communities and rights of Indigenous peoples and implement policies that require consideration of the particular importance of marine resources for these groups.

4. Create inclusive governance by incorporating Indigenous and local knowledge and interests, particularly those of women and youth, in planning and decision-making processes.

5. Promote integrity across ocean governance and ocean industries, enforce transparency and accountability in public service and public finance, and take robust action against corruption.

6. Enhance domestic revenue administration through modernized, progressive tax systems, improved tax policy, and more efficient tax collection.

7. Promote international cooperation to combat child labor and forced labor and eliminate trafficking in persons and contraband along supply chains in the ocean economy.

SUMMARY AND PROGRESS

The benefits of a healthy ocean supporting a healthy economy are well documented. However, cultural norms, institutional constraints, policies, and laws ignore or discount the value—environmental and social—of the ocean. One of the biggest challenges is the lack of financial investment. The ocean is currently underinvested in, and of all the SDGs it has received the smallest share of investment in the last five years. Between 2013–18, an overage of $1.5 billion of official development assistance a year was allocated in support of the sustainable ocean. This represents less than 1 percent of all global official development assistance. Beyond development assistance, the finance available for biodiversity and conservation of the ocean is very small. Governments and the private sector do not seem to understand the positive impact of sustainable ocean investments.

Still, there has been progress. Erna Solberg, Norway's prime minister and one of the leaders of the High-Level Panel for the Ocean, is optimistic about international efforts going forward (see *Expert Voice* 14.1).

EXPERT VOICE 14.1

Erna Solberg on Protecting the Ocean

Erna Solberg is Norway's prime minister. She discusses why she believes that ocean science can boost jobs and well-being:

> I grew up on the west coast of Norway and my parents taught me how to fish when I was a little girl. I caught my first mackerel in a boat on the Hardangerfjord, and had it fried for dinner. Such memories become part of you. The ocean is central to Norway's history and culture, economy and diet. We need it to weather existential threats—from the COVID-19 crisis to climate change. As the country's prime minister, it is my job to ensure that our relationship with it is sustainable: protection, production and prosperity go hand in hand.
>
> We cannot do this alone. In September, I went back to the west coast and spent time picking up plastic waste with volunteers. The plastic was from all over, much brought to Norway on ocean currents. Because the biggest threats to the ocean are now global, its safeguarding must be, too.
>
> Perhaps the most notable international achievement to protect the ocean is the marine-protected area around the Ross Sea in Antarctica. This 2016 agreement was hard won. More than two dozen nations agreed to preserve ecosystems in Antarctic waters, with 70 percent off-limits to fishing.
>
> Protection is important to let damaged waters regenerate, but we need more: we can manage the ocean for its vast capacity to drive economic growth and equitable job creation, sustain healthy ecosystems, and mitigate climate change. We need to allow sustainable industries, such as offshore wind turbines and seaweed cultivation, to develop. This demands political will, scientific insight and international.
>
> That is why, almost three years ago, I set up the High Level Panel for a Sustainable Ocean Economy (the Ocean Panel), which I co-lead with the president of Palau in the western Pacific.
>
> This month, all 14 countries on the panel agreed to sustainably manage 100 percent of their Exclusive Economic Zones (national waters) by 2025, utilizing the ocean without sacrificing its health. Together, the coastlines of Ocean-Panel members comprise almost 40 percent of all national coastlines worldwide. The agreement covers some of the world's busiest shipping lanes, most-productive fishing grounds and most-enticing tourist destinations.
>
> Why are we making this commitment? The ocean covers 70 percent of the planet. It can transport goods more carbon-efficiently than air and provide protein more sustainably than land. By one calculation, $1 invested in a healthier ocean will reap a $5 return, but countries must work together to realize that reward. If we fail, it will be costly. Across sectors such as fishing, shipping and tourism, declining ocean health as a result of overexploitation, pollution and climate change could cost the global economy more than $400 billion each year by 2050.
>
> Countries need to hold each other accountable and craft mutually reinforcing policies, such as sharing data and technology to help monitor illegal fishing and pollution. Norway no longer approaches oil, transport, fisheries, aquaculture and minerals separately, but monitors and manages ocean activities across sectors. This helps to establish common data standards, metrics and goals. It also facilitates coordination across government boundaries.
>
> Solutions will vary by country. Norway and Chile have booming aquaculture industries and Fiji has tropical tourist beaches. Portugal and Japan have notable seafood cultures, whereas Indonesia has fisheries and reefs for diving.

But lasting, meaningful progress requires international cooperation. Otherwise, planning will be ad hoc and ineffective, as we have seen in marine sanctuaries that are "paper parks." These are marked as protected on a map but in fact are not. The "high seas," which cover half of Earth's surface, are essentially unprotected.

It will take time for the benefits to become apparent. In 2017, I joined with philanthropist Bill Gates to launch the Coalition for Epidemic Preparedness Innovations, aimed at developing vaccines for pandemics. Only now are we seeing that work pay off, in accelerated COVID-19 vaccine development and distribution.

Critics might counter that more than 14 countries benefit from and exploit the ocean, some of which are bigger contributors to ocean problems and have more wherewithal to solve them. Still, we must start somewhere, and together we are bringing sustainability to nearly 30 million square kilometers of ocean. We are hopeful our ranks will grow over time, because the benefits of better management are beyond doubt. Sceptics might suggest that the Ocean Panel is just a way for countries to burnish their images. But if sustaining the ocean boosts a country's image, that can only be a good thing.

The proof of our efforts will be in ocean services: fishing, recreation, emissions absorption and the creation of decent jobs. Ocean protection and productivity are inseparable.

Source: Solberg, E. "Norway's Prime Minister: Ocean Science Can Boost Jobs and Wellbeing." *Nature*, 588, (2020). https://www.nature.com/articles/d41586-020-03302-4

The 2020 COVID-19 pandemic has caused a dramatic disruption of the global economy, and in particular has had major effects on tourism: Because of the lockdowns and travel ban implemented in most countries, tourism lost $2.1 trillion in 2020, and as many as one hundred million jobs may have been lost. This sudden and massive hit on the tourism industry offers the opportunity to reinvent tourism as an eco-friendly experience.

Given the crucial role of maritime shipping, maintaining global supply chains will be critical to support recovery from the COVID-19 pandemic and future crisis. However, ocean workers and sectors have been largely absent from economic stimulus packages in response to the COVID-19 pandemic.

Unfortunately, the response to the COVID-19 pandemic has caused a surge in production and consumption of protective equipment, much of which contains single-use plastic that has entered waterways and the ocean.

Making progress on SDG 14 will help advance other SDGs. Addressing marine pollution, for example, could advance targets in SDG 2 Zero Hunger, SDG 3 Good Health and Well-Being, SDG 6 Clean Water and Sanitation, and SDG 12 Responsible Consumption and Production. Figure 14.1 highlights the interconnections between SDG 14 and the other SDGs.

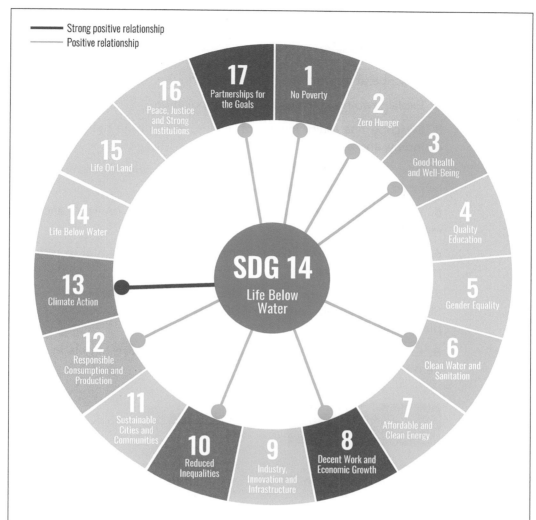

SDG 1: Healthy and productive oceans benefit small-scale fishers, improve tourism revenue and increase potential for blue carbon markets.

SDG 2: Seafood, whether farmed or caught in the wild, is globally important as a source of protein, omega-3 fatty acids, vitamins, calcium, zinc, and iron for one billion people. Sustainable fisheries and aquaculture backed by healthy oceans and coasts are a necessary prerequisite to achieve food security and improved nutrition.

SDG 3: Contamination of coastal zones or seafood with pollutants can cause health problems. Reducing and preventing marine pollution will thus help reduce pollution related deaths and illnesses.

SDG 6: Ocean sustainability directly links to sustainable water management. Preventing marine pollution contributes to improving water quality and vice versa.

SDG 10: Healthy oceans and coasts provide a sustainable resource base for income growth in low-income populations.

SDG 12: Achieving sound management of chemicals throughout their lifecycle will also help minimize marine pollution.

SDG 13: Restoring and protecting the health of oceans, coasts and marine resources contributes to strengthening the resilience and adaptive capacity of both the natural and human systems to climate change.

SDG 17: Achievement of SDG 14 will benefit particularly from the mobilization of financial aid, strengthened technology exchange, capacity building, better policy coherence and multi-stakeholder partnerships.

FIGURE 14.1 Interconnections in SDG 14 and the Other Goals
The ocean connects to all of the SDGs in some way; this figure highlights several of the stronger connections.

QUESTIONS FOR DISCUSSION AND ACTIVITIES

1. Describe at least three challenges facing the ocean.

2. Explain two major causes of marine pollution.

3. Describe one major impact of marine pollution on the environment, economy, and equity.

4. Why do you think governments and the international community have invested so little into the ocean?

5. Explain the interconnections between activities on land and their impacts on the ocean.

6. In what ways is responsible consumption (SDG 12) vital to reducing pollution and overfishing?

7. Visit this site to track plastic debris, then discuss which targets for SDG 12 would also help with SDG 14: https://theoceancleanup.com/plastic-tracker/

8. Explain the potential for the ocean to help realize targets in SDG 2 Zero Hunger and SDG 13 Climate Action.

TERMS

aquaculture
atmospheric pollution
bioaccumulation
biomagnification
blue carbon
coral bleaching
dead zones
ecosystem services
eutrophication
Great Pacific Garbage Patch
gyres
harmful algal blooms
hypoxia

Illegal, Unreported and Unregulated fishing (IUU)
mariculture
marine debris
Marine Protected Area (MPA)
ocean acidification
pollution sink
sediment
stormwater runoff
thermohaline circulation
Total Maximum Daily Load (TMDL)
tragedy of the commons

RESOURCES USED AND SUGGESTED READINGS

Allison, E., Kurien, J., and Ota, Y. "The Human Relationship with Our Ocean Planet." Washington, DC: World Resources Institute (2020). https://www.oceanpanel.org/blue-papers/relationshipbetween-humans-and-their-ocean-planet

Blackford, M. G. *Making Seafood Sustainable: American Experiences in Global Perspective*. Philadelphia: University of Pennsylvania (2012).

Blasiak, R., Wynberg, R., Grorud-Colvert, K., et al. "The Ocean Genome and Future Prospects for Conservation and Equity." *Nature Sustainability* 3: 588–96 (2020). https://doi.org/10.1038/s41893-020-0522-9

Costello, C., Cao, L., Gelcich, S., et al. "The Future of Food from the Sea." *Nature* 588: 95–100 (2020). https://doi.org/10.1038/s41586-020-2616-y

Duce, R., Galloway, J., and Liss, P. "The Impacts of Atmospheric Deposition to the Ocean on Marine Ecosystems and Climate." *World Meteorological Organization Bulletin* 58 (1): 61–66 (2009). https://public.wmo.int/en/bulletin/impacts-atmospheric-deposition-ocean-marine-ecosystems-and-climate

Green, D., and Payne, J. (eds.). *Marine and Coastal Resource Management: Principles and Practice*. New York: Earthscan (2017).

High Level Panel for A Sustainable Ocean Economy. "Ocean Equity." World Resources Institute (2021, February). https://oceanpanel.org/ocean-action/ocean-equity.html

Hoegh-Guldberg, O., et al. "The Ocean as a Solution to Climate Change: Five Opportunities for Action." Washington, DC: World Resources Institute (2019). https://devoceanpanel.pantheonsite.io/sites/default/files/2019-09/19_HLP_Report_Ocean_Solution_Climate_Change_final.pdf

Jambeck, J., Geyer, R., Wilcox, C., Siegler, T., Perryman, M., Andrady, A., Narayan, R., and Law, K. "Plastic Waste Inputs from Land into the Ocean." *Science* 347 (6223): 768–71 (2015, February). https://www.science.org/doi/10.1126/science.1260352

Komatsu, T., Ceccaldi, H.-J., Yoshida, J., Prouzet, P., and Henocque, Y. (eds.). *Oceanography Challenges to Future Earth Human and Natural Impacts on Our Seas*. Cham, Switzerland: Springer Nature (2019).

Koundouri, P. (ed.). *The Ocean of Tomorrow: The Transition to Sustainability Volume 2*. Cham, Switzerland: Springer Nature (2021).

Leal Filho, W., Azul, A. M., Brandli, L., Lange Salvia, A., and Wall, T. (eds.). *Life Below Water* Cham, Switzerland: Springer Nature (2020).

Lubchenco, J., Haugan, P. M., and Pangestu, M. "Five Priorities for a Sustainable Ocean Economy: Unleash the Ocean's Potential to Boost Economies Sustainably While Addressing Climate Change, Food Security and Biodiversity." *Nature* 588: 30–32 (2020). https://doi.org/10.1038/d41586-020-03303-3

McGraw, B. "The Many Sides of Sustainability in Aquaculture," The Fish Site (2017). https://thefishsite.com/articles/the-many-sides-of-sustainability-in-aquaculture

Nathan, A. "The Law of the Seas: A Barrier to Implementation of Sustainable Development Goal 14." *Sustainable Development Law & Policy* 16 (2): Article 8 (2017). http://digitalcommons.wcl.american.edu/sdlp/vol16/iss2/8

Organisation for Economic Cooperation and Development. "The Ocean Economy in 2030. Directorate for Science, Technology and Innovation Policy Note" (2016, April). https://www.oecd.org/futures/Policy-Note-Ocean-Economy.pdf

Organisation for Economic Cooperation and Development. *Rethinking Innovation for a Sustainable Ocean Economy*. Paris: OECD Publishing (2019). https://doi.org/10.1787/9789264311053-en

Ramkumar, M., James, R. A., Menier, D., and Kumaraswamy, K. (eds.). *Coastal Zone Management: Global Perspectives, Regional Processes, Local Issues*. Amsterdam: Elsevier (2019).

Rossi, S. *Oceans in Decline. Cham*, Switzerland: Springer Nature (2019).

Stuchtey, M., Vincent, A., Merkl, A., Bucher, M., et al. "Ocean Solutions That Benefit People, Nature and the Economy" (p. 21). Washington, DC: World Resources Institute (2020). www.oceanpanel.org/ocean-solutions.

US Environmental Protection Agency. "Chesapeake Bay" (2021). https://www.epa.gov/nutrient-policy-data/addressing-nutrient-pollution-chesapeake-bay

US Environmental Protection Agency. "TMDL" (2021). https://www.epa.gov/tmdl/overview-total-maximum-daily-loads-tmdls

US National Oceanic and Atmospheric Administration. "Garbage Patches" (2021). https://marinedebris.noaa.gov/info/patch.html

World Bank and UN Department of Economic and Social Affairs. *The Potential of the Blue Economy: Increasing Long-Term Benefits of the Sustainable Use of Marine Resources for Small Island Developing States and Coastal Least Developed Countries*. Washington, DC: World Bank (2017). https://openknowledge.worldbank.org/bitstream/hand

Terrestrial Ecosystems and Biodiversity

LEARNING OBJECTIVES

After reading this chapter, you should be able to:

- Describe the patterns and trends in ecosystem change and biodiversity

- Identify and explain direct and indirect drivers of ecosystem and biodiversity loss

- Describe the impacts of land degradation

- Identify and explain the various targets and goals associated with SDG 15

- Explain some of the strategies for achieving SDG 15

- Explain how the three Es are present in SDG 15

- Explain how SDG 15 strategies connect to other SDGs

Everglades National Park in Florida was established in 1947 to conserve the natural landscape and prevent further degradation of its land, plants, and animals. It was the first national park to be based on a conservationist ideology to protect and preserve an ecosystem.

One of the biggest challenges to protecting this distinctive ecosystem is invasive species, plants, and animals. In recent years, the nonnative melaleuca tree and snakes from around the world have been found in and around the national park. The snakes were most likely either pets that had been accidentally or intentionally released into the wild. The most notorious is the Burmese python, which can grow to about seventeen feet long and weigh up to 150 pounds. These snakes have no natural predators to control their population and are indiscriminate eaters. Scientists estimate there could be between 100,000 and 300,000 pythons in the park.

Nonnative Burmese pythons now occupy a wide variety of habitats in the park, including uplands, freshwater wetlands, and the saline coastal fringe. These snakes compete directly with other top predators such as the American alligator and, in some instances, very large pythons have been recorded eating alligators. Predation from the python is already changing the delicate balance of the national park's food chain. Twenty-five different bird species, including the endangered wood stork, have been found in the digestive tracts of pythons in Everglades National Park. Scientists report that pythons have hunted the native marsh rabbits to near extinction.

In 2021, a record number 2,000 pythons were humanely removed from the Everglades. Hunters also caught the longest Burmese python ever captured in Florida. The female, who was carrying about sixty eggs, measured almost nineteen feet long. Other efforts to control the python population include legalized python hunting competitions, where hunters are paid for each python they kill (Photo 15.1). Invasive species are not just a nuisance; they are dramatically transforming the Everglades ecosystem.

PHOTO 15.1 A Captured Burmese Python in the Everglades, Florida
A National Park Service Ranger holds a captured Burmese Python. This is part of ongoing effort to control the python population that has reduced native species of birds and marsh rabbits.

TERRESTRIAL ECOSYSTEMS AND BIODIVERSITY: UNDERSTANDING THE ISSUE

Nature is the foundation for all dimensions of human health and well-being. On land, terrestrial ecosystems sustain the quality of the air, fresh water, and soils, regulate the climate, and provide pollination to the numerous plants—including food crops—that support humans. Terrestrial ecosystems are critical to our existence, and yet human activity has altered almost 75 percent of the earth's land surface.

Prior to agriculture and increased human population, human activity had limited influence on terrestrial ecosystems. Instead, the most influential factors were climate, drainage, soil conditions, and large-scale changes such as glaciers, sea levels, storms, and fires. Today, however, human activity significantly shapes the patterns of ecosystems, although this varies in intensity across various scales and geographies. Human activities can have complex and often contradictory impacts on ecosystems. For example, agriculture and forestry have increased production over the last several decades, thus increasing the supply of food and other materials important for people. However, these changes decreased soil fertility, water quality, and pollinator diversity.

According to the 2019 Global Assessment Report on Biodiversity and Ecosystem Services, biodiversity loss is accelerating and ecosystems continue to be degraded. The report stated that the deteriorating health of ecosystems is affecting the foundations of our economies, livelihoods, food security, health, and quality of life worldwide. An estimated two billion hectares of land on earth have been degraded by human activity including pollution. Deforestation continues to decline, driven mainly by agricultural expansion and urbanization. Desertification also poses major challenges in many regions of the world. The report estimates one million species are threatened with extinction and biodiversity is declining faster than at any other time in human history. Wildlife trafficking, habitat loss, and overexploitation are some of the causes of biodiversity loss. Of course, these processes are distributed unequally across space, affecting people and ecosystems differently. At the root of every issue is one common thread: the unsustainable use of land and resources.

Ecosystems

An **ecosystem** is defined as "a biological community of interacting organisms and their physical environment." Despite the brevity of this definition, it encapsulates an extremely complex set of interactions where organisms are interacting with one another, organisms are affecting their environment, and the environment in turn is affecting the organisms that inhabit it.

The concept can be seen in examples such as basic food/life systems. All ecosystems are made up of **food chains** or **food webs** that begin with energy—sunlight—that is converted into organic matter by plants. Animals that eat plants synthesize a portion of the plant material in their own bodies for energy. Thus energy is passed along from one organism to another. A food chain or food web is organized into different segments or levels. Each level is defined by the point of energy transfer from the environment to an organism and then from that organism to another. The following are the **trophic levels** that represent a basic framework for all ecosystems:

1. The first level are the producers of organic energy. These include algae and plants. This is followed by a series of consumers.

2. Herbivores, or primary consumers, make up the second level. These include cattle, buffalo, fish, squirrels, and other plant-eating fish and animals.

3. Carnivores, or second consumers, such as wolves, lions, and snakes are the third level. These are meat eaters.

4. Specialized carnivores, or tertiary or apex consumers, are the fourth level. Hawks, eagles, and sharks are among this group that eats other carnivores.

A food chain is a linear sequence of organisms that pass nutrients and energy to another; most ecosystems are composed of many interconnected food

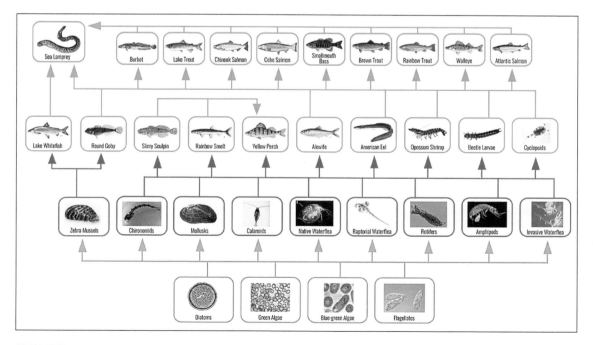

FIGURE 15.1 A Food Web

Most ecosystems are composed of many interconnected food chains that are better conceptualized as food webs.

chains that are better conceptualized as food webs. This is because food chains are often oversimplified, and are not always representative of real-world ecological communities. For instance, an organism can sometimes eat multiple types of prey or be eaten by multiple predators, including ones at different trophic levels. Many prefer the term "food web" as a more accurate representation of these relationships (see Figure 15.1).

There are three basic classes of ecosystems: terrestrial, saltwater, and freshwater. In this chapter, we focus on terrestrial ecosystems, although there are interconnections among all three ecosystems. The terrestrial class is divided into five groups based on the general structure and composition of the vegetation: forest, savanna, grassland, desert, and tundra. Each of these is differentiated geographically according to moisture and temperature into biomes. A **biome** is a large biogeographical unit characterized by a particular combination of vegetation and animals that are generally associated with a general climatic type. Map 15.1 shows the Earth's principal terrestrial biomes.

Ecosystem Services

Ecosystem services are the benefits that humans obtain from ecosystems. These benefits can be both direct and indirect. Some ecosystem services involve the direct provision of material and nonmaterial goods to people and depend on the presence of particular species of plants and animals; for example, food, timber, and medicines. Other ecosystem services come indirectly from the functioning of ecosystem processes. For example, the formation of soils and soil fertility that sustains crop and livestock production depends on the ecosystem processes of decomposition and nutrient cycling by soil micro-organisms.

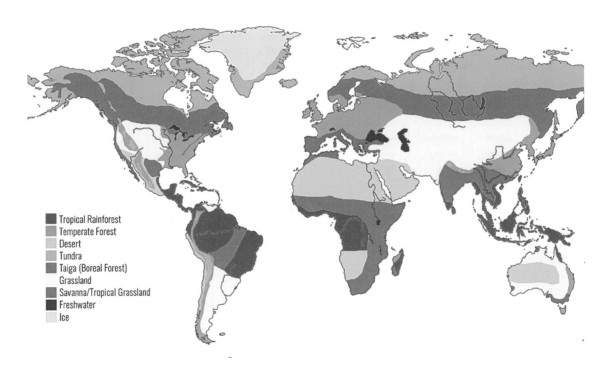

Tropical Rainforest
Temperate Forest
Desert
Tundra
Taiga (Boreal Forest)
Grassland
Savanna/Tropical Grassland
Freshwater
Ice

MAP 15.1 Earth's Major Land Biomes

There are four major categories of ecosystem services, as shown in Figure 15.2:

1. *Provisioning services* include the water produced as a service of the hydro-logical cycle, and also the plant and animal materials used as food and to make clothing, and the natural resources used to produce energy. These services allow humans to exist.

2. *Regulating services* are those necessary for our sustained habitation of the earth, such as the purification of water as it migrates through the soil. These services also include climate regulation, carbon sequestration, flood control, biological regulation of infectious disease, and the soil fertility and pollination necessary for food production, among other ser-vices. These services are the ones most likely taken for granted by most humans—the hidden services that are essential to the continued quality and abundance of many provisioning services.

3. *Cultural services* encompass the nonmaterial benefits of nature. These benefits include those obtained from recreation in greenspace, the eco-nomic benefits generated from people visiting greenspaces, and the aes-thetic and spiritual experience felt when observing or being immersed in the natural environment.

4. *Supporting services* are those services that are necessary to produce all other ecosystem services. These are services such as soil formation and nutrient and water cycling on which the provisioning, regulating, and cultural services are dependent.

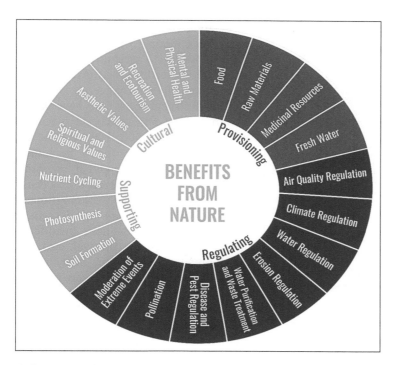

FIGURE 15.2 Ecosystem Services
This refers to the many benefits that nature can provide to society. Seeing ecosystems as a service provided to humans may help to value their contributions. However, this does reinforce an anthropocentric perspective that to be "of value," ecosystems must provide humans some benefit.

Biodiversity

Biodiversity includes diversity *within* and *among* species and ecosystems. Scientists estimate there are about one million species that have been identified, but there are millions of others that have not yet been identified. Some of the upper estimates suggest there could be as many as 100 million different species coexisting with us on this planet.

There are several measures of biodiversity. First is species diversity, both in richness and abundance. Species richness is a count of the species within an ecosystem, while species abundance refers to the number of individuals in a species in that ecosystem (see Figure 15.3). A second measurement is habitat size; the smaller the habitat, the less diversity there will be. A third measure is genetic diversity within a species. Genetic diversity refers to the number of genetic characteristics in the makeup of a species, and can affect a population's ability to adapt to changing environments.

More biodiverse communities are more productive. In more biodiverse communities, some species are "redundant," which means they could be lost without much effect on the structure and functioning of the ecosystem. **Redundancy** is an insurance against the loss of ecosystem function. An ecosystem that has more biodiversity and more redundant species is better able to adapt to change and avoid collapse.

Changes in biodiversity can influence the quality and supply of ecosystem services. Biodiversity loss has been accelerating for a number of reasons including habitat loss due to agricultural expansion, pollution, desertification, invasive alien species, and climate change. According to some experts, human actions have already driven at least 680 vertebrate species to extinction since 1500. Habitat loss and deterioration, largely caused by human actions, have reduced global terrestrial habitat integrity by 30 percent. As a result, around 9 percent of the world's estimated six million terrestrial species—more than 500,000 species—have insufficient habitat for long-term survival, and may be doomed to extinction, many within decades, unless their habitats are restored.

A 2019 **Intergovernmental Science-Policy Platform on Biodiversity (IPBES)** Report found the global rate of species extinction is already at least tens to hundreds of times higher than the average rate over the past ten million years and is increasing. Some estimates are that between 10,000 and 100,000 species are becoming extinct each year. Of the 8,300 animal breeds known 8 percent

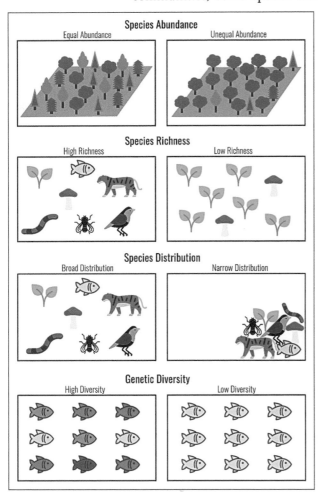

FIGURE 15.3 Measures of Biodiversity
Measures of biodiversity include species abundance, species richness, species distribution, and genetic diversity.

are extinct and another 22 percent are at risk of extinction. Of the 80,000 tree species less than 1 percent have been studied for potential use. Local declines of insect populations such as wild bees and butterflies have been reported, and insect abundance has declined rapidly in some places, although the full extent of insect declines is not known. The report also noted that the average abundance of native species in most major terrestrial biomes has fallen by at least 20 percent, potentially affecting ecosystem processes and hence nature's contributions to people. An earlier 2016 IPBES study shows that 40 percent of pollinator species, such as bees and butterflies, face a risk of extinction, with potential devastating consequences for food and livelihoods because 75 percent of our food crops are dependent on pollinators.

Fewer and fewer varieties and breeds of plants and animals are being cultivated, raised, traded, and maintained around the world. Reductions in the diversity of agricultural crops means that agricultural systems are less resilient against future climate change, pests, or pathogens.

There are numerous examples of declines in the natural world (Figure 15.4). In addition to direct human activity, climate change is also projected to increase the number of species under threat as it will shift the boundaries of biomes and many species will not be able to migrate or adapt.

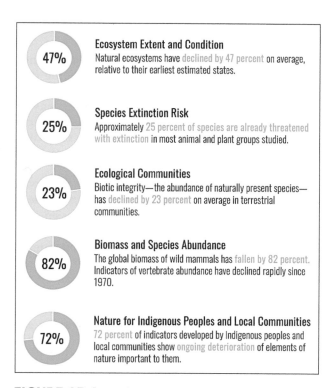

47% **Ecosystem Extent and Condition**
Natural ecosystems have declined by 47 percent on average, relative to their earliest estimated states.

25% **Species Extinction Risk**
Approximately 25 percent of species are already threatened with extinction in most animal and plant groups studied.

23% **Ecological Communities**
Biotic integrity—the abundance of naturally present species—has declined by 23 percent on average in terrestrial communities.

82% **Biomass and Species Abundance**
The global biomass of wild mammals has fallen by 82 percent. Indicators of vertebrate abundance have declined rapidly since 1970.

72% **Nature for Indigenous Peoples and Local Communities**
72 percent of indicators developed by Indigenous peoples and local communities show ongoing deterioration of elements of nature important to them.

FIGURE 15.4 Declines in Nature
Examples of declines in nature involve the extent and condition of ecosystems, those species at risk of extinction, the decline in ecological communities, the decline in biomass and species abundance, and the deterioration of ecosystems in Indigenous communities.

REASONS FOR ECOSYSTEM CHANGE AND BIODIVERSITY LOSS

Humans impact ecosystems and biodiversity in direct and indirect ways. A **direct driver** of change influences ecosystem processes and can be identified and measured, while an **indirect driver** operates more diffusely, by altering one or more direct drivers.

Direct drivers of change in ecosystems are land-use change, pollution, direct exploitation of organisms, and invasion of alien species. Many experts also consider climate change is now a direct driver of change. Globally, land-use change is a direct driver with the largest relative impact on terrestrial ecosystems.

Indirect drivers of change can include population change (such as the increase in global population or urbanization), sociocultural factors (such as whether ecosystems are considerable valuable), political factors (such as taxes or subsidies on pesticides, or the existence of conservation laws), economic factors (such as financial incentives to produce certain foods or products), and technology changes (such as the increase in agricultural cultivation since the 1950s). Generally, changes in ecosystems are often caused by multiple interacting drivers.

Land-Use Change

Land-use change is driven primarily by three human activities: agriculture, forestry, and urbanization.

Agricultural expansion is the most widespread form of land-use change, with more than one-third of the terrestrial land surface today being used for cropping or animal husbandry. This expansion, along with a doubling of urban areas over the last thirty years, has come mostly at the expense of forests (largely old-growth tropical forests), wetlands, and grasslands. Agriculture that results in monocropping economies such as tobacco, sugarcane, banana, and pineapple not only changes land use but it also accelerates the loss of biodiversity.

Although the pace of agricultural expansion has varied from country to country, the greatest losses have occurred primarily in the tropics, home to the highest levels of biodiversity on the planet. For example, between 1980 and 2000 more than 250 million acres (100 million hectares) of tropical forest were lost due to cattle ranching in Latin America and oil palm plantations in Southeast Asia among others.

Agricultural expansion is also degrading shrublands, grasslands, and savannas through overexploitation and poor management. Productivity hotspots like areas along rivers where nutrients and access to water are high are being converted to cropland, leaving behind the less productive, dry, and nutrient-poor areas.

Forestry involves the harvesting of natural vegetation for human use. It can involve trees being cut down for residential and commercial use. Changes in the forest can result in habitat loss, degradation, and fragmentation. In addition, by harvesting wood, forestry also removes a carbon sink because trees and other vegetation store carbon.

Urbanization is also a major driving force behind land-use change. The process of urbanization entails the change of land from nonurban uses (such as open space or agriculture) to urban uses (such as residential, industrial, or commercial). Land-use change due to urbanization can reduce biodiversity, damage wildlife habitats, or introduce the emission of pollutants. The conversion of land typically accelerates in areas experiencing more rapid urbanization.

Agriculture, forestry, and urbanization are accompanied by expanded infrastructure such as roads, bridges, and hydroelectric dams that can generate positive economic effects and can improve quality of life, but can accelerate deforestation, habitat fragmentation, biodiversity loss, and also cause population displacement and social disruption, including for Indigenous peoples and local communities. Each of these main drivers of land-use change can impact ecosystems and biodiversity by causing changes that include:

- *Reduction*: loss in area or coverage of an ecosystem due to burning, agricultural development, urbanization, and lumbering;
- *Fragmentation*: occurs when ecosystems are broken down from large, continuous areas into smaller parcels. Agriculture, cities, highways, and dams can destroy segments or create constrictions;
- *Substitution*: involves replacing one set of organisms in an ecosystem with another. This can include the deliberate conversion of land (the replacement of native grasses with wheat or corn) or the unintended introduction of invasive species;
- *Simplification*: occurs when ecosystems become less diverse (both in types of species and numbers of organisms);
- *Contamination*: involves the incorporation of pollutants into the ecosystem through air, water, or soil; and
- *Overgrowth*: pollution or contamination can lead to overgrowth of ecosystems, such as when excess nitrogen and phosphorus induce rapid growth in algae, resulting in eutrophication.

The three major drivers of land-use change have resulted in the most pressing challenges to ecosystem health and biodiversity: deforestation, desertification, and pollution.

Deforestation

Forests cover nearly 31 percent of our planet's land area and account for ten billion acres (four billion hectares). From the air we breathe, to the water we drink, to the food we eat—forests sustain us. Forests are home to more than 80 percent of all biodiversity and ecosystem services. However, they vary geographically. Forty-five percent of the world's forests are in the tropical areas, followed by the boreal (27 percent), temperate (16 percent), and subtropical (11 percent) areas (Figure 15.5). Regionally, Europe accounts for 25 percent of the world's forest area, followed by South America (21 percent), North and Central America (19 percent), Africa (16 percent), Asia (15 percent), and Oceania (5 percent). More than half (54 percent) of the world's forests is located in only five countries—the Russian Federation, Brazil, Canada, the United States of America, and China. Map 15.2 shows the global distribution of forests.

Deforestation is the conversion of forest to other land uses, such as agriculture and infrastructure. In the course of activities such as mining or farming, trees and other vegetation are cleared away. In contrast to deforestation, forest

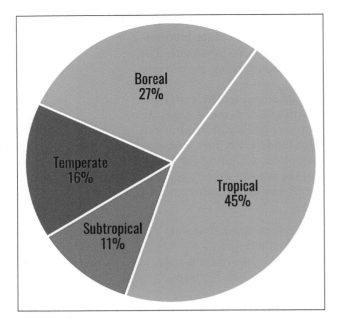

FIGURE 15.5 Global Forest Areas by Climate Domain
The proportion of global forest areas by climate domain. Tropical rainforests constitute the largest proportion of global forests, followed by boreal forests.

area can increase when trees are planted on land that was not previously forested (**afforestation**) or when trees grow back on abandoned agricultural or other land (natural forest expansion).

According to a 2020 UN Food and Agricultural Organization (FAO) report on forests, the world has lost a net area of 440 million acres (178 million ha) of forest since 1990, which is an area about the size of the country of Libya. Lost forests mean the disappearance of livelihoods in rural communities, increased carbon emissions, diminished biodiversity, and the degradation of land. The good news is that the rate of net forest loss decreased substantially over the period 1990–2020 due to a reduction in deforestation in some countries, plus increases in forest area in others through afforestation and the natural expansion of forests. While forest loss remains high, 2021 data show that the proportion of forests in protected areas and under long-term management plans increased or remained stable at the global level and in most regions of the world.

Geographically, South America, and Eastern and Southern Africa are regions where forest loss is highest. In South America, Brazil has experienced the largest loss of forest with Paraguay and Peru also seeing forest decline. The state of Rondônia in western Brazil—once home to about 208,000 square kilometers (51.4 million acres) of forest—has become one of the most deforested parts of the Amazon. In the past three decades, clearing and degradation of the state's forests have been rapid: 4,200 square kilometers cleared by 1978; 30,000 by 1988; and 53,300 by 1998. By 2003, an estimated 67,764 square kilometers of rainforest—an area larger than the state of West Virginia—had been cleared.

Deforestation follows a fairly predictable pattern. The first clearings that appear in the forest are in a fishbone pattern, arrayed along the edges of roads. Over time, the fishbones collapse into a mixture of forest remnants, cleared areas, and settlements. This pattern follows one of the most common deforestation trajectories in the Amazon. Legal and illegal roads penetrate a remote part of the forest, and small farmers migrate to the area. They claim land along the road and clear some of it for crops. Within a few years, heavy rains and erosion deplete the soil, and crop yields fall. Farmers then convert the degraded land to cattle pasture and clear more forest for crops. Eventually the small landholders, having cleared much of their land, sell it or abandon it to large cattle holders, who consolidate the plots into large areas of pasture.

In South America, forests have been replaced by plantation forests that consist almost entirely of introduced species such as soybean. In the Amazon basin, the growth of soybean farming and cattle ranching have

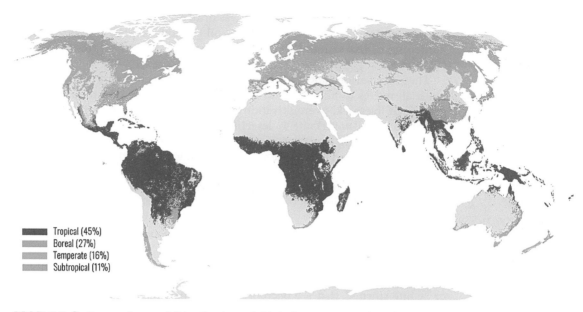

MAP 15.2 Proportion and Distribution of Global Forest Area by Climate Domain (2020)
This map shows the distribution of world's tropical, subtropical, boreal, and temperate forests.

resulted in significant deforestation. Forestry extraction—or timber harvesting—also plays a role in deforestation; between 1990 and 2015, 290 million hectares in native forest cover were lost. Other countries that have seen the highest loss of forest area between 2010–20 are the Democratic Republic of the Congo, Indonesia, Angola, Tanzania, Myanmar, Cambodia, and Bolivia.

In Europe and East Asia—particularly China—there has been a net gain in forest area. A small part of this is reforest planting; most of the forest areas are naturally regenerating. Australia, India, and the United States have also recorded net gains in forest area over the last ten years.

Desertification

Drylands encompass around 38 percent of the Earth's land area, covering much of North and South Africa, western North America, Australia, the Middle East, and Central Asia. Drylands are ecosystems that are characterized by a lack of water. They include scrublands, shrublands, grasslands, savannas, semideserts, and true deserts. Drylands are home to many iconic images of biodiversity: the lions and wildebeest of the Serengeti, reindeer in the Mongolian rangelands, and the prairie dogs of the tall grass prairies in the United States. Botswana, Burkina Faso, Iraq, Kazakhstan, the Republic of Moldova, and Turkmenistan are countries where 99 percent of their areas are classified as drylands. About two billion people live in areas that are at risk for or already experiencing desertification.

Desertification is the term used to describe land degradation in arid, semiarid, and other drylands as a result of drought, deforestation, or agriculture. Desertification does not literally mean the expansion of deserts, but it is a more general term for land degradation in water-scarce parts of the world.

It is a transformation from land that was once arable (drylands or grasslands) into an area that may have significant soil erosion, or even a compaction or hardening of the soil and its inability to retain water or regrow vegetation (Photo 15.2).

There are various and sometimes interconnected causes of desertification, but it all relates to how the land is (or is not) managed. Deforestation, overgrazing of livestock, overcultivation of crops, and excessive or inappropriate irrigation can all contribute to desertification. Added to this is climate change, which is increasing land surface temperatures, and altering precipitation patterns, both of which can accelerate desertification. For example, in areas that are naturally dry, a drought can reduce vegetation cover, particularly if that land is being used by high numbers of livestock. As plants die off due to lack of water, the soil becomes bare and is more easily eroded by wind, and eventually by the rains.

The consequences of desertification are numerous. Areas may see an increase in wildfires, or a declining water table, as more groundwater is needed to irrigate crops or provide water for human or animal consumption. There can also be an increase in sand and dust storms, known as "haboobs" or "white storms," that are becoming more frequent in the Middle East and Northern Africa. Desertification can also lead to a loss of biodiversity in plants and animals, as many have limited ability to adapt quickly to desertification. Degradation can also open up opportunities for invasive species. Finally, desertification can have significant consequences for food security and livelihoods. In some countries in East Africa, such as Somalia, Kenya, and Ethiopia, more than half the population are pastoralists that rely on healthy grazing lands for the livelihoods.

Deforestation and desertification can be interconnected. The Amazon rainforest, for example, generates its own rainfall. Clearing the rainforests may cause a drying in the local climate, which adds to the risk of desertification. These processes are also connected to wider indirect socioeconomic drivers. Demand for meat in Europe can lead to deforestation in South America, as

PHOTO 15.2 Soil Erosion Has Led to Desertification in Huangshatou National Desert Park, located in Hainan, China
Green trees have been planted as desertification control.

land is converted to increase the production of livestock for exports. Over time, this area may experience desertification.

Pollution and Contamination

A main cause of land degradation is pollution. Generally, pollution comes from three major sources: agriculture, urbanization and industry, as listed in Table 15.1. Globally, one-fifth of the Earth's land area (more than seventy-seven million square miles or two billion hectares) is degraded, an area nearly the size of India and the Russian Federation combined. Land degradation is undermining the well-being of some 3.2 billion people, driving species to extinction and intensifying climate change.

Agricultural pollution includes crop fertilizers in stormwater runoff, smoke from field burning and forest fires, blowing dust, and soil erosion from logging. Pollutants are released to the environment accidentally, for example from oil spills or leaching from landfills, or intentionally, through use of fertilizers and pesticides, irrigation with untreated wastewater, or land application of sewage sludge.

Urbanization and industry contribute pollution including from industrial activities (air and water pollution), car exhaust, acid and other chemicals in stormwater runoff, sewage treatment plant leakages and released gases, and chemicals that run off suburban lawns. Regardless of the way in which pollutants enter the environment, they impact ecosystems.

Soil pollution can change the balance of ecosystems, causing the disappearance of predators or competing species or allowing the emergence of new pests and diseases. Soil pollution can also spread antimicrobial resistant bacteria and genes. Arsenic is a metallic pollutant that is used in industrial manufacturing, including those conducted on mining lands. When plants take in arsenic, it can disrupt metabolic processes and lead to cell death. Plants that take up pollutants often pass these along to organisms that consume these plants, affecting the food web. Agricultural pesticides and fertilizers may be responsible for reducing honey bee populations. Pollutants have been found everywhere in the world—in polar bears, in newborn infants, and even on the top of Mount Everest.

TABLE 15.1	Major Land Use Sources of Pollution
Agricultural	• Fertilizers • Pesticides • Herbicides
Urban	• Plastics • Organic solid wastes (food scraps, wood, and debris) • Inorganic: fertilizers and pesticides from lawns, antifreeze and oils from street surfaces • Municipal wastes including wastewater
Industrial	• Medical wastes • Chemicals • Oil spills • Mining and mining wastes • Persistent organic pollutants • Inorganic pollutants such as polychlorinated biphenyls (PCBs) and dioxins

Source: By author, compiled from several sources.

In organisms, pollutants can cause death or other health problems. In 1962 Rachel Carson published her book, *Silent Spring*, writing about the deleterious effects of pesticides. Carson argued that pesticides such as DDT (dichloro-diphenyl-trichloroethane), dieldrin, endrin, and parathion were destroying birdlife. The image of a "birdless" (or silent) spring provides a telling metaphor and her book marked a shift in environmental concern and a growing awareness that human activity could have a significant impact on the natural world and, in turn, ourselves. More than fifty years later, however, pollution—and particularly man-made chemicals—remain a threat to humans and ecosystems.

Man-made or synthetic chemicals are especially problematic. The European Chemicals Agency estimates there are more than 144,000 man-made chemicals in existence, although probably only 40,000 or so are actively being produced today. More alarming is most of these man-made chemicals have not been studied or screened for human or ecosystem health safety. Some are considered toxic; some are considered hazardous. **Toxicity** is the ability of a chemical substance to produce injury once it reaches a susceptible site in or on the body. Toxic chemicals are poisonous, such as lead, cyanide, or dioxin. A **hazardous substance** may be toxic, flammable, or reactive in a way that may produce injury when used (e.g., gasoline or sulfuric acid). The US Environmental Protection Agency (EPA) lists more than 85,000 chemicals on its inventory of substances that fall under the Toxic Substances Control Act (TSCA).

There are three categories of chemicals that are of the most concern today: (1) very persistent chemicals that break down slowly or not at all and bioaccumulate or biomagnify in the bodies of wildlife and people; (2) **endocrine disrupting chemicals** that interfere with the hormone systems of animals and people; and (3) chemicals that cause cancer, reproductive problems, or damage DNA.

Exploitation of Organisms

Many plants and animals are being harvested at unsustainable levels. **Overexploitation** refers to harvesting of a renewable resource that is taken at levels beyond its capacity to reproduce its population. This can occur through excessive hunting, fishing, trapping, or logging. Some "charismatic species" such as elephants and rhinoceros are threatened by unsustainable harvesting, but the majority of endangered species are insects, birds, reptiles, and plants. Unfortunately, overexploitation continues despite the fact that there are international agreements to regulate and protect endangered species. The *SDGs and the Law* 15.1 box discusses the Convention on International Trade in Endangered Species of Wild Fauna and Flora (CITES).

Illicit poaching and trafficking of wildlife and plants contributes to the problem. Currently, nearly 7,000 species of animals and plants have been reported in illegal trade involving 120 countries. According to a 2020 UN Report on World Wildlife Crime, wildlife crime is a lucrative global business, with high demand driving high prices. Figure 15.6 shows the share of illegally trafficked wildlife that was seized by officials. For example, pythons are illegally taken for their use live as pets, their skins to make handbags and shoes, their meat as a food, and their organs as a traditional medicine. Another species that has

SDGs AND THE LAW 15.1

The Convention on International Trade in Endangered Species of Wild Fauna and Flora

The Convention on International Trade in Endangered Species of Wild Fauna and Flora, known as CITES, was signed on March 3, 1973, and entered into force on July 1, 1975. Today almost all countries in the world are parties to this international legally binding agreement.

CITES is an international agreement between governments that aims to ensure that international trade in specimens of wild animals and plants does not threaten their survival. CITES regulates international trade in specimens of species of wild fauna and flora based on a system of permits and certificates issued under certain conditions.

At the time when CITES was under consideration, in the 1960s, international discussion of the regulation of wildlife trade for conservation purposes was something relatively new. With hindsight, the need for CITES is clear.

Annually, international wildlife trade is estimated to be worth billions of dollars and to include hundreds of millions of plant and animal specimens. The trade is diverse, ranging from live animals and plants to a vast array of wildlife products derived from them, including food products, exotic leather goods, wooden musical instruments, timber, tourist curios, and medicines. Levels of exploitation of some animal and plant species are high and the trade in them, together with other factors such as habitat loss, is capable of heavily depleting their populations and even bringing some species close to extinction. Many wildlife species in trade are not endangered, but the existence of an agreement to ensure the sustainability of the trade is important to safeguard these resources for the future.

Because the trade in wild animals and plants crosses borders between countries, the effort to regulate it requires international cooperation to safeguard certain species from overexploitation. CITES was conceived in the spirit of such cooperation. Today, it accords varying degrees of protection to more than 37,000 species of animals and plants, whether they are traded as live specimens, fur coats, or dried herbs.

Source: The Convention on International Trade in Endangered Species of Wild Fauna and Flora. "What Is CITES?" (2021). https://cites.org/eng/disc/what.php

seen a dramatic increase in trafficking is pangolins. Currently pangolins are the most trafficked wildlife, valued for their scales and blood, which are thought to be a healing tonic, and their meat, which is considered a high-end delicacy. Pangolins are illegally exported from Nigeria and Democratic Republic of Congo primarily to countries in Asia such as Vietnam and China.

Nearly 6,000 different species of fauna and flora have been seized between 1999 and 2018, with nearly every country in the world playing a role in the illicit wildlife trade. Trafficking in illegal wildlife is not limited to pythons and pangolins but includes corals, plants, and birds. One of the markets that has increased dramatically is that for illicit rosewood.

Overexploitation can result in extinction of the species, ecosystem changes and degradation, and resource destruction.

Invasive Species

For centuries, humans have transported plants, animals, and other organisms beyond their natural ranges, intentionally or unintentionally. Not all

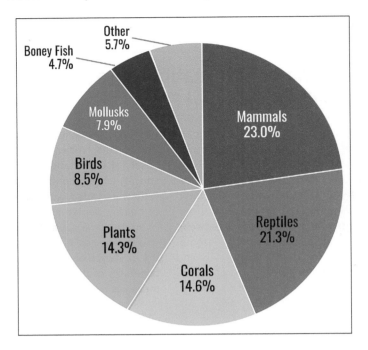

FIGURE 15.6 Illegally Trafficked Wildlife
This figure shows the share of illegally trafficked wildlife that was seized by officials by taxonomic category. While mammals and reptiles constitute the largest share of seizures, many people are unaware in the trafficking of corals, plants, and mollusks. Corals, for example, are harvested for the aquarium trade.

introduced species cause problems in their new locations. However, some can cause substantial and irreversible ecological damage. This is the case with the pythons in the Everglades discussed at the start of this chapter.

Invasive, nonnative species (INNS) are considered a critical driver of biodiversity loss worldwide.

Nearly one-fifth of the Earth's surface is at risk of plant and animal invasions, impacting native species, ecosystem functions, and economies and human health. The rate of introduction of new invasive alien species appears higher than ever before and one UN report indicated that since 1970 there has been a 70 percent increase in numbers of invasive alien species across twenty-one countries with detailed records.

Invasive species can be introduced by mistake, or by design. They can be a cultivated species that escapes and colonizes new areas, or they can be a native species that becomes dominant because of the destruction of a predator species in that ecosystem. Often, invasive species increase aggressively because they may lack existing predators in that ecosystem. Kudzu, a woody vine, is notorious in the United States as an invasive species that is nearly impossible to eradicate (Photo 15.3). It was introduced from China and Japan and was widely planted in the 1930s to stop soil erosion in the Dust Bowl. But kudzu began to spread to all parts of the United States. Another notorious example is the cane toad that was introduced in Australia in 1935 to combat cane beetles that were wreaking havoc on sugar cane crops.

PHOTO 15.3 Kudzu Taking over Forest in Atlanta, Georgia
Kudzu is distinctive in its ability to climb to the forest canopy to get access to sunlight. In doing so, it often crowds out, covers over, or outcompetes some of the native plant communities.

The cane toads quickly became prolific invaders and, instead of regulating cane beetles, became voracious eaters of native insects. They number in the millions and have proven difficult to control in part because they are poisonous to predators.

SUSTAINABLE ECOSYSTEMS AND BIODIVERSITY

Over the next decade, it will be a major challenge to reverse the degradation of ecosystems while also meeting the increasing demands for their services. Table 15.2 lists the targets for SDG 15.

There exist many international organizations and agreements to help finance, support, and implement the targets for SDG 15. For example, the UN Convention on Biological Diversity (CBD) established in 1992 provides a global framework for international cooperation on science, technology, and innovation related to the **conservation** and sustainable use of biodiversity. According to the 2019 CBD report, one trillion US dollars is a conservative estimate of the amount of finance required to assist in addressing current environmental challenges through ecosystem restoration.

In 2019, the UN General Assembly proclaimed 2021–30 to be the UN Decade on Ecosystem Restoration. It is intended that this will build on prior and existing initiatives. Over the next ten years, the CBD aims to enhance global, regional, national, and local commitments and actions to prevent, halt, and reverse the degradation of ecosystems; increase understanding of the multiple benefits of successful ecosystem restoration; and apply this knowledge in education systems and within all public- and private-sector decision making. Within these more general goals, there are

TABLE 15.2 SDG 15 Targets and Indicators to Achieve by 2030

Target	Indicators
15.1 Ensure the conservation, restoration, and sustainable use of terrestrial and inland freshwater ecosystems and their services, in particular forests, wetlands, mountains, and drylands, in line with obligations under international agreements	**15.1.1** Forest area as a proportion of total land area **15.1.2** Proportion of important sites for terrestrial and freshwater biodiversity that are covered by protected areas, by ecosystem type
15.2 Promote the implementation of sustainable management of all types of forests, halt deforestation, restore degraded forests, and substantially increase afforestation and reforestation globally	**15.2.1** Progress toward sustainable forest management
15.3 Combat desertification, restore degraded land and soil, including land affected by desertification, drought, and floods, and strive to achieve a land degradation-neutral world	**15.3.1** Proportion of land that is degraded over total land area
15.4 Ensure the conservation of mountain ecosystems, including their biodiversity, to enhance their capacity to provide benefits that are essential for sustainable development	**15.4.1** Coverage by protected areas of important sites for mountain biodiversity **15.4.2** Mountain Green Cover Index
15.5 Promote fair and equitable sharing of the benefits arising from the utilization of genetic resources and promote appropriate access to such resources, as internationally agreed	**15.6.1** Number of countries that have adopted legislative, administrative, and policy frameworks to ensure fair and equitable sharing of benefits
15.7 Take urgent action to end poaching and trafficking of protected species of flora and fauna and address both demand and supply of illegal wildlife products	**15.7.1** Proportion of traded wildlife that was poached or illicitly trafficked
15.8 Introduce measures to prevent the introduction and significantly reduce the impact of invasive alien species on land and water ecosystems and control or eradicate the priority species	**15.8.1** Proportion of countries adopting relevant national legislation and adequately resourcing the prevention or control of invasive alien species
15.9 Integrate ecosystem and biodiversity values into national and local planning, development processes, poverty reduction strategies, and accounts	**15.9.1** Progress toward national targets established in accordance with Aichi Biodiversity Target 2 of the Strategic Plan for Biodiversity 2011–20
15.A Mobilize and significantly increase financial resources from all sources to conserve and sustainably use biodiversity and ecosystems	**15.A.1** Official development assistance and public expenditure on conservation and sustainable use of biodiversity and ecosystems
15.B Mobilize significant resources from all sources and at all levels to finance sustainable forest management and provide adequate incentives to developing countries to advance such management, including for conservation and reforestation	**15.B.1** Official development assistance and public expenditure on conservation and sustainable use of biodiversity and ecosystems

Target	Indicators
15.C Enhance global support for efforts to combat poaching and trafficking of protected species, including by increasing the capacity of local communities to pursue sustainable livelihood opportunities	**15.C.1** Proportion of traded wildlife that was poached or illicitly trafficked

Source: United Nations Sustainable Development Goals. https://sustainabledevelopment.un.org/sdg15

many strategies and actions that can advance the targets of SDG 15. Both direct and indirect drivers of ecosystem change and biodiversity loss must be addressed. Table 15.3 lists some specific solutions and several others are considered in more detail.

Expand Protected Areas

The creation or expansion of protected ecosystems involve developing incentives for protecting biodiversity, including habitat protection and working to protect genetic diversity of plants and animals. Instruments such as protected areas place legal restrictions on human access and use within their boundaries and impose penalties on offenders. Examples of protected areas include national parks, national forests, and national conservation areas. Research suggests that protected areas are effective at stalling deforestation.

Increase Reforestation

Reforestation refers to the process of restocking or regrowing forests. This can be a naturally occurring process (such as after a forest fire or disease) or intentional (programs that deliberately plant seedlings or trees).

There have been regional efforts to fight desertification in Africa through large-scale reforestation. The Great Green Wall, launched in 2007 by the African Union, aims to restore Africa's degraded landscapes in the Sahel. Once complete, the wall will be the largest living structure on the planet—an 8,000 kilometer stretch across the entire width of the continent.

The Great Green Wall is now being implemented in twenty countries across Africa and more than eight billion dollars have been mobilized and pledged for its support. Between 2007 and 2020, several achievements have been recorded in most of the Great Green Wall member states, with some countries being more successful than others. For example, Ethiopia produced more than 5.5 billion plants and seeds, restored more than one million hectares of land, and created more than 200,000 jobs; Senegal planted more than eighteen million trees and restored 800,000 hectares of degraded land; and Niger planted 146 million trees. By 2030, the ambition of the initiative is to restore 100 million hectares of currently degraded land, sequester 250 million tons of carbon, and create ten million green jobs.

At the local or urban scale, reforestation could include enhancing and expanding green spaces such as urban forests, community gardens, parks, and open space. For example, planting trees in cities can have multiple benefits that include regulating the urban heat island, improving air quality, providing habitat, increasing property values, and resulting in energy savings. The *Solutions* 15.1 box highlights a tree planting program in New York City.

TABLE 15.3 Solutions and Actions for Achieving SDG 15	
Solutions to address indirect drivers of biodiversity loss	• Celebrate a culture of restoration • Incorporate environmental and socioeconomic impacts, including externalities, into public and private decision making • Shift behaviors to reduce ecosystem degradation • Raise awareness of benefits of ecosystem restoration
Solutions to reduce direct pressures	• Redirect fossil fuel, agricultural, forestry, and fishing subsidies to conservation and restoration of ecosystems • Reduce pollution and remediate contaminated areas • Reduce invasive species • Ban pesticides • Establish and expand preservation areas • Amend legislative and policy frameworks to promote restoration • Include biodiversity protection, biodiversity offsetting, river basin protection, and ecological restoration in regional planning
Solutions to improve the status of ecosystems	• Improve and expand the levels of financial support for conservation and sustainable use through a variety of innovative options, including through partnerships with the private sector • Amend legislative and policy frameworks to promote restoration • Establish and expand preservation areas • Prevent extinctions • Conserve gene pool • Restore ecosystems
Solutions to enhance the benefits of ecosystems and biodiversity	• Increase financial support for R&D • Undertake long-term scientific research on the implementation and benefits of ecosystem restoration • Advance habitat mapping • Improve the documentation of nature (e.g., biodiversity inventory and other inventories) and the assessment of the multiple values of nature, including the valuation of natural capital by both private and public entities • Integrate Indigenous knowledge and traditional practices into ecosystem restoration initiatives

Source: By author, compiled from several sources.

Implement Cash for Conservation

During the last several decades, there has been an emphasis on applying monetary value to nature. As a result, there have been efforts to develop more robust systems for valuing ecosystem services—at local, national, and regional scales to protect and manage ecosystems.

One instrument many countries have used is **integrated conservation and development projects (ICDP)**. The main goal of an ICDP is to conserve biodiversity while also furthering the social and economic needs of rural communities. ICDPs are a coordinated set of development incentives to influence indirect drivers of change, such as values and behaviors around conservation. For example, a rural community near a national park could be given money to protect against harvesting the timber, prevent poaching, or expand tourism to the park. These programs have been popular in Africa, where few if any countries have the financial resources to manage and protect the national parks against poaching and other forms of degradation.

The World Wildlife Fund and Conservation International are organizations that have invested into these types of programs and they are found in many countries around the world. However, ICDPs have be criticized for having limited success in achieving both conservation and development objectives.

Two other recent conservation strategies are **Payments for Ecosystem Services (PES)** and Biodiversity Offset Programs. These are programs that

⊜ SOLUTIONS 15.1

Planting One Million Trees in New York City

In 2007, New York City launched its Million-TreesNYC program to plant and care for one million new trees by 2020. In November 2016, the city celebrated the planting of its one millionth tree, planted four years ahead of schedule. The million new trees expand the city's tree canopy by more than 20 percent.

Several aspects of this successful project are notable. New York City's approach to tree planting directly targets equity issues by bringing trees to those areas where the health benefits may be most needed, including neighborhoods with the highest asthma rates and fewest trees. This approach sets New York's program apart from those of many cities that seek to increase the tree canopy citywide, but do not necessarily ensure that communities in need are targeted. Nota-

bly, the MillionTreesNYC program is a successful public-private partnership between NYC Parks and Bette Midler's New York Restoration Project and relied on more than 50,000 volunteers. The city allocated more than $350 million toward the effort, but more than $30 million was also raised from private sources. Because fundraising has been so successful, the city planted an additional 150,000 new trees each year until 2020.

Sources: "About MillionTreesNYC: I'm Counting on You." Million Trees NYC (2018, April 3). http://www.milliontreesnyc.org/html/about/about.shtml.

The City of New York, "PlaNYC A Greener, Greater New York" (2011). http://www.nyc.gov/html/planyc/downloads/pdf/publications/planyc_2011_planyc_full_report.pdf

offer direct payments for the restoration of ecosystems and biodiversity. Unlike protected areas, which use negative incentives to induce behavioral change, PES and Biodiversity Offset Programs aim to promote biodiversity conservation and the provision of ecosystem services through positive incentives in the form of payments to landowners *not* to convert plots of land with high conservation value. PES is an upgraded, more direct version of ICDPs. PES schemes are concentrated predominantly in Latin America and China.

In 2005 United Nations developed the program **Reducing Emissions from Deforestation and Forest Degradation (REDD/REDD+)**. REDD+ aims to fight deforestation and forest degradation while fostering innovative and collaborative approaches to address climate change. REDD+ is based on payment for ecosystem services that is tied to the Paris Agreement on climate change. The program has established a financial value for the carbon stored in forests and offers incentives for countries to keep their forests standing. REDD+ provides technical and financial support for countries to develop forest and land management national plans, increase the use of data and monitoring, and work with Indigenous peoples and local communities in creating these plans. Currently sixty-five countries participate in the REDD+ program. One example of the REDD+ program in Kenya is discussed in *Solutions* 15.2.

In theory, all these programs deliver environmental conservation while providing socioeconomic outcomes such as poverty reduction. However, it is not clear how effective these programs are. Academics are divided in their assessment on these mechanisms. Some see these as win-win solutions that tackle conservation needs or climate change while supporting economic development. However, others evaluate these as having weak or no impacts.

A Critical Perspective on Cash for Conservation

Some scholars argue that PES and Biodiversity Offset programs may not be providing net conservation gains or the social benefits expected. For example, PES

SOLUTIONS 15.2

The Impact of REDD+ in Kenya

Target 15.2: Promote the implementation of sustainable management of all types of forests, halt deforestation, restore degraded forests, and substantially increase afforestation and reforestation globally.

Target 15.9 Integrate ecosystem and biodiversity values into national and local planning, development processes, poverty reduction strategies, and accounts.

Kenya loses significant forest area each year through deforestation. Demand for fuelwood, timber products, and charcoal, as well as population and infrastructure pressures and the conversion of forest to agricultural land have led to dramatic climate change impacts in the country. However, the country's forest cover has been improving in recent years, in part due to the REDD+ projects under way in Kenya, which have educated women and youth across the country on the importance of forest preservation.

The government of Kenya, in collaboration with the UN-REDD Programme, has embarked on a process to develop Free, Prior and Informed Consent (FPIC) Guidelines for Kenya—the first of their kind in Africa. Indigenous peoples and forest-dependent communities, including women and youth, are the central drivers of this multistakeholder process. Their engagement is essential to the success of REDD+ in Kenya because the majority of remaining forests are located where they live, and also because they have played a major historical and cultural role in the sustainable management of forests for centuries.

"The REDD+ project has been a life-changer because previously, communities were left out of the conversation around tree planting efforts," says Tecla Chiumba, treasurer with the National Alliance of Community Forest Associations in Kenya. She says that women and youth are now involved in forest conservation, tree planting both in forests and on their own farms, as well as in developing and implementing innovative wood product ventures to generate income.

"Forest communities are very happy, and there are now a lot of women and youth active in conservation," she says. "Replanting efforts and REDD+ activities have not only meant preserving forests, but have also allowed us to get firewood from the trees on our farms, not in faraway forests, and it has brought water closer to us. This saves us time and dramatically improves our lives as women."

Utilizing the knowledge they have acquired through the UN-REDD Programme's support, forest communities in Kenya will continue planting trees, conserving what they have and creating nonwood products such as beehives and mushrooms to conserve the forest while also generating income and employment, according to Chiumba. "We have realized the importance of trees. Unless we plant trees, we won't have water and we can't sustain life."

Source: United Nations. *10*[th] *Consolidated Annual Progress Report of the UN-REDD Programme Fund.* "Impact Kenya" (p. 63) (2018). https://www.unredd.net/documents/programme-progress-reports-785/2018-programme-progress-reports/17258-un-redd-consolidated-2018-annual-report.html

programs are typically voluntary. Landowners that choose to participate might do so because there is no other use for their land that could be better compensated, yet their land may not have been at risk. It can also be difficult to predict exactly where deforestation will occur, and selecting the right lands to protect may result in deforestation in an unprotected forest nearby. Some researchers say that these programs do not offer enough money to protect these areas in the long term. Other scholars have argued that in places where small-scale or subsistence farmers have weak land tenure, they may be evicted from their land or denied access to it as governments or others "buy" or take their land. Organizations such as Friends of the Earth and No REDD+ accuse the program of facilitating land grabs in the name of climate change mitigation, calling this "carbon colonialism."

💬 EXPERT VOICE 15.1

Dr. Ginger Allington on the Mongolian Rangeland

What does a cashmere sweater have to do with the Mongolia Plateau rangeland? Dr. Ginger Allington can connect the dots (see Photo 15.4).

Allington travels to Mongolia to study the plateau's rangeland—a vast grassland that feeds grazing animals like goats and sheep—and the many factors that shape it. The international demand for cashmere has increased the number of grazing cashmere goats in the Mongolian Plateau and incentivized local herders. These goats eat plants all the way down to the ground, destroying plant communities and causing erosion. Allington studies trends like these to better understand changing rangeland conditions and help local decision makers identify potential steps for recovery.

To get information on the plant communities she studies, Allington plans her fieldwork in the summertime during the growing season. Given the remote locations of her fieldwork, Allington camps in the rangeland. While she's in the field, Allington studies perennial grasses, which maintain their root system from year to year and can help build soil health, and annual grasses, herbs, and shrubs, which seed every year and have an easier time taking root where perennial plants have been degraded. By comparing the differences in the composition of plant communities and soils, Allington can better understand the status of the rangeland's health.

Allington combines the data she collects on the ground with data from NASA satellites to make estimates over broader areas of land. When Allington is in Mongolia, she can usually be found in the field collecting data to better understand the health of the rangeland. She works closely with regional stakeholders to learn about the socioeconomic issues tied to the land, as well as researchers on the ground who provide invaluable feedback based on their extensive time in the field. Climate, wildlife, cultural practices, market forces, and broad-scale policies are all factors that affect the health of the Mongolian Plateau rangeland ecosystem.

PHOTO 15.4 In the Field with Ginger Allington Ginger Allington's collaborators from China collect photosynthesis data at a research site in China.

"To understand the drivers of degradation and how the landscapes are responding, you have to know what's happening with the people living there," Allington said. "And you have to try to understand things at larger spatial scales."

Allington works primarily with data from the joint NASA-US Geological Survey Landsat series of satellites. The data set includes thirty-five years of normalized difference vegetation index (NDVI). NDVI essentially measures the "greenness" of plants from space and provides a measure of plant health based on how the plant reflects light at certain frequencies. This measurement helps Allington determine the density of green on a patch of land and better understand how these arid rangeland areas change over time.

"Satellite data is so useful in a place like the Mongolian Plateau because it's huge," Allington said. "We're collecting data on ground level with tape measures—we can't do that over all of Mongolia. But what we can do is collect data from representative sites and try to link that to observations from satellites so that we can make estimates over broader areas. The really great thing about the Landsat mission is that we have data over space combined with a great archive of data over time. So we can look for signals of big shifts by looking at the trajectory, the trends of an area."

Because her work studying the land is tied closely to the socioeconomics of the area, Allington collaborates closely with regional stakeholders and policy makers from the onset of her work to ensure the system she designs aligns with their needs and learns from their firsthand experience with the region. She also works with nongovernmental organizations to help develop sustainable protocols for herd management and rangeland practices that influence herder decision making and benefit herders, wildlife, and the rangeland.

_____

Source: Ecker, M. NASA Earth Applied Sciences. "A Photo Tour with Scientist Ginger Allington" (2021, January 25). https://appliedsciences.nasa.gov/our-impact/people/destination-mongolia-understanding-plateau-rangeland-ecosystem

Work by critical theorists including political ecologists criticizes these cash for conservation programs as a form of "eco-colonialism," where high-income countries that have already overexploited their forests and natural resources pay lower-income countries not to do so. Geographer Kathleen McAfee says this merely offsets bad practices in the Global North at the expense of the Global South. She also notes this may deny countries in the Global South the economic opportunities they would have if they were free to develop their land how they see fit.

Many political ecologists endorse projects that use a socioecological systems framework instead, an idea discussed next.

Apply a Socioecological Systems Perspective

We know that humans are not separate from nature. People and the natural world are intertwined in a complex **social-ecological system (SES)**. Political economy and political ecology scholars use the concept of social-ecological systems to evaluate the biological system and the social/human system as mutually dependent and nested across scales and they often use a political ecology approach. Any social-ecological system is an open system, with a number of influences impinging on it, such as population growth, technological change, effects of capital markets, and trade. Political change and pressures of globalization were also considered major influences on the system. Human well-being, social concerns, and equity must be a part of solving any complex environmental challenge. For example, understanding that many Indigenous communities have a different but valuable understanding of their ecosystems than planners or policy makers means that they must be included in an effective land management program. Dr. Ginger Allington is a scholar who uses the SES framework to understand changes to the Mongolian grasslands and is discussed in the *Expert Voice* 15.1 box. Many SES scholars are working to help decision makers craft more effective plans for ecosystem management.

Any strategy to improve environmental conservation while also furthering the social and economic needs of community will require improvements in science and research, and they will depend on governments to have the political will to commit financial resources for these strategies. Indeed, research suggests that a critical determinant of success in achieving the SDG 15 targets is the financing that is committed to maintaining biodiversity.

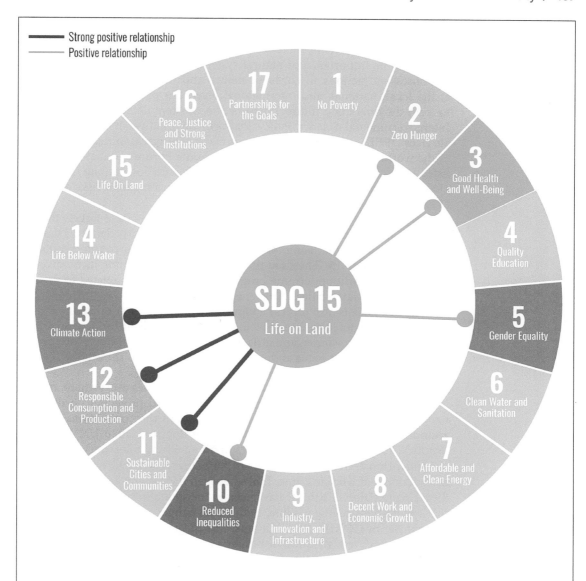

Strong positive relationship
Positive relationship

SDG 2: Combatting desertification, restoring degraded land, and reducing the impact of invasive species as well as fair and better access to genetic resource enable sustainable agriculture. On the other hand, the interconnections can be counterproductive: the extension of agricultural areas can lead to an increase in agricultural income but can also increase deforestation.

SDG 5: Working to advance targets in SDG 5 may leverage progress towards SDG 15 ensuring women's land rights.

SDG 10: Conservation and biodiversity is an issue of inequality for indigenous peoples—for example, those who cause land degradation are often not held accountable. In addition, indigenous peoples are often among those who are considered "left behind" and SDG 15 is an important goal to them.

SDG 12: SDG 15 has strong interlinkages to SDG 12, particularly around reducing meat consumption and promoting bio-based production and supply chains.

FIGURE 15.7 Interconnections in SDG 15 and the Other Goals
Advancing targets in SDG 15 may also advance targets in many other SDGs. This figure highlights several of the stronger connections.

SUMMARY AND PROGRESS

Over the last thirty years, there has been growing awareness and understanding about the importance of natural systems. There have been many policy responses and actions to conserve nature and manage it more sustainably at the local, national, and international level. But progress is not sufficient to stem the direct and indirect drivers of ecosystem deterioration. The 2019 Intergovernmental Science-Policy Platform on Biodiversity and Ecosystem Services (IPBES) presented a bleak summary of progress in nearly all regions of the world. Further, the 2021 World Economic Forum's Global Risk Report listed human environmental change and biodiversity loss among the top five global risks in terms of impact; four of the eight worst global risks are linked to ecosystem changes.

In addition, there is some criticism that other SDGs have targets that fail to factor in the impacts on ecosystems and biodiversity. Sandra Diaz and colleagues caution that some of the other SDGs may be counterproductive to targets in SDG 15. They say that current negative trends in biodiversity and ecosystems will undermine progress toward the goals related to poverty, hunger, health, water, cities, climate, oceans, and land (SDGs 1, 2, 3, 6, 11, 13, 14, and 15). Diaz is also critical about the lack of attention on equity, noting that land or resource tenure insecurity, as well as declines in nature, have greater impacts on women and girls, who are most often negatively impacted. However, the current focus and wording of the targets of these goals obscure or omit these specific consequences.

The COVID-19 outbreak has also affected economies everywhere, making it even more challenging to invest in programs and projects to protect the natural world. The pandemic has reminded us of the connections between **zoonotic diseases** and human health. According to a 2020 UN Report on World Wildlife Crime, three-quarters of all emerging infectious diseases are zoonotic, transferred from animals to humans, facilitated by environmental destruction and wildlife crime. Links between the global health crisis and the illegal exploitation of wildlife have been in the spotlight since it was suggested that wet markets selling wildlife, in this case pangolins, could have facilitated the transfer of COVID-19 to humans. The illegal trade in wildlife, which by definition does not go through proper sanitary and phytosanitary controls, can potentially lead to the spread of zoonoses, such as SARS-CoV-2 that caused the COVID-19 pandemic. Trafficked wild species—pangolins, birds, turtles, tigers, bears, and many more—and the resulting products offered for human consumption therefore pose even greater risks of infection. When wild animals are poached from their natural habitat, butchered, and sold illegally, the potential for transmission of zoonotic diseases is increased.

Failing to achieve SDG 15 will have ripple effects: Twenty-two targets under other SDGs are being undermined by substantial negative trends in terrestrial and ocean ecosystems. Numerous reports stress the need to emphasize the interconnections between targets within the SDGs, as such synergies may help achieve multiple targets. Biodiversity and ecosystem functions and services play a direct role in several SDGs including those on water and sanitation, climate action, and life below water. Nature also plays an important and complex role in the achievement of the SDGs related to poverty, hunger, health, and well-being and sustainable cities. Figure 15.7 highlights just a few of these connections. For example, protecting biodiversity has a direct impact on gender equality (SDG 5), as biodiversity loss and degraded ecosystems can perpetuate gender inequalities by increasing the time spent by women and children on collecting valuable resources including fuel, food, and water, and reducing the time for education (SDG 4 Quality Education) and income-generating activities.

QUESTIONS FOR DISCUSSION AND ACTIVITIES

1. Describe the patterns and trends in ecosystem change and biodiversity.

2. Describe the impacts of land degradation.

3. Go to the World Bank's interactive map on deforestation (https://www.geospatialworld.net/blogs/world-banks-interactive-map-shows-deforestation-severe/). Describe the geography of deforestation. What accounts for the differences between the Global North and the Global South?

4. Why is it important to address indirect drivers to protect ecosystems and biodiversity?

5. Which indirect drivers would you prioritize and why?

6. What type of land-use change is most prevalent in your area and why?

7. View the documentary "Extinction: The Facts" (https://www.pbs.org/show/extinction-facts/?mc_cid=8ac310759b&mc_eid=UNIQID). Answer the following:

   I. What is the "insect apocalypse" and how might it affect food production?

   II. Why does biodiversity loss lead to pandemics?

   III. How does consumer consumption affect biodiversity?

8. What are the benefits/limitations to international agreements such as CITES or CBD?

9. Based on Table 15.3, which strategies would you prioritize for investment and why?

10. Discuss the connection between SDG 3 Good Health and SDG 15.

11. Discuss two other SDGs that could be leveraged to best support the targets for SDG 15.

TERMS

afforestation
biodiversity
biome
conservation
deforestation
desertification
direct driver of change
ecosystem
ecosystem services
endocrine disrupting chemicals
food chains
food webs
hazardous substance
indirect driver of change
integrated conservation and
    development projects (ICDP)

Intergovernmental Science-Policy
    Platform on Biodiversity (IPBES)
invasive, nonnative species
    (INNS)
overexploitation
payments for ecosystems/
    environmental services (PES)
Reducing Emissions from
    Deforestation and Forest
    Degradation (REDD+)
redundancy
reforestation
social-ecological system (SES)
toxicity
trophic levels
zoonotic diseases

RESOURCES USED AND SUGGESTED READINGS

Brondizio, E. S., Settele, J., Díaz, S., and Ngo, H. T. (eds.). Global Assessment Report on Biodiversity and Ecosystem Services of the Intergovernmental Science-Policy Platform on Biodiversity and Ecosystem Services. Bonn, Germany: IPBES Secretariat (2019). https://doi.org/10.5281/zenodo.3831673

Díaz, S. Settele, J. Brondízio, E. S., Ngo, H. T., Guèze, M., Agard, J., Arneth, A., Balvanera, P., Brauman, K. A., Butchart, S. H. M., Chan, K., Garibaldi, L. A., Ichii, K., Liu, J., Subramanian, S. M., Midgley, G. F., Miloslavich, P., Molnár, Z., Obura, D., Pfaff, A., Polasky, S., Purvis, A., Razzaque, J., Reyers, B., Roy Chowdhury, R., Shin, Y. J., Visseren-Hamakers, I. J., Willis, K. J. and Zayas, C. N. (eds.). *"Global Assessment Report Summary for Policy Makers."* Bonn, Germany: IPBES Secretariat (2020). https://ipbes.net/sites/default/files/2020-02/ipbes_global_assessment_report_summary_for_policymakers_en.pdf

Díaz, S. Settele, J. Brondízio, E. S., Ngo, H. T., Guèze, M., Agard, J., Arneth, A., Balvanera, P., Brauman, K. A., Butchart, S. H. M., Chan, K., Garibaldi, L. A., Ichii, K., Liu, J., Subramanian, S. M., Midgley, G. F., Miloslavich, P., Molnár, Z., Obura, D., Pfaff, A., Polasky, S., Purvis, A., Razzaque, J., Reyers, B., Roy Chowdhury, R., Shin, Y. J., Visseren-Hamakers, I. J., Willis, K. J. and Zayas, C. N. (eds.). *Intergovernmental Science-Policy Platform on Biodiversity (IPBES): Summary for Policymakers of the Global Assessment Report on Biodiversity and Ecosystem Services of the Intergovernmental Science Policy Platform on Biodiversity and Ecosystem Services.* Bonn, Germany: IPBES Secretariat (2019). https://ipbes.net/sites/default/files/inline/files/ipbes_global_assessment_report_summary_for_policymakers.pdf

Filho, W. L., Azul, A. M., Brandli, L., Alvia, A. L., and Wall, T. (eds.). *Life on Land (Encyclopedia of the Sustainable Development Goals)*. Cham, Switzerland: Springer Nature (2020).

Gajbe, P. *Biodiversity Conservation*. Nagpur, India: AK Scientific Publishing (2021).

Hobohm, C. (ed.). *Perspectives for Biodiversity and Ecosystems*. Cham, Switzerland: Springer Nature (2020).

Kareiva, P., Tallis, H., Ricketts, T., Daily, G., and Polasky, S. *Natural Capital: Theory and Practice of Mapping Ecosystem Services*. New York: Oxford University Press (2011).

Kolbert, E. *Field Notes from a Catastrophe*. New York: Bloomsbury (2015).

Lovejoy, T. (ed.). *Biodiversity and Climate Change: Transforming the Biosphere*. New Haven, CT: Yale University Press (2019).

Mcafee, K. "The Post- and Future Politics of Green Economy and REDD+," in B. Stephan and R. Lane (eds.). *The Politics of Carbon Markets* (pp. 237–60). New York: Routledge (2015).

McKibben, B. *The End of Nature*. New York: Bloomsbury (2003).

Potts, S. G., Imperatriz-Fonseca, V. L., and Ngo, H. T. (eds.). *Intergovernmental Science-Policy Platform on Biodiversity (IPBES): Summary for Policymakers of the Assessment Report of the Intergovernmental Science-Policy Platform on Biodiversity and Ecosystem Services (IPBES) on Pollinators, Pollination and Food Production*. Bonn, Germany: IPBES Secretariat (2016). https://www.ipbes.net/assessment-reports/pollinators and https://doi.org/10.5281/zenodo.3402856

Ramawat, K. G. (ed.). *Sustainable Development and Biodiversity*. Cham, Switzerland: Springer Nature (2021).

Sale, P. *Our Dying Planet: An Ecologist's View of the Crisis We Face*. Berkeley: University California Press (2011).

Spicer, J., and Gaston, K. *Biodiversity: An Introduction*. Malden, MA: Blackwell (2004).

United Nations Convention on Biodiversity "The Year in Review" (2019). https://www.cbd.int/article/2019-12-20-16-57–49

United Nations Environment Programme. "United Nations Decade on Ecosystem Restoration" (n.d.). https://wedocs.unep.org/bitstream/handle/20.500.11822/31813/ERDStrat.pdf?sequence=1&isAllowed=y

United Nations Food and Agriculture Organization. "Global Forest Resources Assessment 2020: Main report." Rome: United Nations FAO (2020). https://doi.org/10.4060/ca9825en

United Nations Office on Drugs and Crime. *World Wildlife Crime Report: Trafficking in Protected Species* (p. 27). Vienna, Austria: United Nations Office on Drugs and Crime (2020). https://www.unodc.org/documents/data-andanalysis/wildlife/2020/World_Wildlife_Report_2020_9July.pdf

United States Environmental Protection Agency. Toxic Release Inventory (2022, October). https://www.epa.gov/toxics-release-inventory-tri-program

Walsh, S. J., Riveros-Iregui, D., Acre-Nazario, J., and Page, P. H. (eds.). *Land Cover and Land Use Change on Islands: Social and Ecological Threats to Sustainability*. Cham, Switzerland: Springer Nature (2020).

Wilson, E. O. *The Diversity of Life*. New York: W. W. Norton & Company (1992).

Wilson, E. O. *The Future of Life*. New York: Alfred Knopf/Vintage (2002).

World Economic Forum. "The Global Risks Landscape, 2021" (2021). https://www.weforum.org/agenda/2021/01/these-are-the-worlds-greatest-threats–2021/

World Resources Institute and the Millennium Ecosystem Assessment Panel. *Ecosystems and Human Well-Being: Synthesis*. Washington, DC: Island Press (2005).

Zdruli, P., Pagliai, M., Kapur, S., and Cano, A. F. (eds.). *Land Degradation and Desertification: Assessment, Mitigation and Remediation*. New York: Springer (2010).

Peace, Justice, and Human Rights

LEARNING OBJECTIVES

After reading this chapter, you should be able to:

- Explain the concept of human rights

- Describe the trends around war and conflict and forced migration

- Explain the difference between voluntary and forced migration

- Explain how corruption and a weak rule of law undermine sustainability

- Identify and explain the various targets and goals associated with SDG 16

- Explain how the three Es are present in SDG 16

- Explain how SDG 16 strategies connect to other SDGs

Years of war and conflict in Syria have created the largest refugee crisis in the world. Globally, Syria remains a main country of origin of refugees, making up a quarter of all refugees in the world. There are now 5.5 million Syrian refugees. Syrians have found asylum in 127 countries, but the vast majority fled to the neighboring countries including Turkey, Lebanon, Jordan, Iraq, and Egypt. In Jordan, 128,000 have found sanctuary in sprawling refugee camps such as Za'atari, shown in Photo 16.1. In addition to the 5.5 million Syrian refugees, there are an estimated 6.7 million internally displaced persons (IDPs) in Syria.

After a decade of conflict, destruction and displacement, the situation for Syrians—both refugees and IDPs—remains precarious. More than fourteen million Syrians need humanitarian assistance and 90 percent of the population is living in poverty and food insecurity. Syrian refugees face many challenges: Poverty rates for Syrian refugees exceed 60 percent in some countries, while unemployment and uneven access to basic services, such as education, persist. As more refugees remain in poverty, potential risks such as early marriage, sexual and gender-based violence, child labor, and exploitation are liable to worsen. Additionally, years of displacement, exposure to violence, loss of loved ones, lack of jobs, and deepening poverty continue to have a severe impact on people's mental health.

Ongoing war and violence make it difficult to imagine peace and justice in Syria.

PHOTO 16.1 Za'atari Refugee Camp in Jordan
Za'atari refugee camp, close to the border with Syria, was opened
in 2012. The camp covers more than two square miles and is home
to 85,000 refugees, half of them children.

WHAT ARE HUMAN RIGHTS?

Human rights are a cornerstone of the foundation of the United Nations.
Human rights are standards that allow all people to live with dignity, freedom,
equality, justice, and peace. Every person has these rights simply because they
are human beings. They are guaranteed to everyone without distinction of any
kind, such as race, color, sex, language, religion, political or other opinion,
national or social origin, property, birth, or other status.

Human rights reflect the minimum standards necessary for people to live
with dignity. Human rights give people the freedom to choose how they live,
how they express themselves, and what kind of government they want to support. Human rights also guarantee people the means necessary to satisfy their
basic needs, such as food, housing, and education, so they can take full advantage
of all opportunities. Finally, by guaranteeing life, liberty, equality, and security,
human rights protect people against abuse by those who are more powerful.

Human rights are part of international law, contained in treaties and declarations that spell out specific rights that countries are required to uphold.
Countries have incorporated human rights in their own national, state, and
local laws.

The modern human rights era can be traced to the founding of the United
Nations and the 1948 Universal Declaration of Human Rights (UDHR). The
UDHR was the first international document that spelled out the "basic civil,
political, economic, social and cultural rights that all human beings should
enjoy." When it was adopted, the UDHR was not legally binding, although it
carried great moral weight. To give the human rights listed in the UDHR the
force of law, the United Nations drafted two treaties, the International Covenant on Civil and Political Rights (ICCPR) and the International Covenant
on Economic, Social, and Cultural Rights (ICESCR). Together, these three are
known as the International Bill of Human Rights.

The International Bill of Human Rights outlines rights that include the
right to equality and freedom from discrimination; the right to life, liberty,
and personal security; freedom from torture and degrading treatment; the
right to a free trial; the right to participate in government; the right to education; the right to work; and the right to health, food, and housing.

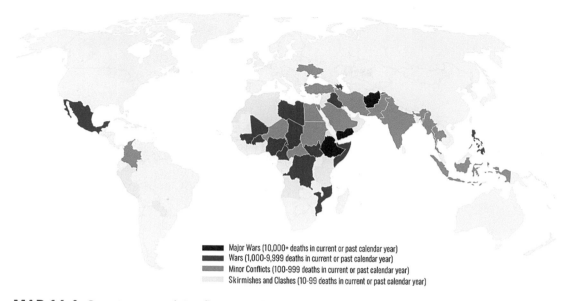

Major Wars (10,000+ deaths in current or past calendar year)
Wars (1,000-9,999 deaths in current or past calendar year)
Minor Conflicts (100-999 deaths in current or past calendar year)
Skirmishes and Clashes (10-99 deaths in current or past calendar year)

MAP 16.1 Ongoing Armed Conflicts (as of March 2021)
Syria, Afghanistan, Mexico, and Yemen are countries where conflict has caused the most deaths over the last five years.

Conflict, war, weak or corrupt institutions, and limited access to justice threaten the foundation of peaceful societies and undermine sustainable development. This chapter considers the context for understanding the movement for peace, justice, and strong institutions within the broader context of human rights.

WAR AND CONFLICT

Every seven minutes, somewhere in the world, a child is killed by violence. Every day, one hundred civilians, including women and children, are killed in armed conflicts, despite protections under international law. In 2021, the number of people fleeing war, persecution, and conflict exceeded 89 million, the highest level ever recorded. In that same year, the United Nations tracked 357 killings and 30 enforced disappearances of human rights defenders, journalists, and trade unionists in forty-seven countries.

Violent conflict reverses development and violates the most fundamental of human rights: the right to life. Armed violence and insecurity have a destructive impact on a country's development, affecting economic growth, and may cause long-standing grievances among communities. Armed conflict can involve the use of armed force between two or more organized armed groups, governmental or nongovernmental, within a country or between countries. Globally, two billion people live in countries where development is affected by fragility, conflict, and violence. Map 16.1 shows active conflicts as of 2021. Syria, Afghanistan, Mexico, and Yemen are countries where conflict has caused the most deaths over the last five years. Yemen has been steeped in crisis for decades, and the combination of famine, war, and other health emergencies has made it one of the worst humanitarian crises in the world.

Afghanistan has endured four decades of conflict and nearly twenty years of direct US military engagement, hundreds of thousands of deaths, and the distinction of being the world's deadliest conflict. Conflict among Eritrea,

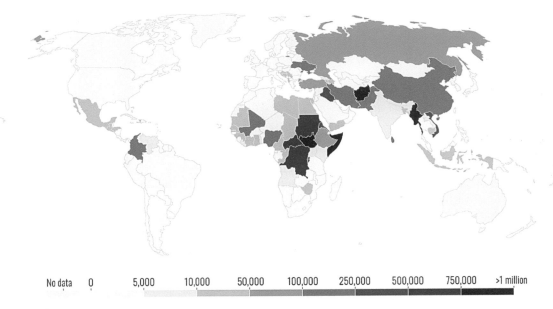

No data 0   5,000   10,000   50,000   100,000   250,000   500,000   750,000   >1 million

MAP 16.2 Refugee Population by Country or Territory of Origin (2017)
The number of refugees reached 26.5 million worldwide. Countries experiencing conflict or violence—such as Afghanistan, Burma, South Sudan, Syria, Ukraine, and Ethiopia—are the source of large numbers of refugees. Most refugees flee to neighboring countries; as a result, sub-Saharan Africa and the Middle East are home to the most refugees.

Ethiopia, and Sudan has caused more than 17,000 deaths since 2020. In Ethiopia escalating violent conflict in Ethiopia's Tigray region threatens to undermine reforms and destabilize not only in Ethiopia but also for the entire Horn of Africa. For decades, systemic and long-standing socioeconomic development challenges kept Ethiopia classified high on fragile state and conflict watchlists. The armed conflict in Ukraine has caused 7 million have fled their homes into Europe, and millions more have been forced to move inside the country seeking safety.

Many experts say funding to fragile states has increased somewhat, but it is not at the level needed to prevent violent conflict. Most of that funding focuses on humanitarian aid in places with active wars like Syria and Yemen. Compare the amount of US funding for defense versus foreign assistance. Military funding is the second-largest item in the US federal budget after Social Security and was estimated to be $733 billion (almost 20 percent of the overall budget) in 2021. Foreign assistance, however, constitutes less than 1 percent of the federal budget, of which only 11 percent is spent supporting political stability, democratic institutions, justice, and peacebuilding.

Forced Migration

In 2020, 89.4 million people were displaced worldwide. There are three main groups of people forcibly displaced, also referred to as forced migrants. The first is **refugees**, defined as someone who is unable or unwilling to return to their country of origin owing to a well-founded fear of being persecuted for reasons of race, religion, nationality, membership of a particular social group, or political opinion. In 2020, the number of refugees reached 26.5 million worldwide (Map 16.2 shows the number of refugees for 2017).

The second group of forcibly displaced people is **asylum seekers**, defined as people outside their country of origin and receiving international protection, but awaiting the outcome of their claim to refugee status. At the end of 2020 there were 4.1 million asylum seekers globally (up from 3.5 in 2018).

The third and largest group, at fifty-five million in 2020, is people displaced to other areas within their own country, a category commonly referred to as **Internally Displaced People or IDPs**. IDPs are predominately concentrated in two regions—the Middle East and North Africa (MENA) and West and Central Africa. The MENA region recorded more than twelve million IDPs as a result of conflict and violence at the end of 2019. Almost all of them lived in just three countries—Syria, Yemen, and Iraq—and around five million were children. Nadia Murad details the experience of the Yazidi in Iraq in the *Expert Voice* 16.1 box.

Sub-Saharan Africa is home to another nineteen million IDPs due to conflict and violence. Around four out of ten internally displaced persons live in the region, half being children. Within the region, 4 million displaced children live in Eastern and Southern Africa and a further 5.4 million in West and Central Africa. While conflict and violence are major causes of internal displacement, increasingly environmental disasters have caused many to leave their homes. In Bangladesh, India, and the Philippines, most of the recent internal displacements were caused by environmental disasters.

In South America, the ongoing crisis in Venezuela had displaced almost five million Venezuelans by 2020; many remain inside Venezuela, about 3.9 million have fled to neighboring countries. This has become one of the largest displacement crises in the world. People continue to leave Venezuela to escape violence, insecurity, and threats as well as lack of food, medicine, and essential services. There are about 2.5 million Venezuelans living in countries within Latin America and the Caribbean.

Humanitarian assistance is often needed by those forced to leave their homes. This consists of actions intended to save lives, alleviate suffering, and maintain human dignity during and after man-made crises and disasters caused by natural hazards. Humanitarian aid can include providing shelter, food, water and sanitation, medical care, and transportation. Such assistance is crucial because many of those forcibly displaced have increased chances of poverty, food insecurity, and mental health issues.

Today, more children than ever before are displaced. UNHCR reports that in 2020, thirty-five million children were displaced: twenty million children were internally displaced within their own countries, fourteen million were refugee children, and around one million were children seeking asylum. Approximately 28.5 million primary school age children who are out of school live in conflict-affected areas. Millions more are on the move as seasonal **migrants** or internal migrants and each day more are displaced by climate-induced disasters. Displaced children are among the most vulnerable in the world. They have limited access to education, health care, clean water, or protective services. Unfortunately, these trends have increased despite international law designed to protect them, the UN Convention on Refugees. The convention outlines basic minimum standards for the treatment of refugees, which include the right to the courts, primary education, work, and documentation.

Increasingly, environmental disasters are forcibly displacing people. In 2021, storms, floods, and wildfires displaced more than thirty million people within their countries. In 2020, for example, cyclones, monsoon rains, and floods displaced five million in Bangladesh, Bhutan, India, and Myanmar.

EXPERT VOICE 16.1

Nadia Murad on Conflict and Displacement

Nadia Murad along with thousands of other women, who are members of Iraq's Yazidi minority, were kidnapped in the village of Kocho in Sinjar. While held hostage, they were subjected to abuse at the hands of ISIS fighters. She escaped after three months' captivity. In 2016, Nadia Murad was appointed Goodwill Ambassador for the Dignity of Survivors of Human Trafficking by the UN Office on Drugs and Crime. In 2018, Ms. Murad received the Nobel Peace Prize together with Dr. Denis Mukwege. The following is an interview Question & Answer with Ms. Murad.

Question: Wartime sexual violence has become a significant aspect of today's conflicts, affecting people across age and gender. Perpetrators often go unpunished and survivors do not get the support and redress that they need. What needs to be done to change this situation and where do we start?

Ms. Murad: We need courts that will bring perpetrators like ISIS to accountability through a fair trial. When we speak of accountability, we must speak of justice for survivors. Perpetrators must be prosecuted for their crimes as part of a complete truth and reconciliation process—even if those perpetrators are state actors.

We must put an end to impunity. We believe this is the best way to end the violence. Accountability challenges the idea that certain groups of people are ultimately without rights and anything can be done to them. This attitude is profoundly damaging to the prospect of lasting peace and is evident, not just in those that carry out sexual violence, but those who refuse to prosecute it.

Question: Many Yazidi families in Iraq remain displaced, some are still missing, homes have been destroyed, and public services in places like Sinjar are not available. How do such communities move forward?

Ms. Murad: We must educate the world about the Yazidis and encourage all people to imagine the Sinjar region as a place that can be prosperous and peaceful. Rebuilding can be a slow process, but the more people learn about our culture, the more they understand how much we have in common and we can imagine a new future for the region together.

We must do more than merely endure. We must work toward a future in which the entire world, not just the Yazidis, say "never again."

Not only does everyone have an interest in peace, but by rebuilding Sinjar we are emphasizing that stopping a genocide includes addressing its aftermath. Victimized communities everywhere should be helped and given the chance to heal. Stopping a gruesome, immediate attack is necessary but we cannot leave seriously injured people to die or fend for themselves. In many ways, that is as brutal as the initial attack. The Yazidis have survived previous genocides and I know we will overcome this one.

Question: What is your message to people who have suffered the atrocities of war?

Ms. Murad: My message to people who have suffered the atrocities of war is to believe that things can be different. Understanding that peace is possible is ultimately about standing up for your own human rights. That is why we all benefit from peaceful societies which honor those rights. This vision of peace depends on justice, healing, and support for victimized communities. We must see the reality of this, educate our children to expect it, and not be distracted by hatred.

Source: United Nations. "Goal of the Month: Exclusive Interview with Nadia Murad, Nobel Peace Prize winner and Goodwill Ambassador" (2019). https://www.un.org/sustainabledevelopment/blog/2019/04/gotm-nadia-murad-interview/

HUMAN TRAFFICKING

Another violation of human rights is trafficking. The United Nations defines **human trafficking** as an act—recruitment, transportation, transfer, harboring, or receipt of persons—carried out by means of threat, deception, or coercion for the purpose of exploitation. Human trafficking and modern-day slavery are umbrella terms—often used interchangeably. Nearly forty million people annually are impacted by trafficking.

Human trafficking has many forms. It includes the practices of forced labor, debt bondage, domestic servitude, forced marriage, sex trafficking, child sex trafficking, and the recruitment and use of child soldiers. The traffickers often use violence or fraudulent employment agencies and fake promises of education and job opportunities to trick and coerce their victims.

As discussed in Chapter 8 Decent work, the most common forms of exploitation are forced labor, which, according to the International Labor Organization (ILO), impacts twenty-five million people a year—16 million in private-sector exploitation, 4 million in state-sanctioned forced labor, and 4.8 million in sex trafficking and forced marriage, which enslaves 15.4 million individuals. The ILO estimates that forced labor generates $150 billion in illegal profits each year.

Men, women, and children of all ages and from all backgrounds can become victims of this crime, which occurs in every region of the world. While human trafficking spans all demographics, trafficked persons most often come from positions of vulnerability. They may come from a low socioeconomic background; be homeless; be a political, cultural, or ethnic minority; have a history of sexual abuse, rape, or domestic violence; or have been subject to conflict or political turmoil.

For example, decades of China's one-child policy created extremely skewed gender ratios because parents historically favored boys over girls. As a result there are currently thirty-five million more men than women, and this discrepancy has exacerbated the practice of single men purchasing trafficked women from remote provinces within China, and also from other Southeast Asian countries and North Korea. In this case, a major reason behind trafficking of women is gender inequality, where women are treated as property to be traded or sold.

A Critical Perspective on the Terms We Use

Dr. Giorgia Serughetti, a postdoctoral research fellow at the Department of Sociology at the University of Milano-Bicocca, Italy, has examined the issue of trafficking of Nigerian women and girls into the European Union.

A growing number of Nigerian women and girls have been landing on the coast of Sicily since 2013. Public authorities and humanitarian agencies have described this migration flow as part of an increase in trafficking for sexual exploitation.

Serughetti's research on Nigerian women and girls arriving in Italy by sea illustrates that the conditions and practices involved in smuggling and trafficking can overlap and be difficult to disentangle. Because European policy distinguishes between human trafficking (forced) and migrant smuggling (voluntary), how a person is labeled has tremendous impact on how they are treated within the EU system. Based on their responses, they will be classified as asylum seekers, unaccompanied foreign minors, victims of trafficking, people with vulnerabilities, or irregular migrants, the last of which results in deportation. There is much to be gained from being classified as trafficked, but much to lose from being considered smuggled. One is deserving of protection, the other is not. However, Serughetti argues this distinction is problematic, and that migration research suggests the need to rethink the categories of **forced migration** and **voluntary migration**.

Serughetti says that a migrant's reason for leaving their country of origin is far more complicated than the dominant narrative of women and girls who have been deceived or forced to move against their will. She found that although some young Nigerians who left their country had been tricked with false employment prospects, increasingly prospective migrants are aware that the work most likely available to them in Italy and other countries is sex work. To make matters more complex, even if a migrant is aware that they will be a sex worker, their migration may still turn into an exploitative experience, as many may face unexpectedly harsh conditions of work in the destination country or a higher than anticipated migratory debt.

Serughetti concludes that it is the duty of state institutions to guarantee protection, both by alleviating vulnerability and supporting resilience, and recognizing the ability of migrants to determine their own lives. State and humanitarian actors fail to fulfil their duty of protection not only when they base their legislation, policies, and practices on misconstrued concepts of vulnerability and employ rigid and stereotypical labels but also when they fail to acknowledge agency and respect people's—and especially women's—choices.

WEAK RULE OF LAW AND A LACK OF ACCESS TO JUSTICE

The **rule of law** and development are highly interconnected. At the national scale, the rule of law is necessary to create a society that ensures due process, protects the environment, and creates better economic opportunities. A country with a weak rule of law may mean its people are more susceptible to human rights violations.

Access to justice is a basic principle of the rule of law. In the absence of access to justice, people are not able to have their voices heard, exercise their rights, challenge discrimination, or hold decision makers accountable. Lack of access to justice means that conflicts remain unresolved and people cannot obtain protection and redress. Institutions that do not function according to legitimate laws are prone to arbitrariness and abuse of power, and are less capable of delivering public services to everyone. To exclude and to discriminate not only violates human rights but also causes resentment and animosity and could give rise to violence.

One of the major obstacles in accessing justice is the cost of legal advice and representation. In many countries in the Global South, informal justice systems resolve a majority of disputes over personal security and local crime; protection of land, property, and livestock; and resolution of family and community disputes. Informal justice systems are also often referred to as "traditional," "Indigenous," "customary," or "nonstate" justice systems.

Corruption is the abuse of entrusted power for private gain. This includes corruption in public and business sectors, from local to international levels. It extends from petty corruption felt by citizens every day, to kleptocracy and high-level grand corruption damaging entire societies. Examples include bribery and embezzlement, theft, tax evasion, and profiting from illicit flows (such as drugs or illegal arms). Bribery is one of the most common forms of corruption. Map 16.3 shows the percentage of those who reported having paid a bribe in the last year to access public services such as education, health, police, and so forth. In countries such as India, Yemen, Liberia, and Vietnam, at least two out of three survey respondents report having paid a bribe within

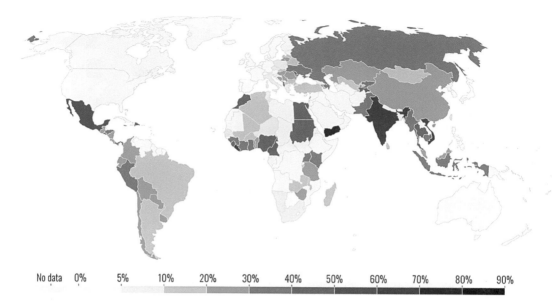

MAP 16.3 Bribery Rates (2017)
The map shows the percentage of those who reported having paid a bribe in the last year to access public services such as education, health, police, and others. Yemen, Liberia, Vietnam, Mexico, and India have among the highest rates of bribery.

the past year. In Japan, the United Kingdom, and Sweden, only one in a hundred (or fewer) reports having paid a bribe.

Corruption is a global problem but impacts low-income countries the most. The World Economic Forum reports that corruption cost low- and middle-income countries $1.26 trillion. The amount of money lost due to corruption could be used to lift those who are living on less than $1.25 a day above $1.25 for at least six years. For example, in war-torn Afghanistan, of the $8 billion donated in recent years, as much as $1 billion has been lost to corruption.

Although corruption is difficult to measure, the nongovernmental organization Transparency International publishes a "Corruption Perceptions Index," the most widely used indicator of corruption worldwide and shown in Map 16.4. The Corruption Perceptions Index scores 178 countries on their degree of corruption—10 is the cleanest possible and 0 indicates endemic corruption. The five countries with the highest scores (and thus perceived as most "clean") are Denmark, New Zealand, Finland, Singapore, and Sweden. At the other extreme, the countries with the lowest scores (and highest perceived corruption) are Somalia, Syria, South Sudan, Yemen, and North Korea.

Corruption is often a part of daily life for many: Having to pay bribes to access public services, for example, shows that small-scale corruption can have a big impact as the constant demands for bribes push people further into poverty. Government and politics scholar Eric Uslaner notes that "corruption not only thrives under conditions of high inequality and low trust, but in turn it leads to more inequality."

A 2015 UN Global Compact report noted corruption is a major obstacle to economic and social development. Its political costs can include the destruction of public order and the erosion of societal trust in the institutions. In economic terms, corruption depletes wealth, contributes to further inequality, and hinders entrepreneurship. According to the report there are some estimates that indicate the direct cost of corruption far exceeds $1 trillion per year.

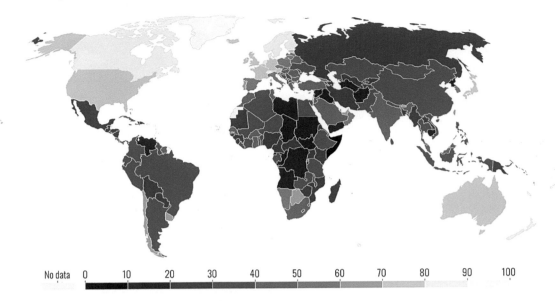

MAP 16.4 Corruption Perceptions Index (2018)
The most recent Corruption Perceptions Index scores 178 countries on their degree of corruption—10 is the cleanest possible and 0 indicates endemic corruption. Countries where the rule of law is weaker may have higher levels of corruption. The five countries with the highest scores (and thus perceived as most "clean") are Denmark, New Zealand, Finland, Singapore, and Sweden. At the other extreme, the countries with the lowest scores (and highest perceived corruption) are Somalia, Syria, South Sudan, Yemen, and North Korea.

Corruption in the Justice Sectors

It is particularly difficult to end corruption if the justice sector—the judiciary, prosecution, and the police—are corrupt because these are the institutions that should embody the principles of independence, impartiality, integrity, and equality. A 2020 UN report from the Office of Drugs and Crime noted a corrupt act during one step of the criminal justice chain can severely harm the whole process or even nullify its essence and erode public trust in law and order. The report also found challenges in claiming rights and enforcing contracts in court proceedings can create an atmosphere of legal uncertainty and ultimately deter business, entrepreneurial spirit, and investment. Disrespect for the equal application of the law undermines the legitimacy of public institutions and contributes to impunity.

The establishment of an effective, independent justice system that safeguards human rights and is transparent and objective is fundamental to economic stability and peace. At the center of the justice system are the police. Police play a vital role in society to maintain peace and security and the rule of law, and rely on the trust of the people and the communities they serve. Communities must believe that the police are generally free of corruption and misconduct, and that any violations have consequences. This also extends to how the police interact and treat the people they service on a daily basis. However, some reports found that in many countries police stations, courthouses, and prisons are frequently dilapidated, and key legal records and other necessary materials are often missing. Typically, the independence of the judiciary is weak and their salaries low and often unpaid, which can be a fertile ground for corruption.

In addition to the justice sector, there is growing interest in better understanding how corruption may be impacting the prison system, but to date little exists on this issue. Some statistics indicate that 30 percent of prisoners held in detention have yet to be sentenced, a clear violation of due process. Corruption can pose a severe security threat to prison staff and prison management, for example, trafficking of mobile phones, drugs, or weapons into and inside of prison walls in exchange for bribes. This risk is further increased if organized crime groups or other prisoners manage to come into a position of power and control.

Detention and imprisonment not only limit the freedom of movement but also creates a situation in which inmates depend on prison authorities for almost all their day-to-day needs, an imbalance of power that can lead to abuse. Experts suggest that around the world many prisons suffer from extreme overcrowding, lack of food, absence of adequate medical care, and poor sanitation. This became more evident during the COVID-19 pandemic. The fact that prisons are relatively restricted systems, and therefore less visible to public scrutiny, may mean internal and external auditing, monitoring, and inspection mechanisms are less transparent.

Finally, corruption in customs is also a problem. Inefficient and corrupt customs services are unable to ensure the equal treatment of importers and exporters in paying excise and customs duties and are likely to fail in stopping illicit contraband and trafficking in drugs, firearms, and wildlife, among others.

Missing and Disappeared People

Corruption, the lack of rule of law, and a weak justice system may also be responsible for the killings, enforced disappearances, torture, arbitrary detention, and kidnappings committed against journalists, trade unionists, and human rights activists. This group of people, referred to as **missing and disappeared people**, serves as proxy measurement of fundamental freedoms (such as freedom of opinion, freedom of expression and access to information, the right to peaceful assembly, and freedom of association) because their killings, enforced disappearance, torture, arbitrary detention, and kidnappings have a chilling effect on the exercise of these fundamental freedoms. Some disappearances occur at the hands of government; some at the hands of criminal gangs and militant groups; and still others are detained awaiting a credible government investigation. Map 16.5 shows the total number of persons held in detention who have not yet been sentenced. In 2021, 66 journalists were missing, and a record 275 were imprisoned according to data from the Committee to Protect Journalists. One is journalist Daysi Lizeth Mina Huamán from Peru. She covered news related to the development of the Congress of the Republic's electoral process in the complementary elections in district Santa Rosa. She was last seen waiting for a bus on January 26, 2019, on her way to meet her boyfriend after voting in Peru's congressional elections and filing a report for television broadcaster Cable VRAEM in the central city of Ayacucho. About a week after the disappearance, family members found her identity card and other personal documents along the side of a road between the bus stop and her destination. She is still missing.

According to the Committee to Protect Journalists, the number of journalists in state custody is highest in China, Turkey, Egypt, Saudi Arabia, and Belarus. In China, forty-seven journalists were in prison, serving long sentences; many of those recently incarcerated were arrested for their coverage of COVID-19 challenging Beijing's official narrative of its handling of the disease. Globally, thirty-two journalists were killed in 2020, most signaled out in retaliation for their reporting. Mexico and Afghanistan were the deadliest countries for journalists. Mexico has long been the most dangerous country

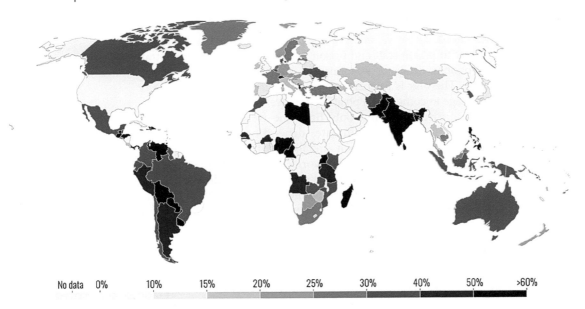

MAP 16.5 Unsentenced Detainees as a Percentage of Overall Prison Population (2017)
The map shows the total number of persons held in detention who have not yet been sentenced.
In 2021, 66 journalists were missing, and a record 275 were imprisoned according to data from the
Committee to Protect Journalists.

in the Western hemisphere as journalists who work there navigate a complex
and dangerous web of criminal drug-trafficking gangs and entrenched official
corruption.

SOLUTIONS TO ENDING WAR AND CONFLICT, ENSURING JUSTICE AND HUMAN RIGHTS

According to the UNDP, there is no single pathway to building resilience
in the face of fragility, conflict, and violence. Ultimately, different societies
will adopt different trajectories in how they strengthen state-society relations. At the same time, there must be respect for all international norms
and principles, in particular the human rights framework. There remains
a need for stronger integration of governance, conflict prevention, and
peacebuilding interventions. Table 16.1 lists the targets associated with
SDG 16. SDG 16 and the entire SDG agenda embrace the core elements of a
social contract between state and society. Some of the key solutions for this
include to strengthen the rule of law and a fair justice system, strengthen
peacebuilding infrastructure, and promote opportunities for inclusive and
resilient societies.

Strengthen the Rule of Law

As discussed, there are many examples where governments need to improve
and enhance the rule of law. The **rule of law** is defined by the United Nations
as "a principle of governance in which all persons, institutions and entities,
public and private, including the State itself, are accountable to laws that are
publicly promulgated, equally enforced and independently adjudicated, and

TABLE 16.1   SDG 16 Targets and Indicators to Achieve by 2030 or Earlier

Target	Indicators
16.1 Significantly reduce all forms of violence and related death rates everywhere	**16.1.1** Number of victims of intentional homicide per 100,000 population, by sex and age **16.1.2** Conflict-related deaths per 100,000 population, by sex, age, and cause **16.1.3** Proportion of population subjected to physical, psychological, or sexual violence in the previous twelve months **16.1.4** Proportion of population that feel safe walking alone around the area they live
16.2 End abuse, exploitation, trafficking, and all forms of violence against and torture of children	**16.2.1** Proportion of children aged 1–17 years who experienced any physical punishment and/or psychological aggression by caregivers in the past month **16.2.2** Number of victims of human trafficking per 100,000 population, by sex, age, and form of exploitation **16.2.3** Proportion of young women and men aged 18–29 years who experienced sexual violence by age eighteen
16.3 Promote the rule of law at the national and international levels and ensure equal access to justice for all	**16.3.1** Proportion of victims of violence in the previous twelve months who reported their victimization to competent authorities or other officially recognized conflict resolution mechanisms **16.3.2** Unsentenced detainees as a proportion of overall prison population
16.4 Significantly reduce illicit financial and arms flows, strengthen the recovery and return of stolen assets, and combat all forms of organized crime	**16.4.1** Total value of inward and outward illicit financial flows (in current US dollars) **16.4.2** Proportion of seized, found, or surrendered arms whose illicit origin or context has been traced or established by a competent authority in line with international instruments
16.5 Substantially reduce corruption and bribery in all their forms	**16.5.1** Proportion of persons who had at least one contact with a public official and who paid a bribe to a public official, or were asked for a bribe by those public officials, during the previous twelve months **16.5.2** Proportion of businesses that had at least one contact with a public official and that paid a bribe to a public official, or were asked for a bribe by those public officials during the previous twelve months
16.6 Develop effective, accountable, and transparent institutions at all levels	**16.6.1** Primary government expenditures as a proportion of original approved budget, by sector (or by budget codes or similar) **16.6.2** Proportion of the population satisfied with their last experience of public services
16.7 Ensure responsive, inclusive, participatory, and representative decision making at all levels	**16.7.1** Proportion of positions (by sex, age, persons with disabilities, and population groups) in public institutions (national and local legislatures, public service, and judiciary) compared to national distributions **16.7.2** Proportion of population who believe decision making is inclusive and responsive, by sex, age, disability, and population group

(continued)

Target	Indicators
16.8 Broaden and strengthen the participation of developing countries in the institutions of global governance	**16.8.1** Proportion of members and voting rights of developing countries in international organizations
16.9 Provide legal identity for all, including birth registration	**16.9.1** Proportion of children under five years of age whose births have been registered with a civil authority, by age
16.10 Ensure public access to information and protect fundamental freedoms, in accordance with national legislation and international agreements	**16.10.1** Number of verified cases of killing, kidnapping, enforced disappearance, arbitrary detention and torture of journalists, associated media personnel, trade unionists, and human rights advocates in the previous twelve months **16.10.2** Number of countries that adopt and implement constitutional, statutory, and/or policy guarantees for public access to information
16.A Strengthen relevant national institutions, including through international cooperation, for building capacity at all levels, in particular in developing countries, to prevent violence and combat terrorism and crime	**16.A.1** Existence of independent national human rights institutions in compliance with the Paris Principles
16.B Promote and enforce nondiscriminatory laws and policies for sustainable development	**16.B.1** Proportion of population reporting having personally felt discriminated against or harassed in the previous twelve months on the basis of a ground of discrimination prohibited under international human rights law

Source: United Nations Sustainable Development Goals. https://sustainabledevelopment.un.org/sdg16

which are consistent with international human rights norms and standards." Figure 16.1 highlights the key aspects of the rule of law.

Actions to support the rule of law include constitutional reform, power sharing, federalism and decentralization, human rights, gender equality, and public accountability. Working toward enhancing or enforcing the rule of law may also serve to reduce armed violence and improve citizen security. There are many ways to strengthen the justice system. These include improving legal aid systems, increasing women's access to justice, improving the quality of police services and community outreach, improving coordination with informal justice providers, and taking an active role in addressing any legacy of mass human rights abuses.

A strong civil society supports a strong rule of law. **Civil society** is "the arena, outside of the family, the state, and the market, which is created by individual and collective actions, organizations and institutions to advance shared interests." Governance is improved when there are multiple means for people to have a say in decision making. For example, countries with strong civil societies share a high propensity to participate, a high degree of tolerance of different ethnic and religious groups, and high public trust in nonprofit organizations. According to CIVICUS, a global alliance of civil society organizations, the most recent rankings place New Zealand first, followed by Canada and Australia. Denmark, Norway, Sweden, the Netherlands, Switzerland, Iceland, and the United States round out the top ten.

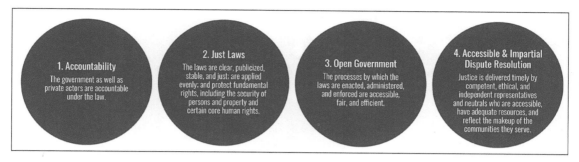

FIGURE 16.1 The Four Elements of the Rule of Law
The Rule of Law has four key elements: accountability, just laws, open government, and accessible and impartial dispute resolution.

In addition to these efforts, those countries that have not ratified or signed important human rights agreements, such as the Convention against Torture, or the protocol to abolish the death penalty should do so (see Maps 16.6 and 16.7). The United States has not signed or ratified this convention because it is still legal to apply the death penalty in many US states. The United States is one of the few countries in the Global North where the death penalty is still legal.

Strengthen Conflict Prevention and Peacebuilding Infrastructure

Conflict prevention and conflict resolution have a long history in international relations and peace studies in both theory and practice.

MAP 16.6 Membership of the Convention against Torture and Other Cruel, Inhuman, or Degrading Treatment or Punishment
There are numerous countries that have neither signed nor ratified this convention, including the United States.

Signed and ratified
Signed but not ratified
Neither signed nor ratified

MAP 16.7 Membership of the Second Optional Protocol to the International Covenant on Civil and Political Rights

International conflict such as war between countries can benefit from the practices of traditional diplomacy to manage conflict. These practices may try to prevent or mitigate violence by using threats of armed force (deterrence, coercive diplomacy, defensive alliances such as NATO); economic sanctions and other nonmilitary threats and punishments, such as the withdrawal of foreign aid; and direct military force to establish demilitarized zones.

A report by the National Research Council noted that conflict, as a concept, has also broadened over the last twenty years to move beyond conventional armed conflict between countries, to include ethnic conflicts and violence *within* a country that may threaten international peace and security. For example, from 1990 to 2010 there was a decrease in the frequency and death toll of international wars; however, there was an increase in subnational ethnic and religious conflicts.

When internal conflicts involve violations of universal norms such as self-determination, human rights, or democratic governance, the international community often attempts to prevent, conclude, or resolve these conflicts. However, there has also been increasing criticism that the international community and the United Nations in particular are too slow to act to prevent or manage these conflicts.

Although peacekeeping missions still sometimes physically separate adversaries to prevent further violence, increasingly they also provide humanitarian relief, resettle refugees, and rebuild infrastructure.

The report also described the numerous tools that **peacebuilding** organizations can use, including truth telling and reconciliation, electoral and constitutional design, autonomy arrangements within federal governance structures, laws and policies to accommodate linguistic and religious differences, training for law enforcement officials in following the rule of law, and the development and support of institutions of civil society. Figure 16.2 shows the numerous actions to promote peace. Table 16.2 highlights strategies in conflict resolution.

One example of a restorative justice approach was implemented in South Africa in the years after apartheid. Apartheid was a system of legally enforced racial segregation in South Africa between 1948 and 1994. In 1994 the Truth and Reconciliation Commission was established by the new South African government to help heal the country and bring about a reconciliation of its people by uncovering the truth about human rights violations that had occurred during apartheid. Abuses included torture, killings, disappearances and abductions, and severe ill treatment suffered at the hands of the apartheid state. The commission, chaired by Archbishop Desmond Tutu, investigated human rights abuses committed from 1960 to 1994, allowing victims the opportunity to tell their stories, granting amnesty, constructing an impartial historical record of the past, and drafting a reparations policy. The commission took the testimony of more than 22,000 victims and held public hearings at which victims gave testimony about violations of human rights. The commission was not without limitations and criticisms. Not everyone with a grievance had a chance to be heard.

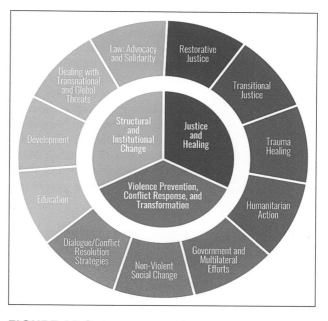

FIGURE 16.2 Actions toward Peacebuilding
These can involve structural and institutional changes, as well as actions to prevent violence and conflict. Finally, to create lasting peace, there must be justice and healing.

Additionally, the commission did not focus on the policies or political economy of apartheid. Failing to examine the larger system of racialized power allowed many who benefited to escape responsibility. However, it still remains a model for how a divided society with a violent past might work through that past and move forward.

TABLE 16.2 Strategies and Tools for Conflict Resolution	
Strategy	**Tools That Feature the Strategy**
Power politics	• Threats of force • Defensive alliances • Economic sanctions • Bargaining as a trade-off of interests • Power mediation
Conflict transformation	• Problem-solving workshops • Alternative dispute resolution • Reconciliation by truth commissions
Structural prevention	• Electoral system design • Autonomy • Legal guarantees of free speech and association • Civilian control of military organizations
Normative change	• Invocation of human rights norms

Note: These strategies and tools are often used in combination; moreover, the conceptual distinctions among them are sometimes blurred in use.
Source: Stern, P. C., and Druckman, D. (eds.). *International Conflict Resolution After the Cold War* (p. 5). Washington, DC: National Academy Press (2000).

Peacebuilding efforts aim to manage, mitigate, resolve, and transform central aspects of conflict through official diplomacy, civil society peace processes, and informal dialogues, negotiations, and mediations. Peacebuilding can be a direct process with organizations that focus on addressing the factors driving conflict (such as Mediators beyond Borders) or it can take the form of a broader set of efforts that support economic development, humanitarian assistance, governance, security, and justice, where organizations may not necessarily use the term "peacebuilding" but their efforts contribute to a broader peace nonetheless (such as Mercy Corp that works with people experiencing drought, war or famine). Many experts have noted that peace processes are more sustainable when civilian led, especially when women are involved. An example of inclusive governance in Mauritania is explored in *Solutions* 16.1.

 SOLUTIONS 16.1

Inclusive Governance in Mauritania

Target 16.7: Ensure responsive, inclusive, participatory, and representative decision making at all levels.

Target 16.8: Broaden and strengthen the participation of developing countries in the institutions of global governance.

The region of Hodh Ech Chargui in Mauritania borders on Mali, a country that is facing several crises (interethnic conflicts, terrorism, transnational organized crime, and climate change). Local communities have been impacted by the humanitarian crisis in Mali. Since 2012, they have hosted more than 50,000 Malian refugees. Hodh Ech Chargui is also one of the most vulnerable regions in Mauritania, with a lack of basic services and infrastructure.

This situation is fueling tension between host communities and refugees, as well as within local communities. Conflicts arise mainly over access to basic social services and natural resources.

Despite these challenges, local communities are participating in inclusive local decision-making mechanisms. They have created dozens of village committees. These committees include a range of marginalized groups such as women, religious leaders, young people, migrants, disabled persons, and refugees. Local populations recognize these committees as their representatives.

Members of the village committees contact local authorities directly to address ongoing security, peace, and rights-related issues. Local governments, as well as humanitarian and development actors, recognize the committees as an important partner and involve them in relevant interventions in the villages. The village committees, for example, are increasingly involved in preventing conflicts between refugees and the local population. The committees have prioritized the protection of civilians, including children and women, and the reduction of all forms of violence and exploitation.

The village committees have also played an instrumental role in the registration of refugees living in the nearby Mbera camp, ensuring messages on refugee registration and protection are communicated as widely as possible. The registration exercise will ultimately ensure that all refugee children born in Mauritania receive birth certificates issued by the Islamic Republic of Mauritania.

Women, in particular, utilize the committees to solve daily problems and to organize income-generation activities. The village committees receive training in conflict prevention, gender-based violence, and sustainable management of natural resources.

Source: The Global Alliance for Reporting Progress on Peaceful, Just and Inclusive Societies. "Enabling the Implementation of the 2030 Agenda through SDG 16+: Anchoring Peace, Justice and Inclusive" (p. 33) (2019). https://www.sdg16hub.org

Peacebuilding should include investments that support the development and application of credible and inclusive national and local capacities for the peaceful settlement of disputes. It should also involve the strengthening of social cohesion as a deterrence to exclusion and violence, including violent extremism; and the facilitation of consensual approaches to critical development challenges through multistakeholder dialogue. Initiatives for creating infrastructures for peace include:

- Create early warning and response systems to emerging tensions;
- Promote dialogue and the building of consensus around contested issues to help prevent or deescalate conflict;
- Enhance the coordination of local, national, and regional peace groups;
- Provide advice and support for the establishment of government departments and institutions responsible for national peacebuilding, dialogue, and mediation;
- Design policies and regulations that support the establishment of peace infrastructures;
- Empower national and local mediators and facilitators, also known as "insider mediators," to serve as credible intermediaries and confidence builders to improve relationships; and
- Ensure active participation of women and youth, as well as marginalized groups, in national and local peace infrastructure and initiatives.

Increase Inclusion and Participation

It is vital that as a country strengthens or establishes a rule of law that it ensures that participation in this process is as diverse as possible, making efforts to include underrepresented groups, especially women and youth, Indigenous peoples, and ethnic groups. Resilient states and societies must be capable of respecting and promoting the rights of minorities, and those marginalized. For example, to address sexual and gender-based violence, increasing women's participation in formulating policy and actions is vital. According to the UNDP, inclusive political processes are those that improve citizen participation, voice, and accountability through electoral processes, parliamentary and political development, constitutional processes, civic engagement, and women's political participation.

A recent UNDP report noted that resilient states and societies are capable of respecting and promoting the rights of minorities, marginalized, and underrepresented groups, especially women and youth, Indigenous peoples, and ethnic groups. There is now good evidence of a statistically significant association between national security and the security and safety of women. UNDP suggests that any attempt to build resilient governance must empower young women and men as key agents of change in their societies and communities. It is crucial to expand women's political participation and leadership in sectors such as justice and security, providing services for survivors; tackling impunity in cases of sexual and gender-based violence; and engaging people at the community level in awareness-raising and prevention activities.

Resilient states and societies also manage diversity and multicultural demographics by strengthening inclusion and tolerance, and through the peaceful management of conflict and differences in opinions, lifestyles, and beliefs. Professor Adele N. Norris discusses how the debate about the removal of Confederate statues in the United States is linked to efforts at making a more inclusive America (see *Expert Voice* 16.2).

EXPERT VOICE 16.2

Adele Norris on the Meaning of the Confederate Flag

Adele N. Norris is Senior Lecturer in Sociology and Social Policy in the Faculty of Arts and Social Sciences, University of Waikato, Hamilton, New Zealand. She reflects on the Confederate flag as a symbol that continues to represent violence, fear, and intimidation among Black people:

Among the extraordinary uprisings in the wake of George Floyd's murder by US law enforcement, Mississippi retired the Confederate emblem from the state flag on June 30, 2020. Under the banner of Black Lives Matter, young activists, led by student-athletes, played a pivotal role in the removal of the Rebel army's battle emblem that occupied the state's flag for 126 years while shedding light on its white supremacy symbolism.

I was a senior at Mississippi State University when voters, in a state-wide election, chose to keep the flag in 2001. My professor at the time held a mock election in one of my agricultural economics classes. Immediately, my classmates began declaring, "It's our heritage. Who's it hurting?" As the only Black student in the class, I became acutely aware that my feelings toward the flag didn't count. The mock election resulted in one vote in favor of changing the flag. I remembered the relief that apparently overcame the class and my professor afterward, although the "heritage" that my classmates were in fear of losing was never articulated or discussed: Mississippi's relationship with the Confederate flag.

The Confederate emblem is directly linked to the Confederate cause during the Civil War. After the war, the flag came to symbolize the preservation of the "old ways" where Confederate veterans and supporters engaged in acts of terror in their efforts to re-establish the South. The flag was waved and displayed in activities that ranged from assaults, lynchings, and the general terrorizing of Black southerners.

Always present during lynchings, one of the most extreme forms of violence prevalent in the South, the Confederate banner was used to strike fear and intimidation among Black people. Not only did the flag serve as a reminder to the Negro of his fate, particularly if he stepped out of his place, but the flag also represented "Southern justice" where White people could freely inflict pain and death upon Black Americans without repercussions.

Rhetoric that valorizes the "old ways" is not relegated to the past, as the divisiveness around retiring the Confederate emblem revealed. Many politicians seeking Mississippi's support evoke this heritage.

Mississippi heritage is one openly hostile to Black racial justice. Black Lives Matter, after all, exist in large measure because of a heritage that celebrates the holding of Black Americans captive to dehumanizing conditions. The mantra "our heritage" speaks to both the progress the United States has made and the challenges it continues to face in terms of anti-Black state violence. The removal of the Confederate emblem does not lessen the legacy of racism. Its removal should signal the extensive work needed to challenge the heritage and culture that valorizes white supremacy.

Source: Norris, A. "Mississippi Removes the Confederate Emblem from State Flag: Confronting Heritage." Palgrave: Social Science Matters Blog (2020). https://www.palgrave.com/gp/blogs/social-sciences/norris

SUMMARY AND PROGRESS

Peace, stability, human rights, and effective governance based on the rule of law are important conduits for sustainable development. Governance institutions and decision-making processes rooted in a human rights-based approach to development are vital (Figure 16.3). Yet in many countries there are increased barriers to inclusion and participation, and this trend risks excluding the needs and voices of the most vulnerable in society, including women, children, youth, refugees, asylum seekers, and internally displaced and stateless people. Some experts have noted that SDG 16 has the highest number of targets (10) and the lowest number of means of implementation (2), making the pursuit of peace, justice, and good governance a formidable challenge. As of 2021, UN reports say that little substantial progress has been made on the targets for SDG 16.

The COVID-19 pandemic threatens to amplify and exploit fragilities across the globe. There has been a call for governments to be transparent, responsive, and accountable in their COVID-19 response, and to ensure their emergency measures are not only legal and necessary but also nondiscriminatory. However, we also know that the pandemic has exacerbated many existing inequalities. For example, 60 percent of countries have prison overcrowding, risking the spread of the disease. Violent extremist groups, including the Islamic State, are also leveraging the pandemic to their advantage by legitimizing their ability to provide services to communities facing economic and food insecurity.

Governments, civil society, and communities must work together to implement lasting solutions to reduce violence, deliver justice, combat corruption, and ensure inclusive participation at all times. Making progress on targets in SDG 16 will also positively address targets in other SDGs; a few are highlighted in Figure 16.4.

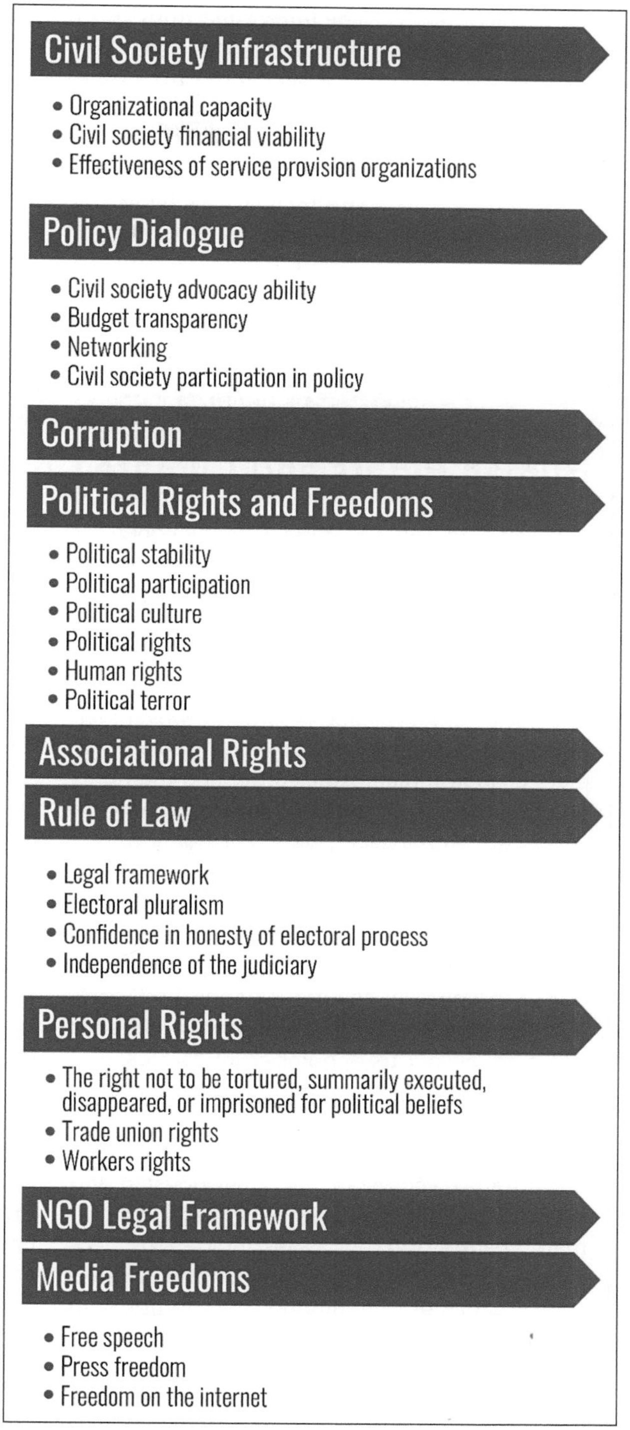

Civil Society Infrastructure

- Organizational capacity
- Civil society financial viability
- Effectiveness of service provision organizations

Policy Dialogue

- Civil society advocacy ability
- Budget transparency
- Networking
- Civil society participation in policy

Corruption

Political Rights and Freedoms

- Political stability
- Political participation
- Political culture
- Political rights
- Human rights
- Political terror

Associational Rights

Rule of Law

- Legal framework
- Electoral pluralism
- Confidence in honesty of electoral process
- Independence of the judiciary

Personal Rights

- The right not to be tortured, summarily executed, disappeared, or imprisoned for political beliefs
- Trade union rights
- Workers rights

NGO Legal Framework

Media Freedoms

- Free speech
- Press freedom
- Freedom on the internet

FIGURE 16.3 How to Ensure a More Equitable Future
The above highlights ways to create institutional structures that will ensure a more equitable future.

The *Interconnections* 16.1 box explores the linkages between human rights and climate change (SDG 13).

Working toward a society where people are free from the threat of violence and able to focus on other issues is a vital component of a sustainable future. Taken together, targets covered under SDG 16 encourage partnerships and collaborative efforts to implement the policies needed to build peaceful, just, and inclusive societies.

INTERCONNECTIONS 16.1

Human Rights and Climate Change

According to the UN High Commissioner for Human Rights, climate change poses a threat not just to human life, but to all life. It already affects the human rights of countless persons and the impacts are only getting worse. The UDHR guarantees that all human beings are entitled to a social and international order in which their rights and freedoms can be fully realized. Climate change threatens this order and the rights and freedoms of all people. Climate change has an impact on, among others, the rights to life, self-determination, development, health, food, water and sanitation, adequate housing, and a range of cultural rights.

The negative impacts of climate change are disproportionately felt by persons and communities who are already in a disadvantageous situation owing to a number of factors. For example, persons, communities, and countries that occupy and rely upon low-lying coastal lands, tundra and Arctic ice, arid lands, and other fragile ecosystems for their housing and subsistence face the greatest threats from climate change. In many cases Indigenous peoples, women, children, migrants, and persons with disabilities are among the groups and individuals disproportionately affected by climate change.

Climate change may cause the displacement of Indigenous peoples and the potential loss of their traditional lands, territories, and resources threatens their cultural survival, traditional livelihoods, and right to self-determination.

Indigenous peoples have long lived in fragile ecosystems that are uniquely sensitive to the effects of a changing climate. Extreme weather events, drought, melting ice, sea-level rise, and species shifts are affecting Indigenous territories, increasing the vulnerability of Indigenous peoples. They are also directly affected by environmental destruction, such as deforestation, land degradation, and excessive exploitation of mineral resources, which are having a negative impact on the local economies, subsistence lifestyles, food security, access to water, and cultures of Indigenous peoples, who often rely heavily on land and natural resources to meet their livelihood needs.

The Local Communities and Indigenous Peoples Platform was established under the UN Framework Convention on Climate Change. The platform serves to strengthen the knowledge and practices of Indigenous peoples to address climate change, to facilitate the exchange of experience and sharing of best practices on mitigation and adaptation, and to enhance the engagement of local communities and Indigenous peoples under the convention.

Source: United Nations Office of the United Nations High Commissioner for Human Rights, Frequently Asked Questions on Human Rights and Climate Change. New York and Geneva, Switzerland (2021). https://www.ohchr.org/Documents/Publications/FSheet38_FAQ_HR_CC_EN.pdf

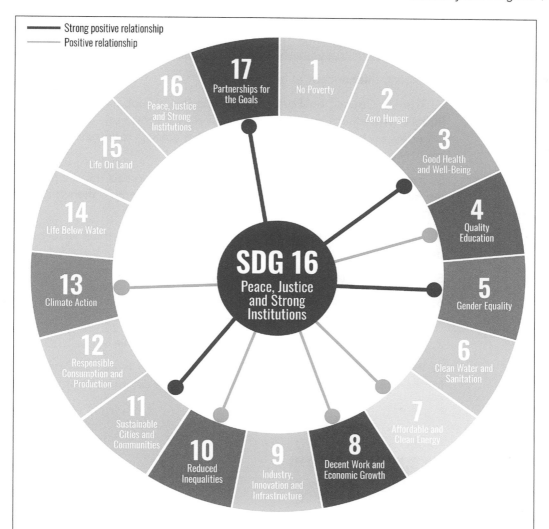

FIGURE 16.4 Interconnections in SDG 16 and the Other Goals
SDG 16 connects to all the SDGs in some way; this figure highlights several of the stronger connections.

SDG 3: Reducing violence and insecurity has direct impacts on improving human health and well-being. SDG 16 is critical to discussion and government's effort to rebuild following the COVID-19 crisis.

SDG 4: A quality education should include knowledge about human rights and a culture of peace.

SDG 8: Promoting rights, justice and the rule of law increases economic productivity and investment allowing economies to prosper. In addition, good governance increases the capacity for tax and revenue collection which can be invested in other services.

SDG 10: Fostering peace and inclusive societies can help reduce inequalities. In addition, promoting rights, justice and the rule of law reaches the vulnerable and furthest behind.

SDG 11: Fostering peace can assist in cities reducing high levels of crime, violence and insecurity, particularly among those communities that are marginalized (such as those who live in slums).

SDG 13: As highlighted in Box 16.10, the negative impacts of climate change are a threat to human life and freedoms of all people.

SDG 17: There are several overlaps between SDG 16 and SDG 17, leading many to say that both SDG 16 and SDG 17 provide the governance framework for the other SDGs.

QUESTIONS FOR DISCUSSION AND ACTIVITIES

1. Explain the concept of human rights.

2. The US Council on Foreign Relations has a Global Conflict Tracker (https://www.cfr.org/global-conflict-tracker/?category=us). Select several countries and describe the types of conflict among countries and within countries.

3. Explain the difference between voluntary and forced migration. Why are forcibly displaced persons considered more vulnerable than voluntary migrants?

4. Explain how corruption and a weak rule of law undermine sustainable development.

5. What are some of the causes of a weak rule of law?

6. How would you assess the status of the rule of law in the United States? What are its strengths and weaknesses?

7. Is peacebuilding an idealistic goal or is it achievable?

8. Which strategies for peace and the rule of law would you prioritize for the Global North; which strategies would you prioritize for the Global South and why?

TERMS

asylum seeker
civil society
corruption
forced migration
human rights
human trafficking
humanitarian assistance

internally displaced person (IDP)
migrant
missing and disappeared people
peacebuilding
refugees
rule of law
voluntary migration

RESOURCES USED AND SUGGESTED READINGS

Alliance for Peacebuilding. "Seven Ways the 117th Congress Can Build Peace in Conflict Affected and Fragile States" (2021, February). https://allianceforpeacebuilding.app.box.com/s/10pub13ja3n4sp47u b6qoflilf7tsvek

Butcher, C., and Hallward, M. (eds.). *Understanding International Conflict Management*. New York: Routledge (2019).

CIVICUS. "CIVICUS Enabling Environment" (2021). https://www.civicus. org/index.php/what-we-do/defend/civicus-enabling-environment and http://www.civicus.org/eei/

Committee to Protect Journalists. "2020 Attacks on the Press." Interactive Map of Countries and Regions (2020). https://cpj.org/attacks-on-press-2020-journalists-killed-jailed/

Fleming, S. "Corruption Costs Developing Countries $1.26 Trillion Each Year—Yet Half of EMEA Think It's Acceptable." World Economic Forum (2019, December). https://www.weforum.org/agenda/2019/12/corruption-global-problem-statistics-cost/

Global Alliance (2021, May). https://sdg.iisd.org/news/global-alliance-report-shares-three-key-findings-on-sdg–16/

Internal Displacement Monitoring Centre. *Global Internal Displacement Database* (2020). https://data.unicef.org/resources/lost-at-home-risks-faced-by-internally-displaced-children/

Meuleman, L. *Metagovernance for Sustainability: A Framework for Implementing the SDGs.* London and New York: Routledge (2019).

Meyer, C. K., and Boll, S. "Editorial: Categorising Migrants: Standards, Complexities, and Politics." *Anti-Trafficking Review* 11: 1–14 (2018). www.antitraffickingreview.org

Myers, E., and Hume, L. *Peacebuilding Approaches to Preventing and Countering Violent Extremism.* Washington, DC: Alliance for Peacebuilding (2018).

National Research Council. *International Conflict Resolution after the Cold War.* Washington, DC: The National Academies Press (2000). https://doi.org/https://doi.org/10.17226/987

Owsaik, A., Greig, J. M., and Diehl, P. *International Conflict Management.* Boston and New York: Polity (2019).

Paehlke, R. *Democracy's Dilemma: Environment, Social Equity, and the Global Economy.* Cambridge, MA: MIT Press (2003).

Serughetti, G. "Smuggled or Trafficked? Refugee or Job Seeker? Deconstructing Rigid Classifications by Rethinking Women's Vulnerability." *Anti-Trafficking Review* 11: 16–35 (2018). www.antitraffickingreview.org

United Nations. *Manual on Human Rights Monitoring* (2012). http://www.ohchr.org/EN/PublicationsResources/Pages/MethodologicalMaterials.aspx

United Nations. *World Trends in Freedom of Expression and Media Development.* Paris and the United Nations (2015). http://www.unesco.org/new/en/world-media-trends

United Nations. *"COVID-19: UN Chief Calls for Global Ceasefire to Focus on 'the True Fight of Our Lives.'"* UN News (2020, March 23). https://news.un.org/en/story/2020/03/1059972

United Nations, Department of Economic and Social Affairs, Population Division. *Trends in International Migrant Stock: The 2017 Revision.* New York: UN DESA (2017, December). https://www.un.org/en/development/desa/population/migration/data/estimates2/data/UN_MigrantStockTotal_2017.xlsx

United Nations Development Program. "Support to the Implementation of Sustainable Development Goal 16" (2016). https://www.undp.org/content/undp/en/home/librarypage/sustainable-development-goals/undpsupport-to-the-implementation-of-the-2030-agenda/

United Nations Global Compact. *A Practical Guide for Collective Action against Corruption.* New York: United Nations (2015). https://d306pr3pise04h.cloudfront.net/docs/issues_doc%2FAnti-Corruption%2FCollectiveActionExperiencesGlobal.pdf

United Nations, Office of Drugs and Crime. "Integrity in the Criminal Justice System" (2021). https://www.unodc.org/unodc/en/corruption/criminal-justice-system.html

United Nations Office of the High Commissioner for Human Rights. The Advocates for Human Rights. "Human Rights Basics" (2021). https://www.theadvocatesforhumanrights.org/human_rights_basics

United Nations Office of the High Commissioner for Human Rights. "Universal Declaration of Human Rights" (2022). https://www.ohchr.org/en/universal-declaration-of-human-rights

Uslaner, E. *Corruption, Inequality and the Rule of Law.* New York: Cambridge University Press (2008).

Uslaner, E. *"Corruption and Inequality."* CESifo DICE Report, ISSN 1613–6373, ifo Institut - Leibniz-Institut für Wirtschaftsforschung an der Universität München, München 9 (2): 20–24 (2011).

Yayboke, E., and Hume, E. *"Ending Violent Conflicts Requires Preventing Them in the First Place."* Center for Strategic and International Studies (2020, October 1). https://www.csis.org/analysis/ending-violent-conflicts-requires-preventing-them-first-place

Collaborative Governance and Partnerships

LEARNING OBJECTIVES

After reading this chapter, you should be able to:

- Describe how the lack of data hinders sustainability

- Explain how debt distress inhibits sustainable development

- Explain the role of the private sector in implementing sustainability

- Explain corporate social responsibility

- Identify and explain the various targets and goals associated with SDG 17

- Discuss how the three Es are present in SDG 17

- Explain how SDG 17 strategies connect to other SDGs

In 2019, Samsung partnered with the UN Environment Programme (UNDP) to offer the Global Goals app on their Galaxy smartphone. The idea is to mobilize Galaxy users to take small actions toward the achievement of the SDGs. The Global Goals app was installed on more than sixty million Galaxy devices. When a Galaxy user watches an in-app advertisement, it automatically donates to UNDP at no cost to the user. Users can also directly donate to specific goals in the app. By 2020, the app had raised $1 million in donations to the UNDP.

During the pandemic, Galaxy users, through the app, were able to direct increased funding to the WHO's COVID-19 Solidarity Response Fund and UNDP's ongoing effort to build resilience for those deeply affected by the pandemic.

COLLABORATIVE GOVERNANCE AND PARTNERSHIPS FOR SUSTAINABILITY

Large-scale progress in sustainability requires engaging across different sectors and diverse partnerships that include the various UN programs and offices, other international organizations, national governments, local communities, and the private sector. Business agility, financial capital, program management, and technological capability are resources and expertise required to achieve sustainability. Some of these skills and expertise are more prominently developed in the private sector.

The challenges are considerable: support for implementing the sustainability has been steady but fragile, the level of financial resources needed is not sufficient, crucial data are still lacking, and there is a need to enhance international collaboration. António Guterres, Secretary-General of the United Nations summarized, "To deliver on the promise of a prosperous and peaceful future, development actors will have to find new ways of working together and leveraging genuine partnerships that make the most of expertise, technology and resources for sustainable and inclusive growth. Hundreds of stakeholders representing government, civil society, private sector and the scientific community are engaged in advancing the SDGs." Sustainable development will involve collaborative governance and partnerships.

It is hoped that collaborative governance and diverse partnerships will help solve some key challenges, including improving access to technology, coordinating policies to help countries manage their public debt, and promoting investment and international trade for low-income countries. However, there are still challenges with regard to the ease in which remittances can be transferred, access to data, and the problem of debt in the Global South.

Remittances

Remittances are money transfers in the form of either cash or goods that migrant workers send back to their country of origin to support their families and communities. Remittances are of the most direct links between migration and development. In low-income households, remittances may buy basic goods, pay for housing, education, and health care. In higher-income households, they may provide capital for small businesses and other entrepreneurial ventures. In some countries, banks have raised overseas financing using future remittances as collateral.

Remittances are an increasingly important source of investment in many economies in the Global South. In 2019, international remittances to low- and middle-income countries totaled $554 billion, more than the total of foreign direct investment (FDI). Remittances have also exceeded official development assistance (ODA) by a factor of three since the mid-1990s (Figure 17.1). The top five recipient countries include India, China, Mexico, Philippines, and Egypt.

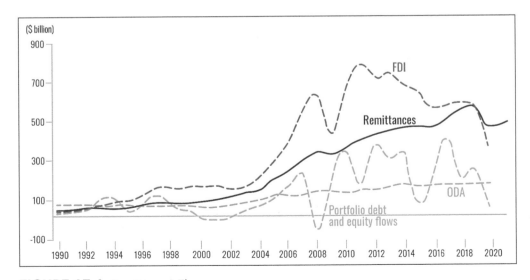

FIGURE 17.1 Remittance Flows
Remittances have increased steadily, overtaking FDI and ODA as major sources of capital investment into low- and middle-income countries.

According to the International Monetary Fund (IMF), remittances are especially important for low-income countries and account for nearly 4 percent of their gross national income (GNI), compared with about 1.5 percent of GNI for middle-income countries. The World Bank estimates remittances may represent more than 30 percent of GNI in Tonga, Haiti, and South Sudan.

Remittances may be sent using cash, check, money, or forms of credit/debit, but they usually go through a service that charges a transaction fee. For smaller remittances—under $200—which is more typical for low-income migrants—fees typically average 7 percent and can be as high as 15–20 percent in countries with less remittance transactions. For some migrants, lack of access to a bank account means they must pay higher fees for remittance transfers. There can also be a currency-conversion fee added. Because of these reasons, some have called for a reform of remittance sending services.

Data and Monitoring

High-quality, open, and timely data are vital in enabling governments, development partners, international organizations, the private sector, and the general public to make informed decisions and wise investments. However, many national statistical systems have faced serious challenges in tracking progress on the sustainability because it requires an unprecedented amount of data and statistics at all levels.

To address the needs of "those left behind," we first need to know who and where they are. The truth is, there is remarkably poor knowledge on who is being "left behind," particularly within countries but also between them. Even where data appears rigorous and comprehensive, certain groups are often missing, such as ethnic minorities or regional groups. Indigenous populations and slum-dwellers are consistently left out of data sets. It is still impossible to know with certainty how many disabled children are in school in many countries. Issues of most concern to women are also poorly covered by existing data. For example, UNICEF reports that only just more than half all countries report data, of varying quality, on intimate partner violence; data is rarely collected from women fifty and over; and little is available on the division of money and labor within households.

In 2019, most countries reported that they were carrying out a national statistical plan. However, many also said that they lacked sufficient funding for full implementation. Only 25 percent of plans were fully funded in sub-Saharan Africa (nine out of thirty-six countries), compared with 95 percent in Europe and North America. Over the past three years, countries in sub-Saharan Africa benefited most, receiving $885 million, a significant increase in funding, but international funding for data and statistics is about half the level it needs to be, according to the UN Task Force on Financing for Sustainable Development. Continued and increased technical and financial support is needed to ensure that countries in developing regions are better equipped to monitor progress of their national development agendas.

The COVID-19 crisis is demonstrating how critical data can effectively guide decision making at each step of the pandemic response. At the same time, the pandemic is also calling attention to the fact that even the most basic health, social, and economic data are often absent. In addition, the most vulnerable populations who need help the most remain invisible or underserved.

To achieve sustainable development, governments and their national statistics offices need better funding and training. Traditional data-collection techniques such as household surveys, censuses, and registers should be made more frequent, rigorous, and universal.

Despite the importance of data, there are a number of challenges that include privacy, data security, ownership, access, and inequality. Some progress has been made in developing better gender statistics and financial statistics, but there are large gaps in the production and usage of data, particularly in low-income countries that may lack statistical systems, have insufficient coordination among data producers at the country level, or lack financial resources.

Debt and Debt Distress

Governments borrow money all the time. They do so to mobilize financing for growth and development. A government can borrow from three types of creditors. It may borrow from private creditors, such as commercial banks, that charge interest on those loans. Governments can borrow with a bilateral creditor, often another country such as the United States, Belgium, or the Netherlands, which will offer lower interest rate loans than commercial banks. Low-income countries can borrow from multilateral creditors such as the World Bank, the African Development Bank, and others that charge little or no interest on the loans. When governments borrow, this is known as **public debt** or public-sector debt (in contrast to private debt that is taken out by individuals, for example home mortgages).

Many countries in the Global South have borrowed money to build the expansive infrastructure needed to industrialize and modernize. Borrowing money gave governments the resources needed to build roads, water networks, and sanitation systems and to invest in health systems, education, and other infrastructure.

Debt is not necessarily bad if it is short term and a country can pay back its loans. Debt service payments include the principal plus interest. However, in some cases, if the interest on that original loan is high or variable, or a loan comes due at a time when its economy is not doing well, a country may have difficulty servicing debt. A country can have difficulty paying back loans if tax revenues are shrinking and there is a decline in GNI; official development assistance is stagnating; exports are declining, in value and volume; or if remittances are falling. These factors may aggravate a country's balance of payments.

A country experiences **debt distress** when debt servicing has become so large that a country uses a high proportion of its foreign exchange earnings to service this debt and still searches for more loans to enable it to meet urgent and pressing domestic obligations. A country with a high debt burden may need to seek debt relief or reschedule the debt, which involves changing the original debt agreement in some way (such as extending the payment time frame or lowering interest rates). Some countries attempt to "restructure" or modify their debt arrangements but may be forced to implement **austerity measures**—which consist of cuts to spending in the public sector such as health, education, social services, and other public goods. Austerity measures generally follow neoliberal principles and, unfortunately, for many countries experiencing debt distress, the impact of austerity measures resulted in declining incomes, increased poverty and malnutrition, and increased rates of mental depression.

For example, in the Gambia, public debt was determined to be unsustainable in 2018, following sharp increases in debt resulting from economic declines combined with governance issues. A large share of the Gambia's external debt is owed to nontraditional creditors, including bilateral creditors and other institutions. The restructuring is ongoing, with coordination issues slowing progress. Unsustainable debt burdens have the potential to siphon off the resources these countries need immediately to fund sustainability efforts. The Gambia is just one of many countries in the Global South that have experienced cycles of debt crisis and debt distress, in some cases for more than thirty years.

Debt distress imposes a number of constraints on a country's growth, hindering sustainable development by draining away limited resources and reducing financial resources for investment in health, education, and other social projects. A debt crisis occurs when government spending continues to fall causing increased levels of poverty, inequality, while debt stays high. Meanwhile, high interest rates on private loans mean speculators continue to take large profits out of the country.

According to the World Bank, today more than half the world's seventy-six lower-income countries struggle with high levels of debt. A 2020 UN Report on the World Economic Situation noted that high levels of debt are pervasive and reached a record $744 billion in 2019. Thirty-eight countries in sub-Saharan Africa are in debt distress; so too are Afghanistan, Bangladesh, Nepal, Haiti, Honduras, Cambodia, and Myanmar. Even a relatively modest volume of debt can pose great problems for the economy of a small low-income country. Elevated debt levels not only pose financial risks but also reduce an economy's resilience to shocks, which in turn may lead to further deterioration in economic activity.

Although debt distress has been associated with the Global South, in fact the United States, the United Kingdom, France, and Germany are the most indebted countries in terms of total dollars. European countries such as Greece and Spain have experienced recent debt Crises. The US national debt is $19.2 trillion.

However, the debt owed by Mongolia, Angola, Sudan, and Lebanon, while significantly lower in total dollars, constitutes a higher debt burden because their economies are not generating sufficient revenues to service their debt— in other words, they may not be able to pay off these loans without putting the burden onto their people, and this often means the poor.

In 2000, the United Nations launched the Heavily Indebted Poor Countries Initiative (HIPC) to deal with debt distress. Initially the aim was to cancel enough of an indebted country's debts so that the total debt would fall to a level viewed by the IMF and World Bank as "sustainable" (in other words, it would have a level of manageable debt). HIPC worked by combining debt relief and low-interest rates loans to cancel or reduce external debt payments. To be included in the HIPC initiative, a country had to be in debt distress and meet certain criteria, including having developed a poverty reduction strategy. In theory, debt relief would free up resources for social spending, particularly in health and education. Map 17.1 shows the countries initially part of the HIPC initiative.

In 2005 there were efforts to accelerate progress on debt reduction. This scheme cancelled all debts from loans made before the end of 2004 by the IMF, and before the end of 2003 by the World Bank and African Development Bank. In addition, many governments went beyond the amount of debt they

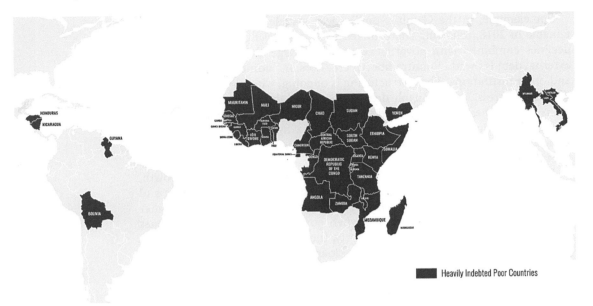

MAP 17.1 Heavily Indebted Poor Countries
Countries involved in the Heavily Indebted Poor Countries initiative that launched in 2000.

needed to cancel based on HIPC calculations. Private-sector creditors were asked to take part in HIPC cancellation, but it was voluntary. To date, the HIPC and its successor programs have relieved thirty-seven countries of more than $100 billion in debt.

However, this program has also been criticized. Critics note that many of the countries that went through the program and were able to reduce debt are now back in debt distress. Tim Jones found that despite having had significant amounts of debt canceled a decade previously, Ghana is losing around 30 percent of government revenue in external debt payments each year. His report concluded that in Ghana, as with many other reindebted countries, the underlying causes of the return to a debt crisis are the continued dependence on commodity exports, as well as borrowing and lending not being responsible enough, meaning that new debts do not generate sufficient revenue to enable them to be repaid.

Although the HIPC initiative has had some success, continued debt and debt distress are not a foundation for sustainable development.

INCREASING COLLABORATIVE GOVERNANCE AND MOBILIZING PARTNERSHIPS

Achieving sustainable development will require mobilizing political will and deepening partnerships between government, the private sector, and civil society. It will also require increasing international cooperation and improving coherence between policies and initiatives both domestically and internationally. Table 17.1 lists the targets for SDG 17.

TABLE 17.1   SDG 17 Targets and Indicators to Achieve by 2030 or Earlier

Target	Indicators
FINANCE	
17.1 Strengthen domestic resource mobilization, including through international support to developing countries, to improve domestic capacity for tax and other revenue collection	**17.1.1** Total government revenue as a proportion of GDP, by source **17.1.2** Proportion of domestic budget funded by domestic taxes
17.2 Developed countries to implement fully their official development assistance commitments, including the commitment by many developed countries to achieve the target of 0.7 percent of ODA/GNI to developing countries and 0.15 to 0.20 percent of ODA/GNI to least developed countries; ODA providers are encouraged to consider setting a target to provide at least 0.20 percent of ODA/GNI to least developed countries	**17.2.1** Net official development assistance, total and to least developed countries, as a proportion of the Organization for OECD's Development Assistance Committee donors' GNI
17.3 Mobilize additional financial resources for developing countries from multiple sources	**17.3.1** FDIs, official development assistance, and South-South Cooperation as a proportion of total domestic budget **17.3.2** Volume of remittances (in US dollars) as a proportion of total GDP
17.4 Assist developing countries in attaining long-term debt sustainability through coordinated policies aimed at fostering debt financing, debt relief, and debt restructuring, as appropriate, and address the external debt of highly indebted poor countries to reduce debt distress	**17.4.1** Debt service as a proportion of exports of goods and services
TECHNOLOGY	
17.5 Adopt and implement investment promotion regimes for least developed countries	**17.5.1** Number of countries that adopt and implement investment promotion regimes for developing countries, including the least developed countries
17.6 Enhance North-South, South-South, and triangular regional and international cooperation on and access to science, technology, and innovation and enhance knowledge sharing on mutually agreed terms, including through improved coordination among existing mechanisms, in particular at the UN level, and through a global technology facilitation mechanism	**17.6.1** Number of science and/or technology cooperation agreements and programs between countries, by type of cooperation **17.6.2** Fixed internet broadband subscriptions per one hundred inhabitants, by speed
17.7 Promote the development, transfer, dissemination, and diffusion of environmentally sound technologies to developing countries on favorable terms, including on concessional and preferential terms, as mutually agreed	**17.7.1** Total amount of approved funding for developing countries to promote the development, transfer, dissemination, and diffusion of environmentally sound technologies

(continued)

Target	Indicators
17.8 Fully operationalize the technology bank and science, technology, and innovation capacity-building mechanism for least developed countries by 2017 and enhance the use of enabling technology, in particular information and communications technology	**17.8.1** Proportion of individuals using the internet
CAPACITY BUILDING	
17.9 Enhance international support for implementing effective and targeted capacity building in developing countries to support national plans to implement all the sustainable development goals, including through North-South, South-South, and triangular cooperation	**17.9.1** Dollar value of financial and technical assistance (including through North-South, South-South, and triangular cooperation) committed to developing countries
TRADE	
17.10 Promote a universal, rules-based, open, nondiscriminatory, and equitable multilateral trading system under the World Trade Organization, including through the conclusion of negotiations under its Doha Development Agenda	**17.10.1** Worldwide weighted tariff-average
17.11 Significantly increase the exports of developing countries, in particular with a view to doubling the least developed countries' share of global exports by 2020	**17.11.1** Developing countries' and least developed countries' share of global exports
17.12 Realize timely implementation of duty-free and quota-free market access on a lasting basis for all least developed countries, consistent with World Trade Organization decisions, including by ensuring that preferential rules of origin applicable to imports from least developed countries are transparent and simple, and contribute to facilitating market access	**17.12.1** Average tariffs faced by developing countries, least developed countries, and Small Island Developing States
SYSTEMIC ISSUES *Policy and Institutional Coherence*	
17.13 Enhance global macroeconomic stability, including through policy coordination and policy coherence	**17.13.1** Macroeconomic Dashboard
17.14 Enhance policy coherence for sustainable development	**17.14.1** Number of countries with mechanisms in place to enhance policy coherence of sustainable development
17.15 Respect each country's policy space and leadership to establish and implement policies for poverty eradication and sustainable development	**17.15.1** Extent of use of country-owned results frameworks and planning tools by providers of development cooperation
MULTISTAKEHOLDER PARTNERSHIPS	
17.16 Enhance the global partnership for sustainable development, complemented by multistakeholder partnerships that mobilize and share knowledge, expertise, technology, and financial resources, to support the achievement of the sustainable development goals in all countries, in particular developing countries	**17.16.1** Number of countries reporting progress in multistakeholder development effectiveness monitoring frameworks that support the achievement of SDGs

Target	Indicators
17.17 Encourage and promote effective public, public-private, and civil society partnerships, building on the experience and resourcing strategies of partnerships	**17.17.1** Amount of US dollars committed to public-private and civil society partnerships
DATA, MONITORING, AND ACCOUNTABILITY	
17.18 Enhance capacity-building support to developing countries, including for least developed countries and Small Island Developing States, to significantly increase the availability of high-quality, timely, and reliable data disaggregated by income, gender, age, race, ethnicity, migratory status, disability, geographic location, and other characteristics relevant in national contexts	**17.18.1** Proportion of sustainable development indicators produced at the national level with full disaggregation when relevant to the target, in accordance with the Fundamental Principles of Official Statistics **17.18.2** Number of countries that have national statistical legislation that complies with the Fundamental Principles of Official Statistics **17.18.3** Number of countries with a national statistical plan that is fully funded and under implementation, by source of funding
17.19 Build on existing initiatives to develop measurements of progress on sustainable development that complement GDP, and support statistical capacity building in developing countries	**17.19.1** Dollar value of all resources made available to strengthen statistical capacity in developing countries **17.19.2** Proportion of countries that (a) have conducted at least one population and housing census in the last ten years; and (b) have achieved 100 percent birth registration and 80 percent death registration

Source: United Nations Sustainable Development Goals (2022). https://sustainabledevelopment.un.org/sdg17

Among the numerous targets, there are seven broad solutions that may advance sustainability and sustainable development. These are:

1. Improve public administration and the quality of government (this has a strong overlap with targets in SDG 16 Peace, Justice and Strong Institutions, particularly those targets on justice and strong institutions);

2. Ensure policy coherence;

3. Improve access to and capacity of technology and data (with some overlap with SDG 9 on innovation);

4. Reduce debt;

5. Engage multistakeholder partnerships;

6. Increase ODA; and

7. Promote trade.

1. Improve Governance and Public Administration

Many scholars have noted that quality of government must include inclusive political institutions, accountability (both political and for public institutions), effective rule of law (as discussed in Chapter 16 Peace, Justice, and Human Rights), property rights, low levels of corruption, an impartial use of political authority, and merit-based bureaucracy. In an ideal world, countries would be willing and able to manage their economic assets for the collective well-being of their citizens. Too often, however, internal weakness and external pressure combine against the will, capacity, and means of developing nations to achieve these goals.

Research by Aaron Maltais and colleagues argues that the quality of government is one of the most important ways to improve economic and social conditions; however, higher rates of corruption lead to less stringent environmental regulation and a lower commitment to sustainability.

It is also important to pay attention to scale, as governance happens at the city, subregional, and even NGO scale. There is an increased role for local communities and civil society to take actions and develop plans. Many local governments have taken the initiative to implement the sustainability systematically, in some cases going further than national governments. For example, New York City, under its sustainability plan called *OneNYC 2050*, is mobilizing efforts to combat climate change, fight injustice, and ensure a safer, healthier city for all its constituents. Similarly, the city of Bristol in the United Kingdom established the Bristol SDG Alliance, which is working to raise awareness about the SDGs. Many urban scholars say that, ultimately, all sustainable development must be implemented locally, meaning that the role of local governments and partnerships will be essential.

2. Ensure Policy Coherence

Many of the challenges faced by countries are global in nature and cannot be adequately addressed by domestic policies and investments alone—for example climate change, international trade, and finance. This is one reason that national policies need to be complemented by more effective international cooperation.

Target 17.14 calls for enhancing policy coherence for sustainable development. In the context of the SDGs, policy coherence refers to (1) representing the interests of the poorest in policy processes that affect them and (2) ensuring that a policy in one area does not undermine policy objectives in another. It promotes consistency between global goals and national contexts, among international agendas, and between economic, social, and environmental policies. Compounding the challenge is that there are no single solutions. This is an area where there is not yet sufficient research to understand interactions between SDG goals and targets and how policy could promote positive interactions while mitigating negative ones. Aaron Maltais and colleagues suggest that there are some enabling factors that will help with policy integration and coherence. These include (1) political will that sends a clear message on the importance of coherence; (2) incentivizing policy makers to see cross-sectoral solutions; and (3) strong leadership.

3. Increase Access and Capacity in Technology and Data

Technology and good data are essential to achieving the SDGs. There are three reasons for this, say Aaron Maltais and his colleagues. First, technology transfer provides developing countries the opportunities to absorb the most productive technological advancements in our economies and to accelerate progress. Second, access to technology is critical for supporting local innovation systems and fully realizing local competitive advantages. Third, environmental conditions, particularly climate change, require that countries in the Global South adopt a different development path than the Global North followed. Meeting the ambitions of both the SDGs and the Paris Agreement requires that as middle- and low-income countries grow economically, they simultaneously move rapidly to clean energy and technology.

Digital technologies, robotics, artificial intelligence (AI) and automation, biotechnology, and nanotechnology all have far-reaching impacts and present both opportunities and challenges. On the one hand, new materials, such as digital technologies, biotechnology, and nanotechnologies, and AI all hold great promise for a range of high-efficiency water and renewable energy systems. There are several examples. The "It's Our Forest Too" app is a forest-monitoring smartphone application used to monitor and report on illegal logging and biodiversity loss by Indigenous communities. It has been used in Cambodia to allow activists to discretely monitor and report illegal activity within forests. "No Food Waste" is a youth and technology-driven surplus food recovery network that collects surplus food from weddings, restaurants, and food industries in India and donates the food to those hungry, thereby combating both hunger and food waste.

Additionally, big data, data science, and geospatial analysis can be applied to assist in many areas of sustainable development. For example, data on spending patterns on mobile phone services could provide proxy indicators of income levels, allowing a better picture of poverty. Crowd sourcing or tracking of food prices listed online could help monitor food security in near real-time. Sensors connected to water pumps can track access to clean water, and smart meters allow utility companies to increase or restrict the flow of electricity, gas, or water to reduce waste and ensure an adequate supply at peak periods. Collecting big data on the patterns in global postal traffic can provide indicators such as economic growth, remittances, and trade. Satellite remote sensing can track encroachment on public land or parks and forests, and maritime vessel tracking data can reveal illegal, unregulated, and unreported fishing activities. The integration of geospatial information and statistical data will be particularly important in increasing useful data for decision making. The *Solutions* 17.1 box discusses the tremendous potential for open-source mapping to empower communities to become more resilient.

While these new technologies offer promise, these must be used within a Life Cycle Analysis (LCA) approach; otherwise it could result in increased energy consumption and associated pollution and wastes (such as e-waste, nano-waste, and chemical wastes). An LCA requires incorporating environmental considerations into the design of these technology systems from the start and will depend on an extraordinary level of international cooperation. Many countries may also need to find new development pathways that incorporate these technologies and to rethink employment and income distribution issues.

In general, technology needs to be transferred from the Global North to the Global South. However, there are many barriers. These can include a lack of technology to support the adoption of new technologies, lack of knowledge on the operation or management of new technologies, lack of access to financing

SOLUTIONS 17.1

Open-Source Mapping

Target 17.6: Enhance North-South, South-South, and triangular regional and international cooperation on and access to science, technology, and innovation and enhance knowledge sharing on mutually agreed terms, including through improved coordination among existing mechanisms, in particular at the UN level, and through a global technology facilitation mechanism.

It may surprise you to learn that the world is not entirely mapped. There are many rural and remote areas that are not mapped. Even some urban areas are not comprehensively mapped. Our world is also constantly changing—buildings go up, buildings come down, new roads are paved, new suburbs emerge. Mapping the world is dynamic and the more we map, the more we can provide information to help decision makers more effectively implement sustainability.

In the world of mapping, there are platforms such as GoogleMaps or ESRI, but these may require license fees that make it too expensive for nonprofits or low-income communities to use. In recent years, there has been a movement to make data and mapping software freely available. OpenStreetMap (OSM) is a community of mappers that maintains data that can be used to make maps. This data is free and available to anyone to use (think the "wikis" of maps).

The "open source" nature of the data allows anyone to contribute vital information, which can be leveraged across many fields of interest. Researchers use new high-resolution satellite imaging technology to measure poverty, consumption expenditures, and asset wealth from high above the Earth—providing systematic information about where poor households are more likely to be found when other data is not available.

One initiative that uses OSM is YouthMappers, a global community of students, researchers, educators, and scholars that use public geospatial technologies to highlight and directly address development challenges worldwide (see Photo 17.1). To date, there are more than 250 college chapters of YouthMappers. On college campuses around the world, students volunteer

PHOTO 17.1 YouthMappers from the State University of Zanzibar in Tanzania use cell phones and field papers to map community assets
Khamis Juma Madai, chapter president of SUZA YouthMappers at the State University of Zanzibar, and chapter members Marhat Abubakar Awadh and Fatma Mohammed Hamza collected data during the Resilience Academy Internship Program in October 2020 at Nungwi, Zanzibar. During the eight-week internship program, students collected data related to the tourism industry using GPS receivers and mobile phones, including attributes and features of the road network.

their time to organize, collaborate, and implement mapping activities that respond to actual development needs around the globe—creating and using geospatial data and information that is made publicly available through open platforms. YouthMappers focus on remotely mapping communities—particularly the most vulnerable—that have not been formally mapped before. The resulting maps and data are publicly available and enable international and local nongovernmental organizations and individuals to improve their response to crises affecting these areas. Many mapping projects support the work of organizations such as the American Red Cross, the World Bank, the US Department of State, and the Peace Corps.

A YouthMappers chapter in Sierra Leone and researchers at Arizona State University teamed up to map buildings and roads to support the design of mini-grid power distribution networks in rural areas. This data will be used for electrification planning by the Ministry of Energy, national utilities, and other Sierra Leone energy-sector stakeholders. Distribution networks in Sierra Leone are not well mapped, especially in rural locations, making this data valuable for all energy-sector planners.

Another mapping project focused on Tanzania, a country with a high incidence of female genital mutilation/cutting, early marriage, and gender-based violence. NGOs on the ground need better road and residential area data to facilitate their outreach work. YouthMappers worked with Tanzania Development Trust to produce maps that will be used to help activists to better protect girls at risk, empower local mappers to map and develop their communities, and enable villages to develop village land-use plans to mitigate land disputes and better develop their resources.

YouthMappers is not only using cutting-edge geospatial technology but they are also an example of how wider collaborative partnerships can be effective at making change. The work of YouthMappers is also supported by partners that include Critigen, Esri, GIS Certification Institute, Hexagon Geospatial, Humanitarian OpenStreetMap Team, Mapbox, Mapillary, MAXAR, Missing Maps, Radiant Earth Foundation, Teach OSM, the American Geographical Society, the Open Source Geospatial Foundation, and the University Consortium for Geographic Information Science.

To learn more about how you can participate in YouthMappers, or start a chapter on your campus, go to https://www.youthmappers.org/

or weak domestic financial institutions, and a lack of government policies that support technology transfer or insufficient infrastructure. Effective technology transfer depends on having the knowledge and capacity to use that technology and to be sure it fits local contexts. Many Global South countries will need external financial support to build infrastructure for these technologies, or to access and implement them.

4. Reduce Debt: Debt Swaps and Social Impact Bonds

Rising public debt levels and heightened debt vulnerabilities have been a cause for concern, particularly in many of the world's lowest-income countries. This is one reason why SDG 17 specifically targets low-income countries in attaining long-term debt sustainability through coordinated policies aimed at fostering debt financing, debt relief, and debt restructuring to reduce debt distress.

Debt swaps are one mechanism. Debt conversion or debt swap mechanisms cancel part of a country's debt and redirects those monies that would have been used for debt servicing to investments in sustainable development. They have a long history, with examples that include debt-for-nature swaps. More recently, debt swaps have been proposed to reduce the debt burden of participating countries while channeling more funds into SDG investments.

The Caribbean region is heavily indebted and periodically exposed to devastating hurricanes. High debt has brought about a period of fiscal consolidation that continues to restrict the capacity of governments to sustain social spending and invest in much-needed climate-resilient infrastructure, while not relieving the debt burden.

In this context, the Economic Commission for Latin America and the Caribbean (ECLAC) has proposed a swap of some of the region's external debt for debtor-country commitments to make annual payments into a new Caribbean Resilience Fund. The proposal is being developed by a regional task force that is now seeking to engage three countries of the region in a first phase of the project. ECLAC proposes that the Green Climate Fund (GCF) buy up some of the external private debt of participating countries, and that it also negotiate a discounted value of the country obligations to certain bilateral and official creditors. GCF would then pay down the discounted country obligations to those creditors over a period of years. For their part, the Caribbean countries would commit to pay into the new Caribbean Resilience Fund the amount that they would have paid as debt servicing to its former creditors. The participating countries would thus exchange the affected debt obligations for an obligation to pay an amount annually into the new regional fund—an amount that would have otherwise been used to pay the debt servicing on the discounted debt. Annual payments would possibly be bolstered by donor contributions of various donors. The fund would be used to finance sustainable investments in the participating countries.

The UNDP has also introduced innovative mechanisms for financing, such as financial aggregation, impact bonds, pay-for-success systems, or equity-based investments. These financial tools allow the private sector to mobilize resources while meeting socioeconomic objectives. Financial aggregation is a process in which an entity bundles together multiple small-scale assets and then seeks financing, or refinancing, from investors on the basis of future cash flows. This bundling of small loans creates investment products that meet the needs of large institutional investors such as global pension funds and insurance funds. In low-income countries, aggregation also allows financial institutions, from banks to microfinance lenders, to make the most of limited balance sheets.

A **social impact bond** (SIB) (also known as a social benefit good) is a type of financial security that provides capital to the public sector to fund projects that have positive social outcomes. These are relatively new; the first was launched in 2010. Currently, more than eighty-nine SIBs have been contracted globally with approximately $322 million in investments in industries ranging from health care to education.

SIBs combine the principles and practices of key trends: public-private partnerships, outcomes-based contracts, and "payment by results" funding of social interventions. They are different from conventional bonds in that they do not offer a fixed rate of return to investors; instead, the repayment of the bonds depends on the success of the project that has been funded using the bonds. If a project is successful, the investors are repaid their investment plus a dividend based on the outcomes that meet or exceed predetermined targets. Advocates say SIBs represent a transformation in the way government can deliver public services because they can generate huge potential savings for the taxpayer, increase revenues for charities and social enterprises, and provide returns to social investors.

One example of an SIB involves conservation in Africa. In 2015, UNDP, the Global Environment Facilities, the Zoological Society of London, and other partners began developing the world's first rhinoceros impact bond to address illegal hunting and poaching, with a goal to launch a $25 million to $35 million Rhino Bond. The initial capital will be used to develop evaluation metrics and prepare proposed conservation sites in South Africa, Kenya, Zimbabwe, and Namibia. Under the model, investors receive returns as payments against performance, such as the identification of conservation sites, improved protection of rhinos, and reduced poaching activities.

Because SIBs are new, they are still being evaluated. Despite a lack of evidence of their effectiveness, there is increasing enthusiasm for SIBs. There are estimated to be between 89–160 SIB projects already operating or under development in nineteen countries, and they are being adapted to finance international development activity. Scholars Steven Sinclair and his colleagues argue that SIBs have at best a limited (and as yet unproven) applicability to particular types of technical social policy interventions, but they are singularly unsuited to deal effectively with wicked social problems that require transformational interventions, such as those found in the SDGs. Sinclair points out that to date, most SIBs have been small scale and not particularly innovative.

Finally, blended financing is another possible option. Blended financing is where public money or philanthropic funds are used to mobilize private capital—through development banks or national aid agencies, for example. This not only catalyzes increased funding but it also creates investment track records and helps to improve local finance sectors and market conditions.

5. Engage in Cross-Sector Collaboration and Multistakeholder Partnerships

A 2021 Brookings Institution Report indicated that while the United Nations is the body with foremost official responsibility for shepherding SDG advances, they form only one piece of a broader puzzle of action required for SDG success. Most of the practical actions for the SDGs will be developed and implemented outside the walls of UN conference rooms. SDG success requires groups of people getting together, around the world, to identify and implement new forms of cooperation. **Cross-sector collaborations** are alliances of organizations that together have a role in solving a problem and achieving a shared goal. These alliances can include businesses, nonprofits, government, education, health care, labor, faith institutions, philanthropy, and even communities (see *Solutions* 17.2).

One of the types of cross-sector collaboration is multistakeholder partnerships. **Multistakeholders** are partnerships between public, private, and civil society actors. Every country and local community needs its own forums to work through multistakeholder debates and priorities. The benefits of engaging in multistakeholder partnerships are that they are more

SOLUTIONS 17.2

Cross-Sector Collaboration

Cross-sector collaborations are alliances of organizations that together have a role in solving a problem and achieving a shared goal. A broad view of sectors includes business, nonprofits, government, education, health care, labor, faith institutions, philanthropy, and even communities. Research shows that increasing the diversity of stakeholders strengthens the quality of problem solving and decision making, as well as the likelihood that solutions will be implemented. This is important particularly when dealing with "wicked" complex problems such as climate change, inequality, or poverty.

Cross-sector collaboration is not new. For decades individuals and organizations have been building alliances within and across sectors to achieve shared goals. What has changed in the last decade is that more individuals and institutions have become focused on this way of working. Public-private partnerships are on the rise. Philanthropic and federal government grants are increasingly requiring recipients to work in cross-sector collaborations. SDG 17 focuses explicitly on improving cross-sector collaborations in Targets 17.16 and 17.17.

There are four different partnership models of cross-sector collaboration:

1. *Joint project*: short-term, one-time collaborative effort or single project
   Example: The TV White Space Partnership was a joint project between Microsoft, the government of the Philippines, and the US Agency for International Development (USAID), facilitated by Resonance, to extend internet access to remote and underserved coastal communities in the Philippines.

2. *Joint program*: multiple projects or deliverables around a single focus area with a small set of partners
   Example: The Last Mile Initiative was a joint program partnership between USAID and Qualcomm, Dialog Telecom, and the National Development Bank of Sri Lanka to extend internet connectivity and services to rural consumers.

3. *Multistakeholder initiatives*: partners and resources align to drive system change on a common agenda; often requires a governance structure
   Example: The Global Alliance for Vaccines and Immunization works with civil society organizations, manufacturers, governments, donors, companies, and more to help vaccinate almost half the world's children against deadly and debilitating infectious diseases.

4. *Collective impact*: long-term commitments to a common agenda, with many actors and independent workstreams
   Example: The Platform for Accelerating the Circular Economy works with global leaders and their organizations to accelerate the transition to a circular economy.

Sources: Gold, A. "A Framework for Learning about Cross-Sector Collaboration." Independent Sector (2016). https://independentsector.org/resource/a-framework-for-learning-about-cross-sector-collaboration-2/#:~:text=Cross%20sector%20collaboration%20is%20an,not%20yet%20labeled%20or%20codified

Resonance Global. "A Guide to Cross-Sector Collaboration" (n.d.). https://www.resonanceglobal.com/a-guide-to-cross-sector-collaboration#common partnershipmodels

flexible and adaptative in their ability to address complex problems, which may provide advantages over traditional forms of governance and international cooperation.

With regard to the SDGs, there are two key roles that stakeholders can play: holding governments accountable for their actions or lack thereof (by tracking implementation or engaging in advocacy activities) and, second, implementing the SDGs by providing services. The latter sometimes happens in close collaboration or even on behalf of governments. In practice, there are additional roles that stakeholders can play, such as providing inputs to policy making.

Most countries are still in the early stages of creating a framework for multistakeholder engagement for the SDGs. A recent Canadian study suggested five principles of effective multistakeholder engagement, as shown in Figure 17.2.

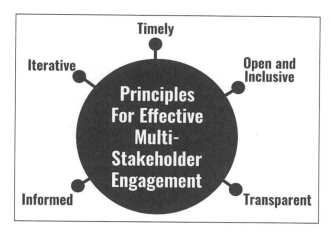

FIGURE 17.2 Engaging with Multiple Stakeholders To engage most effectively with stakeholders, a partner must be informed, transparent, open and inclusive, and respond in timely ways. Engaging with multiple stakeholders is also an iterative process, where a plan, process, or action is analyzed, changed, or adapted in a continuous process.

1. Effective multistakeholder engagement should be timely. This means that stakeholders are given sufficient time frames for their engagement that are well established and communicated in advance of actual engagement opportunities.

2. Engagement is open and inclusive. Extra efforts are often needed to ensure the presence and participation of people and groups that are most often left behind. Providing information in local languages and making use of a range of methods for engagement, including both offline and online options, are also important. Inclusivity can be bolstered by working with representative organizations for nonstate actors, such as civil society platforms and business associations.

3. Multistakeholder engagement should be transparent, with information on engagement processes and plans being clear and widely communicated.

4. Effective multistakeholder engagement requires that participants are informed about the purpose of engagement, how their inputs will be used, and the overall expected outcomes. Documentation should be provided ahead of all interactions with ample time, clear deadlines, and appropriate tools to provide feedback. There should also be follow-up reports and documentation on how inputs have been considered.

5. Effective multistakeholder engagement should be iterative. Engagement is not a singular process or event but rather a continuous process where multiple opportunities for ongoing engagement exist for different stakeholders. Dialogue should be two-way between those hosting the consultations and those in attendance with mechanisms for engagement institutionalized to provide long-term dialogue for continuous input from nonstate.

Engage Businesses in Corporate Sustainability

Public resources will only go so far in securing SDG achievement; mobilizing and strengthening private-sector and philanthropic partners will be essential. A 2020 UN Global Compact report found all UN leaders felt that cross-sector alliances, networks, and partnerships are essential to accelerating the progress on the SDGs and a majority believe that business will be the critical partner in the United Nation's ability to deliver the SDGs. However, that same report said that most UN-business partnerships involve just one business partner, and that these typically last five years or less. The report also noted that nearly two-thirds (59 percent) of UN leaders said their ability to engage in innovative, transformational partnerships is limited by resource constraints that prompt their agencies to focus on securing financial donations; a third of leaders surveyed report that the majority of existing relationships with the private sector are based on philanthropic or in-kind donations. To secure investment needed to achieve the SDGs, they envision partnerships with innovative financing mechanisms gaining prevalence; nearly three-fourths are considering participating in a project using alternative finance.

Corporate accountability and interest in sustainability has expanded rapidly in recent years. **Corporate social responsibility (CSR)**, introduced in Chapter 12, is seen as important for the legitimacy of an organization in the global economy. Social accountability initiatives offer businesses an international framework and set of principles designed to help establish and promote action on corporate sustainability-related issues. Two examples include the Ten Principles of the United Nations Global Compact (UN Global Compact) and the UN Guiding Principles on Business and Human Rights (UN Guiding Principles). The UN Global Compact is described as a network that brings together UN agencies, corporations, governments, civil society groups, labor representatives, and NGOs. The UN Global Compact provides a framework to guide businesses toward corporate responsibility and sustainability. It may be one of the most influential initiatives worldwide to encourage firms to participate voluntarily in corporate responsibility activities.

To join the UN Global Compact, a company must make a voluntary pledge to operate under the 10 Principles of Sustainability (see Table 17.2). After committing to these principles, businesses must report annually on progress in operating responsibly and ways they have supported society. Currently, more than 12,000 companies and organizations based in 160 countries are involved in this compact, including corporations such as American Airlines, PayPal, Kraft Heinz, Lloyds Banking Group, Siemens, and AstraZeneca Pharmaceuticals.

Despite these initiatives, some in the public sector remain suspicious of the private sector, being uncomfortable with the idea that there can be profits in sustainability. However, some business leaders counter that there are public-sector legal policies or procedures that delay or limit partnership opportunities.

Those in the public or nonprofit sector point to the fact that the UN Global Compact is not a legally binding code of conduct, so there is no mandate to enforce or even measure how a business or organization performs. Business scholar Jette Knudsen says companies with a poor reputation may simply choose to join without intending to implement its principles in their organizational structures, assuming that the mere association with the United Nations allows them to create a positive image. In allusion to the blue UN flag, this practice is derogatorily referred to as "blue washing." Economist Jeffrey Sachs offers his criteria for CSR in the *Expert Voice* 17.1 box.

TABLE 17.2 The 10 Principles of the UN Global Compact	
Principles	**Principle derived from**
Principle 1: Businesses should support and respect the protection of internationally proclaimed human rights	Universal Declaration of Human Rights
Principle 2: Make sure that they are not complicit in human rights abuses	Universal Declaration of Human Rights
Principle 3: Businesses should uphold the freedom of association and the effective recognition of the right to collective bargaining	International Labour Organizations' Declaration on the Fundamental Principles of Rights at Work
Principle 4: The elimination of all forms of forced and compulsory labor	International Labour Organizations' Declaration on the Fundamental Principles of Rights at Work
Principle 5: The effective abolition of child labor	International Labour Organizations' Declaration on the Fundamental Principles of Rights at Work
Principle 6: The elimination of discrimination in respect of employment and occupation	International Labour Organizations' Declaration on the Fundamental Principles of Rights at Work
Principle 7: Businesses should support a precautionary approach to environmental challenges	Rio Declaration on Environment and Development
Principle 8: Undertake initiatives to promote greater environmental responsibility	Rio Declaration on Environment and Development
Principle 9: Encourage the development and diffusion of environmentally friendly technologies	Rio Declaration on Environment and Development
Principle 10: Businesses should work against corruption in all its forms, including extortion and bribery	United Nations Convention Against Corruption

Source: The United Nations Global Compact. "The Ten Principles of the UN Global Compact" (2021). https://www.unglobalcompact.org/what-is-gc/mission/principles

6. Increase Official Development Assistance

According to the UN Conference on Trade and Development (UNCTAD), achieving sustainable development will require $5 trillion to $7 trillion in annual investment. Currently, funding is nowhere near that level.

One of the most important flows of finance for development is **official development assistance (ODA)**. ODA is defined as government aid that promotes and specifically targets the economic development and welfare of low-income countries. Loans and credits for military purposes are excluded. ODA can take the form of grants, where financial resources are provided to developing countries free of interest and with no provision for repayment; or soft loans, which have to be repaid with interest, albeit at a significantly lower rate than if developing countries borrowed from commercial banks. Aid may be provided bilaterally, from donor to recipient, or channeled through a **multilateral organization** such as the World Bank. Aid includes grants, soft loans, and the provision of technical assistance. The OECD maintains a list of 150 countries that are eligible for official ODA.

In 2021, ODA totaled $179 billion. The source of nearly all the money for ODA comes from high-income Global North countries such as the

EXPERT VOICE 17.1

Jeffrey Sachs Proposes Criteria for Corporate Social Responsibility

Dr. Jeffrey D. Sachs is University Professor and Director of the Center for Sustainable Development at Columbia University. He is also Director of the UN Sustainable Development Solutions Network. In the following text, he reflects on CSR and the SDGs:

Around the world, companies are being called upon to align their business practices with the Sustainable Development Goals (SDGs). These calls are coming from many places: outside investors such as pension funds, active shareholders within companies, consumers, suppliers, and other stakeholders in society. There is still no single agreed set of standards to assess whether a company is in alignment with the SDGs.

Thousands of businesses around the world are already aligning their corporate reporting with the SDGs, though still without agreed international standards. Annual corporate and sustainability reports increasingly take the SDGs as the baseline for assessing company performance. Yet there is a tendency of companies to highlight the bright areas of their alignment without taking a serious look at ways that the company's performance needs to improve. As investors, shareholders, and other stakeholders call on companies for greater corporate social responsibility, the need for more objective and comprehensive benchmarking with the SDGs will grow.

Various benchmarking efforts are now underway around the world to try to develop an agreed framework for measuring company performance vis-à-vis the SDGs. It will take time to reach a global consensus and to implement new reporting standards.

I propose four high-level criteria to assess a company's performance vis-à-vis the SDGs.

1. A sound product line: The first criterion is whether the company's goods and services are helpful, harmful, or neutral vis-à-vis the Sustainable Development Goals.
2. Sustainable production practices: The second criterion is the company's direct environmental impact as the result of its own production activities. Alignment with the SDGs therefore requires that companies measure, monitor, and reduce their own direct adverse environmental impacts.
3. Sustainable global value chain. In our globalized economy, it's not good enough for a company to focus only on its own production processes while neglecting the companies up and down its global supply chain. Each company should have a reasonable measure of co-responsibility for the alignment of the entire value chain with the SDGs.
4. Good corporate citizenship: The fourth criterion is the company's corporate citizenship, meaning its behavior vis-à-vis governments and local communities in both source and host countries on a range of corporate citizenship practices. These include tax behavior, corporate reporting, compliance with local laws, payment of bribes, and responsibility to local communities.

Source: Sachs, J. "The SDGs and Corporate Social Responsibility." *Insight/CSR Compass* 53: 1–4 (2019, February).

United States, Germany, the United Kingdom, Japan, France, Sweden, and the Netherlands. In 2021, for example, the United States contributed $42 billion, Germany $32 billion, the United Kingdom $16 billion, and Japan $17.6 billion. A long-standing UN target is that high-income countries should devote 0.7 percent of their gross national income to ODA (represented in SDG Target 17.2). ODA has consistently hovered between 0.2 percent and 0.4 percent of GNI. Only a few countries—Luxembourg, Norway, and Sweden—contribute more than 1 percent of their GNI, while the United States contributes less than 0.2 percent of its GNI. More recently non-OECD countries have contributed to ODA, including the United Arab Emirates and China.

ODA is just one source of development financing and is by no means the largest. Other sources include domestic resources, remittances, and foreign direct investment. However, ODA is crucial because it has been a stable source of external financing, particularly compared to private flows, which are more sensitive to economic shocks or crisis. However, ODA financing is influenced by political leadership, public support, and decisions and coordinated action that prioritize development. Even though ODA may not be able to offset drops in other flows, its dependability is crucial to enable long-term planning within developing countries.

A Critical Perspective on Official Development Assistance

The perceived failure of aid to make a difference in ending poverty has been the subject of political economy scholars for some time. Poverty persists seven decades after the establishment of the World Bank, despite spending several trillions of dollars in aid. This has led some to argue that the failure of aid to make a difference in ending poverty questions the legitimacy of aid.

Some political economists viewed ODA as an instrument through which wealthy countries seek to dominate poor countries. For example, Teresa Hayter claims in her book *Aid as Imperialism* that aid provided by the World Bank and OECD countries serves first and foremost the interests of Western nations and their multinational corporations. The work of Dambisa Moyo has been critical of foreign aid in Africa, noting that development assistance has fostered dependency, encouraged corruption, and perpetuated the cycle of poverty.

Wil Hout writes that a political economy analysis is valuable in identifying the different "layers" of analysis: Beneath the daily events in every political system, there are the institutional arrangements (the "rules of the game") that impact day-to-day politics by influencing the policy options of politicians. Even more fundamental are so-called structural elements, which relate to the history of the country under discussion, its natural resource endowment, and the power distribution across social groups. Existing governance arrangements work in the interests of the dominant power holders, while subordinate groups (the poor, Indigenous, and other minority groups, in many cases also women) are marginalized and generally fail to get access to the formal decision-making structures.

A political economy framework highlights that ODA is not a neutral form of assistance but is inevitably political in nature. One lesson that can be learned from a serious engagement with political economy analysis is that development should not be understood rather naïvely as a process that will bring about improvement in the lives of all parts of a population over a relatively short time span.

7. Encourage Trade and Foreign Direct Investment

Historically, trade has proven to be an engine for countries' economic growth, development, and poverty reduction. Higher demand for commodities, such as minerals, ores, and fuels, resulted in higher prices in the 2000s, consequently boosting incomes in many resource-exporting countries in the Global South. This rapid growth, fueled in part by trade, contributed to significant reduction of poverty levels.

A growing emphasis on free-trade, duty-free, and quota-free market access has been a key element to economic growth over the last fifty years and is part of the World Trade Organization's guiding principles. It aligns with neoliberal economic ideas. The annual value of total merchandise trade in recent years averages about $20 trillion.

However, there are prolonged trade disputes that threaten to impede efforts to reduce poverty, create decent jobs, broaden access to affordable and clean energy, and achieve sustainability. For example, the uncertainty around the exit of the United Kingdom from the European Union may negatively impact trade. In addition, there has been escalating international trade tensions and a trade war among the two largest economies, China and the United States. Bilateral trade between the United States and China has plummeted, with significant disruptions to international supply chains. The global electronics and automobile sectors, which have extensive supply chains that rely on many production networks, have been hit particularly hard. A continued escalation of trade wars could risk a major economic impact, threatening jobs and growth in all countries. Some estimates are that the US-China trade impasse may result in a net loss of 2.7 million jobs in Asia.

Foreign Direct Investment

Over the past twenty years, neoliberal and even liberal reform economists have regarded **foreign direct investment (FDI)** as a central driver of global economic integration and economic growth. FDI is a category of cross-border investment in which an investor (individual or business) obtains a lasting interest in a business or corporation in another country. Ownership of 10 percent or more of the voting power in an enterprise in one economy by an investor in another economy is evidence of such a relationship. Amazon opening a new headquarters in Vancouver, Canada, or McDonald's opening restaurants in Japan, would be examples of FDI. Another example would be by investing in a different level in the supply chain—for example McDonald's purchasing a large-scale farm in Canada to produce meat for their burgers. Finally, a business could acquire an unrelated business in a foreign country—Audi motors, based in Germany, acquired a clothing line in Italy. Trade and FDI are different manifestations of global production. International trade is the cross-border flow of goods and services, whereas FDI is the cross-border flow of the ability to produce goods and services.

Global FDI is big business. In 2019, global FDI was $1.54 trillion, according to UNCTAD. The largest recipient of FDI was the United States, at $282 billion, followed by China and Singapore. Moderate-and low-income countries received about $685 billion in 2019, but with significant differences among regions. Countries in Asia, the largest recipient region of FDI, received approximately $475 billion of the $685 billion. FDI to Africa was $46 billion, while FDI to Latin America and the Caribbean totaled $147 billion.

FDI offers advantages to both the investor and the foreign host country. Economists say the benefits of FDI include: a more diverse economy, lower labor costs, preferential tariffs, and tax incentives. Additionally, FDI is seen as a key element in international economic integration because it can create links between economies. FDI is also an important channel for the transfer of technology between countries, promotes international trade through access to foreign markets, and can be an important vehicle for economic development for the host country (e.g., stimulating employment or providing access to management expertise, skills, and technology). In low-income countries, FDI is seen as essential for creating and supporting emerging markets where there is a need for private investment in energy, infrastructure, and water to undertake such large-scale projects. There have also been some examples where FDI has reduced the gender pay gap and increased female workforce entry.

Many studies have found that foreign firms tend to provide higher wages, use more technologies, and train their staff more, all of which improve productivity and economic growth within the host country. The presence of foreign competition can lead to the diffusion of technology and knowledge, stimulating domestic firms to improve their technologies, which thus increases their productivity.

A Critical Perspective on Foreign Direct Investment

There are also disadvantages to FDI. Attracting FDI may not be suitable for strategically important industries that could lower the comparative advantage of that country. In addition, foreign investors could behave immorally by avoiding taxes or engaging in corrupt practices; they could sell unprofitable parts of the company or may not reinvest funds in the local community. Some foreign investors may conveniently avoid the burden and costs of health or environmental regulations in their home countries by investing in a country with less stringent regulations. FDI that targets resource extraction often has a large and direct negative environmental effect. In these cases, FDI may cause considerable environmental damage. Some critics note that there is no global investment agreement that regulates FDI.

Nikki Harish and Michael Plouffe have written about the political economy of FDI. They note that while the positive impact of investment on economic growth is well established FDI's relationship with economic growth—as well as development—is more complicated. Their review of the literature suggests that FDI can increase income inequality, effectively leaving the least skilled workers behind. FDI has long been associated with a regulatory "race to the bottom," in which developing countries compete to reduce regulatory burdens on foreign multinational corporations in an effort to attract investment. Any reductions to existing regulations then have the potential to adversely impact the host country's population and resources. They also note that to attract FDI, a government may offer tax incentives to foreign multinational corporations, which potentially brings about more significant negative fiscal effects for the domestic population, as many low-income countries lack welfare or safety nets. Finally, they say the market power of foreign corporations can drive out inefficient domestic companies, due to their potential market dominance as the result of technological expertise.

Despite substantial literature addressing globalization as a positive route to economic growth and integration, Nita Rudra has written that FDI continues to disadvantage the poorest in society, as domestic institutions ignore the welfare of the poor.

SUMMARY AND PROGRESS

The good news is that many countries have integrated the sustainability into policy frameworks and have established institutional and governance mechanisms to evaluate progress on implementation of the SDGs. Partnerships across sectors are forming, while systems for monitoring and evaluation are being put in place. Some critics caution that in many cases governments have no shortages of plans, strategies, and policies. The challenge is often that there are too many projects and partnerships competing for attention and limited government money.

Progress on SDG 17 has been mixed across countries. A 2020 report found that while a majority of governments report on engaging stakeholders in processes related to the implementation and follow-up of the SDGs, many have struggled to set up and maintain open, inclusive, participatory, and transparent processes. Some countries have not yet developed strategies to engage stakeholders, while others are still at an early stage; still other countries have partial or superficial frameworks in place. There are also some unfortunate reversals occurring: Since 2018, ODA has fallen by 3 percent, humanitarian aid fell by 8 percent, and aid to the lowest-income countries and African countries, who need it most, has also declined, rather than increased. These trends were underway prior to the COVID-19 pandemic.

The COVID-19 pandemic is now threatening past achievements, with trade, FDI, and remittances all projected to decline for the next several years. Global merchandise trade is projected to decline by 13 to 32 percent in 2021. FDI is expected to decline by up to 40 percent as a result of delayed investment caused by the shock in global demand, and by a further 5 to 10 percent in 2021. Remittances to low- and middle-income countries—an economic lifeline for many poor households—are predicted to decline by about 20 percent in 2020—almost $100 billion, the steepest decline in recent history. This is largely the result of a fall in the wages and employment of migrant workers, who tend to be more vulnerable than nonmigrants in economic crises in host countries. It is not yet clear whether these impacts will be short term or longer term. The need for urgent action has become clearer in light of the pandemic.

Many experts say that improving global partnerships are absolutely essential for successfully advancing the entire SDG agenda. As has been true of each of the 17 SDGs explored in this text, achievements under any one goal are linked to progress on others, as highlighted in Figure 17.3. For example, cities must coordinate with local organizations and community groups, the private sector, and national governments to advance many of the targets in SDG 11. On the one hand, SDG 17 has strong interconnections to all other SDGs and, on the other hand, partnerships may prove most vital in those goals that require international financing or governance. For many low-income countries, achieving the SDGs will depend on the effectiveness of ODA and public-private partnerships, both of which depend on international collaboration. For example, to achieve targets on health and well-being, many low-income countries will rely on both ODA and public-private partnerships to expand infrastructure and services. Similarly, clean water and sanitation often requires large and expensive sanitation infrastructure, as does transportation and ICT.

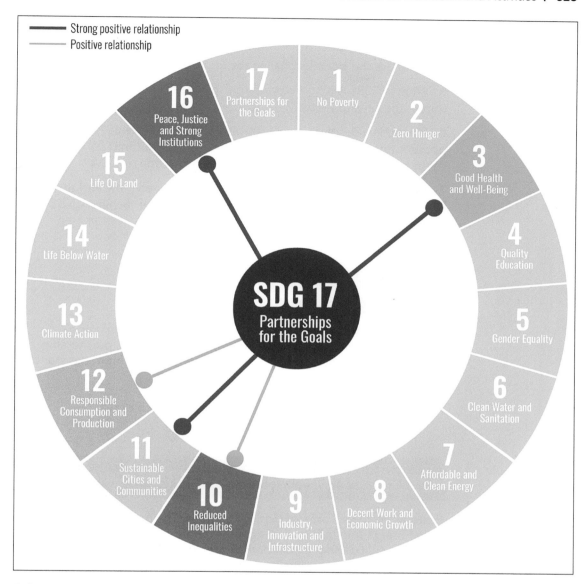

FIGURE 17.3 Interconnections in SDG 17 and the Other Goals
SDG 17 has strong interconnections to all other SDGs, however the role of partnerships may prove most vital in those goals that require international financing or governance such as health, water and sanitation, transportation, and ICT infrastructure.

QUESTIONS FOR DISCUSSION AND ACTIVITIES

1. Explain why the lack of data is impeding efforts at sustainability.

2. Why is debt distress an impediment to sustainability? What do you think the international community should do to address debt distress in the lowest-income countries?

3. Explain the role of the private sector in implementing sustainability. How do we ensure the private sector upholds sustainability principles?

4. What are the benefits/challenges to cross-sector collaboration?

5. What is CSR? Is it "greenwashing," or does it represent a fundamental shift in the private sector?

6. View this video featuring Dambisa Moyo on Foreign Aid and Development (https://www.youtube.com/watch?v=WtleVBnk1Zo). What does she mean by "Dead Aid"? What are the pros/cons to ODA?

7. Many of the strategies for SDG 17 are relatively new and unproven. Which do you think has the best potential for achieving SDG 17 targets?

TERMS

austerity measures
corporate social responsibility (CSR)
cross-sector collaboration
debt distress
debt swap
foreign direct investment

multilateral organization
multistakeholder partnership
official development assistance
public debt
remittance
social impact bond

RESOURCES USED AND SUGGESTED READINGS

Easterly, W. "The Cartel of Good Intentions: The Problem of Bureaucracy in Foreign Aid." *Journal of Economic Policy Reform* 5 (4): 1–28 (2003).

Egan, P. *Globalizing Innovation: State Institutions and Foreign Direct Investment in Emerging Economies.* Cambridge, MA: MIT Press (2017).

Harish, N., and Plouffe, M. *"The Political Economy of Foreign Direct Investment to Developing Countries"* (2020). https://doi.org/10.31219/osf.io/chzpq

Hayter, T. *Aid as Imperialism.* London: Penguin Books (1971).

Hout, W. "Putting Political Economy to Use in Aid Policies." *A Governance Practitioner's Notebook: Alternative Ideas and Approaches* (pp. 1–15). Washington, D.C.: (2015). https://www.oecd.org/dac/accountable-effective-institutions/Governance%20Notebook%201.4%20Hout.pdf

Knudsen, J. "Company Delistings from the UN Global Compact: Limited Business Demand or Domestic Governance Failure?" *Journal of Business Ethics* 103: 331–49 (2011).

Maltais, A., Weitz, N., and Persson, Å. *SDG 17: Partnerships for the Goals. A Review of Research Needs. Technical Annex to the Formas Report Forskning för Agenda 2030: Översikt av forskningsbehov och vägar framåt.* Stockholm: Stockholm Environment Institute (2018).

Organisation for Economic Cooperation and Development. *Better Policies for Sustainable Development 2016: A New Framework for Policy Coherence.* Paris: Organisation for Economic Co-operation and Development *(2016).* http://www.oecd.org/greengrowth/better-policies-for-sustainable-development2016-9789264256996-en.htm

Organisation for Economic Cooperation and Development. *Making Blended Finance Work for the Sustainable Development Goals.* Paris: OECD Publishing (2018). https://www.oecd.org/publications/making-blended-finance-work-for-the-sustainable-development-goals-9789264288768-en.htm

Organisation for Economic Cooperation and Development "Official Development Assistance (ODA)" (2020, April). http://www.oecd.org/dac/financing-sustainable-development/development-finance-standards/What-is-ODA.pdf

Pandya, S. *Trading Spaces: Foreign Direct Investment Regulation, 1970–2000.* New York: Cambridge University Press (2015).

Ratha, D. "What Are Remittances?" *International Monetary Fund* (2021, April). https://www.imf.org/external/pubs/ft/fandd/basics/remitt.htm

Rudra, N. *Globalization and the Race to the Bottom in Developing Countries.* Cambridge: Cambridge University Press (2008).

Sinclair, S., McHugh, N., and Roy, M. J. "Social Innovation, Financialisation and Commodification: A Critique of Social Impact Bonds." *Journal of Economic Policy Reform* 24 (1): 11–27 (2021). https://doi.org/10.1080/17487870.2019.1571415

Toussaint, E., and Millet, D. *Debt, the IMF and the World Bank: Sixty Questions, Sixty Answers.* New York: The Monthly Review (2010).

United Nations. *The Sustainable Development Goals Report 2020.* New York: United Nations (2020). https://unstats.un.org/sdgs/report/2020/The-Sustainable-Development-Goals-Report-2020.pdf

United Nations. *World Economic Situation and Prospects 2020.* New York: United Nations (2020). https://www.un.org/development/desa/dpad/wp-content/uploads/sites/45/WESP2020_FullReport.pdf

United Nations Conference on Trade and Development. *UNCTAD 2020 E-Handbook of Statistics: Foreign Direct Investment.* New York: United Nations (2020). https://stats.unctad.org/handbook/EconomicTrends/Fdi.html

United Nations Economic and Social Affairs, Financing for Sustainable Development Office. *"Financing for Sustainable Development Report 2020."* New York: United Nations (2020). https://developmentfinance.un.org/press-release-financing-sustainable-development-report–2020

United Nations Economic and Social Commission for Asia and the Pacific. *Economic and Social Survey of Asia and the Pacific 2019: Ambitions beyond Growth.* Bangkok: United Nations Economic and Social Commission for Asia and the Pacific (2019). https://www.unescap.org/publications/economic-and-social-survey-asia-and-pacific-2019-ambitions-beyond-growth

United Nations Economic and Social Council. *New Forms of Cooperation and Increased Coherence to Implement the SDGs.* 2016 Development Cooperation Forum Policy Briefs, 15. DCF Secretariat, ECOSOC (2016). https://www.un.org/ecosoc/sites/www.un.org.ecosoc/files/files/en/dcf/dcfbelgium-cooperation-coherence-for-sdgs.pdf

United Nations Inter-agency Task Force on Financing for Development. *Financing for Sustainable Development Report 2019.* New York: United Nation (2019). https://developmentfinance.un.org/fsdr2019

United Nations, the UN Global Compact. *Transforming Partnerships for the SDGs.* New York: United Nations (2018). https://d306pr3pise04h. cloudfront.net/docs/publications%2FTransforming_Partnerships_for_ the_SDGs.pdf

Walker, J., Pekmezovic, A., and Walker, G. *Sustainable Development Goals: Harnessing Business to Achieve the SDGs through Finance, Technology and Law Reform.* Hoboken, NJ: Wiley (2019).

Wayne-Nixon, L., Wragg-Morris, T., Mishra, A., Markle, D., and Shannon, K. "Effective Multi-stakeholder Engagement to Realize the 2030 Agenda." *Good Practice in 2030 Agenda Implementation Series.* Vancouver and Ottawa: British Columbia Council for International Cooperation and Canadian Council for International Cooperation (2019). https://www.bccic.ca/wp-content/uploads/2019/04/Effective_ Engagement_Canada.pdf

Reflections

Our world is marked by tremendous inequality across many dimensions, including economic, political, social, and environmental inequalities. A long history of economic development has led to unsustainable practices, behaviors, and cultural norms that have caused environmental degradation and exploitation, and an uneven distribution of wealth and resources. Moving toward sustainability will not be quick, or easy. It may be the most significant challenge of the twenty-first century.

The concepts of sustainability and sustainable development have become global buzzwords—and, they now form a discourse of change. Sustainability is complex and multidimensional. There are many viewpoints from which to examine sustainability and sustainable development. This textbook has used the UN Sustainable Development Goals (SDGs) as a framework. Yet these seventeen SDGs are not exhaustive or the only vantage point from which to understand sustainability and sustainable development. On the one hand, the SDGs offer a way to examine numerous different areas of sustainability, from poverty to health to biodiversity and human rights. On the other hand, the SDG framework is not without contradictions and limitations. There are many ways the SDGs are missing vital ideas, topics, and actions. It is fitting that we reflect on these in a conclusion of sorts.

WHAT IS MISSING WITHIN THE SUSTAINABLE DEVELOPMENT GOALS FRAMEWORK?

Since the launch of the SDGs in 2015, scholars and development practitioners have criticized the SDGs. Some assert the seventeen SDGs lack coherence by being too vague and too numerous. One claim is that some targets should have been constructed more clearly. For example, SDG 17 has been criticized for having vague or incomplete targets around the important role of the private sector. It offers no measurable targets that address exactly how corporations can help advance sustainable development. Similarly, one way to reduce war and conflict is to end the nearly $40 billion worth of arms sold each year. Yet SDG 16 is silent on this topic, overlooking this multibillion dollar industry. Health scholars have said that health care should have been included as a human right (with direct and explicit targets in SDG 3 and SDG 16). Human rights advocates argue that more should have been included in human rights in SDG 10 and SDG 16, particularly around LGBTQ+ issues.

There are also criticisms that the SDGs do not have meaningful actions or policies associated with them. For example, marine biologists David Osborn and Sam Dupont say that the SDGs do not effectively address ocean acidification because the targets fail to provide clear communication about policy changes that are needed. Although SDG 14 recognizes the gravity of the problem, articulates its target very precisely, and even offers the right indicator for tackling the symptoms of ocean acidification, the *tools for action* are missing or not adapted to the global scale of the problem. They argue that to reach the SDG target on ocean acidification, we need policy changes at the global level (to limit greenhouse gas emissions) and change in individual behaviors through education (to accelerate adaptation by the people affected the most by acidification).

WHAT IS MISSING: SUSTAINABLE DEVELOPMENT GOAL 18 AND MORE?

It may also be that seventeen SDGs are not sufficiently comprehensive and so do not cover the critical issues of sustainability and sustainable development. What other SDGs are missing? What should have been SDG 18 or SDG 19?

In 2019 the Over Population Project held a conference on the missing SDGs, featuring a number of experts from different disciplines. One possible missing SDG is population growth. Jenna Dodson and Patricia Derer, of the Over Population Project, highlighted the role of population growth in driving biodiversity loss and climate change, noting that the SDGs do not address population growth explicitly. They say if population increase continues, it will undermine achievement of the SDGs, particularly SDG 1, SDG 2, SDG 14, and SDG 15. However, others have countered that while population growth is not an explicit goal, it is "implied" in several targets around health and gender.

Other speakers noted that air quality, harmful chemicals, and democratic governments have been neglected and may deserve to be their own goals. Air pollution is particularly noteworthy because about eight million people die annually from air pollution. Yet air quality is included only implicitly under several other SDGs such as SDG 3 Health, SDG 7 Energy, and SDG 13 Climate Change. Molecular biologist Erik Thomson has argued that air should be a separate, independent goal with the following proposed four targets:

- 18.1 reduce air pollution,
- 18.2 reduce indoor air pollution,
- 18.3 end uncontrolled burning of toxic materials, and
- 18.4 update metrics for air-quality assessment.

Biologist Thomas Backhaus has argued that harmful chemicals are neglected within the SDGs. He notes how little we know about chemical safety. Yet the SDGs do not cover chemical safety or include targets for achieving a "nontoxic environment."

The SDGs conference also featured human rights scholar Johan Karlsson Schaffer who argues that SDG 18 should have been "Strengthen Democracy and End Autocracy." Although SDG 16 promotes access to justice and more inclusive societies and institutions, he says it fails to include some essential components such as democracy, parliaments, parties, free elections, and freedom of expression.

Another example of a "missing SDG" concerns culture. Traditions and practices and creative expressions are not found specifically in the SDGs, yet culture and international development have long been recognized as having a reciprocal and interdependent connection. In 2020, the British Council published a report titled "The Missing Pillar—Culture's Contribution to the Sustainable Development Goals." The report argues that arts and culture are important in the development context and that many countries have begun to view culture as an asset in eradicating poverty, addressing social inclusion and inequality, and creating economic growth. The report further says that enhanced trade in cultural goods and services strengthens local and national markets and can contribute to the empowerment and inclusion of all people, irrespective of age, sex, disability, race, ethnicity, origin, religion, or economic status. Even though the SDGs did not include culture as a stand-alone goal, the 2030 agenda includes several explicit references to aspects of culture. They argue that not only should culture be its own goal but also it should be considered an equal pillar alongside environment, economy, and equity.

Another missing stand-alone goal is environmental justice. Mary Menton and colleagues found that although equity and inequalities are fundamentally part of the SDGs, in fact not a single SDG explicitly incorporates principles of environmental justice. As discussed in the "Introduction," environmental justice is integral to sustainability to ensure no one group is disproportionately disadvantaged by public policy. The lack of targets and goals that recognize the right of every person to freely access natural resources and to meaningfully participate in decisions that affect their environment is concerning. Jessica Hope has researched Indigenous land struggles in Bolivia. She argues that the SDGs do not adequately address environmental justice, and so weaken Indigenous struggles for territorial sovereignty. She maintains that the SDG framework is accessible only to large state actors and organizations such as the Bolivian government, which used the SDG framework to instead consolidate power over land to be used for road infrastructure, at the expense of Indigenous groups that opposed the road project. The lack of targets that explicitly use the term "environmental justice," and perhaps the lack of environmental justice as a stand-alone SDG, reduces the transformational capacity of the goals.

WHAT IS MISSING: FINANCING THE SUSTAINABLE DEVELOPMENT GOALS

What is also missing from the SDG agenda is a reliable funding mechanism. Even before COVID-19, financing for the SDGs was insufficient.

Sources of financing for sustainable development come from two broad categories. One is from a country's domestic sources and includes tax revenues or private savings. The second is foreign: private inflows through FDI, remittances, public inflows from loans, and official development assistance. There is little doubt that it will cost trillions of dollars to achieve the SDGs. This cannot be done by government/public financing alone.

Some SDG targets require longer-term investment that must come from governments. However, many of the goals will require complementing public-sector resources with increased private-sector investments, including FDI, and blended finance (public-private partnerships). Blended finance through public-private partnerships is likely to advance some SDGs more than others. This is particularly true with respect to infrastructure investment (energy, water, communication, and transportation), where there is a significant potential for profits. But it is crucial that more resources are mobilized for lower-income countries; in reality all countries need to invest more in the SDGs.

In 2015, the Addis Ababa Action Agenda was designed to provide a foundation for financing the SDGs. The Addis agenda was different from the previous two agreements on financing for development that were stand-alone processes (the 2002 Monterrey Consensus and the 2008 Doha Declaration) because the Addis agenda harmonized with the SDGs. The Addis agenda contains more than one hundred concrete measures, addresses numerous sources of finance, and covers cooperation on a range of issues including technology, science, innovation, trade, and capacity building. The document encourages countries to establish a range of measures aimed at widening the revenue base, improving tax collection, and combating tax evasion and illicit financial flows. High-income countries that signed the Addis agenda also reaffirmed their commitment to official development assistance, particularly for the lowest-income countries, and pledged to increase South-South cooperation. The Addis agenda also had targets for aligning private investment with sustainable development.

Despite the Addis Ababa Action Agenda, there has been a significant finance gap. Djeneba Doumbia and Morten Lykke Lauridsen set out to understand how big the financing gap is. Their research suggests that meeting the SDGs will require the global community to increase development financing from "billions" to "trillions." Their study noted significant investment gaps, with some of the largest funding needs related to economic infrastructure. At up to $950 billion, power infrastructure carries the greatest financing need, followed by climate change mitigation ($850 billion) and transport infrastructure ($770 billion). There are also sizeable investment gaps in social infrastructure, ranging from $140 billion in health to $250 billion in education.

Another 2019 study by Homi Kharas and John McArthur found that the lowest-income countries have the largest SDG financing gaps, not a surprising finding, but certainly one that makes achieving the targets even more challenging. Their study estimated that meeting the SDGs in education, health, roads, electricity, water, and sanitation will require additional private and public *annual* spending of $528 billion for low- and lower-middle-income countries. Overall investments would need to be on the order of 7.2 percent of GDP. Currently, average investment ranges from 2.5 percent GDP in sub-Saharan Africa to 5.7 percent in East Asia. These estimates mean the *annual* financing gap may be as large as $2.5 trillion.

The challenge of financing the SDGs is not just about mobilizing more money but also is more systemic and much deeper than closing any financial gap. Financing is theoretically available, given the size, scale, and level of the global financial system—with gross world product and global gross financial assets estimated at more than $85 trillion and $200 trillion, respectively. Yet the financial flows, mostly private, do not (or cannot) reach the places and people that need them the most. Most private investments are still not channeled toward sustainable development at the scale and speed required. This means that not only is money the issue, so too is the political will to invest it equally across each of the SDGs and to invest in people and places that need it the most. Without this commitment, the SDGs are threatened with failure.

Clearly, more needs to be done to finance the SDG agenda.

WHAT IS MISSING: STRUCTURAL TRANSFORMATION

The shift to sustainable development and the SDG framework is welcomed by mainstream development actors as an important step in addressing environmental degradation. However, for many political economists and political ecologists the SDG framework is an ineffective resolution to tensions between development and the environment. It does not fundamentally transform capitalist structures.

One of the glaring contradictions of the SDG framework is its pro-economic-growth approach. Economist Partha Dasgupta has written that there is a fundamental flaw in thinking that healthy rates of economic growth will result in sustainable development. The problem with measuring growth—and by extension several of the targets of the SDG—is that growth will be measured by GDP. GDP measures goods and services, but it does not measure the way in which a good is produced, and it does not factor in human well-being or the environment. Dasgupta says this can set up the potential for false comfort in the SDGs.

Others are also concerned about the progrowth approach. Hayden Washington and Paul Twomey contend that the SDGs need to be decoupled from economic-growth targets, otherwise they are only partial solutions. Jason Hickel has highlighted the contradictions that he sees in the economic growth targets in SDGs 8 and 9 with the targets to protect the environment in SDGs 6, 13, 14, and 15.

Even more concerning for some is that that the SDGs do not explicitly challenge individuals or society to transform fundamental cultural values that influence how we perceive our relationship with the natural environment. Genuine sustainability will require changing the anthropocentric mindset that has dominated the world for the past several hundred years. The SDGs and the global agenda are seen as "greenwashing" because most efforts to address sustainability under the SDG framework will result in small reforms, not the radical transformation of the market-oriented global capitalist system and the cultural value system that places economic growth ahead of other interests.

Olga Korestkaya and Lauran van Oers, for example, contend the SDGs framework fails to address the structural or root causes of many of the issues. They note that SDG 1 (No Poverty) does not address any of the structural factors that cause poverty, such as overconsumption, increased inequality, labor exploitation, dispossession, and ecosystems' destruction. If SDG 1 addressed structural causes, a number of stronger governmental approaches needed to be proposed, for instance, reducing and eliminating uneven agricultural subsidies, recognizing peasant sovereignty over seeds, stopping all land grabbing, and regulating speculation on agricultural commodities and land reform. Structural solutions call for the dismantling of the international political and economic arrangements that systematically benefit the wealthy and disenfranchise the poor.

Although some political ecologists might contend that because of the shortcomings of the SDG framework it is irrelevant, this book contends that it is still crucial to understand how sustainable development is currently operationalized and implemented in a world that remains oriented to capitalism. It is important for anyone interested in sustainability to understand how this is implemented within the existing political, economic, and social systems, and to what degree progress is realized, all while retaining a critical perspective. This is because we are unlikely to achieve the goals by 2030—certainly another initiative will be launched; we need to learn what worked and what did not.

This book has attempted to balance the SDG framework with a critical perspective. It has introduced key ideas in sustainability and sustainable development through the SDG framework, while also providing a critical perspective on the way in which the SDG framework may not make the fundamental transformations needed. In short, it has had its cake and eaten it too, all while recognizing this paradox.

WHAT IS MISSING: YOU

The SDGs are important, world-changing objectives that will require cooperation among governments, international organizations, and world leaders. The SDG framework implies access and accessibility at the scale of state governments and international organizations but may lack obvious entrée points for smaller-scale communities and individuals. Are you—the individual—missing in the global agenda?

It may seem unlikely that the average person can make an impact, but sustainability involves being aware of how decisions and day-to-day activities impact the environment as well as those who come after us.

Students like you have the power to enact positive change, not just in university communities but also in the world. There are many examples: Teresa Cheng first became involved in labor rights activism at the University of Southern California coordinating the United Students against Sweatshops. The organization went on to win several legal cases for workers' rights. Duke University students pressured the university's board of trustees to reform guidelines on investment, transforming a $6 billion endowment into socially responsible investments. Boston University students organized a sit-in to demonstrate support for gender-neutral housing and eventually won their campaign for gender-neutral housing on campus. On many campuses, Fossil Free and Sunrise student movements are impacting how universities invest their money. To date, more than a hundred universities in the United States, the United Kingdom, and elsewhere have pledged to divest from the fossil fuel industry, including Cornell University, Oxford University, Syracuse University, and the entire University of California system. These student movements have influence beyond the campus, encouraging companies and governments to make similar pledges. Student activism campaigns are alive and well today.

TABLE 18.1	17 Actions You Can Take to Advance Sustainability
1	Get involved with sustainability efforts on your college campus, or join regional or national organizations
2	Make sustainability a focus of an independent project or thesis
3	Combat climate change by renting, borrowing or sharing a bike
4	Walk
5	Unplug your electronic devices and have a "power down" day where you curtail use of electronics
6	Listen to your ecosystem
7	Learn about where your food comes from and how it is grown
8	Take care of nature wherever it is found
9	Watch your water usage—reducing shower time by even a few minutes on a regular basis saves water and energy
10	Recycle, compost, and reuse
11	Be a smarter shopper and buy secondhand
12	Eat less meat and reduce your food waste
13	Vaccinate yourself and your family from disease
14	Appreciate what you have so you consume less
15	Volunteer for and support mentoring and tutoring programs in your area aimed at keeping girls in school
16	Interrogate assumptions about what sustainability means
17	Vote

In addition to activism, keep learning. Many colleges have political economy or political ecology research groups (and some have majors in political economy); join, learn, and then transform! For example, UC Berkeley offers a major in political economy as do Williams College and Kings College London. Or you can go mainstream and learn more through the SDG Academy, an online educational initiative that offers free courses and book clubs (unsdsn.org/sdg-academy).

Beyond activism and education, there are still many actions that you can take to advance sustainability. Table 18.1 lists these actions. Some of these are simple changes to your daily habits; others will require more investment of your time and energy.

"We don't have to engage in grand, heroic actions to participate in change. Small acts, when multiplied by millions of people, can transform the world."

—Howard Zinn, historian, philosopher, and playwright

RESOURCES TO HELP YOU LEARN MORE
AND STAY ENGAGED

Association for the Advancement of Sustainability in Higher Education (AASHE): The go-to resource for campus sustainability information, ideas, initiatives, and challenges is the AASHE. The AASHE has an annual conference, a newsletter, an online guidance center, an assessment and tracking tool, and links to everything that has anything to do with college sustainability. *https://www.aashe.org/*

AASHE List of State and Regional Campus Sustainability Organizations: Along with state coalitions, there are larger regional organizations on AASHE's list. Many are focused on facilitating communication and collaboration between universities and higher-level administration. *https://hub.aashe.org/*

Cooperative Food Empowerment Directive: Offers tools and guidance to student groups to bring cooperatively run, ethically sourced food options to campus. *https://www.cofed.coop/*

Energy Action Coalition: An alliance of thirty organizations that support student clean energy initiatives, including the grassroots Campus Climate Challenge. *https://www.powershift.org/*

Engineers for a Sustainable World: An organization dedicated to integrating sustainability into engineering curricula, and to helping students and faculty implement practical sustainability plans and projects. *http://www.eswusa.org/about-us*

Focus the Nation: A clean energy leadership development organization dedicated to fostering agents of change among college students. *http://thebaumfoundation.org/education/focus-nation*

Fossil Free USA: Offers tools, resources, and guidance to anyone looking to start a campus fossil free organization. *https://gofossilfree.org/USA/*

National Wildlife Federation's Campus Ecology Program: The National Wildlife Federation's student outreach program, offering student consulting on sustainability issues, educational outreach and resources, and a climate action competition. *https://www.nwf.org/Campus-Ecology.aspx*

Net Impact: An alliance of young professionals, grad students, and business majors dedicated to confronting sustainability challenges. *https://www.netimpact.org/home?action*

Roots and Shoots: The Jane Goodall Institute's international environmental and humanitarian program for youth of all ages. *https://www.rootsandshoots.org/*

Sierra Student Coalition: The student arm of the Sierra Club, a progressive environmental organization that was established by famed naturalist John Muir in 1892. *https://www.sierraclub.org/youth*

SustainUS: A nonpartisan, nonprofit organization dedicated to fostering social, economic, and environmental sustainability. *https://sustainus.org/*

United Students for Fair Trade: A cooperatively run student group promoting fair trade principles and models. *https://fairworldproject. org/fair-world-project-fwp-united-students-for-fair-trade-usft-announce-formal-partnership/*

RESOURCES USED AND SUGGESTED READINGS

British Council. "The Missing Pillar—Culture's Contribution to the Sustainable Development Goals" (2020). https://www.britishcouncil. org/sites/default/files/the_missing_pillar.pdf

Dasgupta, P. "What's Missing from the SDGs?" DevEx (2016, May). https://www.devex.com/news/what-s-missing-from-the-sdgs–88207

Doumbia, D., and Lauridsen, M. "Closing the SDG Financing Gap—Trends and Data." *Fresh Ideas about Business in Emerging Markets Compass* 73: 1–8 (2019, October). https://www.ifc.org/ wps/wcm/connect/842b73cc-12b0-4fe2-b058-d3ee75f74d06/EMCompass-Note-73-Closing-SDGs-Fund-Gap. pdf?MOD=AJPERES&CVID=mSHKl4S

Hickel, J. "The Contradiction of the Sustainable Development Goals: Growth Versus Ecology on a Finite Planet." *Sustainable Development* 27: 1–12 (2019, February). https://doi.org/10.1002/sd.1947

Hope, J. "Globalising Sustainable Development: Decolonial Disruptions and Environmental Justice in Bolivia." *Area* (2020, April). https://rgs-ibg.onlinelibrary.wiley.com/doi/epdf/10.1111/area.12626

Kharas, H., and McArthur, J. "How Much Does the World Spend on the Sustainable Development Goals?" *Future Development.* Washington, DC: Brookings Institution (2019). https://www.brookings.edu/blog/ future-development/2019/07/29/how-much-does-the-world-spend-on-the-sustainable-development-goals//

Korestskaya, O., and van Oers, L. "Sustainable Development Goals: Neglected Debate on Economic Growth." *Science for Everyone* (2020, May). https://ontgroei.degrowth.net/sustainable-development-goals-neglected-debate-on-economic-growth/

Menton, M., Larrea, C., Latorre, S., Martinez-Alier, J., Peck, M., Temper, L., and Walter, M. "Environmental Justice and the SDGs: From Synergies to Gaps and Contradictions." *Sustainability Science*: 1–16 (2020).

Osborn, D., Dupont, S., Hansson, L., and Metian, M., "Ocean Acidification: Impacts and Governance." In P. Nunes, L. Svensson, and A. Markandya (eds.). *Handbook on the Economics and Management of Sustainable Oceans* (pp. 396–415). Cheltenham, UK: Edward Elgar Publishing (2017).

The Overpopulation Project. "The Missing SDG—What Should Be Added?" (2019, November). https://overpopulation-project.com/the-missing-sdg-what-should-be-added/

Washington, H. "Questioning the Assumptions, Sustainability and Ethics of Economic Growth." *Journal of Risk Financial Management* 14 (497): 1–15 (2021). https://doi.org/10.3390/jrfm14100497

Washington, H., and Twomey, P. *A Future beyond Growth: Towards a Steady State Economy.* London: Routledge (2016).

GLOSSARY

absolute poverty: The state where one is barely able, or unable, to afford basic necessities. (ch. 1)

access (to health services): The perceptions and experiences of people as to their ease in reaching health services or health facilities in terms of location, time, and ease of approach. (ch. 3)

accessibility (of health services): Aspects of the structure of health services or health facilities that enhance the ability of people to reach a health-care practitioner, in terms of location, time, and ease of approach. (ch. 3)

acidification: Ongoing decrease in pH away from neutral value of 7. Often used in reference to oceans, freshwater, or soils, as a result of uptake of carbon dioxide from the atmosphere (see *ocean acidification* for a specific definition). (ch. 14)

acute undernutrition: Short-term food insecurity due to wars, natural disasters, or displacement of people. (ch. 2)

adaptation (climate change): Adjustment to environmental conditions. Used in the context of climate change to indicate the process of adjusting to current or expected climate change and its effects. (ch. 13)

advanced treatment systems: Any treatment of sewage that goes beyond the secondary or biological water treatment stage and includes the removal of nutrients such as phosphorus and nitrogen and a high percentage of suspended solids. (ch. 14)

aerobic: Having molecular oxygen (O_2) as a part of the environment, or a biological process that occurs only in the presence of molecular oxygen. (ch. 14)

aerobic treatment: A process by which microbes decompose complex organic compounds in the presence of oxygen and use the liberated energy for reproduction and growth. (Such processes include extended aeration, trickling filtration, and rotating biological contactors.) (ch. 14)

afforestation: Planting of new forests on lands that historically have not contained forests. (ch. 15)

agroecology: The science and practice of applying ecological concepts, principles, and knowledge (e.g., the interactions of, and explanations for, the diversity, abundance, and activities of organisms) to the study, design, and management of sustainable agroecosystems. It includes the roles of human beings as a central organism in agroecology by way of social and economic processes in farming systems. Agroecology examines the roles and interactions among all relevant biophysical, technical, and socioeconomic components of farming systems and their surrounding landscapes. (ch. 15)

anaerobic: Absence of molecular oxygen (O_2) as a part of the environment, or a biological process that occurs in the absence of molecular oxygen; bound oxygen is present in other molecules, such as nitrate (NO3–), sulfate (SO4+), and carbon dioxide (CO_2). (ch. 14)

anaerobic decomposition: The reduction of the net energy level and change in chemical composition of organic matter caused by microorganisms in an oxygen-free environment. (ch. 14)

Anthropocene: A proposed term for the present time interval, which recognizes humanity's profound imprint on and role in the functioning of the Earth system. The term has evolved in breadth and diversity, now ranging from a proposed definition of a new geological epoch, a widely used metaphor for global change, a novel analytical framework, a meme about the relationship of society to nature, and the framing for new and contested cultural narratives. Different starting periods have been proposed for the geological definition of the Anthropocene, including early agriculture and domestication, colonial species exchange, the onset of the Industrial Revolution, nuclear bomb deployment in 1945, and the post–World War II period characterized by the great acceleration of global changes and the widespread adoption of fossil fuels. (ch. 13)

anthropocentrism: Interpreting the world exclusively in terms of human values and experiences. Seeing humans as the center of the universe and basing all one's actions on furthering only human interests. Utilizes resources from the ecosystem around us irrespective of the negative consequences. (Intro.)

aquaculture: The farming of aquatic organisms, including fish, mollusks, crustaceans, and aquatic plants, in both inland and coastal areas, and involving some form of intervention in the rearing process to enhance production, such as regular stocking, feeding, protection from predators, and so forth. (ch. 14)

aquifer: A permeable rock layer through which groundwater flows easily. (chs. 6, 14)

asylum seeker: An individual that has been forced to flee their own country for the safety of another but has yet to receive any legal recognition or status. (ch. 16)

atmospheric pollution: Occurs in many forms but can generally be thought of as gaseous and particulate contaminants that are present in the Earth's atmosphere. Chemicals discharged into the air that have a direct impact on the environment are called primary pollutants.

These primary pollutants sometimes react with other chemicals in the air to produce secondary pollutants. Air pollution is separated into two categories: outdoor air pollution and indoor air pollution. (ch. 14)

austerity measures: Refer to strict economic policies implemented by a government to reduce government spending and public debt. Governments implement austerity measures when their public debt is so large that the risk of default or the inability to service the required payments on its obligations becomes a real possibility. (ch. 17)

basic education: Basic education comprises primary education (first stage of basic education) and lower secondary education (second stage). It also covers a wide variety of nonformal and informal public and private activities intended to meet the basic learning needs of people of all ages. (ch. 4)

Belt and Road Initiative: A transcontinental long-term policy and investment program that aims at infrastructure development and acceleration of the economic integration of countries along the route of the historic Silk Road. The initiative was unveiled in 2013 by China's President Xi Jinping and until 2016, was known as OBOR: One Belt One Road. According to the Belt and Road Initiative's official web site, the initiative aims to "promote the connectivity of Asian, European and African continents and their adjacent seas, establish and strengthen partnerships among the countries along the Belt and Road, set up all-dimensional, multi-tiered and composite connectivity networks, and realize diversified, independent, balanced and sustainable development in these countries." (ch. 9)

bioaccumulation: The continuous accumulation of foreign substances, such as pesticides or toxic chemicals, within an organism, relative to that in the environment. (ch. 15)

biocentrism: The belief that all living things have an inherent value and that human beings do not have an inherent value more than other species. As human beings, we are morally obliged to preserve all living beings because of the fact that we all belong to the same planet, holding the same intrinsic values. (Intro.)

biochemical oxygen demand (BOD): A measure of the amount of oxygen required or consumed for the microbiological decomposition (oxidation) of organic material in water. The purpose of this indicator is to assess the quality of water available to consumers in localities or communities for basic and commercial needs. It is also one of a group of indicators of ecosystem health. (ch. 14)

biodiversity (biological diversity): The variability among living organisms from all sources including, terrestrial, marine, and other aquatic ecosystems and the ecological complexes of which they are part; this includes diversity within species, between species, and of ecosystems. (ch. 15)

biodiversity conservation: The management of human interactions with genes, species, and ecosystems so as to provide the maximum benefit to the present generation while maintaining their potential to meet the needs and aspirations of future generations; encompasses elements of saving, studying, and using biodiversity. (ch. 15)

biofortification: Is the process by which the nutritional quality of food crops is improved through agronomic practices, conventional plant breeding, or modern biotechnology (e.g., wheat with zinc or golden rice with vitamin A). (ch. 2)

biofuel: Any fuel that is derived from organic materials (usually plant or animal), often used to refer specifically to organically derived liquid fuels used to replace fossil fuels for transportation, although wood and other biomass burned directly is also a biofuel. (ch. 7)

biomagnification: The process by which a compound (such as a pollutant or pesticide) increases its concentration in the tissues of organisms as it travels up the food chain or food web. (ch. 14)

biomass: Renewable organic material that comes from plants and animals. Biomass sources of energy include wood and wood-processing wastes, agricultural crops (e.g., sugar cane, soybeans, algae), biogenic materials in waste (e.g., paper, cotton, food or yard wastes), animal manure, and human sewage. (ch. 7)

biome: A set of naturally occurring communities of plants and animals occupying an environmental and/or climatic domain, defined on a global scale. Terrestrial biomes include desert, grassland, tropical rainforest, deciduous and coniferous forests, and tundra. (ch. 15)

bioretention: A stormwater management practice that is designed to provide both temporary surficial water storage and runoff retention subsurface in soil media. Runoff is directed to shallow depressions where it is infiltrated, filtered, or evapotranspirated. These systems are typically designed with a soil media selected to promote infiltration and runoff retention and are vegetated with plants picked to withstand both inundation and drought. (ch. 14)

bioswale: Relatively wide, shallow, open channel, typically vegetated with turf grasses, with a slight gradient. These systems are designed to let water flow slowly through the turf grasses. The roughness of the turf slows the runoff velocity and provides some filtration and settling of suspended solids. Runoff volumes can also be reduced through infiltration depending on the porosity of the underlying soils. Swales can be designed with underdrains to convey excess runoff from saturated soils. (ch. 14)

blue carbon: The carbon stored in marine and coastal ecosystems. Blue carbon ecosystems can store ten times more carbon than tropical rainforests. (ch. 14)

British thermal unit: The amount of energy needed to raise one pound of water by one degree Fahrenheit, used primarily in the US energy context. (ch. 7)

brownfield: An abandoned, idled, or underused industrial and commercial facility/site where expansion or redevelopment is complicated by real or perceived environmental contamination. They are found in urban, suburban, and rural areas. (ch. 11)

Brundtland Report: (Also called *Our Common Future*) A report published in 1987 on behalf of the UN World Commission on Environment and Development. It is cited as providing the globally accepted definition of sustainable development: Development that meets the needs of the present without compromising the ability of future generations to meet their own needs. (Intro.)

burden of disease: The impact of a health problem as measured by financial cost, mortality, morbidity, or other indicators. It is often a measurement of the gap between current health status and an ideal situation in which everyone lives into old age, free of disease and disability. (ch. 3)

cap-and-trade: An economic policy instrument in which the state sets an overall environmental target (the cap) and assigns environmental impact allowances (or quotas) to actors that they can trade among each other.

carbon cycle: The flow of carbon (in various forms, e.g., as carbon dioxide [CO_2]) through the atmosphere, ocean, terrestrial and marine biosphere, and lithosphere. (ch. 13)

carbon dioxide (CO_2): A colorless, odorless gas produced by burning carbon and organic compounds and by respiration. It is naturally present in air (about 0.03 percent) and is absorbed by plants in photosynthesis. (ch. 13)

carbon footprint: A measure of the emission of gases that contribute to heating the planet in carbon dioxide (CO_2)-equivalents per unit of time or product; there is no universally accepted definition of the term. (ch. 11)

carbon neutrality: Refers to any process by which the net effect on atmospheric carbon dioxide is zero. (ch. 7)

carbon offset: A reduction in GHG emissions—or an increase in carbon storage (e.g., through land restoration or the planting of trees)—that is used to compensate for emissions that occur elsewhere. (ch. 7)

carbon sequestration: The long-term storage of carbon in plants, soils, geologic formations, and the ocean. Carbon sequestration occurs both naturally and as a result of anthropogenic activities. (ch. 13)

carbon sink: Any process, activity, or mechanism that removes carbon dioxide from the atmosphere. Using vegetation to store carbon dioxide in woods, roots, leaves, and starches, and to release oxygen. Fossil fuels are a carbon sink, but humans release these voluntarily. (ch. 13)

child labor: Work that deprives children of their childhood, potential, and dignity, and that is harmful to physical and mental development. Note that not all work done by children will be classified as child labor. Children's or adolescents' participation in work that does not affect their health and personal development or interfere with their schooling, is generally regarded as being something positive. (ch. 8)

child marriage: Any marriage where at least one of the parties is under eighteen years of age. (ch. 5)

chronic undernutrition: Refers to long-term undernutrition (see also *undernutrition*). (ch. 2)

circular economy: Based on the principles of designing out waste and pollution, keeping products and materials in use, and regenerating natural systems. In a circular economy, waste does not exist. (ch. 12)

civil society: Refers to a wide array of organizations: community groups, nongovernmental organizations, labor unions, Indigenous groups, charitable organizations, faith-based organizations, professional associations, and foundations. (ch. 16)

climate change: As defined in Article 1 of the United Nations Framework Convention on Climate Change (UNFCCC), "a change of climate which is attributed directly or indirectly to human activity that alters the composition of the global atmosphere and which is in addition to natural climate variability observed over comparable time periods." (ch. 13)

climate denial: Rejection of the proposition that climate change caused by human activity is occurring or that it constitutes a significant threat to human welfare and civilization. (ch. 13)

coagulation: In wastewater treatment, coagulation is a process used to neutralize charges and form a gelatinous mass to trap (or bridge) particles, thus forming a mass large enough to settle or be trapped in the filter. (ch. 6)

coal: Fossil fuel composed of carbon and other elements, which is formed from decomposed plant materials that are trapped under the earth and have undergone intense heat and pressure for thousands to millions of years. (ch. 7)

combined sewer overflow (CSO): A discharge of a mixture of stormwater and domestic waste when the flow capacity of a sewer system is exceeded during rainstorms. (chs. 6, 14)

combined sewer systems (CSS): A system that collects rainwater runoff, domestic sewage, and industrial wastewater into one pipe. (ch. 6)

communicable disease: An illness caused by an infectious agent or its toxins that occurs through the direct or indirect transmission of the infectious agent or its products from an infected individual or through an animal, vector, or the inanimate environment to a susceptible animal or human host. (ch. 3)

community-supported agriculture (CSA): A way for the public to create a relationship with a farm and receive a weekly basket of produce. (ch. 2)

Comprehensive Health Services: Health services that are managed so as to ensure that people receive a continuum of health promotion, disease prevention, diagnosis, treatment and management, rehabilitation, and palliative care services, through the different levels and sites of care within the health system, and according to their needs throughout their lives. (ch. 3)

conservation: The careful preservation and/or protection of a natural resource or ecosystem. It involves the planned management to prevent exploitation, destruction, or neglect. (ch. 15)

contagious: Capable of being transmitted from one person to another by contact or close proximity. (ch. 3)

coral bleaching: When water is too warm, corals will expel the algae (zooxanthellae) living in their tissues, causing the coral to turn completely white. Corals can survive a bleaching event, but they are under more stress and are subject to mortality. (ch. 14)

core nations (also Global North): Core nations are dominant capitalist countries, highly industrialized, technological, and urbanized. (Intro.)

corporate social responsibility (CSR): Companies conduct their business in a way that is ethical, taking account of their social, economic and environmental impact, and consideration of human rights. (Intro.; chs. 12, 17)

corruption: Dishonest behavior by those in positions of power, such as managers or government officials for private gain. Corruption can include giving or accepting bribes or inappropriate gifts, double-dealing, under-the-table transactions, manipulating elections, diverting funds, laundering money, and defrauding investors. (ch. 16)

cradle-to-cradle (C2C): Is about seeing garbage (or waste) as an eternal resource, used within a circular system, rather than a linear system. In practical terms, C2C requires products to be designed in such a way to ensure that all materials can be classified into one of two cyclical systems: (1) biological system, where materials that naturally biodegrade can be returned to the ecological system or (2) technological system, where materials that do not naturally biodegrade can be recycled or reused. (ch. 12)

cross-sector collaboration (also multistakeholder partnerships): A term used to describe a process where various community organizations come together to collectively focus their expertise and resources on a complex issue of importance to a community they serve. (ch. 17)

cultural values: Shared social values and norms, which are learned and dynamic, and which underpin attitudes and behavior and how people respond to events and opportunities, and affects the hierarchy of values people assign to objects, knowledge, stories, feelings, other beings, forms of social expressions, and behaviors. (Intro.)

dead zones (also hypoxia): Less oxygen dissolved in the water is often referred to as a "dead zone" because most marine life either dies or, if they are mobile such as fish, leave the area. (ch. 14)

death rate: The ratio of total deaths to total population over a specified period. (ch. 3)

debt bondage (also debt slavery and peonage): Workers are told they can pay off a loan of their own or of a family member by working it off. The work is often difficult and imposed under brutal circumstances. Many find that repayment of the loan is impossible. Then, their enslavement becomes permanent. Debt-bonded labor is used across a variety of industries to produce products for consumption around the world. (ch. 8)

debt default: Missing a debt payment when it is due. Often contracts allow for a "grace period" after a payment is missed before a default is officially declared—often around one month. A default can be a full default—on all payments coming due—or a partial default—on just some of the debt, usually depending on who the creditor is. (ch. 17)

debt distress: When the country is already experiencing difficulties in servicing its debt, as evidenced, for example, by the existence of arrears, ongoing or impending debt restructuring, or indications of a high probability of a future debt distress event (e.g., debt and debt service indicators show large near-term breaches, or significant or sustained breach of thresholds). Debt accumulation beyond sustainable levels constitutes a threat to development outcomes. (ch. 17)

debt service: The amount spent on paying debt principal and the interest over a particular period, usually one year. (ch. 17)

debt swaps (also debt-for-environment swaps): Financial transactions in which a portion of a lower-income country's foreign debt is forgiven in exchange for local investments in environmental conservation measures. According to the Organisation for Economic Co-operation and Development, debt swaps are normally negotiated in the context of debt restructuring of public and publicly guaranteed long-term debt. Debtor countries qualify if they are heavily indebted (according to IMF standards), if they have exhausted other more favorable debt relief instruments (e.g., unconditional debt relief), and if they can convince creditors that they are capable of allocating a sustainable part of the resources that have been budgeted for debt repayment to finance domestic projects that will yield significant environmental benefits at the national, regional, or global level. (ch. 16)

decent work: Work that is productive and delivers a fair income, security in the workplace and social protection for families, better prospects for personal development and social integration, freedom for people to express their concerns, organize, and participate in the decisions that affect their lives, and equality of opportunity and treatment for all women and men. (ch. 8)

decentralized energy system: Energy that is generated close to where it will be used, rather than at an industrial plant and sent through a larger grid. Decentralized systems typically use renewable energy sources, including small hydro, combined heat and power (CHP), biomass, solar, and wind power. A decentralized energy system allows for more optimal use of renewable energy as well as combined heat and power, reduces fossil fuel use, and increases eco-efficiency. (ch. 7)

decentralized wastewater treatment systems: Consists of a variety of approaches for collection, treatment, and dispersal/reuse of wastewater for individual dwellings, industrial or institutional facilities, clusters of homes or businesses, and entire communities. These systems are a part of permanent infrastructure and can be managed as stand-alone facilities or be integrated with centralized sewage treatment systems. They provide a range of treatment options from simple, passive treatment with soil dispersal, commonly referred to as septic or onsite systems, to more complex and mechanized approaches such as advanced treatment units that collect and treat waste from multiple buildings and discharge to either surface waters or the soil. Decentralized systems can cost far less than centralized systems and they work well in rural and suburban settings. (ch. 6)

deforestation: Human-induced conversion of forested land to nonforested land. Deforestation can be permanent, when this change is definitive, or temporary when this change is part of a cycle that includes natural or assisted regeneration. (ch. 15)

degenerative disease (also noncommunicable diseases): The result of a continuous process based on degenerative cell changes, affecting tissues or organs, that will increasingly deteriorate over time. (ch. 3)

degraded lands: Land in an area that results from persistent decline or loss of biodiversity and ecosystem functions and services that cannot fully recover unaided within decadal timescales. (ch. 15)

degrowth: A political ecology term that is used for both a movement as well as a set of theories that critiques the paradigm of economic growth as unsustainable. The term refers to an economic situation during which the economic wealth produced does not increase or even decrease, but it is not a negative growth rate. Degrowth is based on the awareness that in a finite world with limited resources, global production, and consumption cannot continue in a linear way. As a practice, degrowth would include tools such as Life Cycle Analysis, or agroecology. (ch. 8)

deindustrialization: Decline in industrial activity in a region or economy. This process has been most pronounced in the economies of the Global North. (ch. 11)

desalination: Removing the salts from water to make it drinkable. (ch. 6)

desertification: Land degradation in arid, semiarid, and dry subhumid areas resulting from various factors, including climatic variations and human activities. Desertification does not refer to the natural expansion of existing deserts. (ch. 15)

diarrhea: The passage of three or more loose or liquid stools per day (or more frequent passage than is normal for the individual). It is usually a symptom of an infection in the intestinal tract, which can be caused by a variety of bacterial, viral, and parasitic organisms. Infection is spread through contaminated food or drinking water, or from person-to-person as a result of poor hygiene. It can last several days and can leave the body without the water and salts that are necessary for survival. Severe dehydration and fluid loss are the main causes of diarrhea deaths. (ch. 3)

Direct Potable Reuse: The supply of highly treated reclaimed water directly to a drinking water treatment plant or distribution system, with or without an engineered storage buffer. Refers to recycled or reclaimed water that is safe for drinking. (ch. 6)

disability: Inability to adequately or independently perform routine daily activities, such as walking, bathing, and toileting; the negative aspects of the interaction between a person with a health condition and his or her context (environmental and personal factors). (ch. 3)

disinfection (water): The removal, deactivation, or killing of pathogenic microorganisms. Disinfection can be attained by means of physical (UV light or heat) or chemical disinfectants (such as chlorine or soaps). (ch. 6)

Doha Agreement: The 2012 agreement that set a goal of reducing GHG emissions by 18 percent compared to 1990 levels for participating countries. (ch. 13)

drivers (direct): Both non-human-induced and anthropogenic, that affect nature directly. Direct anthropogenic drivers are those that flow from human institutions and governance systems and other indirect drivers. They include positive and negative effects, such as habitat conversion, human-caused climate change, or species introductions. Direct non-human-induced drivers can directly affect anthropogenic assets and quality of life (e.g., a volcanic eruption can destroy roads and cause human deaths). (ch 15)

drivers (indirect): Human actions and decisions that affect nature diffusely by altering and influencing direct drivers as well as other indirect drivers. They do not physically impact nature or its contributions to people. Indirect drivers include economic, demographic, governance, technological, and cultural ones, among others. (ch. 15)

drivers of change: Refer to those external factors that affect nature and, as a consequence, affect the supply of nature's contributions to people. The Intergovernmental Science-Policy Platform on Biodiversity and Ecosystem Services (IPBES) conceptual framework includes drivers of change as two of its main elements: indirect drivers, which are all anthropogenic, and direct drivers, both natural and anthropogenic. (ch. 15)

ecocentrism: Considers that ecological systems, both living and nonliving components, have inherent value. Decisions are made based on an analysis of systems and ecosystems. Ecocentrism sees that the ecosystem as a whole is the center of life and humans are just functional parts of this cosmos. (Intro.)

ecological footprint: A measure of the amount of biologically productive land and water required to support the demands of an individual, a population, or productive activity. Ecological footprints can be calculated at any scale: for an activity, a person, a community, a city, a region, a nation, or humanity as a whole. (ch. 11)

ecosystem: An interconnected geographic area including all the living organisms (people, plants, animals, and microorganisms), their physical surroundings (such as soil, water, and air), and the natural cycles that sustain them. Ecosystems can be small scale (a small area in detail such as a single stand of aspen) or large scale (a large area such as an entire watershed including hundreds of forest stands across many different ownerships). (ch. 15)

ecosystem services: The benefits that people receive from ecosystems. Some of these, such as the provisioning services (or goods) like food, timber, and fresh water, are well-known and routinely included in assessments. Others, such as the regulating services of carbon storage and sequestration, watershed protection, storm protection and pollination. Supporting services, that is, the natural processes such as nutrient cycling and primary production, or the cultural services of recreation and spiritual values, are often overlooked because they are to a lesser extent traded in the market and internalized in traditional cost-benefit analyses. (ch. 15)

ecotourism: Responsible travel to natural areas that conserve the environment, sustains the well-being of the local people, and involves interpretation and education. (ch. 8)

endangered species: A species at risk of extinction in the wild. (ch. 15)

endemic disease: The constant presence of a disease or infectious agent within a given geographic area or population group; may also refer to the usual prevalence of a given disease within such area or group. (ch. 3)

endocrine disrupting chemicals (EDCs): Chemicals that mimic, block, or interfere with hormones in the body's endocrine system. EDCs have been associated with a diverse array of health issues such as cancerous tumors, birth defects, and other developmental disorders. (ch. 15)

energy: The ability to do work, in this context referring primarily to heat energy and kinetic energy. (ch. 7)

energy portfolio: Refers to the different mixes of energy sources that provide energy to a location such as a city, state, region, or country. (ch. 7)

environmental hazard: A substance, state, or event that has the potential to threaten the surrounding natural environment that adversely affects people's health. (ch. 11)

environmental justice: The fair treatment and meaningful involvement of all people regardless of race, color, national origin, or income with respect to the development, implementation, and enforcement of environmental laws, regulations, and policies. (Intro.)

Environmental Kuznets Curve: The environmental Kuznets curve suggests that economic development initially leads to a deterioration in the environment, but after a certain level of economic growth, a society begins to improve its relationship with the environment and levels of environmental degradation reduces. (ch. 12)

epidemic disease: The occurrence of more cases of disease than expected in a given area or among a specific group of people over a particular period. (ch. 3)

epidemiology: The study of the distribution and determinants of health-related states or events in specified populations, and the application of this study to the control of health problems. (ch. 3)

equality: Aims to ensure that everyone gets the same things to enjoy full, healthy lives. Like equity, equality aims to promote fairness and justice, but it can only work if everyone starts from the same place and needs the same things. (Intro.)

equal pay: Men and women in the same employment performing equal work must receive equal pay, unless any difference in pay can be justified. (ch. 5)

equity: The state, quality, or ideal of being just, impartial, and fair. The concept of equity is synonymous with fairness and justice. (Intro.)

equity in health: (1) The absence of systematic or potentially remediable differences in health status, access to health care, and health-enhancing environments, and treatment in one or more aspects of health across populations or population groups defined socially, economically, demographically, or geographically within and across countries. (2) A measure of the degree to which health policies are able to distribute well-being fairly. (ch. 3)

eradication: Permanent reduction to zero of the worldwide incidence of infection caused by a specific pathogen, as a result of deliberate efforts, with no risk of reintroduction. Documentation of eradication is termed certification. (ch. 3)

eutrophication: An enrichment of water by nutrients that causes structural changes to the ecosystem, such as increased production of algae and aquatic plants, depletion of fish species, general deterioration of water quality, and other effects that reduce and preclude use. (ch. 14)

external costs or negative externalities: Disadvantageous impacts paid for by the consumer for the actions of firms and individuals. For example, the environmental and social costs of the current economic system, where these external costs are passed on to the consumer. (Intro.)

extreme poverty: The World Bank defines extreme poverty as individuals living on less than $.190 per day ($694/year). The United Nations defines extreme poverty as characterized by severe deprivation of basic human needs, including food, safe drinking water, sanitation facilities, health, shelter, education, and information. It depends not only on income but also on access to services. (ch. 1)

famine: An acute episode of extreme hunger that results in excess mortality due to starvation or hunger-induced diseases. (ch. 2)

filtration: A treatment process for removing solid (particulate) matter from water by means of porous media such as sand or a man-made filter; often used to remove particles that contain pathogens. (ch. 6)

financial inclusion: Individuals and businesses have access to useful and affordable financial products and services that meet their needs—transactions, payments, savings, credit, and insurance—delivered in a responsible and sustainable way. (ch. 9)

floodplain: The flat or nearly flat land along a river or stream or in a tidal area that is covered by water during a flood. (ch. 14)

fluorinated gases: Man-made gases that include hydrofluorocarbons, perfluorocarbons, sulfur hexafluoride, and nitrogen trifluoride. Fluorinated gases are the most potent and longest-lasting greenhouse gases emitted by human activities. (ch. 13)

food chain: Is a linear sequence of organisms through which matter and energy in the form of food are transferred from organism to organism. (ch. 15)

food deserts: Locations without easy access to fresh, healthy, and affordable foods. (ch. 2)

food insecurity: People are at risk of, or worried about, not being able to meet their preferences for food, including in terms of raw calories and nutritional value. In the Food and Agriculture Organization of the United Nations (FAO) definition, all hungry people are food insecure, but not all food-insecure people are hungry. (ch. 2)

food security: The World Food Summit of 1996 defined "food security" as existing "when all people at all times have access to sufficient, safe, nutritious food to maintain a healthy and active life." (ch. 2)

food swamps: Easy access to less healthy, energy-dense foods, particularly if they are convenient and cheap, may swamp out healthier choices. (ch. 2)

food web: Important ecological concept representing feeding relationships within a community and implying the transfer of food energy from its source in plants through herbivores to carnivores; normally, food webs consist of a number of food chains integrated together. (ch. 15)

forced labor: Work that is performed involuntarily and under the menace of any penalty. It refers to situations in which persons are coerced to work through the use of violence or intimidation, or by more subtle means such as manipulated debt, retention of identity papers, or threats of denunciation to immigration authorities. (ch. 8)

forced marriage: A marriage that takes place without the consent of one or both people in the marriage. Consent means that an individual has given full, free, and informed agreement to marry the intended spouse and to the timing of the marriage. Forced marriage may occur when family members or others use physical or emotional abuse, threats, or deception to force an individual to marry without his/her consent. (ch. 8)

forced migration (also involuntary migration): A migratory movement that, although the drivers can be diverse, involves force, compulsion, or coercion. This term has been used to describe the movements of refugees, displaced persons (including those displaced by disasters or development projects), and, in some instances, victims of trafficking. At the international level the use of this term is debated because of the widespread recognition that a continuum of agency exists rather than a voluntary/forced dichotomy and that it might undermine the existing legal international protection regime. (ch. 16)

foreign direct investment (FDI): An investment made by a firm or individual in one country into business interests located in another country. Generally, FDI takes place when an investor establishes foreign business operations or acquires foreign business assets in a foreign company. (ch. 17)

formal economy: All those types of employment that offer regular wages and hours, which carry with them employment rights, and on which income tax is paid—in contrast to the informal sector. However, these two sectors are by no means entirely separate. (ch. 8)

fossil fuels: Energy resources that are derived from organic materials of the distant past that are trapped and transformed within the Earth's crust, consisting of coal, oil, and natural gas. (ch. 7)

Fourth Industrial Revolution: Builds on the Third Industrial Revolution that relied on electronics and information technology. The Fourth Industrial Revolution is considered the digital revolution that has been occurring since the middle of the twentieth century. It is characterized by a fusion of technologies that is blurring the lines between the physical, digital, and biological spheres, and includes artificial intelligence (AI), robotics, the Internet of Things, genetic engineering, and quantum computer. Examples of the products and services developed include GPS systems that suggest the fastest route to a destination, voice-activated virtual assistants such as Apple's Siri, and Facebook's ability to recognize your face and tag you in a friend's photo. (ch. 9)

Fourth World: Fourth World follows the First World, Second World, and Third World classification of nation-state status; however, unlike the former categories, Fourth World is not spatially bounded, and is usually used to refer to the perceived nonrecognition and exclusion of ethnically and religiously defined peoples. (ch. 10)

functional redundancy: The occurrence in the same ecosystem of species filling similar roles, which results in a sort of "insurance" in the ecosystem, with one species able to "replace" a similar species from the same functional niche. (ch. 15)

gender equality: Means that all genders are free to pursue whatever career, lifestyle choice, and abilities they want without discrimination. Their rights, opportunities, and access to society are not different based on their gender. Gender equality does not necessarily mean that everyone is treated exactly the same, but that women's and men's rights, responsibilities, and opportunities will not depend on whether they are born male or female. (ch. 5)

gender equity: The process of being fair to women and men. To ensure fairness, strategies and measures must often be available to compensate for women's historical and social disadvantages that prevent women and men from otherwise operating on a level playing field. Equity leads to equality. (ch. 5)

Gender Inequality Index (GII): Measures gender inequalities in three important aspects of human development—reproductive health, measured by maternal mortality ratio and adolescent birth rates; empowerment, measured by proportion of parliamentary seats occupied by females and proportion of adult females

and males aged twenty-five years and older with at least some secondary education; and economic status, expressed as labor market participation and measured by labor force participation rate of female and male populations aged fifteen years and older. The GII helps to better expose differences in the distribution of achievements between women and men. It measures the human development costs of gender inequality. Thus the higher the GII value, the more disparities between females and males and the more loss to human development. (ch. 5)

genetic diversity: The variation at the level of individual genes, which provides a mechanism for populations to adapt to their ever-changing environment. The more variation, the better the chance that at least some of the individuals will have an allelic variant that is suited for the new environment and will produce offspring with the variant that will in turn reproduce and continue the population into subsequent generations. (ch. 15)

genetically modified organism (GMO): Organism in which the genetic material (DNA) has been altered in a way that does not occur naturally by mating and/or natural recombination. (ch. 2)

geothermal energy: Energy drawn from the heat within the interior of the earth. One of only three energy resources not originally drawn from the power of the sun. (ch. 7)

Gini coefficient or Gini index: Measures the extent to which the distribution of income (or, other indicators such as educational attainment, life expectancy, etc.) among individuals or households deviates from a perfectly equal distribution. A Gini index of 0 represents perfect equality, while an index of 100 implies perfect inequality. (Intro., ch. 10)

Global Gender Gap Index: Measure gender-based gaps in access to resources and opportunities in countries rather than the actual level of the available resources and opportunities in those countries. It is calculated in four key areas: health, education, economy, and politics. The index measures women's disadvantage compared to men. (ch. 5)

Global Multidimensional Poverty Index (MPI): Measure of poverty that captures deprivations in education and access to basic infrastructure in addition to income or consumption at the $1.90 international poverty line. (ch. 1)

Global North and Global South: Terminology that distinguishes not only between political systems or degrees of poverty but also between the benefactors of global capitalism and those who have not benefited. (Intro.)

global warming: Refers to the long-term rise in global temperatures due mainly to the increasing concentrations of GHGs in the atmosphere. It is part, but not all, of the changes due to climate change. See *climate change*. (ch. 13)

grassed swales: A term to describe a vegetated, open runoff channel planted with grasses or turf. Similar terms include grassed channel, dry swale, wet swale, biofilter, or bioswale. Such systems are designed to treat and attenuate stormwater runoff. As runoff flows along the channels, the vegetation in the channel promotes filtration, settling, and infiltration of runoff into the underlying soils. The specific design features and methods of treatment differ in each of these designs but are improvements on the traditional drainage ditch. The designs incorporate modified geometry and other features for use of the swale as a treatment and conveyance practice. (ch. 15)

gray infrastructure: Refers to human engineered structures such as dams, seawalls, roads, pipes, or other systems. With regard to stormwater management, gray infrastructure is a network of water retention and purification infrastructure (pipes, ditches, culverts, retention ponds) meant to slow the flow of stormwater during rain events to prevent flooding and reduce the amount of pollutants entering waterways. (ch. 6)

graywater: Any washwater that has been used in a home or business, except water from toilets. This water is considered to be more reusable, especially for landscape irrigation purposes. (ch. 14)

Great Pacific Garbage Patch: An accumulation of chemicals, plastic waste, and debris in the North Pacific Ocean. (ch. 14)

greenfield: Previously undeveloped land such as forests, meadows, or other natural lands. (ch. 15)

green growth: A means to foster economic growth and development while ensuring that natural assets continue to provide the resources and environmental services on which our well-being relies. (ch. 8)

greenhouse gases (GHGs): Gases that contribute to the greenhouse effect by absorbing infrared radiation produced by solar warming of the Earth's surface. Major GHGs include carbon dioxide (CO_2), methane (CH_4), nitrous oxide (NO_2), and water vapor. Elevated levels of GHGs have been observed in the recent past and are related, at least in part, to human activities such as burning fossil fuels. (ch. 13)

greenhouse effect: A natural process that warms the Earth's surface. When the Sun's energy reaches the Earth's atmosphere, some of it is reflected back to space and the rest is absorbed and reradiated by greenhouse gases. (ch. 13)

green infrastructure: Green infrastructure refers to the natural or seminatural systems (e.g., riparian vegetation) that provide services such as wildlife habitat, flood protection, drinking water source protection, air quality, and urban heat island reduction. In regard to stormwater management, green infrastructure includes bioswales, forests, floodplains, wetlands, and soils that provide flood protection with equivalent or similar benefits to conventional (built) gray infrastructure (e.g., water treatment plants). (chs. 6, 15)

green job: Jobs in businesses that produce goods or provide services that benefit the environment or conserve natural resources. Jobs in which workers' duties involve making their establishment's production processes more environmentally friendly or use fewer natural resources. (ch. 8)

Green Revolution: Period of increased food crop productivity that started in the 1960s due to a combination of high rates of investment in crop research, infrastructure, and market development and appropriate policy support, and whose environmental impacts have been mixed: on one side saving land conversion to agriculture, on the other side promoting an overuse of inputs and cultivation on areas otherwise improper to high levels of intensification, such as slopes. (ch. 2)

green roof: Also known as eco-roofs or rooftop gardens, green roofs are engineered soil media systems that are planted on rooftops and designed to reduce runoff, combined sewer overflows, and urban heat island impacts and provide other ecological and human benefits such as aesthetics, wildlife habitat, and aesthetics. The soil media mix and vegetation are planted over existing roof structures and consist of a waterproof, root-safe membrane that is covered by a drainage system, lightweight growing medium, and plants. Green roofs reduce rooftop and building temperatures, filter pollution, lessen pressure on sewer systems, and reduce the heat island effect. (ch. 15)

gross national income (GNI): The income of a nation calculated based on goods and services produced, plus income earned by citizens and corporations headquartered in that country. (ch. 1)

gyres (oceanic): A large system of rotating ocean currents. There are five major gyres: the North and South Pacific Subtropical Gyres, the North and South Atlantic Subtropical Gyres, and the Indian Ocean Subtropical Gyre. (ch. 14)

habitat: The place or type of site where an organism or population naturally occurs. Also used to mean the environmental attributes required by a particular species or its ecological niche. (ch. 15)

haboob: An intense dust storm, with winds up to thirty miles per hour. (ch. 15)

hard infrastructure: Tangible or built infrastructure, is the physical infrastructure of roads, bridges, tunnels, railways, ports, and harbors. (ch. 9)

harmful algal blooms (HABs): Occur when colonies of algae grow out of control and produce toxic or harmful effects on people, fish, shellfish, marine mammals, and birds. The human illnesses caused by HABs, though rare, can be debilitating or even fatal. (ch. 14)

hazardous child labor: Work in dangerous or unhealthy conditions that could result in a child being killed or injured or made ill as a consequence of poor safety and health standards and working arrangements. Such work can result in permanent disability, ill health, and psychological damage. (ch. 8)

hazardous substance: Those substances that are known or suspected to cause cancer or other serious health effects, such as reproductive effects or birth defects, or adverse environmental effects. (ch. 15)

health: The state of complete physical, mental, and social well-being and not merely the absence of disease or infirmity. (ch. 3)

health in all policies: A policy or reform designed to secure healthier communities, by integrating public health actions with primary care and by pursuing healthy public policies across sectors. (ch. 3)

health indicator: A measure that reflects, or indicates, the state of health of persons in a defined population, for example, the infant mortality rate. (ch. 3)

health service: Any service (not limited to medical or clinical services) aimed at contributing to improved health or to the diagnosis, treatment, and rehabilitation of sick people. (ch. 3)

health system: (1) All the activities whose primary purpose is to promote, restore, and/or maintain health; (2) the people, institutions, and resources, arranged together in accordance with established policies, to improve the health of the population they serve, while responding to people's legitimate expectations and protecting them against the cost of ill-health through a variety of activities whose primary intent is to improve health. (ch. 3)

heat island effect: Describes built up areas that are hotter than nearby rural areas. Heat islands can affect communities by increasing summertime peak energy demand, air conditioning costs, air pollution and greenhouse gas emissions, heat-related illness and mortality, and water quality. (ch. 11)

hidden hunger: See *micronutrient deficiency*; a nutritional deficiency caused by lack of balance in an otherwise full diet. (ch. 2)

higher education: Education beyond basic; college or university. (ch. 4)

horizontal inequality: Inequalities among groups of people within an area; horizontal inequality is seen when people of similar origin, intelligence, and so forth, still do not have equal success and have different status, income. (ch. 10)

Human Development Index (HDI): A statistic composite index of life expectancy, education, and per capita income indicators, which are used to rank countries into four tiers of human development (very high development, high development, medium development, and low development). (Intro.)

human immunodeficiency virus (HIV/AIDS): A virus that attacks the body's immune system. If HIV is not treated, it can lead to AIDS (acquired immunodeficiency syndrome). There is currently no effective cure. Once people get HIV, they have it for life. (ch. 3)

humanitarian assistance: The actions of governments and nongovernmental organizations to alleviate suffering and maintain human dignity in the face of disasters and human crises. (ch. 16)

human rights: Rights inherent to all human beings, regardless of race, sex, nationality, ethnicity, language, religion, or any other status. Human rights include the right to life and liberty, freedom from slavery and torture, freedom of opinion and expression, the right to work and education, and many more. (ch. 16)

human trafficking: The recruitment, transportation, transfer, harboring, or receipt of people through force, fraud, or deception, with the aim of exploiting them for profit. Traffickers often use violence or fraudulent employment agencies and fake promises of education and job opportunities to trick and coerce their victims. (ch. 16)

hunger A condition in which a person cannot eat sufficient food to meet basic nutritional needs for a sustained period, leading to physical discomfort. (ch. 2)

hydroelectricity: The production of electricity by turning a turbine with flowing water. (ch. 7)

hydrologic cycle: Movement or exchange of water between the atmosphere and earth. (ch. 6)

hypoxia: Low dissolved oxygen levels in coastal and oceanic waters. Condition created when coastal waters or estuaries become polluted by nutrient overload, creating a lack of oxygen. The most direct result is fish kills, which lead to habitat loss and loss of ecological biodiversity. (ch. 14)

illegal, unreported and unregulated (IUU) fishing: According to the Food and Agriculture Organization of the United Nations (FAO), this is a broad term that includes fishing and fishing-related activities conducted in contravention of national, regional, and international laws; nonreporting, misreporting, or under reporting of information on fishing operations and their catches; fishing by "stateless" vessels; fishing in convention areas of Regional Fisheries Management Organizations (RFMOs) by nonparty vessels; and fishing activities that are not regulated by states and cannot be easily accounted for and monitored. (ch. 14)

immunity, herd: The resistance of a group to invasion and spread of an infectious agent, based on the resistance to infection of a high proportion of individual members of the group. The resistance is a product of the number susceptible and the probability that those who are susceptible will come into contact with an infected person. (ch. 3)

income distribution: The smoothness or equality with which income is dealt out among members of a society. If everyone earns exactly the same amount of money, then the income distribution is perfectly equal. (ch. 10)

infectious disease: (Communicable disease) disorders caused by organisms—such as bacteria, viruses, fungi, or parasites. (ch. 3)

informal economy: Diversified set of economic activities, enterprises, jobs, and workers that are not regulated or protected by the state. The concept originally applied to self-employment in small, unregistered enterprises. It has been expanded to include wage employment in unprotected jobs. People who work in the informal sector do not declare their income and pay no taxes on them. This includes illegal activities, such as drug pushing and smuggling. It also includes cleaning car windshields at traffic lights or doing construction work, that is, legal work. (ch. 8)

information and communication technology (ICT) (also information technology (IT)): Refers to all communication technologies, including the internet, wireless networks, cell phones, computer software, video-conferencing, social networking, and other media applications and services enabling users to access, retrieve, store, transmit, and manipulate information in a digital form. (ch. 9)

integrated conservation and development projects (ICDP): Biodiversity conservation projects with rural development components. This is an approach that aspires to combine social development with conservation goals. (ch. 15)

integrated pest management: The use of pest and environmental information in conjunction with available pest control technologies to prevent unacceptable levels of pest damage by the most economical means and with the least possible hazard to persons, property, and the environment. (ch. 15)

intergenerational justice: The concept or idea of fairness or justice between generations. (ch. 13)

Intergovernmental Panel on Climate Change (IPCC): An intergovernmental body of the United Nations that is dedicated to providing the world with objective, scientific information relevant to understanding human-induced climate change and its effects. The IPCC was established in 1988 by the World Meteorological Organization and the UN Environment Programme. (ch. 13)

Intergovernmental Science-Policy Platform on Biodiversity (IPBES): Founded in 2012, the Intergovernmental Science-Policy Platform on Biodiversity and Ecosystem Services is an intergovernmental organization established to improve the interface between science and policy on issues of biodiversity and ecosystem services. (ch. 15)

Internally Displaced Person (IDP) Someone who is forced to flee their home but has not sought shelter across any international borders. (ch. 16)

International Investment Agreement (IIA): A type of treaty between countries that addresses issues relevant to cross-border investments, usually for the purpose of protection, promotion, and liberalization of such investments. IIAs are designed to protect the investments of foreign investors in the state hosting the investment. (ch. 17)

International Labour Organization (ILO): The ILO is a UN agency whose mandate is to advance social and economic justice through setting international labor standards. The main aims of the ILO are to promote rights at work, encourage decent employment opportunities, enhance social protection, and strengthen dialogue on work-related issues. The ILO was founded in 1919 and is the first and oldest specialized agency of the United Nations. (ch. 8)

International Monetary Fund (IMF): The IMF is an international organization which was created by the United States and the United Kingdom at the end of World War II. Its original purpose was to give loans to countries suffering from short-term economic crisis, in the expectation that they would quickly recover, and the loans would be able to be repaid. Over time, the IMF has shifted to lending to countries in longer-term economic crisis that are unable to pay their debts. These IMF loans repay the original lenders, while the debt remains with the country. In return for such loans, the IMF usually insists on changes to government policy such as cuts in government spending, increases in regressive taxes such as Value-Added Taxes (VAT), privatization of state-owned companies, and removal of regulations on businesses. The IMF is run by its member governments according to a voting formula designed to ensure the United States and Europe control the institution. The head of the IMF has always been a European, following a deal agreed to between the United States and Europe, and the head of the World Bank has always been an American. (ch. 17)

invasive, nonnative species (INNS): Plants and animals that have been purposefully or accidentally introduced, mainly by human activity. (ch. 15)

Jevons paradox: Occurs when technological progress or government policy increases the efficiency with which a resource is used (reducing the amount necessary for any one use), but the rate of consumption of that resource rises due to increasing demand. (ch. 12)

kilowatt hour: A unit used to measure energy, especially electrical energy in commercial applications; a measure of electrical energy equivalent to a power consumption of 1,000 watts for one hour. (ch. 7)

Kyoto Protocol: A 1997 amendment to the international treaty on climate change, assigning mandatory targets for the reduction of greenhouse gas emissions to signatory nations. Canada and the European Union are included in the more than 140 member countries. The United States did not ratify the Protocol. (ch. 13)

land extensification: Involves increasing the total area of land under cultivation to meet growing demands for agricultural production. (ch. 2)

landfill: A system of garbage and trash disposal in which waste is buried between layers of earth. Late-twentieth-century landfills are well-engineered and managed facilities for the disposal of solid waste. (ch. 12)

land intensification: Activities undertaken with the intention of enhancing the productivity or profitability per unit area of agricultural land use. Activities may include increased inputs (such as fertilizers, machinery, or knowledge). (ch. 2)

land use: The human use of a piece of land for a certain purpose (e.g., irrigated agriculture or recreation). Land use is influenced by, but not synonymous with, land cover. (ch. 15)

land-use change: Refers to a change in the use or management of land by humans, which may lead to a change in land cover. (ch. 15)

last mile: Last mile, in supply chain management and transportation planning, is the last leg of a journey comprising the movement of people and goods from a transportation hub to a final destination. In ICT, the last mile refers to the final leg of the telecommunications networks that deliver telecommunication services to customers. It is the portion of the telecommunications network chain that physically reaches the customer's premises. (ch. 9)

Leadership in Energy and Environmental Design (LEED): A designation given to buildings that meet requirements for energy and water efficiency in both the construction and operation of the building. (ch. 7)

Life Cycle Assessment (LCA) (also Life Cycle Analysis): A tool used for comparing the environmental impact of products during the whole life cycle of a product: from raw material extraction through production, use, and waste treatment to final disposal. (ch. 12)

liquid petroleum gas (LPG): Lightweight fuels such as propane, ethane, and butane, which can be easily stored in liquid form and used for purposes such as cooking gas. (ch. 7)

livestock exclusion fencing: Fencing that keeps livestock away from rivers and streams. (ch. 14)

load: The quantity of sediment transported by a current. It includes the suspended load of small particles and the bedload of large particles that move along the bottom. (ch. 14)

malaria: An intermittent and remittent fever transmitted to people through the bites of infected female Anopheles mosquitoes that invade the red blood cells. The parasite is transmitted by mosquitoes in many trop-

ical and subtropical regions. It is preventable and curable. (ch. 3)

malnutrition: Refers to deficiencies, excesses, or imbalances in a person's intake of energy and/or nutrients. The term malnutrition covers two broad groups of conditions. One is "undernutrition," which includes stunting (low height for age), wasting (low weight for height), underweight (low weight for age), and micronutrient deficiencies or insufficiencies (a lack of essential vitamins and minerals). The other condition is overweight, obesity, and diet-related noncommunicable diseases (e.g., heart disease, stroke, diabetes, and cancer). (ch. 2)

mariculture: A branch of aquaculture involving the culture of organisms in a medium or environment that may be completely marine (sea), or sea water mixed to various degrees with fresh water, including brackishwater areas. (ch. 14)

marine debris: Any persistent solid material that is manufactured or processed and directly or indirectly, intentionally or unintentionally, disposed of or abandoned into the marine environment or the Great Lakes. (ch. 14)

Marine Protected Areas (MPA): A broad term for a park or other protected area that includes some marine or lake areas. The International Union for the Conservation of Nature (IUCN) defines a protected area as "a clearly defined geographical space, recognized, dedicated and managed, through legal or other effective means, to achieve the long term conservation of nature with associated ecosystem services and cultural values." (ch. 14)

maternal morality: Refers to death of the mother due to complications from pregnancy or childbirth. (ch. 3)

megacity: A city of at least ten million residents. (ch. 11)

metacity: A city of at least twenty million residents. (ch. 11)

methane: A colorless odorless flammable gaseous hydrocarbon that is a product of biological decomposition of organic matter and of the carbonization of coal. Methane is a far more potent greenhouse gas than carbon dioxide. (ch. 13)

microfiltration: Using a device with a filter media to physically prevent biological contamination from passing through. Ceramic and solid block carbon are commonly used to provide microfiltration. (ch. 14)

microfinance (also microcredit): Is a type of banking and offers financial services to low-income clients, including consumers and the self-employed, who would otherwise lack access to banking and related services. (ch. 6)

micronutrient deficiency: Inadequacies in intake of vitamins and minerals; micronutrients enable the body to produce enzymes, hormones, and other substances that are essential for proper growth and development. (ch. 2)

micronutrients: Substances that are only needed in very small amounts but essential to organisms to produce enzymes, hormones, and other substances fundamental for proper growth and development. (ch. 2)

microplastic: Extremely small pieces of plastic debris (less than five millimeters long) in the environment resulting from the disposal and breakdown of consumer products and industrial waste. (ch. 12)

migrant: An individual that has chosen to leave their home voluntarily most often in search of employment. (ch. 16)

Millennium Development Goals (MDGs): The eight international development goals for the year 2015 that had been established following the Millennium Summit of the United in 2000, following the adoption of the UN Millennium Declaration. (Intro.)

mini-grid/micro-grid: A mini grid, also sometimes referred to as a "micro grid or isolated grid," can be defined as a set of electricity generators and energy storage systems interconnected to a distribution network that supplies electricity to a localized group of customers. They involve small-scale electricity generation (10 kW to 10MW) that serves a limited number of consumers using a distribution grid that can operate in isolation from large scale electricity transmission networks. (ch. 7)

minimum dietary energy requirement: Minimum intake of food energy to sustain a human metabolism and to drive muscles. Older people and those with sedentary lifestyles require less food energy; children and physically active people require more. (ch. 2)

missing and disappeared people: A missing person is a person who has disappeared and whose status as alive or dead cannot be confirmed as their location and condition are not known. Some missing people may go missing through a voluntary disappearance. Others have been subjected to enforced disappearance that is considered to be the arrest, detention, abduction, or any other form of deprivation of liberty by agents of the state or by persons or groups of persons acting with the authorization, support, or acquiescence of the state, followed by a refusal to acknowledge the deprivation of liberty or by concealment of the fate or whereabouts of the disappeared person, which place such a person outside the protection of the law. (ch. 16)

missing women: Refers to the observation that in parts of the Global South notably in India and China—the ratio of women to men is atypically low. The term indicates a shortfall in the number of women relative to the expected number of women in a region or country and is theorized to be caused by sex-selective abortions, female infanticide, and inadequate health care and nutrition for female children. The phrase was coined by Amartya Sen. (ch. 5)

mitigation (climate change): With regard to climate change it is the process or result of making something less severe, dangerous, painful, harsh, or damaging; mitigation typically involves actions to reduce greenhouse gas emissions. (ch. 13)

moderate food security: Includes those who struggle or worry about the ability to access or afford a healthy, nutritious balanced diet, not only those who struggle to meet their energy needs. (ch. 2)

modern slavery: The severe exploitation of other people for personal or commercial gain. Types of modern slavery can include forced labor, organ trafficking, child labor, and child exploitation. (ch. 8)

monoculture: The agricultural practice of cultivating a single crop over a whole farm or area. (ch. 15)

morbidity: Detectable, measurable clinical consequences of infections and disease that adversely affect the health of individuals. Evidence of morbidity may be overt (e.g., the presence of blood in the urine, anemia, chronic pain, or fatigue) or subtle (e.g., stunted growth, impeded school or work performance, or increased susceptibility to other diseases). (ch. 3)

mortality rate: A measure of the frequency of occurrence of death in a defined population during a specified interval of time. (ch. 3)

mortality rate, child: A ratio expressing the number of deaths among children under five years of age reported during a given period per 1,000 children in this age group. (ch. 3)

mortality rate, infant: A ratio expressing the number of deaths among children under one year of age reported during a given period divided by the number of births reported during the same period. The infant mortality rate is expressed per 1,000 live births. (ch. 3)

mortality rate, neonatal: A ratio expressing the number of deaths among children from birth up to but not including twenty-eight days of age divided by the number of live births reported during the same period. The neonatal mortality rate is expressed per 1,000 live births. (ch. 3)

multilateral organization: Refers to an alliance of multiple countries and/or international organization pursuing a common goal, such as treaties, conventions, protocols, and other binding instruments related to sustainability. Multilateral organizations include the International Monetary Fund, World Bank, United Nations, and World Trade Organization. (ch. 17)

multistakeholder partnerships (also called cross-sector collaborations, or public private partnerships): Initiatives voluntarily undertaken by governments, intergovernmental organizations, major groups, and other stakeholders, whose efforts are contributing to the implementation of intergovernmentally agreed development goals and commitments, such as the Sustainable Development Goals. (ch. 17)

native landscaping: Landscaping that is designed to use native plants adapted to the specific geographic location of their origin. (ch. 15)

natural capital: The stock of renewable and nonrenewable resources (e.g., plants, animals, air, water, soils, minerals) that combine to yield a flow of benefits to people. (ch. 12)

natural gas: Fossil fuel energy resource comprised primarily of methane for a variety of uses, including electrical generation, home heating, cooking, and transportation. (ch. 7)

nature-based solutions (NBS): Actions to protect, sustainably manage, and restore natural or modified ecosystems, that address societal challenges effectively and adaptively, simultaneously providing human well-being and biodiversity benefits. (ch. 6)

Neglected Tropical Diseases (NTDs): Communicable diseases that are found primarily in tropical and subtropical countries and affect more than one billion people. (ch. 3)

nitrous oxide: A colorless gas with a sweetish odor, prepared by heating ammonium nitrate. It produces exhilaration or anesthesia when inhaled and is used as an anesthetic and as an aerosol propellant. (ch. 13)

noncommunicable disease: The result of a continuous process based on degenerative cell changes, affecting tissues or organs, which will increasingly deteriorate over time. (ch. 3)

nongovernmental organization (NGO) A non-for-profit, voluntary citizens' group, which is organized on a local, national, or international level to address issues in support of the public good. (Intro.)

nonpoint source: Diffuse runoff (e.g., without a single point of origin or not introduced into a receiving stream from a specific outlet). (ch. 6)

nonrenewable energy resource: Term for any energy resource that will not regenerate within a short period of time; includes all fossil fuels and nuclear energy. (ch. 7)

nuclear power: The production of electricity through the fission of heavy elements, primarily uranium. (ch. 7)

obesity: A person who is too heavy for his or her height. (ch. 2)

ocean acidification: Reduction in the pH of the ocean over an extended period, typically decades or longer, which is caused primarily by uptake of carbon dioxide from the atmosphere but can also be caused by other chemical additions or subtractions from the ocean. Anthropogenic ocean acidification refers to the component of pH reduction that is caused by human activity. (ch. 14)

official development assistance (ODA): Defined by the OECD Development Assistance Committee (DAC) as government aid that promotes and specifically targets the economic development and welfare of developing countries. The DAC adopted ODA as the "gold standard" of foreign aid in 1969 and it remains the main source of financing for development aid. (ch. 17)

off-the-grid: Not connected to or served by publicly or privately managed utilities (such as electricity, gas, or water). (ch. 7)

oil (or petroleum): Fossil fuel formed from phytoplankton that is liquid to semisolid at room temperature; the world's most consumed energy resource. (ch. 7)

organic agriculture: Any system that emphasizes the use of techniques such as crop rotation, compost, or manure application, and biological pest control in preference to synthetic inputs. Most certified organic farming schemes prohibit all genetically modified organisms and almost all synthetic inputs. Recognition and certification of organic agriculture may vary significantly across countries. (ch. 2)

organic matter: The organic component of the soil consisting in living organisms, dry plants, and residues of animal origin. In a mass unit, this organic component is the most chemically active of the soil. Such a component stores several essential elements, stimulates the proper structure of the soil, is a source with capacity for the exchange of cations and regulates the pH changes, supports the relationship between air and water in the soil, and is a huge geochemical storage of carbon. (ch. 15)

overexploitation: Harvesting species from the wild at rates faster than natural populations can recover. Includes overfishing and overgrazing. (ch. 15)

pandemic: An epidemic occurring over a very wide area (several countries or continents) and usually affecting a large proportion of the population. (ch. 3)

Paris Agreement: A legally binding international treaty on climate change. It was adopted by 196 parties at the Conference of the Parties (COP 21) in Paris, on December 12, 2015, and entered into force on November 4, 2016. Its goal is to limit global warming to well below 2, preferably to 1.5 degrees Celsius, compared to preindustrial level. (ch. 13)

pathogenic Of a bacterium, virus, or other microorganism, causing disease. (ch. 3)

Payments for Ecosystem Services (PES): A term used to describe a process whereas a beneficiary or user of an ecosystem service makes a direct or indirect payment to a provider of that service. PES involve a series of payments to land or other natural resource owners in return for a guaranteed flow of ecosystem services or certain actions likely to enhance their provision over and above what would otherwise be provided in the absence of payment. (ch. 15)

peacebuilding: A broad range of measures implemented in the context of emerging, current, or post-conflict situations, and that are explicitly guided and motivated by a primary commitment to the prevention of violent conflict and the promotion of a lasting and sustainable peace. (ch. 16)

people-centered care: Care that is focused and organized around the health needs and expectations of people and communities rather than on diseases. People-centered care extends the concept of patient-centered care to individuals, families, communities, and society. Whereas patient-centered care is commonly understood as focusing on the individual seeking care—the patient—people-centered care encompasses these clinical encounters and also includes attention to the health of people in their communities and their crucial role in shaping health policy and health services. (ch. 3)

petroleum: Fossil fuel formed from phytoplankton that is liquid to semisolid at room temperature; the world's most consumed energy resource. (ch. 7)

poaching: Animal killing or trapping without the approval of the people who control or own the land. (ch. 15)

point source pollution: Pollution from a single identifiable source of air, water, heat, light, or noise. For example, a pipe dumping chemicals into a river. (ch. 6)

political ecology: A field within environmental studies focusing on power relations as well as the coproduction of nature and society. Theoretical inspirations are taken from different sources such as Marxism, political economy, and poststructuralism. Contributions to this field tend to question the status of powerful actors (e.g., governments, businesses, conservation organizations) and what is taken for granted in leading discourses. Political ecology embraces complex systems thinking and points to the limitations of adopting a narrow definition of the natural environment and treating local and global environments as independent from one another. (Intro.)

pollution sink: Vehicle for removal of a chemical or gas from the atmosphere-biosphere-ocean system, in which the substance is absorbed into a permanent or semipermanent repository, or else transformed into another substance. (ch. 14)

postindustrial city/postindustrial economy: An urban economy that has transitioned from one based predominantly on the secondary sector (e.g., manufacturing) to one based primarily on the tertiary sector (e.g., services). (ch. 11)

poverty: A state of economic deprivation. Its manifestations include hunger and malnutrition, limited access to education, and other basic services. (ch. 1)

poverty line (also poverty threshold): The estimated minimum level of income needed to secure the necessities of life in a particular country. (ch. 1)

primary education: The first years of compulsory schooling are called elementary, primary, or grammar schools and typically consist of grades 1–5. (ch. 4)

primary health care: Health care provided by a medical professional (e.g., as a general practitioner, pediatrician, or nurse) for people making an initial approach to a doctor or nurse for treatment. It is considered health care at a basic rather than specialized level. (ch. 3)

primary sector: That portion of a region's economy devoted to the extraction of basic materials (e.g., mining, lumbering, agriculture). (ch. 9)

protected area: A clearly defined geographical space, recognized, dedicated, and managed, through legal or other effective means, to achieve the long-term conservation of nature with associated values to people. (ch. 15)

public debt (also public-sector debt): Public debt is an obligation of a government to pay certain sums to the holders at some future time. Public debt is distinguished from private debt, which consists of the obligations of individuals, business firms, and nongovernmental organizations. (ch. 17)

public private partnerships (PPP): Involve collaboration between a government agency and a private-sector company that can be used to finance, build, and operate projects, such as public transportation networks, parks, and convention centers. Financing a project through a public-private partnership can allow a project to be completed sooner or make it a possibility in the first place. Public-private partnerships often involve concessions of tax or other operating revenue, protection from liability, or partial ownership rights over nominally public services and property to private-sector, for-profit entities. (ch. 9)

pull factors: Those that attract people to migrate to a new home and include better opportunities. Pull factors are reasons to migrate and are often the positive aspects of a different place that encourage people to migrate or immigrate to seek a better life. Generally, a combination of push-pull factors helps determine migration or immigration. (ch. 11)

purchasing power parity (PPP): The PPP calculation tells you how much things would cost if all countries used the same currency. It is the rate at which the currency of one country would have to be converted into that of another country to buy the same amount of goods and services in each country. For example, the price of a McDonald's Big Mac may vary from one country to another, even though the product is the same. PPP is based on an economic theory that states the prices of goods and services should equalize among countries over time. (ch. 1)

push factors: With regard to migration, these are factors that push people away from their home and include inequalities, war and conflict, and lack of opportunities. Push factors are reasons to leave. In many cases, push factors force an individual to leave out of desperation. Generally a combination of push-pull factors helps determine migration or immigration. (ch. 11)

rain garden: Depressed area of the ground planted with vegetation, allowing runoff from impervious surfaces such as parking lots and roofs the opportunity to be collected and infiltrated into the groundwater supply or returned to the atmosphere through evaporation and evapotranspiration. Rain gardens are typically cheaper to build and design than bioretention or bioinfiltration cells because they are often built without specific performance standards and without the assistance of a certified professional to design them. (ch. 15)

REDD+: Mechanism developed by parties to the UN Framework Convention on Climate Change (UNF-CCC), which creates a financial value for the carbon stored in forests by offering incentives for developing countries to reduce emissions from forested lands and invest in low-carbon paths to sustainable development. Developing countries would receive results-based payments for results-based actions. REDD+ goes beyond simply deforestation and forest degradation, and includes the role of conservation, sustainable management of forests, and enhancement of forest carbon stocks. (ch. 15)

redundancy (also species or genetic redundancy): In ecology, the basic concept of redundancy is that if a species is removed, and the community remains constant, then that species was redundant. (ch. 15)

reforestation: The establishment of a forest through artificial plantings or natural regeneration. (ch. 15)

refugees: Persons who flee their country due to well-founded fear of persecution due to reasons of race, religion, nationality, membership of a particular social group, or political opinion, and who are outside of their country of nationality or permanent residence and due to this fear are unable or unwilling to return to it. (ch. 16)

relative poverty: Relative poverty is a state of living in which people can afford necessities but are unable to meet their society's average standard of living. (ch. 1)

remediation: Any action taken to rehabilitate ecosystems after their degradation. (ch. 15)

remittance: A payment of money that is transferred to another party. The term is most often used to describe a sum of money sent by an international migrant working abroad to his or her family back home. (ch. 17)

renewable energy resource: Term for any energy resource that is renewed by natural systems in a short amount of time, which applies to all forms of energy consumed by humans except fossil fuels and nuclear energy. (ch. 7)

resilience: The capacity of a system to absorb disturbance and reorganize while undergoing change so as to still retain essentially the same function, structure, identity, and feedbacks. (Intro.)

resilience hubs: Community-serving facilities augmented to support residents and coordinate resource distribution and services before, during, or after a natural hazard event. (ch. 9)

richness (biodiversity): The number of distinct biological entities (typically species, but also genotypes, taxonomic genera or families) within a given sample, community, or area. (ch. 15)

riparian area: Vegetated ecosystems along a waterbody through which energy, materials, and water pass. Riparian areas characteristically have a high water table and are subject to periodic flooding and influence from the adjacent waterbody. These systems encompass wetlands, uplands, or some combination of the two, although they will not in all cases have all the characteristics necessary for them to be classified as wetlands. (ch. 15)

Rule of Law: A principle of governance in which all persons, institutions, and entities, public and private, including the state, are accountable to laws that are publicly promulgated, equally enforced, and independently adjudicated, and that are consistent with international human rights norms and standards. It requires, as well, measures to ensure adherence to the principles of supremacy of law, equality before the law, accountability to the law, fairness in the application of the law, separation of powers, participation in decision making, legal certainty, avoidance of arbitrariness, and procedural and legal transparency. (ch. 16)

secondary education: Involves education starting at age eleven and can involve middle school and high school and other terms such as sixth-form, vocational schools, preparatory schools, and so forth. Secondary education is considered the second and final phase of basic education. (ch. 4)

secondary sector: That portion of a region's economy devoted to the processing of basic materials extracted by the primary sector. (Intro.)

sediment: Topsoil, sand, and minerals washed from the land into water, usually after rain or snow melt. (ch. 14)

sediment trap: A structure or vegetative barrier designed to collect soil material transported in runoff and to reduce water flow velocity and therefore scouring and erosion. Sediment traps mitigate siltation of natural drainage features. (ch. 14)

single-use plastics: Goods that are made primarily from fossil fuel–based chemicals (petrochemicals) and are meant to be disposed of right after use—often, in mere minutes. Single-use plastics are most commonly used for packaging and serviceware, such as bottles, wrappers, straws, and bags. (ch. 12)

slum: Unplanned, illegal, informal housing that lacks one or more of the following: (1) durable housing of a permanent nature that protects against extreme climate conditions; (2) sufficient living space, which means not more than three people sharing the same room; (3) easy access to safe water in sufficient amounts at an affordable price; (4) access to adequate sanitation in the form of a private or public toilet shared by a reasonable number of people; and (5) security of tenure that prevents forced evictions. (ch. 11)

slum upgrading (also slum improvement): Providing a package of basic services: clean water supply and adequate sewage disposal to improve the well-being of the community. In addition, slum upgrading may also include legalizing and "regularizing" the properties in situations of insecure or unclear tenure. (ch. 11)

smart grid: An electricity supply network that uses digital communications technology to detect and react to local changes in usage. (ch. 7)

smog: Photochemical smog is air pollution caused by motor vehicles and industrial processes and may cause health or irritation to humans, animals, and plants. (ch. 7)

social costs of carbon (SCC): An estimate, in dollars, of the economic damages that would result from emitting one additional ton of greenhouse gases into the atmosphere. The SCC puts the effects of climate change into economic terms to help policy makers and other decision makers understand the economic impacts of decisions that would increase or decrease emissions. (ch. 13)

social-ecological systems (SES): Linked systems of people and nature. The term emphasizes that humans must be seen as a part of, not apart from, nature—that the delineation between social and ecological systems is artificial and arbitrary. (ch. 15)

social impact bonds (SIB): According to the OECD, a social impact bond is an innovative financing mechanism in which governments enter into agreements with social service providers, such as social enterprises or nonprofit organizations, and investors to pay for the delivery of predefined social outcomes. SIBs derive their name from the fact that their investors are typically those who are interested in not just the financial return on their investment but also in its social impact. (ch. 17)

Social Institutions and Gender Index (SIGI): Measures discrimination against women in social institutions by considering laws, social norms, and practices. (ch. 5)

social protection: Consists of policies and programs designed to reduce poverty and vulnerability throughout one's life. Social protection includes benefits for children and families, maternity, unemployment, employment injury, sickness, old age, disability, and health protection. (chs. 1, 8)

soft infrastructure: Infrastructure that makes up institutions that help maintain the economy. These usually require human capital and help deliver services that are required to maintain the economic, health, and cultural and social standards of a population. Examples include the health-care system, financial and banking systems, governmental systems, law enforcement, and education systems. (ch. 9)

soil fertility: The capacity of a soil to receive, store, and transmit energy to support plant growth. It is the component of overall soil productivity that deals with its available nutrient status, and its ability to provide nutrients out of its own reserves and through external applications for crop production. (ch. 15)

solar energy: Any form of energy that directly uses the heat of solar rays or the photovoltaic effect. (ch. 7)

solid waste: The US Environmental Protection Agency defines solid waste as any garbage or refuse, sludge from a wastewater treatment plant, water supply treatment plant, or air pollution control facility and other discarded material, resulting from industrial, commercial, mining, and agricultural operations, and from community activities. (ch. 12)

species: An interbreeding group of organisms that is reproductively isolated from all other organisms, although there are many partial exceptions to this rule in particular taxa. Operationally, the term "species" is a generally agreed fundamental taxonomic unit, based on morphological or genetic similarity, that once described and accepted is associated with a unique scientific name. (ch. 15)

sprawl: Unplanned and uncontrolled spreading of urban development on undeveloped land. (ch. 13)

stakeholders (also multistakeholder partnerships, cross-sector collaboration): In reference to sustainability, the term is used in its broadest sense to include all nongovernmental actors that can contribute to the 2030 agenda, such as individuals, civil society actors, youth and women organizations, Indigenous peoples, movements and networks, academia, the private sector, trade unions, and institutions with an accountability function, such as human rights institutions, parliamentarians, or supreme auditing institutions. In addition, the framework considers local and regional governments as stakeholders, given their dual role as government actors ("duty bearers") and actors that need to be included in national engagement practices. (ch. 17)

stand-alone energy systems: An off-the-grid electricity system for locations that are not fitted with an electricity distribution system and includes wind, solar or diesel, or biofuel generators. (ch. 7)

stormwater runoff: Water that originates from rain, including snow and ice melt. Stormwater can soak into the soil, be stored on the land surface in ponds and puddles, evaporate, or contribute to surface runoff. Stormwater runoff also picks up and carries with it many different pollutants that are found on surfaces such as sediment, nitrogen, phosphorus, bacteria, oil and grease, trash, pesticides, and metals. These pollutants come from a variety of sources, including pet waste, lawn fertilization, cars, construction sites, illegal dumping and spills, and pesticide application. Researchers have found that as the amount of paved surfaces (impervious cover) in the watershed increases, stream health declines accordingly. (ch. 14)

stunting: A child who has very low height-for-age; is also often underweight. (ch. 2)

subsistence agriculture: Farming system emphasizing production for use rather than for sale. (ch. 2)

Sustainable Development Goals (SDGs): A collection of seventeen interlinked global goals designed to be a "blueprint to achieve a better and more sustainable future for all." The SDGs were set up in 2015 by the UN General Assembly and are intended to be achieved by the year 2030. (Intro.)

sustainable industrial development: Industrial development that includes all countries and all peoples, as well as the private sector, civil society organizations, and multinational development institutions, and offers equal opportunities and an equitable distribution of the benefits of industrialization to all stakeholders. The term "sustainable" addresses the need to decouple the prosperity generated from industrial activities from excessive natural resource use and negative environmental impacts. Sustainable industrial development implies that no one is left behind and all parts of society benefit from industrial progress, which also provides the means for tackling critical social and humanitarian needs. (ch. 9)

sustainable procurement: The integration of Corporate Social Responsibility principles into a company's procurement processes and decisions. This includes both sustainably developed products and materials, as well as responsibility for supplier conduct. It consists of balancing environment (e.g., energy use, waste reduction), economy (e.g., product quality), and ethics (e.g., working conditions, respect for diversity) within procurement practices. (ch. 17)

sustainability: Living on Earth while maintaining a balance of all natural ecosystems, including humans.

Recognizing that resources are limited, so that they are sustained throughout time and for future generations. (Intro.)

syndemic: A set of closely intertwined and mutual enhancing health problems that significantly affect the overall health status of a population within the context of a perpetuating configuration of harmful social conditions. (ch. 3)

systems thinking: A system is a cohesive conglomeration of interrelated and interdependent parts; ability to understand how changing one part of the system affects other parts and the whole system. (Intro.)

tertiary sector: That portion of a region's economy devoted to service activities (e.g., transportation, retail and wholesale operations, insurance). (Intro.)

thermohaline circulation: Describes the movement of ocean currents due to differences in temperatures and salinity in different regions of water. Temperature and salinity change the density of water, resulting in the water to move accordingly. Deep ocean currents are driven by differences in the water density, which is controlled by temperature (thermo) and salinity (haline), hence the name "thermohaline circulation." This process is also sometimes referred to as the "ocean's conveyor belt system." (ch. 14)

Three "Es": The environment, the economy, and equity (human well-being) form the three "Es" of sustainability. (Intro.)

Total Maximum Daily Load (TMDL): A calculation of the highest amount of a pollutant that a waterbody can receive and safely meet water-quality standards set by a government. (ch. 14)

toxicity: The ability of a substance to cause poisonous effects resulting in severe biological harm or death after exposure to, or contamination with, that substance. The quantities and length of exposure to toxic substances necessary to cause these effects can vary widely. (ch. 15)

tragedy of the commons: The exploitation of a common resource when competing parties share it, each pursuing their own interest. Term originated by Garrett Hardin, originally in relation to English common fields but now applied to many common areas, such as the ocean and fisheries. (ch. 12)

transformative change: A fundamental, system-wide reorganization across technological, economic, and social factors, including paradigms, goals, and values. (Intro.)

trophic level: The level in the food chain in which one group of organisms serves as a source of nutrition for another group of organisms (e.g., primary producers, primary or secondary consumers, decomposers). (ch. 15)

Tuberculosis (TB): An airborne communicable disease caused by the bacterium Mycobacterium tuberculosis. (ch. 3)

undernourishment: Percentage of the population whose food intake is insufficient to meet dietary energy requirements continuously. (ch. 2)

undernutrition: A diet that is both insufficient in terms of energy (caloric) requirements and not diverse enough to meet additional nutritional needs. (ch. 2)

uneven development: The unevenness exists in various dimensions of an economy. The concept of uneven development originally emerged within a the Marxian political economy framework to describe relations between production and labor. It sees uneven development as an endemic feature of capitalist development. The term has expanded in meaning beyond uneven economies, to include uneven resource distribution and use, uneven sociocultural indicators (gender, class, race, ethnic group, age), and more. The term also implies uneven geographical development that involves a number of different metrics (employment rates, income levels, rates of economic growth, and so on), and it has been described at all geographical scales—from intra-urban disparities all the way through subnational regional differences to uneven international development. (Intro.)

Universal Health Coverage (UHC): That all people have access to the health services they need, when and where they need them, without financial hardship. (ch. 3)

urban: Characteristic of the city or city life. In the United States, an area is considered urban when it has more than 2,500 people; other countries have different population thresholds that define urban. (ch. 11)

urban ecosystem: Any ecological system located within a city or other densely settled area or, in a broader sense, the greater ecological system that makes up an entire metropolitan area. (ch. 11)

urban forest canopy: The land surface area that lies directly beneath the crowns of all trees and tall shrubs. (ch. 15)

urban heat island: Urbanized areas that experience higher temperatures than outlying areas. Structures such as buildings, roads, and other infrastructure absorb and reemit the sun's heat more than natural landscapes such as forests and water bodies. Because urban areas have more concentrated structures and more limited greenery, they become "islands" of higher temperatures relative to outlying areas. Daytime temperatures in urban areas are about 1–7°F higher than temperatures in outlying areas and night-time temperatures are about 2–5°F higher. (ch. 11)

urbanization: The increase in the proportion of a population living in urban areas; the process by which a large number of people becomes permanently concentrated in relatively small areas, forming cities. (ch. 11)

urban metabolism: A method to evaluate the flows of energy and materials within an urban system, which can provide insights into the system's sustainability and the severity of urban problems such as excessive social, community, and household metabolism at scales ranging from global to local. (ch. 11)

urban primacy (also primate city): A city that is at least twice as large as the next largest city and more than twice as significant. A primate city will usually have precedence in all other aspects of its country's society such as economics, politics, culture, and education. Primate cities also attract large numbers of internal migrants. (ch. 11)

urban stormwater runoff: Any precipitation in an urban or suburban area that does not evaporate or soak into the ground, but instead collects and flows into storm drains, rivers, and streams. Stormwater runoff is the leading cause of stream impairment in urban areas. (ch. 6)

vector: In health, a vector is a living organism that can transmit infectious pathogens between humans, or from animals to humans. Many vectors are bloodsucking insects, which ingest disease-producing microorganisms during a blood meal from an infected host (human or animal) and later transmit it into a new host, after the pathogen has replicated. Often, once a vector becomes infectious, they are capable of transmitting the pathogen for the rest of their life during each subsequent bite/blood meal. (ch. 3)

vector-borne disease (or disease vector): An animate intermediary in the indirect transmission of an agent that carries the agent from a reservoir to a susceptible host. An example would be mosquitoes that transmit malaria. (ch. 3)

vector control: Vector control is any method to limit or eradicate the mammals, birds, insects, or other arthropods that transmit disease pathogens. (ch. 3)

vegetated swales: A shallow drainage conveyance that has vegetative turf (typically grasses) with relatively gentle side slopes, generally with flow depths of less than one foot. (ch. 15)

vertical inequality: Inequality among individuals or households, often measured by income. (ch. 10)

virtual water: The water "hidden" in the products and services people buy and use every day. Virtual water often goes unseen by the end-user of a product or service, but that water has been consumed throughout the value chain, which makes creation of that product or service possible. By contrast, direct water use is the water that is seen, felt, and used in a given time and location to produce an item or service (e.g., "tap water"). (ch. 6)

voluntary migration: Migration based on one's free-will and initiative. People move for a variety of reasons, including economic, social, cultural, or political reasons.

Voluntary migration occurs as a result of an individual or group of people desiring to influence their circumstances, not the other way around. (ch. 16)

wage gap: The difference between median earnings of men and women relative to median earnings of men. These differences are often systemic. (ch. 5)

waste hierarchy: A set of priorities for the efficient use of resources. The hierarchy ranks the various management strategies from most to least environmentally preferred. The hierarchy places emphasis on reducing, reusing, and recycling as key to sustainable materials management. (ch. 12)

waste stream: The composition of various types of materials in solid waste, such as paper, metals, textiles, electronics, plastics, and organic debris/material. (ch. 12)

wasting: A low weight-for-height and is often a sign of life-threatening undernutrition. (ch. 2)

water conservation: The practice of using water efficiently to reduce unnecessary water usage. (ch. 6)

water footprint: Measures the amount of water used to produce each of the goods and services we use. It can be measured for a single process, such as growing rice; for a product, such as a pair of jeans; for the fuel we put in our car; or for an entire multinational company. The water footprint can also tell us how much water is being consumed by a particular country—or globally—in a specific river basin or from an aquifer. (chs. 6, 11, 14)

water infrastructure: A broad term for systems of water supply, treatment, storage, water resource management, flood prevention, and hydropower. The term also includes water-based transportation systems such as canals. (ch. 6)

Water, Sanitation, and Hygiene (WASH): Universal, affordable and sustainable access to WASH is a key public health issue within international development and is the focus of the first two targets of Sustainable Development Goal 6. (ch. 6)

watershed: A divide that separates neighboring drainage basins. (ch. 14)

well-being (human): A perspective on a good life that comprises access to basic resources, freedom and choice, and health and physical, including psychological, well-being, good social relationships, security, equity, peace of mind, and spiritual experience. Well-being is achieved when individuals and communities can act meaningfully to pursue their goals and can enjoy a good quality of life. The concept of human well-being is used, together with living in harmony with nature, and living well in balance and harmony with Mother Earth. All these are different perspectives on a good quality of life. (ch. 3)

wind power: The creation of energy using the wind to turn rotor blades and thereby turn a crankshaft and produce power. (ch. 7)

World Bank (also called the World Bank Group): The World Bank Group was established in 1944 with the primary mission to end extreme poverty. It is the world's largest development institution. The bank provides low-interest loans, zero- to low-interest credits, and grants to developing countries. These loans and grants support a diversity of investments in such areas as education, health, public administration, infrastructure, financial and private-sector development, agriculture, and environmental and natural resource management. Some World Bank projects are cofinanced with governments, other multilateral institutions, commercial banks, and private-sector investors. (ch. 1)

worst forms of child labor: Slavery and similar issues such as the trafficking of children, debt bondage, serfdom, and children in armed conflict. It can also involve sexual exploitation of children such as prostitution and pornography. (ch. 8)

zero waste: The conservation of all resources by means of responsible production, consumption, reuse, and recovery of products, packaging, and materials without burning and with no discharges to land, water, or air that threaten the environment or human health. (ch. 12)

zoonotic disease (also known as zoonoses): An infectious disease that is transmissible under normal conditions from animals to humans. (e.g., air—influenza; bites and saliva—rabies). These can include direct contact with saliva, blood, feces, or other body fluids of animals, indirect contact such as contact in areas where animals live or objects that have been contaminated with germs (e.g., chicken coops), vector-borne (being bitten by a tick or mosquito), or food-borne (undercooked meat or eggs). (chs. 3, 17)

CREDITS

INTRODUCTION

Figure I.3. The United Nations (2022).

Figure I.4. UN Development Program. "Human Development Index Ranking 2020." https://hdr.undp.org/en/content/latest-human-development-index-ranking

Map I.1. Adapted from UN Development Program. "Human Development Index Ranking 2020." https://hdr.undp.org/en/content/latest-human-development-index-ranking

CHAPTER 1

Figure 1.1. Based on data from the World Bank (2017).

Figure 1.2. Adapted from Oxford Poverty and Human Development Initiative (OPHI). *Global Multidimensional Poverty Index 2018: The Most Detailed Picture to Date of the World's Poorest People*. Oxford: Oxford University Press (2018).

Figure 1.3. Adapted from Oxford Poverty and Human Development Initiative (OPHI). *Global Multidimensional Poverty Index 2018: The Most Detailed Picture to Date of the World's Poorest People*. Oxford: Oxford University Press (2018), p. 47.

Map 1.1. World Bank, World Bank Development Indicators (2018).

Map 1.2. Adapted from Oxford Poverty and Human Development Initiative (OPHI). *Global Multidimensional Poverty Index 2018: The Most Detailed Picture to Date of the World's Poorest People*. Oxford: Oxford University Press (2018).

CHAPTER 2

Figure 2.1. Based on data from the World Food Programme (2017).

Figure 2.3. Based on USDA Economic Research Services Food Security Hierarchy (2021).

Figure 2.4. The United Nations (2021).

Figure 2.5. The United Nations (2021).

Figure 2.6. The United Nations (2021).

Figure 2.7. Based on data from UN Environment (2019).

Map 2.1. Based on data from UN Food and Agriculture Organisation.

Map 2.3. Based on data from the DC Policy Center (2017, March 13). https://www.dcpolicycenter.org/publications/food-access-dc-deeply-connected-poverty-transportation/

Photo 2.1. USAID Africa Bureau, Public domain, via Wikimedia Commons (2011).

CHAPTER 3

Figure 3.1. Based on data from Johns Hopkins University and Medicine, Coronavirus Resource Center (2021, May 28). https://coronavirus.jhu.edu/data/new-cases

Figure 3.2. Based on data from Our World in Data (2020). https://ourworldindata.org/malaria#:~:text=Since%20the%20beginning%20of%20the,2000%20to%20438%2C000%20in%202015

Map 3.1. Based on data from Johns Hopkins University and Medicine, Coronavirus Resource Center (2021, June). https://coronavirus.jhu.edu/map.html

Map 3.2. Based on data from World Health Organization (2019). https://www.who.int/images/default-source/maps/global_hale_2016.png?sfvrsn=64d958f8_0

Map 3.3. Based on data from UNICEF, World Health Organization, and UN Population Fund, "Maternal Mortality Ratio" (2019). https://data.unicef.org/topic/maternal-health/maternal-mortality

Map 3.4. Based on data from UNAIDS: https://www.dw.com/en/unaids-report-hiv-related-deaths-down-a-third-since-2010/a-49605119 and World Health Organization (2020). https://www.who.int/data/gho/data/indicators/indicator-details/GHO/estimated-number-of-people-(all-ages)-living-with-hiv

Map 3.5. Based on data from World Health Organization (2017).

Map 3.6. Based on data from Wikipedia (2015). https://en.wikipedia.org/wiki/Neglected_tropical_diseases#/media/File:Number_of_people_requiring_interventions_against_neglected_tropical_diseases_(NTDs),_OWID.svg

Map 3.7. Based on data from World Health Organization (2018).

Map 3.8. Based on data and map from UNICEF (2015). https://weshare.unicef.org/Package/2AMZIFS7JPS

Photo 3.1. Pascale Deloche/Stone Collection from Getty Images.

Photo 3.2. Acraftsoft from Wikimedia Commons (2020, January). https://commons.wikimedia.org/wiki/File:Bad_Road_Issue_(Benue_State,_Nigeria).jpg

CHAPTER 4

Figure 4.1. Based on data from UNESCO (2019). http://uis.unesco.org/en/topic/out-school-children-and-youth

Figure 4.2. Based on data from Friedman et al. "Measuring and Forecasting Progress towards the Education-Related SDG Targets." *Nature* 580: 636–39 (2020). https://doi.org/10.1038/s41586-020-2198-8 and https://www.nature.com/articles/s41586-020-2198-8#Fig1

Figure 4.3. Based on data from Friedman et al. "Measuring and Forecasting Progress towards the Education-Related SDG Targets." *Nature* 580: 636–39 (2020). https://doi.org/10.1038/s41586-020-2198-8 and https://www.nature.com/articles/s41586-020-2198-8#Fig1

Figure 4.4. Based on data from Friedman et al. "Measuring and Forecasting Progress towards the Education-Related

SDG Targets." *Nature* 580: 636–39 (2020). https://doi.org/10.1038/s41586-020-2198-8 and https://www.nature.com/articles/s41586-020-2198-8#Fig1

Map 4.1. Based on data from UNESCO Institute for Statistics (2019).

Map 4.2. Adapted from UNESCO (2021, January). https://en.unesco.org/news/unesco-figures-show-two-thirds-academic-year-lost-average-worldwide-due-covid-19-school

Photo 4.1. Shaun D. Metcalfe, from Wikimedia Commons (2014). https://commons.wikimedia.org/wiki/File:Malala_Yousafzai_and_Kaliash_Satyarthi_at_the_Nobel_Peace_Prize_ceremony.jpg

Photo 4.2. Shaun D. Metcalfe, from Wikimedia Commons (2009). https://en.wikipedia.org/wiki/Educational_inequality#/media/File:School_children_in_Rhbat,_Nagar.jpg

CHAPTER 5

Figure 5.1. Adapted from Statista. "Average Gender Gap Closed Worldwide as of 2021, by Region" (2021). https://www.statista.com/statistics/1211887/average-gender-gap-closed-worldwide-by-region/

Map 5.1. Based on data from UN Development Programme, Gender Inequality Index (2019). http://hdr.undp.org/en/content/gender-inequality-index-gii

Map 5.2. Based on data from (n.d.). https://www.orchidproject.org/about-fgc/what-is-fgc/?gclid=CjwKCAiAouD_BRBIEiwALhJH6PhEkaRIXH-4mJcRY_BIlyyfrFDgXFrOogQovgQUVqF-Own5wBpQ1hoCtzIQAvD_BwE

Map 5.3. Based on data from US Census Bureau. "2018 American Community Survey" (2018).

Photo 5.1. Sir Francis Canker Photography/Moment Collection from Getty Images (n.d.).

Photo 5.2. planetlight from Wikimedia Commons (2012). https://commons.wikimedia.org/wiki/File:Ceylon_tea_picker_(5530574131).jpg

CHAPTER 6

Figure 6.1. Based on information from the US EPA (2022, August 12).

Figure 6.2. Based on information from the US EPA (2022, August 12).

Figure 6.3. Based on information from the US Geological Survey (2019, October 16).

Figure 6.4. Based on information from the US EPA (2004, September).

Figure 6.5. Based on information from the US EPA (2022, August 12).

Map 6.1. Based on data from UNICEF, WHO/UNICEF JMP Progress on Drinking Water, Sanitation and Hygiene: 2017 Update and SDG Baseline (2017). https://data.unicef.org/wp-content/uploads/infograms/10152/index.html

Map 6.2. Adapted from UNICEF, WHO/UNICEF JMP Progress on Drinking Water, Sanitation and Hygiene: 2017 Update and SDG Baseline (2017). https://data.unicef.org/wp-content/uploads/infograms/10152/index.html

Map 6.3. Based on data from the UN Food and Agricultural Organization (2015).

Map 6.4. Based on information from United Nations, UNDESEA, "Water Scarcity" (2012). https://www.un.org/waterforlifedecade/scarcity.shtml

Photo 6.1. Mike Goldwater/Alamy Stock Photo (2013).

Photo 6.2. Tim Gainey/Alamy Stock Photo (2013).

Photo 6.3. Sustainable Sanitation Alliance, from Wikimedia Commons (2011, July). https://commons.wikimedia.org/wiki/File:Rainwater_harvesting_tank_(5981896147).jpg

Photo 6.4. Purepix/Alamy Stock Photo (2012, January).

CHAPTER 7

Figure 7.1. Based on data from US Energy Information Administration (2022). https://www.eia.gov/energyexplained/us-energy-facts/

Figure 7.2. Based on data from US Energy Information Administration (2022). https://www.eia.gov/energyexplained/us-energy-facts/

Figure 7.3. Based on data from several sources.

Figure 7.4. Based on data from UN Sustainable Development Goals (2022).

Map 7.1. Based on data from Our World in Data (2019). https://ourworldindata.org/energy-access#how-does-per-capita-energy-consumption-vary-across-the-world

Map 7.2. Based on data from Africa Energy Outlook 2019, International Energy Agency.

Map 7.3. Based on data from Africa Energy Outlook 2019, International Energy Agency.

Photo 7.2. Andrew Aitchison/Alamy Stock Photo (2011, June).

Photo 7.3. Fogarty International Center from Wikimedia Commons (2018). https://upload.wikimedia.org/wikipedia/commons/f/f4/Fogarty-nih-50th-symposium-cookstoves_%2841744498535%29.jpg

CHAPTER 8

Figure 8.1. Adapted from *Global Estimates of Child Labour: Results and Trends, 2012–16*. International Labour Office, Geneva (2017), p. 23. https://www.ilo.org/wcmsp5/groups/public/@dgreports/@dcomm/documents/publication/wcms_575499.pdf

Map 8.1. Based on data from International Labour Office, World Social Protection Data Base (2017). https://www.ilo.org/wcmsp5/groups/public/---dgreports/---dcomm/---publ/documents/publication/wcms_605078.pdf

Map 8.2. Based on data from *Global Estimates of Child Labour: Results and Trends, 2012–16*. International Labour Office, Geneva (2017), p. 23. https://www.ilo.org/wcmsp5/groups/public/@dgreports/@dcomm/documents/publication/wcms_575499.pdf

Map 8.3. Based on data from Walk Free, Global Slavery Index (2018). https://www.globalslaveryindex.org/ and https://www.globalslaveryindex.org/2018/data/maps/#prevalence

Photo 8.1. Rob Crandall/Alamy Stock Photo (2006, May).

Photo 8.2. Y. Levy/Alamy Stock Photo (2013, January.)

Photo 8.3. Matthew Williams-Ellis Travel Photography/ Alamy Stock Photo (2017, March).

CHAPTER 9

Figure 9.1. Adapted from America Society of Civil Engineers, 2021 Report Card for America's Infrastructure (2021). https://infrastructurereportcard.org/tag/2020/ and https://infrastructurereportcard.org/wp-content/uploads/2020/12/2021-Grades-Chart.jpg

Figure 9.2. Based on Maryland Resiliency Hub Grant Program and Mercy Corps (2018). https://www.adaptation clearinghouse.org/resources/maryland-resiliency-hub-grant-program.html

Map 9.1. Based on data in Our World in Data (2017). https://ourworldindata.org/grapher/mobile-cellular-subscriptions-per-100-people

Map 9.2. Based on data in Our World in Data (2017). https://ourworldindata.org/internet

Map 9.3. Based on data from World Bank. "Belt and Road Initiative, March 2018" (2018). https://www.worldbank.org/en/topic/regional-integration/brief/belt-and-road-initiative

Photo 9.1. Oliver Förstner/Alamy Stock Photo (2014, November).

Photo 9.2. UPI/Alamy Stock Photo (2007, August).

Photo 9.3. Christophe95 from Wikimedia Commons (2018). https://commons.wikimedia.org/wiki/File:Bridge_construction_in_Luang_Prabang_Province_2.jpg

CHAPTER 10

Map 10.1. Based on data from District of Columbia Department of Health, Annual Epidemiology & Surveillance Report. Washington, DC: Government of the District of Columbia (2020). https://dchealth.dc.gov/sites/default/files/dc/sites/doh/publication/attachments/2020-HAHS-TA-Annual-Surveillance-Report.pdf

Map 10.2. Based on data from UNHCR (2008). https://indicators.ohchr.org/ and https://en.wikipedia.org/wiki/International_Convention_on_the_Elimination_of_All_Forms_of_Racial_Discrimination#/media/File:ICERD-members.PNG

Map 10.3. Based on data from UNHCR Status of Ratification Interactive Dashboard, and https://indicators.ohchr.org/ on May 20, 2021.

Map 10.4. Based on data from Wikipedia. "List of Income Inequality by Country" (2018). https://en.wikipedia.org/wiki/List_of_countries_by_income_equality

Photo 10.1. SOPA Images Limited/Alamy Stock Photo (2021, June).

Photo 10.2. SOPA Images Limited/Alamy Stock Photo (2019, June).

CHAPTER 11

Map 11.1. Based on data from Our World in Data (2018). https://ourworldindata.org/urbanization

Map 11.2. Based on data from Our World in Data (2018). https://ourworldindata.org/urbanization

Map 11.3. Based on data from UN Habitat "SDG 11.1 Training Module: Adequate Housing and Slum Upgrading." Nairobi: UN-Habitat (2018), p. 4. https://unhabitat.org/sites/default/files/2020/06/indicator_11.1.1_training_module_adequate_housing_and_slum_upgrading.pdf

Photo 11.1. ZUMA Press, Inc./Alamy Stock Photo (2020, May).

Photo 11.3. John Ruberry/Alamy Stock Photo (2019, February).

CHAPTER 12

Figure 12.1. Based on information from the US Environmental Protection Agency (2022).

Figure 12.2. Based on data from Our World in Data (2018). https://ourworldindata.org/plastic-pollution#:~:text=100%2C000s%20tonnes.-,How%20much%20plastic%20does%20the%20world%20produce%3F,381%20million%20tonnes%20in%202015

Figure 12.4. Based on various sources.

Figure 12.5. Based on various sources.

Photo 12.1. Stefan Muller, from Wikimedia Commons (2019). https://commons.wikimedia.org/wiki/File:Fast_Fashion_killt_das_Klima.jpg

Photo 12.2. REUTERS/Alamy Stock Photo (2008, April).

Photo 12.3. Aman Luthra (2018).

CHAPTER 13

Figure 13.1. Based on US Environmental Protection Agency (2022).

Figure 13.2. Based on data and information from NASA's Goddard Institute for Space Studies (2019).

Figure 13.3. Based on NASA "Graphic: The relentless rise of carbon dioxide," https://climate.nasa.gov/climate-resources/24/graphic-the-relentless-rise-of-carbon-dioxide/ based on data from the following: Etheridge, D.M., et al. 2010. "Law Dome Ice Core 2000-Year CO_2, CH_4, and N_2O Data." IGBP PAGES/World Data Center for Paleoclimatology Data Contribution Series #2010-070. NOAA/NCDC Paleoclimatology Program, Boulder CO, USA; Lüthi, D., et al.. 2008. EPICA Dome C Ice Core 800KYr Carbon Dioxide Data. IGBP PAGES/World Data Center for Paleoclimatology Data Contribution Series # 2008-055. NOAA/NCDC Paleoclimatology Program, Boulder CO, USA; and Petit, J.R., Jouzel, J. "Vostok ice core deuterium data for 420,000 years." *Pangaea* (1999) https://doi.org/10.1594/PANGAEA.55505

Figure 13.4. Based on IPCC report (2021). https://www.ipcc.ch/working-group/wg1/

Figure 13.5. Based on IPCC report (2021). https://www.ipcc.ch/working-group/wg1/

Map 13.1. Based on data from Wikipedia (2017). https://en.wikipedia.org/wiki/List_of_countries_by_carbon_dioxide_emissions_per_capita#/media/File:CO2_emissions_per_capita,_2017_(Our_World_in_Data).svg

Map 13.2. Based on data from Wikipedia (2017). https://en.wikipedia.org/wiki/List_of_countries_by_carbon_dioxide_emissions_per_capita#/media/File:CO2_emissions_per_capita,_2017_(Our_World_in_Data).svg

Photo 13.1. Shahee Ilyas from Wikipedia (2004). https://en.wikipedia.org/wiki/Maldives#/media/File:Male-total.jpg

Photo 13.2. Imagery from the NASA Worldview application (2019). (https://worldview.earthdata.nasa.gov), part of the NASA Earth Observing System Data and Information System.

CHAPTER 14

Map 14.1. Based on image from National Oceanic and Atmospheric Administration. "Garbage Patches" (2022). https://marinedebris.noaa.gov/info/patch.html

Map 14.2. Based on information from various sources.

Map 14.3. Based on Halpern et al., "Spatial and Temporal Changes in Cumulative Human Impacts on the World's Ocean." *Nature Communications* 6 (7615) (2015). See Figure 1. https://doi.org/10.1038/ncomms8615 https://www.nature.com/articles/ncomms8615#citeas

Map 14.4. Based on data from NOAA (2017). https://www.noaa.gov/media-release/gulf-of-mexico-dead-zone-is-largest-ever-measured

Map 14.5. Based on map from Antarctic and Southern Ocean Coalition "Protecting the Southern Ocean" (2022). https://www.asoc.org/advocacy/marine-protected-areas

Photo 14.1. NOAA/NASA GOES Project (2011).

Photo 14.2. NASA (2022). https://earthobservatory.nasa.gov/ContentWOC/images/yellow_river/yellowriverwoc_tm5_1995261_lrg.jpg

Photo 14.3. Julian Nieman/Alamy Stock Photo (2014, April).

Photo 14.4. J. Roff from Wikimedia Commons (2016 and 2005). https://en.wikipedia.org/wiki/Coral_bleaching#/media/File:Bleachedcoral.jpg (2016) and https://en.wikipedia.org/wiki/Coral_bleaching#/media/File:Lodestone_Reef_Valentines_Day_2016,_Green_Chromis_on_Coral.jpg (2005).

CHAPTER 15

Figure 15.1. Based on Ecology of Ecosystems: Figure 5, by OpenStax College, Biology, CC by 4.0 (2018).

Figure 15.2. Based on information from US Environmental Protection Agency (2022).

Figure 15.3. Based on information from US Environmental Protection Agency (2013).

Figure 15.4. Based on Diaz et al., "Pervasive Human-Driven Decline of Life on Earth Points to the Need for Transformative Change." *Science* 13 (366): 6471 (2019). http://doi.org/10.1126/science.aax3100

Figure 15.5. Based on Diaz et al. "Pervasive Human-Driven Decline of Life on Earth Points to the Need for Transformative Change." *Science* 13 (366): 6471 (2019). http://doi.org/10.1126/science.aax3100

Figure 15.6. Based on UNODC World WISE database (2018).

Map 15.2. Based on a map from FAO and UNEP. The State of the World's Forests 2020. Forests, biodiversity, and people. Rome (2020), p. 12. https://doi.org/10.4060/ca8642en

Photo 15.1. National Park Service (2012).

Photo 15.2. Sipa USA/Alamy Stock Photo (2020, September).

Photo 15.3. Scott Ehardt, public domain, from Wikimedia Commons (2006).

Photo 15.4. Ginger Allington (2019).

CHAPTER 16

Figure 16.1. Based on UN information on governance and rule of law (2011). https://peacekeeping.un.org/sites/default/files/un_rule_of_law_indicators.pdf

Figure 16.2. Based on Kroc Institute of International Peace Studies (n.d.).

Figure 16.3. Based on UN information on governance and rule of law (n.d.). https://www.un.org/ruleoflaw/files/Governance%20Indicators_A%20Users%20Guide.pdf

Map 16.1. Based on map (2021). https://en.wikipedia.org/wiki/List_of_ongoing_armed_conflicts

Map 16.2. Based on data from the World Bank (2017). https://data.worldbank.org/indicator/SM.POP.REFG.OR

Map 16.3. Based on data from Our World in Data (2017).

Map 16.4. Based on data from Transparency International (2018). https://www.transparency.org/en/cpi/2020/index/nzl

Map 16.5. Based on data from UN Statistics Division (2017). https://dataunodc.un.org/data/prison/total%20persons%20held%20unsentenced

Map 16.6. Based on data from UN Human Rights (2020). https://indicators.ohchr.org/

Map 16.7. Based on data from UN Human Rights (2020). https://indicators.ohchr.org/

Photo 16.1. ITAR-TASS News Agency/Alamy Stock Photo (2018).

CHAPTER 17

Figure 17.1. Based on data from the World Bank (2020).

Figure 17.2. Based on multiple sources.

Map 17.1. Based on World Bank information (2018). https://www.worldbank.org/en/topic/debt/brief/hipc

Photo 17.1. YouthMappers (2020).

LISA BENTON-SHORT is professor of geography at George Washington University. From 2010 to 2015, she served as Academic Program Director for Sustainability at George Washington. She led the development of the pan-university sustainability minor for undergraduates. Since 2012, she has team-taught an undergraduate course, Introduction to Sustainability. Much of the structure and content in this book was inspired by teaching this class, interacting and learning from the colleagues with whom she teaches as well as hundreds of students over the years. She has authored numerous books, including *The Presidio: From Army Post to National Park* (1998), *Cities and Nature* (2013), *The National Mall: No Ordinary Public Space* (2016), *Urban Sustainability in the US: Cities Take Action* (2019, with Melissa Keeley), and *A Regional Geography of the United States and Canada: Toward a Sustainable Future* (2019, with John Rennie Short and Chris Mayda).